Lecture Notes in Computer Science **15830**

The series Lecture Notes in Computer Science (LNCS), including its subseries Lecture Notes in Artificial Intelligence (LNAI) and Lecture Notes in Bioinformatics (LNBI), has established itself as a medium for the publication of new developments in computer science and information technology research, teaching, and education.

LNCS enjoys close cooperation with the computer science R & D community, the series counts many renowned academics among its volume editors and paper authors, and collaborates with prestigious societies. Its mission is to serve this international community by providing an invaluable service, mainly focused on the publication of conference and workshop proceedings and postproceedings. LNCS commenced publication in 1973.

Ipek Oguz · Shaoting Zhang ·
Dimitris N. Metaxas
Editors

Information Processing in Medical Imaging

29th International Conference, IPMI 2025
Kos, Greece, May 25–30, 2025
Proceedings, Part II

Editors
Ipek Oguz
Vanderbilt University
Nashville, TN, USA

Shaoting Zhang
Shanghai AI Laboratory
Shanghai, China

Dimitris N. Metaxas
Rutgers University
Piscataway, NJ, USA

ISSN 0302-9743 ISSN 1611-3349 (electronic)
Lecture Notes in Computer Science
ISBN 978-3-031-96624-8 ISBN 978-3-031-96625-5 (eBook)
https://doi.org/10.1007/978-3-031-96625-5

This Springer imprint is published by the registered company Springer Nature Switzerland AG
The registered company address is: Gewerbestrasse 11, 6330 Cham, Switzerland

Preface

It is our pleasure to present the proceedings of the 29th International Conference on Information Processing in Medical Imaging (IPMI 2025), held on the island of Kos, Greece, May 25–30, 2025. The conference was organized by co-chairs Ipek Oguz from Vanderbilt University and Dimitris N. Metaxas from Rutgers University.

Since 1969, IPMI has become widely recognized as a premier international forum for the presentation of cutting-edge methodological advancements in the medical image computing field. The conference is known for its unique format that fosters in-depth paper discussion, interdisciplinary collaboration, and rigorous scientific exchange in a retreat-style setting.

For the IPMI 2025 conference, we received 196 submissions, of which 143 papers went through the double-blind peer review. The majority of the papers was evaluated by three reviewers and few received two reviews. Following the initial review process, each manuscript was further evaluated by two members of the Paper Selection Committee and subsequently discussed by the full committee at the paper selection meeting. Papers with conflicting scores from reviewers (both accept and reject recommendations) were discussed in detail at this meeting. Finally, 51 submissions were accepted for publication and presentation at the conference. Selected papers were invited for a Special Issue of the Machine Learning for Biomedical Imaging (MELBA) journal following the conference. We are deeply grateful to the Paper Selection Committee for their efforts and commitment. We also thank the 88 reviewers listed below, who each provided thoughtful feedback on an average of 5 papers.

The scientific program was punctuated by two excellent scientific keynotes delivered by Leon Axel from NYU Grossman School of Medicine and Shaoting Zhang from Shanghai Jiao Tong University. We are also grateful to Jim Duncan, Julia Schnabel, Herve Lombaert, and Miaomiao Zhang, who participated in the Future Directions for IPMI: A Soul Searching Panel event at the conference, which was a thought-provoking session for our community.

The François Erbsmann Prize is awarded during each IPMI to a young scientist of age 35 or below, the first author of a paper, giving their first oral presentation at IPMI. The prizes winners were determined by the IPMI Board. NVIDIA sponsored the awards for the François Erbsmann Prize, the runner-up award, as well as the best poster award. NVIDIA also led a virtual workshop at IPMI 2025 focusing on the latest upcoming technology trends with an AWS cloud platform hands-on session titled: "Holoscan × MONAI: Next-Gen Multi-Modal AI & Co-pilots". We are grateful for their support of IPMI 2025.

We were honored to organize IPMI 2025 and present the excellent set of scientific contributions documented in these proceedings. IPMI continues to be the premier venue for methodological advancements in medical image computing, and we are excited to pass the baton to the IPMI 2027 team to organize the next exciting meeting in the series.

May 2025

Ipek Oguz
Shaoting Zhang
Dimitris N. Metaxas

Organization

General Chairs

Ipek Oguz	Vanderbilt University, USA
Dimitris N. Metaxas	Rutgers University, USA

Program Chairs

Ipek Oguz	Vanderbilt University, USA
Shaoting Zhang	Shanghai Jiao Tong University, China

Paper Selection Committee

Ismail Ben Ayed	École de Technologie Superieure, Canada
Shireen Elhabian	University of Utah, USA
Sarah Frisken	Brigham and Women's Hospital, USA
Sharon Huang	Pennsylvania State University, USA
Marc Niethammer	University of California San Diego, USA
Baba Vemuri	University of Florida, USA
Carl-Fredrik Westin	Harvard University, USA

Information Processing in Medical Imaging Board

Gary Christensen	University of Iowa, USA
Albert Chung	University of Exeter, UK
Marleen de Bruijne	Erasmus MC, The Netherlands
James S. Duncan	Yale University, USA
Alejandro Frangi	University of Leeds, UK
Polina Golland	Massachusetts Institute of Technology, USA
Richard Leahy	University of Southern California, USA
Alison Noble	University of Oxford, UK
Ipek Oguz	Vanderbilt University, USA
Sebastien Ourselin	King's College London, UK
Stephen M. Pizer	University of North Carolina at Chapel Hill, USA
Jerry Prince	Johns Hopkins University, USA

Stefan Sommer	University of Copenhagen, Denmark
Martin Styner	University of North Carolina at Chapel Hill, USA
Gabor Szekely	ETH Zurich, Switzerland
Chris Taylor	University of Manchester, UK
Andrew Todd-Pokropek	University College London, UK
William M. Wells III	Harvard Medical School, USA

Reviewers

Aasa Feragen
Albert C. S. Chung
Alexis Arnaudon
Alison Noble
Alison Pouch
Alvin Chen
Andrew P. King
Ario Sadafi
Aristeidis Sotiras
Arrate Munoz Barrutia
Baiying Lei
Bennett A. Landman
Bjoern Menze
Carole H. Sudre
Chris Taylor
Christian Desrosiers
Chunfeng Lian
Chuyang Ye
Daniel C. Moyer
Daniel Rueckert
Dou Qi
Dustin Scheinost
Ender Konukoglu
Gary E. Christensen
Gianfranco Cortes
Guotai Wang
Hao Li
Haomiao Ni
Hoel Kervadec
Hong Xu
Hongmin Cai
Ilker Hacihaliloglu
Ilwoo Lyu
Islem Rekik
Jack Noble
Jadie Adams
James S. Duncan
Jan Kybic
Jerry L. Prince
Jiazhou Chen
John Ashburner
Jon Sporring
Jonghye Woo
Jose Dolz
Julia A. Schnabel
Julio Silva-Rodriguez
Junzhou Huang
Kaleem Siddiqi
Karl Rohr
Karthik Gopinath
Krithika Iyer
Linwei Wang
Marco Lorenzi
Maria A. Zuluaga
Martin Styner
Matthew Toews
MattiasPaul Heinrich
Meng Ye
Miaomiao Zhang
Minjeong Kim
Nicha Dvornek
Ninon Burgos
P. Thomas Fletcher
Peirong Liu
Peng Jin
Pengcheng Shi
Pew-Thian Yap
Qingyu Zhao

Reuben Dorent
Reyer Zwiggelaar
Sara Garbarino
Steve Pizer
Suyash P. Awate
Tim Cootes
Tobias Heimann
Tuo Zhang
Ulas Bagci
Vivek Gopalakrishnan
Wenzheng Tao
Xiang Li
Xiaodong Wu
Yi Hong
Yixuan Yuan
Yonggang Shi
Yuan Xue
Yuan Zhou
Yuankai Huo
Yunhe Gao

Contents – Part II

Diffusion Models

Self-supervised Learning

Vision-Language Models

Contents – Part I

Classification/Detection

Registration

Image Enhancement

Segmentation

Computer-Aided Diagnosis/Surgery

Concepts from Neurons: Building Interpretable Medical Image Diagnostic Models by Dissecting Opaque Neural Networks

Shizhan Gong[1(✉)], Huayu Wang[2], Xiaofan Zhang[3,4], and Qi Dou[1]

[1] The Chinese University of Hong Kong, Hong Kong SAR, China
szgong22@cse.cuhk.edu.hk
[2] University of Washington, Seattle, USA
[3] Shanghai Jiao Tong University, Shanghai, China
[4] Shanghai Artificial Intelligence Laboratory, Shanghai, China

Abstract. Deep learning has achieved remarkable success in medical image analysis, yet its translation to clinical settings is often impeded by the opaque nature of these models. While interpretable models like Concept Bottleneck Models (CBMs) maintain good interpretability, they often require manual designed and annotated concepts, or rely heavily on pre-trained vision-language models. In the medical domain, however, the concepts are often too complicated to be described precisely by pure plain text, and the unified foundation models of high performance are still missing. To address these challenges, we propose a novel framework that extracts human-understandable concepts from pre-trained opaque models and then builds surrogate CBMs for interpretable diagnosis. We first employ sparse autoencoders to disentangle learned representations into a limited set of clinically relevant concepts, which are then transformed into plain text with the assistance of domain experts or large language models (LLMs). Utilizing concept activation vectors (CAVs), we can project these concepts into a shared representation space and apply submodular optimization to select the most informative concepts for model inference. The interpretable surrogate CBMs are finally constructed through sparsely decomposing the visual representation into concepts representations. We validate our framework on three medical diagnostic benchmarks: HAM10000, Harvard-FairVLMed, and MIMIC-CXR. The results indicate that our method achieves performance comparable to opaque models while significantly enhancing interpretability, outperforming previous CBMs. The code of this work is publicly available at https://github.com/med-air/CFN.

Keywords: Concept bottleneck models · Mechanistic interpretation · Computer-aided diagnosis · Vision-language models

I. Oguz et al. (Eds.): IPMI 2025, LNCS 15830, pp. 3–18, 2026.
https://doi.org/10.1007/978-3-031-96625-5_1

1 Introduction

Deep learning has attained impressive success in a range of image-related tasks, including classification [27], segmentation [15,16], and registration [5,42]. However, in the realm of medical image diagnosis, merely attaining high accuracy is not enough. Clinicians often express skepticism towards these models due to their lack of transparency, making it risky to fully trust the predictions produced by the networks. This skepticism poses a barrier to the widespread adoption of deep learning in medical contexts. An ideal AI system for medical diagnosis should not only generate predictions but also offer explanations, enabling clinicians to make informed decisions based on the model's insights rather than simply accepting its conclusions. Explanations in computer vision typically fall into two categories: post-hoc explanations and inherently interpretable models. While post-hoc explanations [14,36,38,39] are often criticized for lacking fidelity to the original network [37], interpretable models are designed to modify the reasoning processes of the networks, ensuring that their logic is understandable to humans [8,13,41]. One such example is concept bottleneck models (CBMs) [25].

CBMs reason by compressing information into a lower-dimensional concept bottleneck layer, then applying a linear function to make predictions based solely on this layer. The bottleneck layer is designed so that each dimension corresponds to human-understandable concepts, emphasizing the most relevant data aspects that inform certain decisions. Early CBMs [23,25,26] utilized manually selected concepts and learned bottleneck layers from human annotations, a process that demands significant expertise and effort. Recent advancements in large language models (LLMs) and vision-language models (VLMs) have enabled the automation of concept selection and annotation. For example, CompDL [45] automates concept scoring using vision-language similarity scores derived from CLIP [34], while LaBo [43] and label-free CBM [30] utilize prior knowledge from GPT [7] to generate concepts. However, applying these methods in the medical field presents significant challenges. First, clinical concepts are often harder to describe accurately in plain text compared to those in natural images, and the same concept can manifest diverse symptom patterns across different patient subgroups. It is more desirable to define the concepts by textual and visual features jointly. Second, the concept mining and scoring processes require generalizable VLMs. In medical field, a unified foundation model with well-proven generalizability is still lacking. Lastly, there remains a performance gap between these CBMs and opaque models, indicating that current concept mining strategies may be suboptimal compared to the implicit concepts identified by opaque models.

Recent advancement in mechanistic interpretation makes it possible to extract learned concepts from neurons of pre-trained opaque models. Existing research [1,3,4,31] has demonstrated that well-trained vision encoders can effectively capture structured patterns in latent space, with each neuron associated with human-understandable concepts. Additionally, studies have shown that components like sparse autoencoders (SAEs) [6] can further enhance neuron-concept alignment and tackle the issue of polysemanticity, where a single neuron may correspond to multiple concepts. These findings motivate us to extract concepts from well-trained opaque vision encoders and develop interpretable surrogate CBMs that replicate the reasoning processes of the original opaque models.

In this work, we investigate the potential to dissect opaque models and create surrogate interpretable models for medical image diagnosis. We begin with an opaque classifier that has been pre-trained on specific medical diagnostic tasks. Using this pre-trained model, we apply SAEs to separate the learned representations into a sparse set of human-understandable concepts. These concepts are then transformed into plain text by domain experts or GPT, summarizing and extracting common insights from captions or clinical notes associated with the top-activated images. We then employ concept activation vectors (CAVs) [22] to project the extracted concepts into a shared representation space and use submodular optimization to select a subset of candidate concepts, whose span can effectively define the latent space. For making the diagnostic predictions, we propose to decompose the visual representation of input images into a sparse linear combination of the concept embeddings, with the coefficients indicating the importance of the corresponding concepts. By fine-tuning the final classification head, we demonstrate that the model adheres to the CBMs abstraction while exhibiting reasoning behavior similar to that of the original opaque model.

We conduct experiments on three publicly available medical diagnostic benchmarks, which include dermoscopy images [40], fundus images [29], and chest X-ray images [21]. Our extensive experiments reveal that our models achieve performance comparable to their opaque counterparts and surpass previous CBMs, all while maintaining intrinsic interpretability. We also showcase several model explanations to clarify how they can can aid in understanding the model's reasoning process and support decision-making. Our main contributions include: (1) proposing a novel framework for extracting and selecting clinically meaningful concepts from trained opaque models, (2) presenting a novel form of CBMs by sparsely decomposing the visual representation into concept embeddings within the same representation space, and (3) demonstrating the efficacy and interpretability of our model through comprehensive experiments.

2 Method

Figure 1 illustrates the our proposed framework for dissecting opaque models and building surrogate CBMs. In Sect. 2.1, we introduce how to extract clinically meaningful concepts from opaque models. In Sect. 2.2, we clarify how to project the concept into the representation space and then present a submodular optimization to select representative concepts from large amounts of candidates. In Sect. 2.3, we describe how to build CBMs with the selected concepts.

2.1 Concepts Extraction via Dissecting Opaque Models

Given a pre-trained opaque model, we aim to analyze its visual representation and identify the concepts associated with each neuron. However, interpreting the network's raw features can be challenging [12], due to the phenomenon of

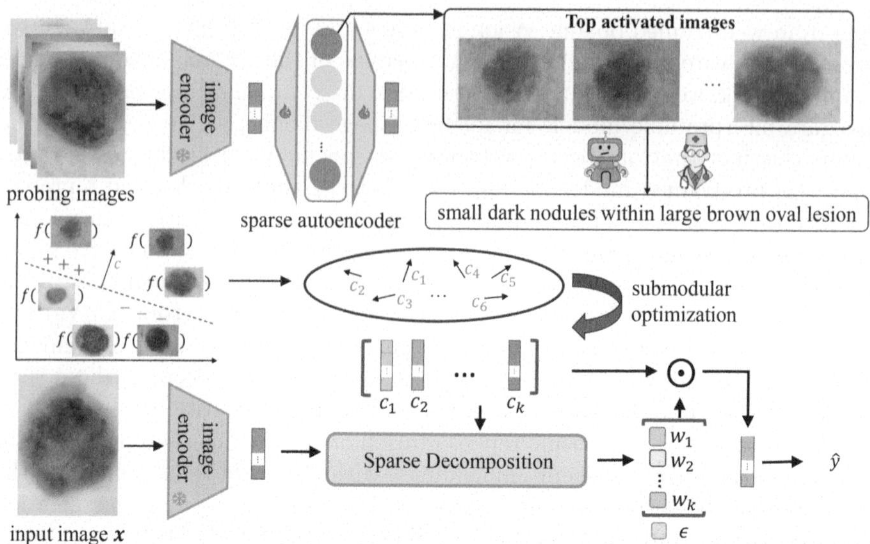

Fig. 1. Overview of our proposed framework: We dissect a pre-trained opaque model to extract learned concepts, project them into representation space, and use submodular optimization to select a representative concept pool. Finally, we apply sparse decomposition to the visual representation to create CBMs.

polysemanticity, where a single neuron may respond to multiple concepts with entirely different meanings. To address the issue, one promising approach for interpreting opaque models is SAEs [6], which decompose the representations into a sparse linear combination of learned dictionary vectors. Existing research has demonstrated that SAEs can effectively untangle learned representations into a limited set of human-understandable concepts [6,10].

Specifically, SAEs project the visual representation $\mathbf{I} \in \mathbb{R}^d$ into a hidden latent space of significantly higher dimension with a linear encoder $\mathbf{W}_E \in \mathbb{R}^{d \times h}$, followed by a ReLU non-linearity function, and finally re-project the representation back to the original space with a linear decoder $\mathbf{W}_D \in \mathbb{R}^{h \times d}$, i.e., $SAE(\mathbf{I}) = \mathbf{W}_D^T ReLU(\mathbf{W}_E^T \mathbf{I})$. The SAE is trained with an L_2 reconstruction loss together with L_1 sparse regularization on the hidden representation:

$$\mathcal{L}_{SAE}(\mathbf{I}) = \|\mathbf{I} - SAE(\mathbf{I})\|_2^2 + \lambda \|ReLU(\mathbf{W}_E^T \mathbf{I})\|_1, \tag{1}$$

where λ is a hyperparameter. In this process, the original representations are transformed into non-negative sparse latent representations. Previous studies [6,35] have shown that neurons in this latent space often align closely with human-understandable concepts. As a result, the representation is effectively disentangled into a non-negative combination of these concepts.

We then describe the concept associated with each neuron using plain text. For each neuron, we collected top m activated images (e.g., $m = 10$). Ideally, experts would summarize the common patterns in these images, but this can be labor-intensive. To streamline this process, we introduce an automated pipeline for naming concepts with the assistance of LLMs. For each image, we can collect relevant clinical notes or use multi-modal LLMs to generate descriptions for the symptoms within the image. We then prompt ChatGPT to summarize these descriptions and output a common concept hidden within the descriptions. This concept would be highly likely to be associated with the neuron.

2.2 Concept Embeddings Generation and Selection

The previous pipeline faces challenges as not all neurons in the hidden space align with human-interpretable concepts, with some acting as uninterpretable shortcuts linking inputs to predictions [22]. Additionally, summarizing a limited number of images to extract concepts may introduce noisy or fake concepts, while the high-dimensional latent space of SAEs often includes redundant or correlated entries, increasing model complexity and reducing interpretability. To mitigate these issues, we propose to rebuild the embeddings of the concepts using the samples screened by LLMs, ensuring that the embeddings faithfully correspond to the summarized concepts (even if these concepts do not precisely align with the original latent representations of the SAEs). Afterward, we utilize submodular optimization [2] to filter out random concepts and select a subset of the most representative concepts that capture the variance within the space.

Specifically, we utilize CAVs [22] to project the concepts into the representation space. For each neuron, we collect n top activated images and n images with zero activations ($n \gg m$). We then prompt ChatGPT to verify if the concept exists in the top activated image and absent from the zero-activated image, by telling the ChatGPT the description of the concept together with the clinical note (or generated description) of each image. We filter out and only keep the images that pass the ChatGPT verification to ensure the derived embedding is faithfully associated to the concept, resulting in n_1 positive cases and n_2 negative cases. We collect the representations for these cases, and train a linear SVM to learn the corresponding CAV, which is the vector normal to the linear classification boundary. To this end, we obtain the concept embeddings $\mathbf{c} \in \mathbb{R}^d$, which is in the same latent space as the visual representations.

Once we obtain the concept pool composed of C concept embeddings $\mathcal{C} = [\mathbf{c}_1, \cdots, \mathbf{c}_C]$, we want to find a subset with a pre-defined cardinality k: $\mathcal{C}_s \subset \mathcal{C}, |\mathcal{C}_s| = k$. We further define the coverage score for the subset as:

$$\text{cover}(\mathcal{C}_s) = \sum_{\mathbf{c}_1 \in \mathcal{C}} \max_{\mathbf{c}_2 \in \mathcal{C}_s} \cos(\mathbf{c}_1, \mathbf{c}_2). \tag{2}$$

The objective is to find the subset $\mathcal{C}_s$ that maximize the coverage score. We note this objective satisfies the diminishing returns property. As a result, we follow [43] and utilize a greedy algorithm to find the optimal solution, which results in a smaller and more representative concept pool for building CBMs.

It is important to note that this process does not guarantee the extracted concepts will precisely align with the original meanings of the SAEs' latent space, as they are derived from a limited set of samples. However, we can ensure that: 1) the embeddings accurately represent the extracted concepts, 2) the concepts are discriminative with respect to the data, and 3) the model's behavior is faithfully interpreted through the extracted concepts.

2.3 Class-Concept Association by Representation Decomposition

Having obtained a filtered concept pool $\mathcal{C}_s = [\mathbf{c}_1, \cdots, \mathbf{c}_k]$, the next step to build CBMs is to assign a score to each concept for a given image. We observe that the concept embeddings reside within the same representation space as the visual representation, allowing us to decompose the visual representation as a weighted sum of concept embeddings. Furthermore, due to the sparsity of the hidden layers in SAEs, a visual representation will assign non-zero scores to only a subset of concepts. Thus, we aim for this visual representation decomposition to be sparse as well. Mathematically, for the visual representation $\mathbf{I}_i$ of an image, we can express it as: $\mathbf{I}_i = \hat{\mathbf{I}}_i + \epsilon_i = \sum_{j=1}^{k} w_j^i \mathbf{c}_j + \epsilon_i$, where ϵ_i is the residue term and w_j^i are coefficients with only $k' \ll k$ of them being non-zero. This can be accomplished by sparse coding algorithms such as orthogonal matching pursuit [32]. We then discard the residue term and retain only the weighted sum as an approximation of the original visual representation. Finally, we fit a linear function as a classification head for making predictions, expressed as: $\hat{y}_i = \arg\max(\mathbf{W}^T \hat{\mathbf{I}}_i + \mathbf{b})$. Note this can be reformulated as:

$$\hat{y}_i = \arg\max\,(\sum_{j=1}^{k} w_j^i \mathbf{W}^T \mathbf{c}_j + \mathbf{b}). \tag{3}$$

Therefore, this approach aligns with the abstraction of CBMs, where the input is first mapped to a score for each bottleneck concept, and the final prediction is determined entirely by these concept scores through a linear function. In this context, the coefficients $[w_1^i, \cdots, w_k^i]$ represents the concept scores, and $\mathbf{W}^T[\mathbf{c}_1, \cdots, \mathbf{c}_k]$ serves as the class-concept weight matrix. In contrast to traditional CBMs that focus on training a sparse class-concept weight matrix to enhance model interpretability, our method emphasizes sparsity on the concept score side by applying sparse decomposition to the visual representation.

3 Experimental Setup

3.1 Dataset

We conduct experiments on three publicly-available dataset, namely:

- **HAM10000** [40]: The dataset contains 10,015 dematoscopic images for skin lesion classifications. The targeted labels contains 7 classes. The dataset is split into train/dev/test set with 8,010/1,000/1,005 images.

Table 1. Comparison between our CBMs and their corresponding opaque models. We report mean accuracy (%) and s.td. based on three random trials.

[1.5pt] Models	Inter.	Dermoscopy	Fundus	X-Ray	Average
Densenet [20]	✗	87.53$_{\pm 0.12}$	74.80$_{\pm 0.51}$	**89.37**$_{\pm 0.01}$	83.90$_{\pm 0.14}$
Ours (Densenet)	✓	**88.33**$_{\pm 0.35}$	**75.07**$_{\pm 0.60}$	89.35$_{\pm 0.07}$	**84.25**$_{\pm 0.26}$
ResNet [18]	✗	87.13$_{\pm 0.66}$	75.12$_{\pm 0.10}$	**89.36**$_{\pm 0.06}$	83.88$_{\pm 0.18}$
Ours (ResNet)	✓	**87.46**$_{\pm 0.26}$	**75.13**$_{\pm 0.08}$	89.25$_{\pm 0.01}$	**83.94**$_{\pm 0.09}$
ConvNext [28]	✗	89.65$_{\pm 0.20}$	73.93$_{\pm 0.16}$	89.30$_{\pm 0.08}$	84.29$_{\pm 0.10}$
Ours (ConvNext)	✓	**89.72**$_{\pm 0.49}$	**75.00**$_{\pm 0.91}$	**89.36**$_{\pm 0.03}$	**84.69**$_{\pm 0.23}$
ViT [11]	✗	85.54$_{\pm 0.11}$	66.08$_{\pm 2.26}$	89.06$_{\pm 0.05}$	80.23$_{\pm 0.76}$
Ours (ViT)	✓	**85.57**$_{\pm 0.17}$	**66.37**$_{\pm 2.16}$	**89.08**$_{\pm 0.05}$	**80.34**$_{\pm 0.77}$

- **Harvard-FairVLMed** [29]: The dataset contains 10,000 fundus images with clinical notes for glaucoma classification. The response variable is a binary variable indicating whether the patient has glaucoma or not. The dataset is split into train/dev/test set with 7,000/1,000/2,000 images.
- **MIMIC-CXR** [21]: The dataset contains 370,955 chest X-ray and the corresponding radiology report. 14 diagnostic labels are extracted from the radiology report with one to be "no findings". We discarded the "no findings" label and combined all non-positive labels ("negative", "not mentioned", "uncertain") into an aggregate "negative" label for simplicity, resulting in 13 binary classification tasks. We followed previous work [9] and excluded samples without the findings section, and divided the dataset into train/dev/test set with 270,790/2,130/3,858 images according to its original split.

3.2 Implementation Details

For concept extraction, we use Llama-3.2-11B-Vision-Instruct to generate descriptions for HAM10000 and use the ground truth clinical notes for the other two datasets. GPT-4o is used for summarizing the concepts. CAVs are calculated using scikit-learn's implementation of SVM with default hyperparameters. For sparse decomposition, we use scikit-learn's implementation of orthogonal matching pursuit. To train the linear classification head, we train the scikit-learn's L-BFGS logistic regression with a hyper-parameter sweep on the L_2 regularization weight. This is achieved through a binary search on the validation set, starting with the range $[1e^6, 1e^4, 1e^2, 1, 1e^{-2}, 1e^{-4}, 1e^{-6}]$ and iteratively halving the interval over eight steps. The size of the concept pool k is tuned from $[500, 600, 700, 800, 1000]$ based on validation accuracy. The opaque models are trained with a batch size of 16 and a learning rate of $5e^{-5}$ over 200 epochs. The SAEs are trained using images from the training set, with a learning rate of $5e^{-4}$ and L_1 sparsity set to $3e^{-5}$. Both models utilize Adam [24] optimizers. All experiments were conducted on NVIDIA GeForce RTX 4090 GPUs.

4 Evaluation

4.1 Comparison with Opaque Models

We begin by comparing the performance of our proposed CBMs with the original opaque models. This comparison involves various backbones for the opaque models, including Densenet [20], ResNet [18], ConvNext [28], and ViT [11]. Specifically, we use ResNet-50 and ConvNext-Base for the MIMIC-CXR dataset, while ResNet-18 and ConvNext-Tiny are employed for the other two datasets. Both Densenet-121 and ViT-B are applied across all datasets. For each opaque model, we dissect the network and build the corresponding CBMs, and compare their performance. The results are shown in Table 1. Our key finding is that the CBMs achieve performance comparable to that of the original opaque models, with most metrics showing improvements. Since our CBMs utilize concepts extracted from the pre-trained opaque models, they can faithfully replicate the original behaviors while providing a reasoning process that is more understandable to humans. This allows for enhanced interpretability without compromising accuracy. Furthermore, our framework is highly flexible, enabling the transformation of opaque models with any backbone into interpretable models, provided they can generate visual embeddings prior to the classification heads.

Table 2. Comparison between our proposed CBMs and several CBMs-type baselines. We report mean accuracy (%) and s.td. based on three random trials.

[1.5pt] Dermoscopy	PCBM [44]	Labo [43]	label-free CBM [30]	DN-CBM [35]	Ours
Accuracy	$74.83_{\pm 0.30}$	$81.39_{\pm 0.01}$	$85.27_{\pm 0.71}$	$86.96_{\pm 0.40}$	$\mathbf{88.33}_{\pm 0.35}$

4.2 Comparison with Baseline CBMs

We then compare our method with several baseline CBMs, including Labo [43], label-free CBM [30], PCBM [44], and DN-CBM [35]. This comparison focuses on the HAM10000 dataset, which has been extensively studied in the literature, providing a wealth of candidate concepts (both expert-selected and automatically generated) for evaluation. Since Labo relies on a pre-trained CLIP model, we use CLIP with ViT-L-14 [11] as its backbone. For the remaining methods, including ours, we fix the backbone to Densenet-121 to ensure a fair comparison. We also utilize the concepts extracted from their original implementations. The results, presented in Table 2, indicate that our proposed CBMs achieve state-of-the-art performance, surpassing all previous baselines by 1.37% compared to the second-best method. Some of the baselines depend heavily on pre-trained vision-language models, which excel with natural images but are less effective for domain-specific tasks [17], limiting their advantage in medical diagnosis. Additionally, the concepts used by these baselines are often manually

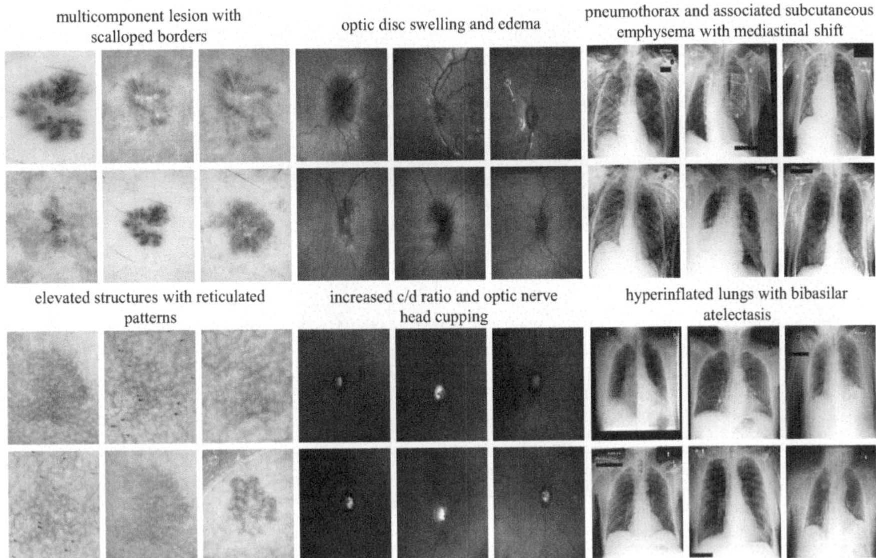

Fig. 2. Illustration of extracted concepts. We show sample concepts summarized by GPT and the corresponding top activated images for each concept.

selected or generated by large language models based on prior knowledge, making them data-independent and model-agnostic. These extracted concepts may not be representative of certain populations or may be difficult for specific networks to discriminate, reducing the efficiency of CBM construction. In contrast, our concepts are derived from models trained on the specific dataset, taking into account both the data distribution and the visual encoder's representation learning capabilities. Our approach is also flexible enough to represent a single clinical symptom that exhibits different patterns across subgroups with multiple concepts, effectively capturing variations among different samples.

4.3 Examples of Explanations

We present several examples of concepts extracted from the trained network. As illustrated in Fig. 2, we showcase two examples of extracted concepts for each dataset, along with the top activated images. For this illustration, we utilize the Densenet-121 version of the models. We observe that the top activated images display a high degree of visual similarity. Moreover, this similarity can be encapsulated by meaningful clinical concepts, indicating that the original opaque models can indeed capture clinically relevant patterns. Given the variability and uniqueness of certain clinical symptoms, some concepts may correspond to multiple clusters with distinct visual characteristics. Including sample images alongside text descriptions can significantly enhance clinical decision-making.

We further provide examples to illustrate the model's reasoning process. In Fig. 3, we present several cases along with their predictions and the top four con-

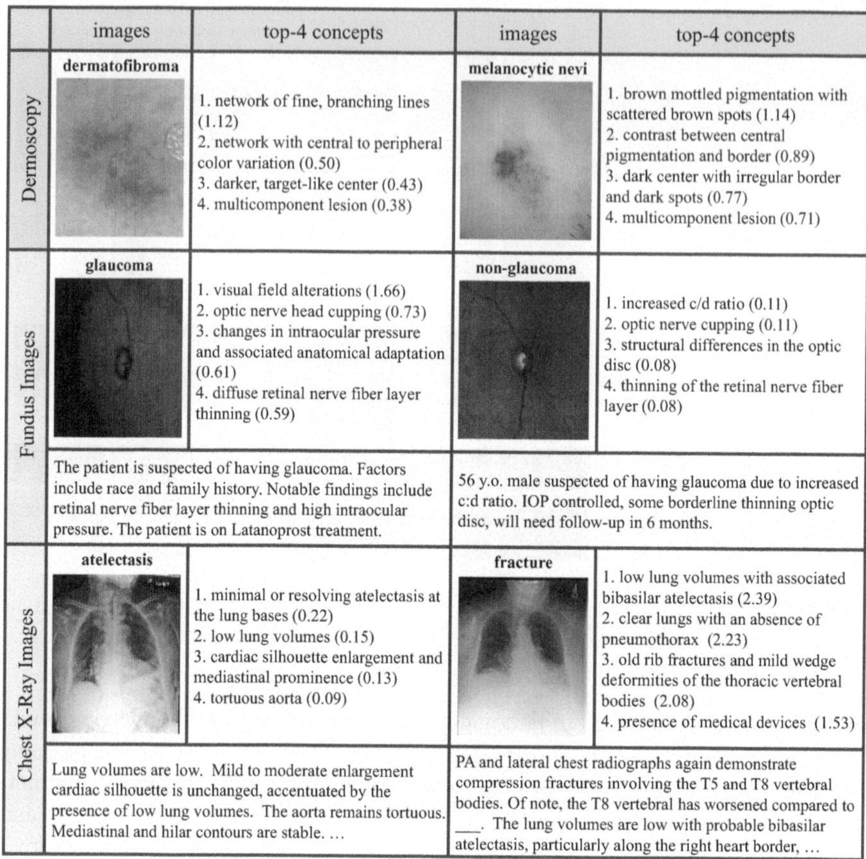

Fig. 3. Examples of explanations provided by our method. For each input image, we show top-4 concepts and their contributions to the logits of the correct labels. We also present the corresponding reports of Harvard-FairVLMed and MIMIC-CXR for references. Some descriptions of the normal findings are omitted.

cepts that contribute most significantly to the logits for those predictions. For the Harvard-FairVLMed and MIMIC-CXR, we also include the corresponding reports for reference. We find that the explanations offer an accurate description of the symptoms depicted in the images, which aligns well with the clinical notes. The concept descriptions are detailed and specific, often more precise than manually selected concepts. Additionally, the concepts are comprised of visual features that can be directly inferred from the images, which is an improvement over GPT-generated concepts that may include a significant proportion of non-visual or nonsensical features [43]. Through CBM modeling, the generated explanations usually establish meaningful causal relationships with the predicted labels, aiding clinicians in better understanding the model's reasoning process.

4.4 Interventions

One advantage of CBMs is their ease of model intervention. Here, we provide two examples of how we can adjust the models to correct potential mispredictions. As shown in Fig. 4, we can successfully rectify the model's incorrect predictions by modifying the concept scores. For instance, in the first image depicting a benign keratosis-like lesion, we identify a concept defined as "a dark center with a lighter periphery or halo, with irregular and slightly raised borders" that does not accurately represent the image. The lesion indeed has raised borders, but the center is actually lighter than the edges. By setting the score for this concept to zero, the prediction is corrected. The second example illustrates a similar correction. These cases demonstrate that, with the aid of explanations, domain experts can intervene in the reasoning process to enhance overall model performance.

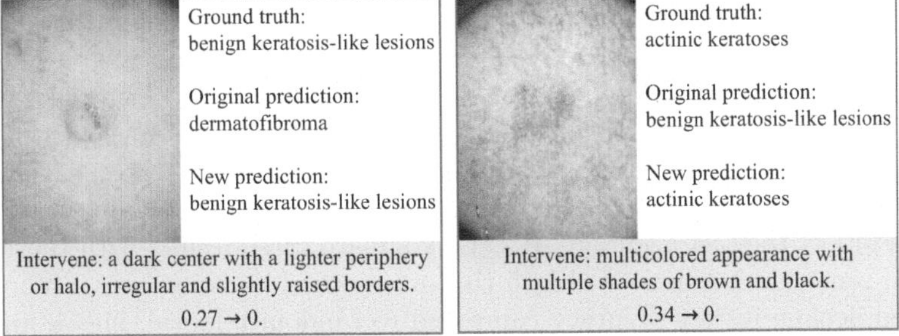

Fig. 4. Examples of two edits we performed on our model predictions. For both cases, we find a concept with high score but does not match the images. By zeroing the incorrect activations, we successfully fix the error.

4.5 Analytical Studies

In this section, we study the faithfulness of the explanation to the model's reasoning process. We also conduct sensitivity analysis to study the effects of sparsity in visual representation decomposition on the model performance. The experiments are conducted on the HAM10000 dataset with Densenet-121 as the backbone.

Deletion. We follow the idea of [33] to examine the importance of the concepts highlighted by the models. Specifically, for each class, our model ranks the concepts based on their importance for predicting that label, as indicated by the class-concept weight matrix. For an image predicted to belong to class y_0, we discard top $x\%$ of the most important concepts for predicting y_0, and then we reapply decomposition and label prediction. By gradually increasing x, we observe how the test accuracy declines. A more rapid decrease indicates that the masked concepts are more crucial for the model to make predictions. We use a random mask as a baseline for comparison. The results, shown in Fig. 5

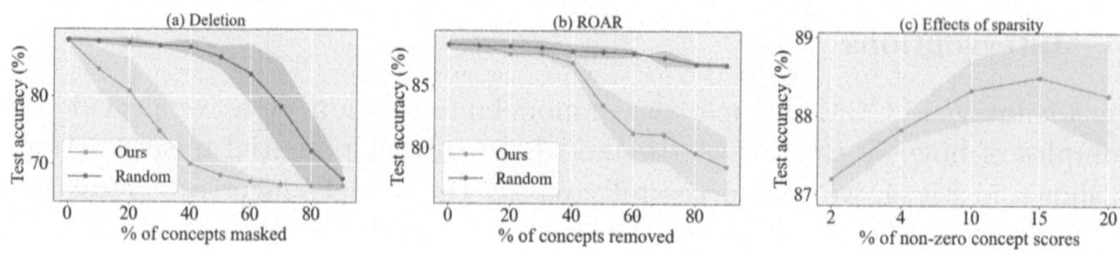

Fig. 5. Test accuracy when (a) masking the top label-related concepts, (b) removing the concepts highlighted by the explanation and retraining the classification head, and (c) changing the sparsity of representation decomposition.

(a), indicate that our method exhibits a more pronounced decline in test accuracy compared to random masks, demonstrating that the concept explanations faithfully reflect the model's reasoning process.

Remove and Retrain. We further apply a variant of Remove and Retrain (ROAR) [19] for explanation analysis. Specifically, for each sample, we identify the top $x\%$ of the concepts that contribute the most to the logits of the predicted labels. We then remove these top concepts from the concept pool and reapply representation decomposition, resulting in a newly fitted representation. The final classification head is retrained using this new set of representations, and we report the test accuracy. We also compare these results with those from random removal. As shown in Fig. 5 (b), our method demonstrates a more pronounced decline in test accuracy compared to random removal, indicating that the explanations effectively highlight the most task-related concepts.

Effects of Sparsity. Finally, we examine the impact of sparsity during the visual representation decomposition on model performance. The orthogonal matching pursuit algorithm features a hyperparameter that governs the number of non-zero coefficients in the fitted linear function. We tune this hyperparameter to observe how test accuracy varies. The results are shown in Fig. 5 (c). We find that the model's performance is generally robust to changes in sparsity, with accuracy slightly improving as sparsity decreases. Overall, higher sparsity reduces model complexity, making it easier to interpret. Consequently, a sparsity level of 10% strikes a favorable balance between model performance and interpretability.

5 Conclusions and Future Work

In this paper, we present a novel pipeline for constructing interpretable medical image diagnostic concept bottleneck models based on pre-trained opaque models. This pipeline involves dissecting the opaque models to extract learned concepts, calculating concept embeddings, selecting the most informative concepts, and building CBMs through visual representation decomposition. It shows good interpretability as well as better performance compared to existing CBMs. However, the current method has several limitations. Firstly, it requires paired clinical reports for concept extraction, which are not always available. Secondly, for data lacking clinical reports, we rely on VLMs to generate reports, which depends heavily on the VLMs' performance and is only applicable to certain modalities that have access to modality-specific VLMs or show similarity to nature images. Thirdly, the model's performance is contingent upon the efficacy of the original opaque models. In the future, we aim to improve our methods and to address these limitations. We would also conduct larger-scale user studies to explore the method's utility in real clinical applications.

Acknowledgments. The work described in this paper was partially supported by a grant from the Research Grants Council of the Hong Kong Special Administrative Region, China (Project Reference Number: T45-401/22-N), and partially supported by National Natural Science Foundation of China (No. 62301311).

References

1. Ahn, Y.H., Kim, H.B., Kim, S.T.: WWW: a unified framework for explaining what where and why of neural networks by interpretation of neuron concepts. In: Proceedings of the IEEE/CVF Conference on Computer Vision and Pattern Recognition, pp. 10968–10977 (2024)
2. Bach, F.: Convex analysis and optimization with submodular functions: a tutorial. arXiv preprint arXiv:1010.4207 (2010)
3. Bai, N., Iyer, R.A., Oikarinen, T., Weng, T.W.: Describe-and-dissect: interpreting neurons in vision networks with language models. arXiv preprint arXiv:2403.13771 (2024)
4. Bau, D., Zhou, B., Khosla, A., Oliva, A., Torralba, A.: Network dissection: quantifying interpretability of deep visual representations. In: Proceedings of the IEEE Conference on Computer Vision and Pattern Recognition, pp. 6541–6549 (2017)
5. Bouza, J.J., Yang, C.H., Vemuri, B.C.: Geometric deep learning for unsupervised registration of diffusion magnetic resonance images. In: Frangi, A., de Bruijne, M., Wassermann, D., Navab, N. (eds.) Information Processing in Medical Imaging. IPMI 2023. LNCS, vol. 13939pp. 563–575. Springer, Cham (2023). https://doi.org/10.1007/978-3-031-34048-2_43
6. Bricken, T., et al.: Towards monosemanticity: Decomposing language models with dictionary learning. Transformer Circuits Thread (2023). https://transformer-circuits.pub/2023/monosemantic-features/index.html
7. Brown, T.B., et al.: Language models are few-shot learners. In: Proceedings of the 34th International Conference on Neural Information Processing Systems. NIPS 2020, Curran Associates Inc., Red Hook, NY, USA (2020)
8. Chen, C., Li, O., Tao, D., Barnett, A., Rudin, C., Su, J.K.: This looks like that: deep learning for interpretable image recognition. In: Advances in Neural Information Processing Systems, vol. 32 (2019)
9. Chen, Z., Song, Y., Chang, T.H., Wan, X.: Generating radiology reports via memory-driven transformer. In: Proceedings of the 2020 Conference on Empirical Methods in Natural Language Processing (EMNLP), pp. 1439–1449 (2020)
10. Cunningham, H., Ewart, A., Riggs, L., Huben, R., Sharkey, L.: Sparse autoencoders find highly interpretable features in language models. arXiv preprint arXiv:2309.08600 (2023)

11. Dosovitskiy, A., et al.: An image is worth 16×16 words: transformers for image recognition at scale. In: International Conference on Learning Representations (2021). https://openreview.net/forum?id=YicbFdNTTy
12. Elhage, N., et al.: Toy models of superposition. arXiv preprint arXiv:2209.10652 (2022)
13. Frosst, N., Hinton, G.: Distilling a neural network into a soft decision tree. arXiv preprint arXiv:1711.09784 (2017)
14. Gong, S., Dou, Q., Farnia, F.: Structured gradient-based interpretations via norm-regularized adversarial training. In: Proceedings of the IEEE/CVF Conference on Computer Vision and Pattern Recognition, pp. 11009–11018 (2024)
15. Gong, S., Lu, W., Xie, J., Zhang, X., Zhang, S., Dou, Q.: Robust cardiac MRI segmentation with data-centric models to improve performance via intensive pre-training and augmentation. In: Camara, O., et al. Statistical Atlases and Computational Models of the Heart. Regular and CMRxMotion Challenge Papers. STACOM 2022. LNCS, vol. 13593, pp. 494–504. Springer, Cham (2022). https://doi.org/10.1007/978-3-031-23443-9_47
16. Gong, S., et al.: Segmentation of tiny intracranial hemorrhage via learning-to-rank local feature enhancement. In: 2024 IEEE International Symposium on Biomedical Imaging (ISBI), pp. 1–5 (2024). https://doi.org/10.1109/ISBI56570.2024.10635853
17. Gong, S., et al.: 3DSAM-adapter: Holistic adaptation of SAM from 2d to 3d for promptable tumor segmentation. Med. Image Anal. **98**, 103324 (2024)
18. He, K., Zhang, X., Ren, S., Sun, J.: Deep residual learning for image recognition. In: Proceedings of the IEEE Conference on Computer Vision and Pattern Recognition, pp. 770–778 (2016)
19. Hooker, S., Erhan, D., Kindermans, P.J., Kim, B.: A benchmark for interpretability methods in deep neural networks. In: Advances in Neural Information Processing Systems, vol. 32 (2019)
20. Huang, G., Liu, Z., Van Der Maaten, L., Weinberger, K.Q.: Densely connected convolutional networks. In: Proceedings of the IEEE Conference on Computer Vision and Pattern Recognition, pp. 4700–4708 (2017)
21. Johnson, A.E., et al.: Mimic-CXR, a de-identified publicly available database of chest radiographs with free-text reports. Sci. Data **6**(1), 317 (2019)
22. Kim, B., et al.: Interpretability beyond feature attribution: quantitative testing with concept activation vectors (TCAV). In: International Conference on Machine Learning, pp. 2668–2677. PMLR (2018)
23. Kim, E., Jung, D., Park, S., Kim, S., Yoon, S.: Probabilistic concept bottleneck models. In: International Conference on Machine Learning, pp. 16521–16540. PMLR (2023)
24. Kingma, D., Ba, J.: Adam: a method for stochastic optimization. In: International Conference on Learning Representations (ICLR), San Diega, CA, USA (2015)
25. Koh, P.W., et al.: Concept bottleneck models. In: International Conference on Machine Learning, pp. 5338–5348. PMLR (2020)
26. Lin, M., Feragen, A., Bashir, Z., Tolsgaard, M.G., Christensen, A.N.: I saw, i conceived, i concluded: Progressive concepts as bottlenecks. arXiv preprint arXiv:2211.10630 (2022)
27. Liu, J., Ge, R., Wan, P., Zhu, Q., Zhang, D., Shao, W.: Multi-task multi-instance learning for jointly diagnosis and prognosis of early-stage breast invasive carcinoma from whole-slide pathological images. In: Frangi, A., de Bruijne, M., Wassermann, D., Navab, N. (eds) Information Processing in Medical Imaging. IPMI 2023. LNCS, vol. 13939, pp. 145–157. Springer, Cham (2023). https://doi.org/10.1007/978-3-031-34048-2_12

28. Liu, Z., Mao, H., Wu, C.Y., Feichtenhofer, C., Darrell, T., Xie, S.: A convnet for the 2020s. In: Proceedings of the IEEE/CVF Conference on Computer Vision and Pattern Recognition, pp. 11976–11986 (2022)
29. Luo, Y., et al.: Fairclip: harnessing fairness in vision-language learning. In: Proceedings of the IEEE/CVF Conference on Computer Vision and Pattern Recognition, pp. 12289–12301 (2024)
30. Oikarinen, T., Das, S., Nguyen, L., Weng, L.: Label-free concept bottleneck models. In: International Conference on Learning Representations (2023)
31. Oikarinen, T., Weng, T.W.: CLIP-dissect: automatic description of neuron representations in deep vision networks. In: The Eleventh International Conference on Learning Representations (2023). https://openreview.net/forum?id=iPWiwWHc1V
32. Pati, Y.C., Rezaiifar, R., Krishnaprasad, P.S.: Orthogonal matching pursuit: Recursive function approximation with applications to wavelet decomposition. In: Proceedings of 27th Asilomar Conference on Signals, Systems and Computers, pp. 40–44. IEEE (1993)
33. Petsiuk, V., Das, A., Saenko, K.: Rise: randomized input sampling for explanation of black-box models. In: Proceedings of the British Machine Vision Conference (BMVC) (2018)
34. Radford, A., et al.: Learning transferable visual models from natural language supervision. In: International Conference on Machine Learning, pp. 8748–8763. PMLR (2021)
35. Rao, S., Mahajan, S., Böhle, M., Schiele, B.: Discover-then-name: task-agnostic concept bottlenecks via automated concept discovery. In: Leonardis, A., Ricci, E., Roth, S., Russakovsky, O., Sattler, T., Varol, G. (eds.) Computer Vision. ECCV 2024. LNCS, vol. 15135, pp. 444–461. Springer, Cham (2024). https://doi.org/10.1007/978-3-031-72980-5_26
36. Ribeiro, M.T., Singh, S., Guestrin, C.: " why should i trust you?" explaining the predictions of any classifier. In: Proceedings of the 22nd ACM SIGKDD International Conference on Knowledge Discovery and Data Mining, pp. 1135–1144 (2016)
37. Rudin, C.: Stop explaining black box machine learning models for high stakes decisions and use interpretable models instead. Nat. Mach. Intell. **1**(5), 206–215 (2019)
38. Scott, M., Su-In, L.: A unified approach to interpreting model predictions. Adv. Neural. Inf. Process. Syst. **30**, 4765–4774 (2017)
39. Selvaraju, R.R., Cogswell, M., Das, A., Vedantam, R., Parikh, D., Batra, D.: Grad-cam: visual explanations from deep networks via gradient-based localization. In: Proceedings of the IEEE International Conference On Computer Vision. pp. 618–626 (2017)
40. Tschandl, P., Rosendahl, C., Kittler, H.: The ham10000 dataset, a large collection of multi-source dermatoscopic images of common pigmented skin lesions. Sci. Data **5**(1), 1–9 (2018)
41. Wolf, T.N., Pölsterl, S., Wachinger, C.: Don't panic: prototypical additive neural network for interpretable classification of Alzheimer's disease. In: Frangi, A., de Bruijne, M., Wassermann, D., Navab, N. (eds) Information Processing in Medical Imaging. IPMI 2023. LNCS, vol. 13939, pp. 82–94. Springer, Cham (2023). https://doi.org/10.1007/978-3-031-34048-2_7
42. Wu, N., Zhang, M.: Neurepdiff: neural operators to predict geodesics in deformation spaces. In: Frangi, A., de Bruijne, M., Wassermann, D., Navab, N. (eds.) Information Processing in Medical Imaging. IPMI 2023. LNCS, vol. 13939, pp. 588–600. Springer, Cham (2023). https://doi.org/10.1007/978-3-031-34048-2_45

43. Yang, Y., Panagopoulou, A., Zhou, S., Jin, D., Callison-Burch, C., Yatskar, M.: Language in a bottle: language model guided concept bottlenecks for interpretable image classification. In: Proceedings of the IEEE/CVF Conference on Computer Vision and Pattern Recognition, pp. 19187–19197 (2023)
44. Yuksekgonul, M., Wang, M., Zou, J.: Post-hoc concept bottleneck models. In: The Eleventh International Conference on Learning Representations (2023)
45. Yun, T., Bhalla, U., Pavlick, E., Sun, C.: Do vision-language pretrained models learn composable primitive concepts? arXiv preprint arXiv:2203.17271 (2022)

BioSonix: Can Physics-Based Sonification Perceptualize Tissue Deformations From Tool Interactions?

Veronica Ruozzi[1(✉)], Sasan Matinfar[2,4], Laura Schütz[2], Benedikt Wiestler[3,4], Alberto Redaelli[1], Emiliano Votta[1], and Nassir Navab[2,4]

[1] Department of Electronics Information and Bioengineering, Politecnico Di Milano, Milan, Italy
veronica.ruozzi@polimi.it

[2] Computer Aided Medical Procedures, Technische Universität München, Munich, Germany

[3] Neuroradiology Department, Klinikum Rechts der Isar, TU Munich, Munich, Germany

[4] Munich Center for Machine Learning (MCML), Munich, Germany

Abstract. Perceptualizing tool interactions with deformable structures in surgical procedures remains challenging, as unimodal visualization techniques often fail to capture the complexity of these interactions due to constraints such as occlusion and limited depth perception. This paper presents a novel approach to augment tool navigation in mixed reality environments by providing auditory representations of tool-tissue dynamics, particularly for interactions with soft tissue. BioSonix, a physics-informed design framework, utilizes tissue displacements in 3D space to compute excitation forces for a sound model encoding tissue properties such as stiffness and density. Biomechanical simulations were employed to model particle displacements resulting from tool-tissue interactions, establishing a robust foundation for the method. An optimization approach was used to define configurations for capturing diverse interaction scenarios with varying tool trajectories. Experiments were conducted to validate the accuracy of the sound-displacement mappings. Additionally, two user studies were performed: the first involved two clinical professionals (a neuroradiologist and a cardiologist), who confirmed the method's impact and achieved high task accuracy; the second included 22 biomedical experts, who demonstrated high discrimination accuracy in tissue differentiation and targeting tasks. The results revealed a strong correlation between tool-tissue dynamics and their corresponding auditory profiles, highlighting the potential of these sound representations to enhance the intuitive understanding of complex interactions.

Keywords: Mixed Reality · Sonification · Biomechanical Modelling · Medical Augmented Reality · Minimal Invasive Surgery

V. Ruozzi and S. Matinfar—Contributed equally to this work.

I. Oguz et al. (Eds.): IPMI 2025, LNCS 15830, pp. 19–33, 2026.
https://doi.org/10.1007/978-3-031-96625-5_2

1 Introduction

Intraoperative decision-making relies heavily on medical imaging; however, interpreting these images is influenced by cognitive biases, as noted in [7]. The challenges are heightened in percutaneous and minimally invasive surgeries, where constrained access, lack of haptic feedback, and reliance on two-dimensional imaging place significant cognitive demands on surgeons. These limitations hinder spatial reasoning, instrument localization, and dynamic anatomical mapping. Robot-assisted surgeries add to the difficulty by introducing trust issues with the system. The complexity intensifies during soft tissue manipulation, such as tumor resection, where tissue deformations deviate from preoperative plans, complicating real-time awareness and precise instrument placement within narrow, deformable pathways. Medical Augmented Reality Systems (MARS) have shown great promise in improving preoperative planning, real-time visualization, and intervention precision. By augmenting patient anatomy, tool pathways, and catheter positions, MARS enhances spatial understanding of tool-target relationships. Its effectiveness in image-guided interventions is evident in orthopedics [34], brain surgery [3], and other areas such as pulmonary and abdominal procedures [8,38].

However, unimodal visual channels impose limitations, including occlusion, depth perception challenges, and frequent shifts of focus between displays, complicating hand-eye coordination. Humans excel at solving complex problems and performing intricate tasks through multisensory processing [10,30]. By integrating visual, tactile, and auditory inputs, they form a holistic understanding of their environment, shaped by technical skills, experience, and intuition. However, current MARS systems fail to fully harness this capability, emphasizing the need for advanced cognitive support systems that bridge human-machine interaction. Delivering real-time, context-sensitive feedback aligned with human perception can enhance trust, augment cognitive abilities, improve precision, and empower surgeons to navigate complex scenarios with greater confidence.

1.1 Challenges in Soft Tissue Modeling and Augmentation

Current systems address soft tissue challenges through two primary strategies: modeling deformations accurately and augmenting these models effectively for the surgeon.

Soft Tissue Deformation Modeling – A major challenge in AR technologies is registering preoperative imaging with the surgical field, made complex by patient positioning and soft tissue deformations. While AR excels in rigid-body contexts like orthopedics, where fixed structures such as bones or the skull facilitate alignment, soft tissue applications face greater difficulties. For example, brain shifts during craniotomies or tissue configuration changes due to patient positioning complicate registration and tool accuracy [11]. Nonetheless, AR has shown promise in guiding soft tissue procedures like tumor ablation. By employing deformable registration techniques and integrating intraoperative imaging for real-time updates, AR systems can adapt to dynamic changes and improve surgical outcomes [1,5].

Most AR systems depend on preoperative imaging, limiting adaptability to real-time intraoperative changes. While some integrate intraoperative data for navigation [5], such cases are uncommon. The lack of dynamic adaptation to anatomical changes, especially those caused by tool navigation, is a significant limitation. This hinders the clinical adoption of AR systems, where real-time anatomical updates and precise targeting are crucial for successful minimally invasive procedures.

Perceptual Augmentation – Even with accurate deformation modeling, MARS faces the challenge of delivering intuitive, real-time feedback to surgeons and achieving seamless integration into the surgical workflow.

Haptic feedback, while essential for tool guidance, faces challenges such as integration issues, safety concerns with bi-directional systems, and replicating tactile sensations. Adoption is further limited by high costs, latency, and operating room constraints [25].

While the human visual system excels at spatial representation, the auditory system efficiently captures dynamic object characteristics with exceptional temporal resolution. By perceiving sonic qualities like pitch, timbre, and rhythm, it enables rapid comprehension of complex data and swift responses to changes, making auditory cues an intuitive and cost-effective medium for surgical tasks. This has led studies [15,24] to explore real-time acoustic sensing for tool-tissue interactions, though the signals often remain imperceptible to humans.

Sonification, the transformation of data into sound, has proven effective for surgical applications by helping surgeons maintain focus, reduce distractions, and optimize outcomes [4,18,19,22,26,37]. Methods for transforming basic data inputs, like multistate signals [20] or spatial mapping [22,37], have proven effective in simple scenarios. However, these approaches are often prescriptive and fail to capture the complexity or provide detailed, textural insights into tool-tissue interactions. The dynamic nature of surgery—shaped by patient-specific anatomy, procedural variability, and diverse strategies—poses challenges for standardizing augmentations. Despite AI and robotics advances, human expertise remains unmatched in integrating sensory inputs and adapting to complexities. Augmentation systems must meet the nuanced demands of surgery to effectively support highly trained surgeons.

Sonification of high-dimensional medical imaging has been explored to preserve data richness [12,13], but usability challenges often limit accessibility, especially for non-musical users. The human auditory system efficiently processes spectro-temporal sound features within fractions of a second, shaping cognitive patterns [23,33]. Subtle sound differences are easily perceived, with those resembling real-world auditory experiences often felt as more intuitive. The effectiveness of these sounds relies on the relationship between sound and gesture, as well as users' prior experiences with everyday objects [6,16,32]. A study proposed model-based sonification as an alternative to simple parametric mapping, transforming high-dimensional medical imaging into physics-driven sound models to intuitively represent tissue textures, with stiffer tissues sounding distinct from softer ones [21]. This approach showed promise in interactive scenarios like retinal surgery [17] and brain tumor localization [27]. However, it relies on preoperative static data, failing to account for soft tissue deformations and the complexities of tool-tissue interactions during surgery.

1.2 Contributions

This paper presents an auditory-based AR tool designed to enhance nuanced decision-making in complex surgical tasks, addressing the limitations of MARS in augmented soft tissue deformation. It advocates for a systematic sonification approach that moves beyond discrete tissue states, focusing on defining tool-tissue interactions to (1) develop a generalizable model adaptable to diverse scenarios and (2) provide detailed insights that enable surgeons to integrate prior knowledge, explore the surgical field, deepen anatomical understanding, and minimize noise and errors. The aim is to deliver perceptually intuitive feedback that enhances cognitive and procedural understanding while supporting decision-making.

As a proof of concept, we propose a methodology for dynamic modeling and sonification of tool-tissue interactions. The framework translates tissue particle displacements—reflecting tissue characteristics and interactions—into excitation forces within a physics-based sound model, enabling physics-driven auditory feedback. Biomechanical numerical simulations, specifically Finite Element Modeling (FEM), are employed to model tool-tissue interactions by probing the anatomical model across different regions and trajectories. These simulations enable the representation of diverse tool-tissue dynamics. We validate the approach through experiments analyzing the correlation between biomechanical displacement and acoustic output, optimizing the sonification model. User studies assess its technical validity, perceptual accuracy, and clinical relevance, providing guidelines for fine-tuning the model across various scenarios.

2 BioSonix Design Framework

The proposed framework, shown in Fig. 1, comprises three modules: (i) the Anatomical Domain Module, which reconstructs anatomy from pre-operative images while preserving tissue spatial information; (ii) the Mapping Module, which translates system physics into the Sonification Domain; and (iii) the Sonification Domain Module, which generates the final audio output. The following sections detail each module, their rationale, and interconnections.

2.1 Anatomical Domain Module

Surgical tool navigation strategies are typically planned using preoperative static volumetric images, such as CT scans. These images enable 3D reconstruction of the anatomical region of interest (ROI) and the identification of safe pathways

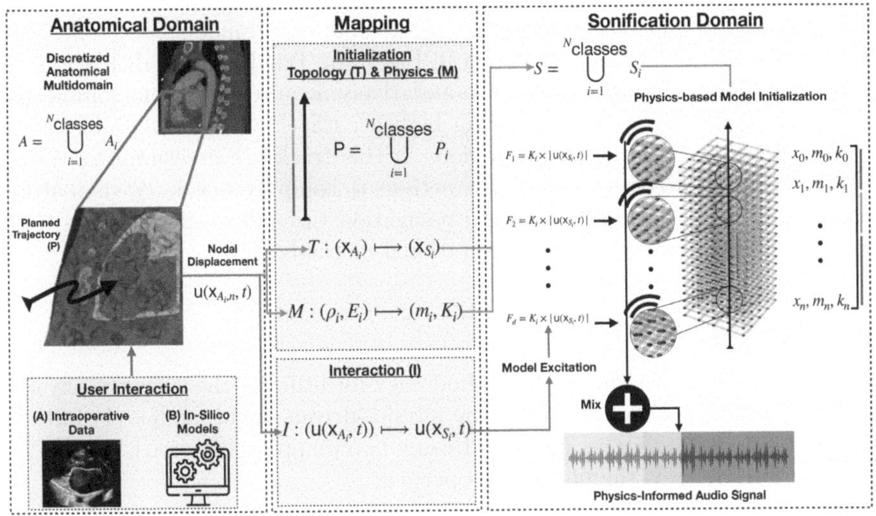

Fig. 1. BioSonix Design Framework – A General Overview

to target structures while avoiding critical tissues. Advanced processing software generates multidomain spatial discretizations of the ROI, preserving tissue classes and defining an anatomical discrete domain, A.

The anatomical domain A is represented as a union of subdomains, each corresponding to a specific tissue class. Mathematically, this can be expressed as:

$$A = \bigcup_{i=1}^{N_{classes}} A_i, \quad \text{with} \quad A_i = \bigcup_{e=1}^{N_{el,i}} V_e^{(i)}. \tag{1}$$

where A_i is the subdomain corresponding to tissue class i, $N_{classes}$ denotes the total number of tissue classes, $V_e^{(i)}$ represents a single volumetric element e within subdomain A_i, and $N_{el,i}$ is the number of volumetric elements in A_i.

In the surgical field, the domain undergoes solicitations from sources such as physiological motions (e.g., heartbeat and respiration) and interactions between surgical tools and anatomical structures. These solicitations generate a 3D displacement field over time, which can be referred to the nodes of the domain A, and expressed as $\mathbf{u}(\mathbf{x}A_i, n, t), \quad i = 1, \ldots, Nclasses, \ n = 1, \ldots, N_{nodes}$, where $\mathbf{x}A_i, n$ are the spatial coordinates of the n-th node in subdomain A_i, and N$nodes$ is the total number of nodes in the anatomical domain A.

The displacement field at time t is calculated as the difference in a node's spatial coordinates between consecutive time frames

$$\mathbf{u}(\mathbf{x}_{A_i,n}, t_i) = \mathbf{x}_{A_i,n}(t_i) - \mathbf{x}_{A_i,n}(t_{i-1}), \tag{2}$$

or for simplicity, we will refer to the nodal displacement as $\mathbf{u}(\mathbf{x}_{A_i}, t)$, omitting the nodal index n and focusing primarily on the tissue class i.

The temporal and spatial dynamics of the anatomical model can be derived through two distinct tracks, as shown in Fig. 1. In **Track A**, the displacement field $\mathbf{u}(\mathbf{x}A_i, t)$ is computed from intraoperatively acquired dynamic volumetric data, such as 3D ultrasound, ensuring high accuracy by capturing all motion sources. However, tool-tissue interactions in this track are deterministic. Conversely, **Track B** uses user-driven interactions to compute $\mathbf{u}(\mathbf{x}A_i, t)$, simulating dynamic behavior induced by surgical navigation through in-silico models that integrate system physics using computational methods.

2.2 Mapping Module

The Mapping Module defines the methodology for utilizing the system's physics in A to: (1) initialize and (2) excite the physics-driven sound model. (1) Initialization: The sound model is initialized using two mapping functions, M and T. The function M maps the physical properties:

$$M : (\rho_i, E_i) \longmapsto (m_i, K_i). \tag{3}$$

Each subdomain A_i is defined by specific density values ρ_i and Young's Modulus E_i, which approximate homogenized tissue properties and linear elastic mechanical behavior. The mapping function M_i converts these values into mass m and stiffness K_i for the mass-spring model underlying the sound domain's physical representation. To represent the spatial distribution of tissue classes, a new coordinate system $\boldsymbol{p}$ is defined along the planned safe trajectory. By projecting the intersected elements $V_e^{(i)}$ of the 3D anatomical domain A onto $\boldsymbol{p}$, the domain is reduced to a 1D coordinate system. This system is divided into segments labeled by tissue classes (i), where each segment's length represents the distance traversed through that tissue. The 1D domain $\boldsymbol{P}$ is defined as:
$\mathbf{P} : \bigcup_{i=1}^{N_{classes}} [0, \ell_i]$,
where ℓ_i is the segment length for tissue class i. The mapping function T then maps nodes from the Anatomical Domain A to the corresponding nodes in the Sound Domain S for each tissue class i:

$$T : (\mathbf{x}_{A_i}) \longmapsto (\mathbf{x}_{S_i}), \tag{4}$$

2) The module for exciting forces (F) applied within the Sound Domain S, in time and in space, is computed by the function I:

$$I : (\mathbf{u}(\mathbf{x}_{A_i}, t)) \longmapsto F(\mathbf{x}_{S_i}, t) = K_i \times |\mathbf{u}(\mathbf{x}_{S_i}, t)|. \tag{5}$$

2.3 Sonification Domain Module

The Sonification Domain Module is built on a physical model that implements mass-spring interactions, functioning as a virtual instrument initialized and excited by physical data. The domain S is defined as:
$S = \bigcup_{i=1}^{N_{4classes}} S_i$,

where S_i represents the subdomain corresponding to each tissue class i. Building on the sound physical model presented in [35], the defined coordinate system $\boldsymbol{p}$ serves as the central axis, extending along 3D orthogonal directions to form a complex network of masses interconnected by linear spring elements. The spatial distribution of tissue classes is propagated from $\boldsymbol{p}$ to neighboring regions through the mapping function T, resulting into subsets of masses and springs specific to each tissue class. Using M, each mass subset is characterized by a mass value m_i, and each spring subset is characterized by a stiffness value K_i. The virtual instrument is excited by defining a number d of *Drivers* at multiple locations around the model ($\mathbf{x}_{S_i}$). The excitation force $F(\mathbf{x}_{S_i}, t)$ applied at each driver is mapped by I. This results in multiple excited regions, generating specific wave patterns that encode the physical properties and dynamics of the anatomical domain. The signals generated from these excitation area are gathered by *Listeners* and integrated into a Mixer, which produces the final audio output with a specific wave pattern encoding the physical properties and dynamics of the anatomical domain.

3 Methods

To establish a robust methodology for the proposed sonification framework, this study employs FEM simulations to model user interactions in a controlled and flexible environment. This section describes FEM-based simulations integrating user interactions into an abstract tissue model via needle insertions. These simulations capture dynamic tool-tissue interactions, facilitating the analysis of their correlation with resulting acoustic signals.

3.1 Model Implementation

Abstract Anatomical Model – The simulations were conducted using the software SOFA Framework[1] with the Cosserat Plugin[2], which is designed for the simulation of tool-tissue interactions. The Cosserat Plugin includes an open-source implementation of needle insertion simulation, based on Cosserat Theory [2]. The interaction between the needle and tissue is modeled using the Complementary Constraints approach [9]. Building upon this existing FEM simulation, an anatomical model was incorporated and defined as a cubic geometry ($25cm \times 25cm \times 25cm$), discretized into 13,465 tetrahedral elements. The bottom face of the cube is fixed, while a spring with a stiffness of $10^2 MPa$ is applied to the nodes on the lateral faces to prevent excessive lateral motion. The non-homogeneous tissue model consisted of three distinct sub-domains, corresponding to three parallel layers (see Fig. 2 A), each representing a specific tissue class. In the presented preliminary study skin, fat and muscle are considered. The constitutive model of the tissue layers were approximated using

[1] https://www.sofa-framework.org/.

[2] https://github.com/SofaDefrost/Cosserat.

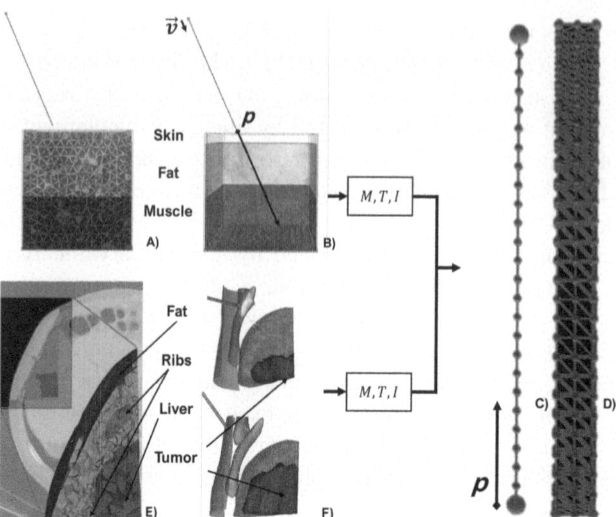

Fig. 2. FEM-based simulations to model tool-tissue dynamics by simulating insertion trajectories. (A) *Anatomical Domain*: skin, fat, and muscle. (B) Tool insertion defined by an entry point, orientation $\boldsymbol{p}$, and a velocity $\boldsymbol{v}$. (C) A 1D spring representing the basic sound model. (D) 3D advanced sound. (Color figure online)

a linear elastic constitutive law, with a co-rotational formulation employed to handle large deformations. Each sub-domain was characterized by its density (ρ) and a Young's Modulus (E), with values derived from literature [31]. A uniform Poisson's ratio of 0.4 was applied across all domains. The needle was modeled as a 17 Gauge, 26 cm-long rigid structure with a Young's modulus of 210 GPa. Each needle insertion was simulated by specifying an entry point on the top surface of the tissue model and an initial tool orientation in space. During the simulation, the needle was pushed from its handle with a constant velocity along the predefined trajectory. Herein, the simulated trajectory was defined as shown in Fig. 2 B and a velocity of $\boldsymbol{v} = \frac{1\ \mathrm{cm}}{s}$ is imposed. The remaining degrees of freedom of the tool were determined at each time step by solving the contact interaction with the tissue through the Complementary Constraint problem.

Toward Realistic Models – Building upon the abstract model, new extensions of it were developed by: i) incorporating additional tissue classes with varying thickness distributions within the multilayer cubic geometry; ii) implementing a new simulation setup to represent a tumor structure within a homogeneous medium; and iii) simulating diverse user interactions by varying trajectory definitions. Figure 3 provides illustrative examples, and the detailed application of these models is described in Sect. 4.1.

Real Anatomical Models – To demonstrate the applicability of the BioSonix design to real surgical tasks, the liver tumor biopsy procedure was considered. The modeling of the *Anatomical Domain A* is shown in Fig. 2 E and F. Start-

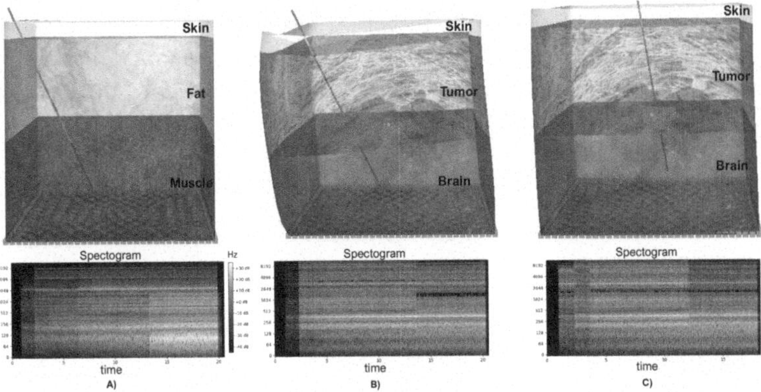

Fig. 3. Illustrative examples of the BioSonix design framework applied to abstract models: (A, B) Stability and responsiveness to tissue class variations; (B, C) Differentiation of dynamic tool-tissue interactions via sound signals.

ing from a public CT dataset[3], a small region of interest (ROI) encompassing subcutaneous fat, liver, and ribs was segmented. The 3D reconstruction into a multi-domain model, reflecting distinct tissue classes, was built using the ANSA v24.0.1 software (Beta CAE Systems, USA). Four different tumor configurations were created. The application of this model is described in Sect. 4.2.

Sound Model – The development of the Biosonix design framework was facilitated by the open-source miPhysics modelling libraries[4], which enabled the implementation of the physics-based sound model with the characteristics described in Sect. 2.3. The $\boldsymbol{p}$-axis was defined as a straight path with the same orientation in space of the simulated trajectory traversing the whole model. The informative nodes of the anatomical model from wich the displacement was extracted ($3u(\mathbf{x}_{A_i}, t)$), were specific to the $\boldsymbol{p}$-axis and were mapped to the sound models by T (light blue dots in Fig. 2 B)). The extracted $\mathbf{u}(\mathbf{x}_{A_i}, t)$, underwent minimal processing: the signals were first normalized within the 0.1 to 0.9 percentile range and subsequently scaled to a new range spanning from 0 to 0.1. For the preliminary benchmark, the sound model was implemented in its simplest form: a 1D string directly projected from the $\boldsymbol{p}$-axis as shown in Fig. 2 C. The spring connections are 10 mm long. The amount of *Drivers* d and *Listeners*, and their location were defined with the method outlined in Sect. 3.2.

3.2 Model Optimization

To provide effective perceptual feedback to the user when touching the target anatomy, it was essential to define both d and position of *Drivers* starting from the target region, where the greatest emphasis is required. In this abstract model,

[3] https://github.com/wasserth/TotalSegmentator.
[4] https://github.com/mi-creative/.

muscle tissue was considered as the target structure. Within this controlled simulated environment, three systematic experiments were conducted to investigate the impact of the mapping functions (Eqs. 4 and 5) on the resulting sound. In Fig. 4 few examples of this systematic approach are illustrated.

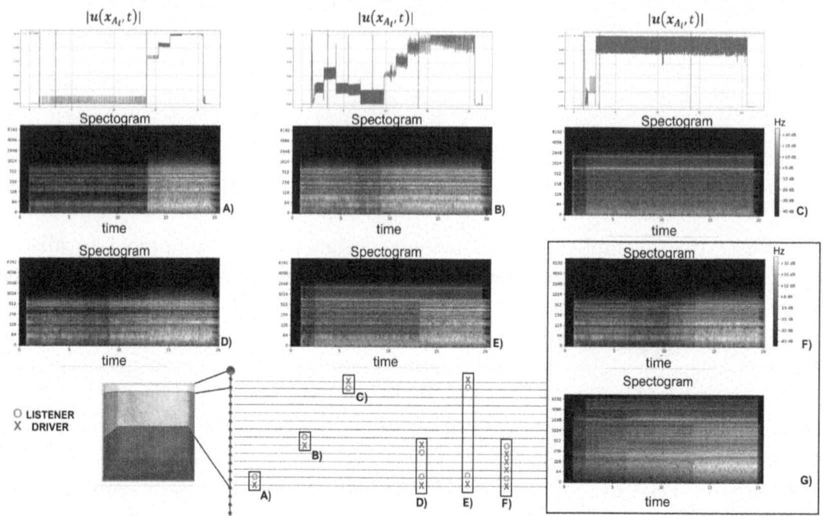

Fig. 4. Systematic experimentation to optimize the sound model: (A, B, C) *Nodal Characterization* – correlation between displacement norms and resulting sounds; (D, E) *Sonic Area Definition* – effects of varying *Driver* distances; (F) *Nodal Contribution Impact* – gradient from activating multiple nodes; (G) Fine-tuned model.

Nodal Characterization – Given that the physics-informed subsets of masses and springs were interconnected into a unified model, it was essential to analyze the contribution of each node to the resulting sound when excited. Figure 4 A), B), C) illustrate the correlation between the norm of displacement data in the *Anatomical Domain* and spectrograms. Notably, transitions corresponding to the tool puncturing the tissue—such as from skin to fat and from fat to muscle—are highlighted by red lines in the displacement plot.

Sonic Area Definition – The Sonic Area was defined as the minimal informative region above the target that contributes significant information to the resulting sound. Figure 4 D and E illustrates the result of selecting a narrow area and a wider area.

Nodal Contributions Impact – With the Sonic Area defined, it was investigated how the final sound was influenced by adding contributing nodes from different locations within the Sonic Area. This aspect was particularly important for creating smooth sound transitions as the tool approached the target. Figure 4 F) illustrates the resulting gradient in the spectrogram.

Fine Tuning – To further enhance the potential of the sound model, we introduced a more advanced design: a 3D topology formed by expanding the 1D structure along its orthogonal directions, centered on the $\boldsymbol{p}$-axis as shown in Fig. 2 D). Additionally, the function M was empirically fine-tuned. The returned values (m_i, K_i), were tested within the physics-based sound model to ensure stability and maximize perceptibility. Figure 4 G) illustrates spectograms obtained by the fine-tuning of configuration Fig. 4 F).

4 User Experiments and Expert Evaluation

To evaluate the method's user relevance and applicability to realistic tasks and clinical settings, we conducted two user studies. In all simulations, a constant insertion velocity of $\boldsymbol{v} = \frac{1\ \mathrm{cm}}{s}$ was applied, and minimal sound processing. Before each task, participants received brief training with three audiovisual examples of tool-tissue interactions(three for the general perceptibility study, and five for the expert evaluation). Each task then included 10 randomized audio-only examples.

4.1 General Perceptibility Study

The main objective of this study was to verify the perceptibility of sounds generated by the abstract models. A total of 41 audio samples were created using the optimized sound model and based on the abstract models (Fig. 2A,B). Six soft tissue classes (fat, brain, muscle, skin, tumor), along with bone tissue to add variability. The study involved three interactive tasks: **(1) Stopping:** detecting tool entry into critical tissue (always at the bottom), **(2) Tumor Detection:** identifying tool contact with a tumor, and **(3) Sorting:** sequencing tissue classes as the tool passed through them. We encourage readers to explore the audiovisual examples provided via this link: https://tinyurl.com/47f2k8ku.

General Perceptibility Results – A total of 22 participants (4 females, 20 males; aged 22–64) took part in the study, with diverse musical backgrounds ranging from minimal exposure to trained musicians. Most reported occasional or regular interaction with music but did not play instruments, ensuring a balanced perspective for auditory task analysis. Accuracy rates were 85.26% for the sorting task (190 responses) and 83% for the tumor detection task (200 cases). Statistics for the stopping task are provided in Table 1.

Table 1. Statistical results of the stopping task include Exact Match (%)—the percentage of exact matches. High-performing values are bolded based on thresholds: Mean values close to 0 ($-0.5 \leq$ Mean $\leq +0.5$), Std. Deviation below 1.0, and Exact Match (%) above 70%. Negative values indicate stopping early, positive values indicate stopping late (in seconds), and zero indicates precise stopping. T# denotes the trajectory number, and the second value indicates the critical tissue layer thickness (in millimeters).

Statistic	Brain-Tumor		Fat-Bone		Fat-Muscle	
	T2, 80mm	**T1, 112mm**	**T1, 112mm**	**T1, 160mm**	**T1, 112mm**	**T1, 160mm**
Mean [Sec]	-1.05	0.55	1.10	-0.71	-1.38	0.75
Std. Deviation	1.19	**0.51**	1.21	**0.90**	3.76	**0.72**
Median	-2.00	1.00	1.00	0.00	0.00	1.00
Exact Match	20.00%	45.00%	15.00%	57.14%	**76.19%**	10.00%
Statistic	**Muscle-Bone**		**Muscle-Fat**		**Tumor-Brain**	
	T2, 112mm	**T1, 112mm**	**T2, 112mm**	**T1, 180mm**	**T1, 112mm**	**T1, 180mm**
Mean [Sec]	-0.70	0.55	-0.60	**0.05**	**-0.05**	**0.00**
Std. Deviation	**0.47**	**0.51**	**0.50**	**0.40**	**0.23**	**0.20**
Median	-1.00	1.00	-1.00	0.00	0.00	0.00
Exact Match	30.00%	45.00%	40.00%	**84.21%**	**94.74%**	**89.00%**

4.2 Expert Evaluation of the Needle Biopsy Task

The simulation setup of Fig. 2 E and F was tested in this user study to assess the relevance of tasks to realistic medical scenarios, particularly the **Tumor Detection** task, which required identifying whether the tool touched the tumor. Twelve samples were generated by combining four tumor configurations with three trajectories. The second user study involved two male participants: a 41-year-old neuroradiologist with limited needle biopsy experience and a 46-year-old cardiologist who performs it weekly. Both had musical backgrounds as casual listeners but no experience playing instruments. Of the 20 cases, 70% were correctly identified.

5 Discussion and Conclusion

BioSonix introduced an optimized method for modeling dynamic tool-tissue interactions with auditory feedback. Experimental evaluations demonstrated its robustness in capturing the behavior of tissue types with similar properties, by leveraging inherent tissue characteristics rather than artificial mappings. The system proved resilient to variations in interaction angles and insertion points, providing granular insights into tool-tissue dynamics. Machine-based evaluations, including spectral analysis, validated the model's stability and reliable psychoacoustic properties, confirming its generalizability across scenarios. An initial user study assessed the effectiveness and perceptibility of sonifications in semi-interactive tasks. Designed for non-clinical users, this step laid the groundwork for future studies involving realistic surgical scenarios. The preliminary expert study demonstrated the method's ability to represent realistic anatomical interactions with minimal training. While both user studies showed relatively high accuracy, participants noted a need for greater contrast between tissue classes, indicating low confidence despite accurate results. Improving tissue distinguishability, though not the primary focus here, remains an area for future exploration. Although methods such as [14,36] could optimize model parameterization or enhance output signals [28,29], this study was essential for identifying the key parameters relevant to such optimizations. The experiments revealed the model's complexity and the influence of its parameters on the resulting sound.

The mean time deviation in the stopping task was ± 1 s, equating to a 10 mm tool pathway distance within a $25cm \times 25cm \times 25cm$ tissue sample. This result is specific to the given tissue configuration and requires further testing to generalize across different dimensions or setups. The best performance occurred in the "Tumor-Brain" case, likely due to the higher contrast and increased thickness, though detailed studies are needed to confirm these factors. As sonification remains a relatively novel modality for interactive tasks, further research is needed to fully explore and realize its potential.

Acknowledgments. This work was supported by MUSA – Multilayered Urban Sustainability Action – project, funded by the European Union – NextGener- ationEU, under the National Recovery and Resilience Plan (NRRP) Mission 4 Component 2 Investment Line 1.5: Strenghtening of research structures and creation of R&D "innovation ecosystems", set up of "territorial leaders in R&D".

References

1. Al-Naser, Y., Halka, F., Alshadeedi, F., Albahhar, M., Athreya, S.: The applications of augmented reality in image-guided tumor ablations: a scoping review. J. Med. Imaging Radiat. Sci. **55**, 125–133 (2024). https://doi.org/10.1016/j.jmir.2023.12.006
2. Antman, S.S.: The Special Cosserat Theory of Rods, pp. 259–324. Springer New York, New York, NY (1995). https://doi.org/10.1007/978-1-4757-4147-6_8
3. Benmahdjoub, M., Thabit, A., Veelen, M.L.C.V., Niessen, W.J., Wolvius, E.B., Walsum, T.V.: Evaluation of AR visualization approaches for catheter insertion into the ventricle cavity. IEEE Trans. Visual. Comput. Graph. **29**, 2434–2445 (2023). https://doi.org/10.1109/TVCG.2023.3247042
4. Black, D., Hansen, C., Nabavi, A., Kikinis, R., Hahn, H.: A survey of auditory display in image-guided interventions. Int. J. Comput. Assist. Radiol. Surg. **12**, 1665–1676 (2017)
5. Bopp, M.H.A., Grote, A., Gjorgjevski, M., Pojskic, M., Saß, B., Nimsky, C.: Enabling navigation and augmented reality in the sitting position in posterior fossa surgery using intraoperative ultrasound. Cancers **16**, 1985 (2024). https://doi.org/10.3390/cancers16111985
6. Chase, E.D., Gerstenberg, T., Follmer, S.: Realism of visual, auditory, and haptic cues in phenomenal causality. In: 2023 IEEE World Haptics Conference (WHC), pp. 306–312. IEEE (2023)
7. Chen, J., Gandomkar, Z., Reed, W.M.: Investigating the impact of cognitive biases in radiologists' image interpretation: a scoping review. Eur. J. Radiol. **166**, 111013 (2023)

8. Doornbos, M.C.J., et al.: Augmented reality implementation in minimally invasive surgery for future application in pulmonary surgery: a systematic review. Surg. Innov. **31**, 646–658 (2024). https://doi.org/10.1177/15533506241290412
9. Duriez, C., Guébert, C., Marchal, M., Cotin, S., Grisoni, L.: Interactive Simulation of Flexible Needle Insertions Based on Constraint Models, pp. 291–299 (2009). https://doi.org/10.1007/978-3-642-04271-3_36
10. Ernst, M.O., Di Luca, M.: Multisensory perception: from integration to remapping. Sens. Cue Integr. **15**, 224–250 (2011)
11. Eves, J., Sudarsanam, A., Shalhoub, J., Amiras, D.: Augmented reality in vascular and endovascular surgery: scoping review. JMIR Ser. Games **10**, e34501 (2021). https://doi.org/10.2196/34501
12. Gionfrida, L., Roginska, A.: A novel sonification approach to support the diagnosis of Alzheimer's dementia. Front. Neurol. **8**, 647 (2017)
13. Hermann, T., Nattkemper, T., Schubert, W., Ritter, H.: Sonification of multi-channel image data. In: Proceedings of the Mathematical and Engineering Techniques in Medical and Biological Sciences (METMBS 2000), pp. 745–750 (2000)
14. Hermann, T., Weger, M.: Data-driven auditory contrast enhancement for everyday sounds and sonifications. In: Proceedings of the 25th International Conference on Auditory Display (ICAD 2019) (2019)
15. Illanes, A., Boese, A., Maldonado, I., Pashazadeh, A., Schaufler, A., Navab, N., Friebe, M.: Novel clinical device tracking and tissue event characterization using proximally placed audio signal acquisition and processing. Sci. Rep. **8**(1), 12070 (2018)
16. Lemaitre, G., Houix, O., Visell, Y., Franinović, K., Misdariis, N., Susini, P.: Toward the design and evaluation of continuous sound in tangible interfaces: the Spinotron. Int. J. Hum. Comput. Stud. **67**(11), 976–993 (2009)
17. Matinfar, S., Dehghani, S., Sommersperger, M., Navab, N.: Ocular stethoscope: auditory support for retinal membrane peeling. In: Linguraru, M.G., et al. (eds.) Medical Image Computing and Computer Assisted Intervention. MICCAI 2024. LNCS, vol. 15006. Springer, Cham (2024). https://doi.org/10.1007/978-3-031-72089-5_41
18. Matinfar, S., Hermann, T., Seibold, M., Fürnstahl, P., Farshad, M., Navab, N.: Sonification for process monitoring in highly sensitive surgical tasks. In: Proceedings of the Nordic Sound and Music Computing Conference 2019 (Nordic SMC 2019) (2019)
19. Matinfar, S., et al.: Surgical soundtracks: automatic acoustic augmentation of surgical procedures. Int. J. Comput. Assist. Radiol. Surg. **13**(9), 1345–1355 (2018)
20. Matinfar, S., et al.: Surgical soundtracks: towards automatic musical augmentation of surgical procedures. In: Descoteaux, M., Maier-Hein, L., Franz, A., Jannin, P., Collins, D.L., Duchesne, S. (eds.) MICCAI 2017. LNCS, vol. 10434, pp. 673–681. Springer, Cham (2017). https://doi.org/10.1007/978-3-319-66185-8_76
21. Matinfar, S., Salehi, M., Dehghani, S., Navab, N.: From tissue to sound: model-based sonification of medical imaging. In: Greenspan, H., et al. (eds.) Medical Image Computing and Computer Assisted Intervention. MICCAI 2023. LNCS, vol. 14228, pp. 207–216. Springer, Cham (2023). https://doi.org/10.1007/978-3-031-43996-4_20
22. Matinfar, S., et al.: Sonification as a reliable alternative to conventional visual surgical navigation. Sci. Rep. **13**(1), 5930 (2023)
23. Moerel, M., De Martino, F., Formisano, E.: Processing of natural sounds in human auditory cortex: tonotopy, spectral tuning, and relation to voice sensitivity. J. Neurosci. **32**(41), 14205–14216 (2012)

24. Ostler-Mildner, D., Wegener, L., Fuchtmann, J., Feussner, H., Wilhelm, D., Navab, N.: The sound of surgery-development of an acoustic trocar system enabling laparoscopic sound analysis. Int. J. Comput. Assist. Radiol. Surg. **19**, 1–9 (2024)
25. Rassi, I.E., Rassi, J.M.E.: A review of haptic feedback in tele-operated robotic surgery. J. Med. Eng. Technol. **44**, 247–254 (2020). https://doi.org/10.1080/03091902.2020.1772391
26. Schütz, L., et al.: Interactive shape sonification for tumor localization in breast cancer surgery. In: Proceedings of the CHI Conference on Human Factors in Computing Systems. CHI 2024, Association for Computing Machinery, New York, NY, USA (2024). https://doi.org/10.1145/3613904.3642257
27. Schütz, L., et al.: A framework for multimodal medical image interaction. IEEE Trans. Visual Comput. Graph. **30**(11), 7419–7429 (2024). https://doi.org/10.1109/TVCG.2024.3456163
28. Schwarz, D.: Corpus-based concatenative synthesis. IEEE Signal Process. Mag. **24**(2), 92–104 (2007)
29. Schwarz, D., Hackbarth, B.: Navigating variation: composing for audio mosaicing. In: International Computer Music Conference (ICMC), pp. 1–1 (2012)
30. Shams, L., Seitz, A.R.: Benefits of multisensory learning. Trends Cogn. Sci. **12**(11), 411–417 (2008)
31. Singh, G., Chanda, A.: Mechanical properties of whole-body soft human tissues: a review. Biomed. Mater. **16**, 062004 (2021). https://doi.org/10.1088/1748-605X/ac2b7a
32. Susini, P., Misdariis, N., Lemaitre, G., Houix, O.: Naturalness influences the perceived usability and pleasantness of an interface's sonic feedback. J. Multimod. User Interfaces **5**, 175–186 (2012)
33. Theunissen, F.E., Elie, J.E.: Neural processing of natural sounds. Nat. Rev. Neurosci. **15**(6), 355–366 (2014)
34. Verhey, J.T., Haglin, J.M., Verhey, E.M., Hartigan, D.E.: Virtual, augmented, and mixed reality applications in orthopedic surgery. Int. J. Med. Robot. Comput. Assist. Surg. **16**, e2067 (2020). https://doi.org/10.1002/rcs.2067
35. Villeneuve, J., Leonard, J.: Mass-interaction physical models for sound and multisensory creation: Starting anew. In: Proceedings of the 16th Sound & Music Computing Conference, pp. 187–194 (2019)
36. Weger, M., Hermann, T., Höldrich, R.: Real-time auditory contrast enhancement. In: Proceedings of the 25th International Conference on Auditory Display (ICAD 2019) (2019)
37. Ziemer, T.: Three-dimensional sonification as a surgical guidance tool. J. Multimod. User Interfaces **17**(4), 253–262 (2023)
38. Zou, X.C., Xu, X.D., Huang, J.B., Chao, H.C., Zeng, T.: The clinical application value of mixed reality in robotic laparoscopic partial nephrectomy. Front. Oncol. **14**, 1478051 (2024). https://doi.org/10.3389/fonc.2024.1478051

Brain

Explainable Deep Model for Understanding Neuropathological Events Through Neural Symbolic Regression

Tingting Dan[1] and Guorong Wu[1,2(✉)]

[1] Department of Psychiatry, University of North Carolina at Chapel Hill, Chapel Hill, NC 27599, USA
{Tingting_Dan,grwu}@med.unc.edu

[2] Department of Computer Science, Department of Statistics and Operations Research (STOR), Carolina Institute for Developmental Disabilities, and the UNC NeuroScience Center, University of North Carolina at Chapel Hill, Chapel Hill, NC 27599, USA

Abstract. Mounting evidence shows that Alzheimer's disease (AD) is characterized by the propagation of tau aggregates throughout the brain in a prion-like manner. Since current pathology imaging technologies can only provide a spatial brain mapping of tau accumulation, computational modeling becomes indispensable in analyzing the spatiotemporal propagation patterns of widespread tau aggregates. To address this challenge, we present a novel physics-informed neural network for AD (coined *PINN4AD*) by conceptualizing the intercellular spreading of tau pathology in a reaction-diffusion model, where each node (brain region) is ubiquitously wired with other nodes while interacting with amyloid burdens. In this context, we formulate the biological process of tau spreading in a principled potential energy transport model that describes the mechanistic role of Aβ-tau interaction in the widespread flow of tau aggregates. The physics principle and mathematics insight allow us to develop an explainable neural network to uncover the spatiotemporal dynamics of tau propagation from the unprecedented amount of longitudinal neuroimages. On top of this, we introduce a symbolic regression module into the *PINN4AD* to further elucidate the analytic expressions underlying Aβ-tau interaction and tau propagation mechanism. We have achieved not only an enhanced prediction accuracy of tau propagation on ADNI and OASIS datasets but also a system-level understanding of the pathophysiological mechanism in AD progression, suggesting great potential for research in AD and AD-related dementias.

Keywords: Physics-informed neural networks · Symbolic regression · Reaction-diffusion model · Alzheimer's disease

1 Introduction

There is an overwhelming consensus that the progression of Alzheimer's disease (AD) is characterized by the spread of tau pathology throughout the brain cortex

I. Oguz et al. (Eds.): IPMI 2025, LNCS 15830, pp. 37–50, 2026.
https://doi.org/10.1007/978-3-031-96625-5_3

[27]. While positron emission tomography (PET) imaging technology allows us to investigate tau aggregation *in-vivo*, the dynamic process of how tau aggregates spread throughout the brain and how toxic amyloid-β proteins contribute to heterogeneous neurodegeneration processes is still largely elusive [16].

The leading hypothesis of tau propagation is that pathological tau aggregates spread directly from cell to cell through anatomical connections in a prion-like manner [14], where the epicenter of misfolded tau seeds induces misfolding of native tau in neighboring cells. Furthermore, increasing evidence shows that the spread of tau pathology in AD follows a stereotypical pattern underlying the topology of white matter fibers [25]. Meanwhile, along with the progression of tau propagation, amyloid-beta (Aβ) plaques, another hallmark of AD, play a critical role in exacerbating tau aggregation and promoting the propagation of misfolded tau. Since Aβ acts as an upstream trigger for tau pathology, the Aβ-tau interaction becomes the gateway to understanding the biological mechanism of neurodegeneration and cognitive decline in AD.

In the past decade, striking efforts have been made to model tau propagation using neuroimaging techniques. For example, a network diffusion model has been proposed to characterize the spreading of tau aggregate under the umbrella of graph heat diffusion [22,28]. Since the graph diffusion process falls into a linear system, the solution is expressed in terms of the eigenvectors of the graph Laplacian, which is associated with the underlying structural brain network. To take into account the interaction of AD biomarkers, more sophisticated system biology approaches have been used to study not only the accumulation and degradation of each biomarker, but also the biomarker-to-biomarker pathways between amyloid, tau, and neurodegeneration process [29]. Recently, machine learning has been introduced to predict tau aggregation by parameterizing the temporal evolution of tau pathology across individuals [23]. For instance, graph convolution network (GCN) has been widely investigated to predict the future tau burden [1,8,9], where regional tau SUVR (standardized update value ratio) is considered as a graph embedding vector.

Both model-based approaches and learning-based methods have their own strengths and limitations. Model-based approaches are rooted in the neuroscientific understanding of the biological processes underlying tau propagation. However, model design and parameter tuning depend heavily on domain expertise, which can restrict their broader applicability. In contrast, machine learning has achieved great success in predicting tau accumulation and clinical outcomes using high-dimensional neuroimaging data. However, the lack of model explainability in current learning-based methods makes it difficult to uncover the underlying mechanisms that could enhance our understanding of AD etiology.

In this context, it is a natural choice to combine the insight of biophysics principles and the power of machine learning, yielding a novel physics-informed neural network (PINN) for tau prediction and uncovering novel mechanisms regarding the tau propagation mechanism as well as the Aβ-tau interaction. Specifically, we characterize the evaluation of tau aggregates in a reaction-diffusion model (RDM) [11,12], where the macro-scale tau propagation underlines the topology of the structural brain network and the regional tau accumulation is influenced by the amyloid burden. Following the notion of PDE-based deep models [3,5], we present a deep neural network architecture that parameterizes tau propagation in the framework of RDM. Furthermore, we integrate the symbolic regression [20] component in the deep model to uncover the analytic expressions of the diffusion function (characterizing the kinetics of tau propagation) and reaction function (characterizing the Aβ-tau interaction), respectively. Taken together, our deep model, coined *PINN4AD*, allows us to discover a novel RDM to explain the propagation of misfolded tau protein in AD. We have evaluated the prediction of future tau accumulation on ANDI and OASIS datasets, where our *PINN4AD* achieves higher prediction accuracy than counterpart deep models.

As a proof-of-concept approach, we explore the system-level understanding of the tau propagation mechanism using discovered analytic expressions of tau spreading and Aβ-tau interactions. Our exploration addresses key neuroscience questions, such as: (1) Is tau spreading a linear, monotonic process, or is it a nonlinear, complex phenomenon? (2) Does amyloid contribute to local tau aggregation (i.e., Aβ-tau interactions occurring within the same brain region) or remote tau aggregation (i.e., amyloid plaque accumulation in one region influencing tau aggregation in a different brain region)?

2 Method

In this section, we delve into the challenge of uncovering novel mechanisms that drive tau propagation. Suppose we have the brain network denoted by $\mathcal{G} = (Z, W)$ with N nodes $Z = \{z_i | i = 1, ...N\}$ and the weighted adjacency matrix $W = [w_{ij}]_{i,j=1}^{N} \in \mathcal{R}^{N \times N}$. We use u_i to represent the tau pathology level associated with node z_i. The backbone of our learning-based approach is a physics-informed deep model rooted in the reaction-diffusion model (RDM). This model is designed to reveal novel insights into tau spreading dynamics. The reaction-diffusion system is mathematically described by the following partial differential equation (PDE):

$$\frac{\partial u(t)}{\partial t} = g(u(t)) + h(u(t), v(t)), \tag{1}$$

where $u(t) = [u_1(t), u_2(t), ..., u_N(t)]$ is the whole-brain concentration level of tau protein at time t. Similarly, $v(t)$ denotes whole-brain amyloid concentration. Following the spirit of RDM, $g(\cdot)$ is called the diffusion operator (characterizing the prion-like tau propagation) and $h(\cdot)$ is the reaction function (modeling the Aβ-tau interaction). In the landscape of machine learning, the diffusion function g is often discretized and parameterized using a graph neural network (GNN), i.e., $g(\cdot) = -\nabla \cdot (\nabla u)$, where the graph Laplacian $\nabla \cdot (\nabla)$ symbolizes the graph diffusion process. Since the mechanistic role of Aβ is still elusive, the reaction function h is implicitly parameterized by a multilayer perceptron (MLP).

Since both GNN and MLP lack model interpretability and transparency in decision-making, the neural network architecture for the RDM in Eq. 1 has limited power to uncover the intrinsic dynamics of tau protein propagation. In this regard, we propose an end-to-end deep model of physics-informed neural symbolic regression to achieve an understanding of reaction and diffusion functions.

2.1 Physics-Informed Neural Symbolic Regression

In this work, we aim to derive an explicit mathematical expression to capture the dynamics of tau propagation through the reaction-diffusion models. This is well-suited to a symbolic regression model [20,30], which can effectively identify patterns and relationships within the data.

Broadly speaking, mathematical expressions can naturally be represented as symbolic expression trees, a specialized form of binary tree where internal nodes denote mathematical operators, and terminal nodes represent input variables or constants. These operators can be unary (e.g., sine, which takes a single argument) or binary (e.g., multiplication, which takes two arguments). This structured representation enables a clear and intuitive depiction of the relationships within an expression. Inspired by this principled approach, we redesign our reaction-diffusion model for tau propagation by integrating a learning component for the latent tree-like structure of analytic expressions. The overview of our deep model is shown in Fig. 1.

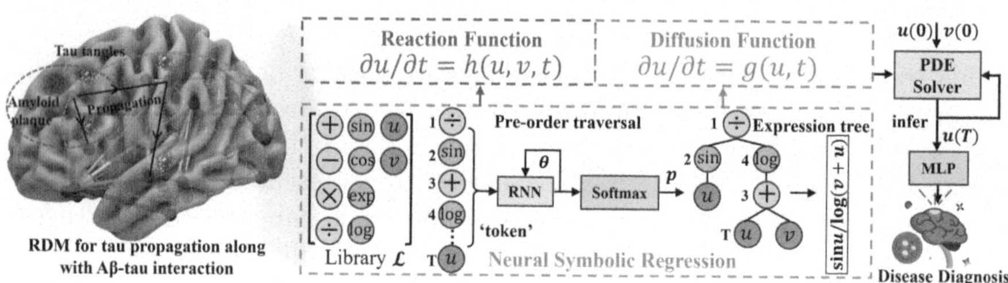

Fig. 1. The overview of physics-informed neural network with neural symbolic regression. Our overarching goal is to model the evolution of tau aggregates using a principled deep model (green dashed box) where the learning mechanism is rooted in a RDM (blue dashed box). Furthermore, we uncover the analytic expressions for the diffusion function g for tau propagation and reaction function h for Aβ-tau interaction using neural symbolic regression that seeks the most accurate expression with the least complexity (shown in the orange dashed box). (Color figure online)

Neural Symbolic Regression. We represent mathematical expressions as sequences, which enables the development of an autoregressive model to generate expressions. To facilitate sequential generation, each expression tree is converted into a sequence of node values, or "tokens", using *pre-order traversal*—a depth-first approach that processes each node before visiting its children, moving from top to bottom. This traversal establishes a one-to-one correspondence between the expression tree and its sequential representation, allowing for systematic generation while preserving the structural integrity of the original expression.

The to-be-discovered analytic expression $\psi(\cdot)$ is represented as a sequence obtained from the pre-order traversal of the underlying expression tree. The k-th token of the sequence is denoted by ψ_k, and the total length of the sequence is given by $|\psi| = K$. Each token is sampled from a predefined library $\mathcal{L}$, which contains possible values, including operators (e.g., $+, -, \times, \div, \sin, \cos, \exp, \log$) and variables (e.g., u, v). This structured representation provides a rich foundation for generating biologically reasonable diffusion and reaction functions. After that, expressions are generated by collecting tokens along the pre-order traversal, progressing sequentially from ψ_1 to ψ_K. The probabilities of selecting each token from the library $\mathcal{L}$ are governed by a categorical distribution parameterized by ϕ. To incorporate the "context" of the expression as it is being constructed, the probability of a particular token being selected is conditioned on all previously selected tokens in the traversal. Therefore, this conditional dependence can be effectively modeled using a recurrent neural network (RNN) with parameters θ, which estimates a probability p in an auto-regressive fashion. At each selection step, the RNN updates its hidden state based on the previously selected tokens, enabling it to capture and leverage the evolving structure of the expression. By doing so, the kth output of the RNN is connected with a softmax layer to predict vector $\psi^{(k)}$, which defines the probability distribution for selecting the kth token ψ_k, conditioned on the previously selected tokens $\psi_{1:(k-1)}$, i.e., $p(\psi_k|\psi_{1:(k-1)};\theta) = \psi^{(k)}_{\mathcal{L}(\psi_k)}$, where $\mathcal{L}(\psi_j)$ is the index in $\mathcal{L}$ corresponding to ψ_k. The likelihood of the entire expression is calculated as the product of the probabilities of its tokens: $p(\psi|\theta) = \prod_{k=1}^{K} p(\psi_k|\psi_1, \psi_2, \ldots, \psi_{k-1};\theta) = \prod_{k=1}^{K} \psi^{(k)}_{\mathcal{L}(\psi_k)}$, as shown in bottom of Fig. 1.

Uncover Reaction Function of Aβ-tau Interaction. Since the observed tau pathology $u(t)$ and amyloid burden $v(t)$ are highly correlated, we model their interaction in a multi-variate function h by neural symbolic regression. To that end, the learned analytic expression of the reaction function is expressed as $h(u, v) \Leftarrow \psi_h(u, v)$.

Characterize Kinetics of Diffusion Function. Tau propagation involves not only the regional change of tau aggregations but also the dynamic region-to-region spreading pathways constrained by the underlying network topology. In this context, we consider the diffusion function g to be dependent on connectomic measurements such as connectivity degrees. Specifically, we first compute the degree of node i, denoted by $D_i = \sum_j W_{ij}$. Then we estimate the average tau burden at the node i^{th} neighbors as $\alpha_i = \frac{\sum_j W_{ij} \cdot u_j}{D_i + \epsilon}$, where $\epsilon = 1e-6$ is added to avoid division by zero. In addition, we compute the peak tau burden within the graph neighborhood of node z_i, denoted by $\beta_i = \max_{j \in neighbor(z_i)}(u_j)$. Finally, we generate an augmented feature embedding of node i^{th} as $\hat{u}_i = [D_i, \alpha_i, \beta_i]$, which describes local topological pattern of tau aggregations including node centric, average characteristics, and peak value of tau burden on each node. Taken together, the diffusion model is defined as $g(u) \Leftarrow \psi_g(\hat{u})$.

Reward Function. Once a pre-order traversal is sampled, the corresponding symbolic expression is instantiated and evaluated using a reward function. In this work, we use the *projection error* η as the reward function. A common fitness measure in genetic programming-based symbolic regression is the root-mean-square error (RMSE). Thus, we define the projection error as: $\eta = \frac{\sum_{i \in N} \|\hat{\psi}(i) - \psi(i)\|_2^2}{N}$, where $\hat{\psi}(i)$ represents the target values of node i, ψ denotes the candidate symbolic expression.

Evaluation Metric. We adopt the *score* metric, as defined in [6], to evaluate the correctness of symbolic expressions. The *score* is computed as: $score = -\frac{\Delta \log(\mathrm{MAE})}{\Delta C}$, where MAE is the mean absolute error between predictions and observed data, C represents the complexity of the expression, and Δ indicates a local change in the expression [6]. A higher *score* signifies that a slight increase in complexity leads to a disproportionately larger increase in MAE, encouraging the resulting expressions to effectively balance simplicity and predictive accuracy. We prioritize models with both low *prediction error* and high *score.*

2.2 *PINN4AD*: An Explainable Deep Model for Understanding Tau Propagation in AD

Current deep learning methods typically discretize g and h using GCNs and MLPs to optimize performance for tasks like disease diagnosis or tau accumulation prediction through graph embeddings. In contrast, our *PINN4AD* has three key components.

- A MLP ϕ_ϑ for projecting the observed regional concentration level of tau pathology to the initial state of tau propagation $u(0)$. The network parameter of this MLP layer is ϑ.
- Deep RDM for tau propagation. We alternate the following two components in the deep model: (1) an RNN of neural symbolic regression to uncover the analytic expression for diffusion function g and reaction function h based on $u(t)$ and amyloid burden $v(t)$, (2) a PDE solver [13] to predict tau aggregate $u(t+1)$. The output of deep RDM is the terminal tau pathology $u(T)$.
- A fully connected layer for down-stream learning task using $u(T)$. For example, MAE between $u(T)$ and observed tau measurement from follow-up tau PET scan is used as the loss function in predicting future tau level.

We propose an end-to-end framework that integrates symbolic regression with PDE-based modeling to achieve interpretable and physically consistent predictions. First, neural symbolic regression, implemented using a *DeepSymbolicRegressor* [19], extracts analytical expressions representing reaction-diffusion processes from the projected tau concentration by ϕ_ϑ. These expressions are subsequently embedded into a PDE framework as initial conditions to model the spatiotemporal evolution of the system. The PDE solver refines the predictions by incorporating physical constraints, i.e., diffusion and reaction dynamics. The output of the PDE solver is then used to iteratively retrain the symbolic regression model. This iterative process refines the symbolic representation while maintaining its interpretability and guarantees robustness by integrating data-driven insights with physical consistency. For the PDE solver, we employ a fused solver based on the Euler method [2], using six time-steps during the feed-forward phase. For backpropagation, we utilize the Adam optimizer [15] with a learning rate of 0.001 to optimize the model parameters.

3 Experiments

In this work, we evaluate the prediction accuracy of future tau accumulation and the learning performance for uncovering the latent propagation mechanism of tau spreading on the Alzheimer's Disease Neuroimaging Initiative (ADNI) dataset [21] and Open Access Series of Imaging Studies-3 (OASIS-3) dataset [18]. The datasets include 163 subjects from ADNI and 77 subjects from OASIS-3, each with longitudinal tau and amyloid-PET scans (ranging from 2–5 visits), along with T1-weighted MRI and diffusion-weighted imaging (DWI) data. To construct the structural connectome, we utilize the Destrieux atlas [10], which divides the whole brain into 160 regions of interest (148 cortical regions - (frontal lobe, insula lobe, temporal lobe, occipital lobe, parietal lobe) and 12 subcortical regions - (hippocampus, caudate, thalamus, amygdala, globus pallidum, and putamen)). We compare our method against several state-of-art approaches, including deep neural network (DNN), vanilla graph convolutional network (GCN), network diffusion model (NDM) [22], graph attention network (GAT) [24], simple and deep graph convolutional networks (GCNII) [4], DeepGCNs [17], graph neural diffusion (GRAND) [3] and recent work TauFlowNet [7]. The evaluation consisted of two key experiments: (1) Assessing the prediction accuracy of future tau accumulation (this task uses the mean squared error (MSE) as the loss function); (2) Disease diagnosis performance in AD (this task is regarded as a classification task and utilizes cross-entropy as the loss function). To ensure robust and reliable performance assessment, we employ 5-fold cross-validation for both experiments.

Prediction Accuracy of Future Tau Accumulation. To evaluate the accuracy of computational methods in predicting future tau accumulation, we use follow-up tau SUVR measurements as the ground truth, and prediction accuracy is quantified using the mean absolute error (MAE). The prediction errors for various methods are presented in the first column of Table 1, with the results on ADNI and OASIS-3 listed in the top and bottom sections, respectively. To assess the robustness of our *PINN4AD*, we further introduced uncorrelated Gaussian noise with standard deviations ranging from 0.02 to 0.1 to the observed SUVR values. The prediction accuracies under different noise levels are shown in subsequent columns of Table 1. Experimental results demonstrate that the proposed method consistently outperforms counterpart approaches across all noise level conditions. Moreover, our *PINN4AD* exhibits remarkable robustness to noise, which benefits from the principled model design as explained below. *First*, symbolic regression is employed to fit the complex relationship between input and output. By generating analytical expressions that capture the underlying trends and patterns in the data, symbolic regression effectively filters out high-frequency noise and random errors. *Second*, our model applies PDE-based inference, which introduces temporal and spatial constraints to enforce consistency and smoothness in the predictions. For instance, even when the initial input conditions ($u(0)$) are perturbed by noise, the pre-determined dynamics of the PDE is able to stabilize the propagation, ensuring that the output remains robust and reliable. The synergistic combination of symbolic regression and PDE modeling creates a powerful framework to the extent that symbolic regression provides a noise-resistant foundation by extracting clear, interpretable expressions, while PDE-based inference further refines the predictions through its robust physical modeling capabilities. Together, these steps ensure reliable performance, even in the presence of noisy or uncertain input conditions.

Table 1. The prediction performance (MAE) between observed and predicted tau SUVR on ADNI dataset (top) and OASIS-3 dataset (bottom) by various methods.

Noise level	-	std=0.02	std=0.04	std=0.08	std=0.1
GCN	0.124 ± 0.08	0.128 ± 0.08	0.130 ± 0.08	0.142 ± 0.08	0.161 ± 0.10
DNN	0.070 ± 0.03	0.072 ± 0.04	0.080 ± 0.04	0.104 ± 0.06	0.112 ± 0.06
NDM	0.113 ± 0.02	0.117 ± 0.03	0.138 ± 0.03	$0.141 \pm \pm \pm 0.03$	0.174 ± 0.04
GRAND	0.259 ± 0.04	$0.259 \pm \pm \pm 0.04$	0.259 ± 0.04	0.259 ± 0.04	0.266 ± 0.04
GCNII	0.207 ± 0.04	0.208 ± 0.04	0.211 ± 0.03	0.211 ± 0.04	0.211 ± 0.04
GAT	0.306 ± 0.05	0.307 ± 0.05	0.304 ± 0.05	0.307 ± 0.05	0.307 ± 0.05
DeepGCNs	0.079 ± 0.02	0.083 ± 0.02	0.088 ± 0.02	0.083 ± 0.02	0.101 ± 0.01
TauFlowNet	0.049 ± 0.02	0.058 ± 0.03	0.066 ± 0.03	0.079 ± 0.04	0.081 ± 0.04
PINN4AD	$\mathbf{0.023 \pm 0.001}$	$\mathbf{0.024 \pm 0.002}$	$\mathbf{0.024 \pm 0.002}$	$\mathbf{0.028 \pm 0.001}$	$\mathbf{0.028 \pm 0.001}$
GCN	0.256 ± 0.030	0.270 ± 0.038	0.274 ± 0.043	0.287 ± 0.040	0.292 ± 0.047
NDM	0.102 ± 0.032	0.110 ± 0.030	0.121 ± 0.034	0.129 ± 0.033	0.133 ± 0.037
DNN	0.076 ± 0.038	0.078 ± 0.041	0.079 ± 0.048	0.112 ± 0.053	0.119 ± 0.063
TauFlowNet	0.063 ± 0.031	0.065 ± 0.035	0.066 ± 0.036	0.070 ± 0.038	0.090 ± 0.042
PINN4AD	$\mathbf{0.057 \pm 0.005}$	$\mathbf{0.057 \pm 0.006}$	$\mathbf{0.060 \pm 0.008}$	$\mathbf{0.062 \pm 0.008}$	$\mathbf{0.063 \pm 0.008}$

Disease Diagnosis Performance in AD. Given baseline amyloid and longitudinal tau scans (note that only use tau as input regarded as the ablation study of comparison methods), We evaluate the prediction accuracy for diagnosing Alzheimer's Disease (AD) using various methods, as shown in Table 2. At a significance level of 0.001, our *PINN4AD* outperforms all other methods in terms of classification accuracy (indicated by '*'). **Verifying the Reliability of the Learned RDM Expressions.** To verify the reliability of the identified diffusion function $\hat{g}$ and reaction function $\hat{h}$, we use the learned expressions to predict future tau accumulation by following the governing equations $\frac{\partial u}{\partial t} = \hat{g}(u(t))$ and $\frac{\partial u}{\partial t} = \hat{h}(u(t), v(t))$, respectively[1]. As shown in Fig. 2, we present the linear fitting results between the observed follow-up tau concentration and the predicted values by $\hat{g} = \psi_g$ (dashed lines) and $\hat{h} = \psi_h$ (solid lines). The results demonstrate promising accuracy, with ($R^2 \geq 0.9$) in most brain regions. Since

[1] We specifically split the diffusion and reaction parts and evaluate them separately. In all other experiments, we use the complete RDM in Eq. 1.

modeling region-to-region spreading characteristics in g is more complex than regional interaction in h, the identified reaction functions slightly outperform the diffusion functions in terms of prediction accuracy. Collectively, the promising line-fitting results indicate that neural symbolic regression holds great potential to achieve not only highly competitive prediction accuracy but also appealing model interpretability, as demonstrated next.

Table 2. Disease diagnosis accuracy in AD.

Input	Methods	Accuracy (ADNI)	Accuracy (OASIS-3)
Tau	**GCN**	0.7707 ± 0.0215 (*)	0.8705 ± 0.0220 (*)
Tau+$A\beta$	**GCN**	0.7707 ± 0.0215 (*)	0.8705 ± 0.0220 (*)
Tau	**GAT**	0.8105 ± 0.0204 (*)	0.8842 ± 0.0205 (*)
Tau+$A\beta$	**GAT**	0.8263 ± 0.0188 (*)	0.8921 ± 0.0117 (*)
Tau	**GCNII**	0.8379 ± 0.0124 (*)	0.9052 ± 0.0114 (*)
Tau+$A\beta$	**GCNII**	0.8405 ± 0.0119 (*)	0.9137 ± 0.0192 (*)
Tau	**DeepGCNs**	0.8403 ± 0.0214 (*)	0.9169 ± 0.0136 (*)
Tau+$A\beta$	**DeepGCNs**	0.8415 ± 0.0105 (*)	0.9209 ± 0.0118 (*)
Tau	**GRAND**	0.8396 ± 0.0108 (*)	0.9249 ± 0.0100 (*)
Tau+$A\beta$	**GRAND**	0.8414 ± 0.0100 (*)	0.9281 ± 0.0096 (*)
Tau+$A\beta$	**TauFlowNet**	0.8434 ± 0.0475 (*)	0.9353 ± 0.0169 (*)
Tau+$A\beta$	***PINN4AD***	**0.8571 ± 0.0178**	**0.9422 ± 0.0086**

Explore the Novel Mechanism of tau Propagation in AD. We investigate the identified reaction-diffusion functions by neural symbolic regression. To do so, we apply the *eval* function in Python to decode the learned expressions, yielding the curves of analytic expressions for the reaction and diffusion functions at each brain region. Taking *Hippocampus* region u_{151} for example, the identified diffusion and reaction functions are explicitly formulated as: $\hat{g}(u) = 1.0703u_{142}$ and $\hat{h}(u, v) = u_{151} - cos(u_{46}) \times cos(v_{45})$. Then, we classified the trends of these functions into five categories: increasing (red), decreasing (blue), stable (orange), oscillating (yellow), and complex[2] (cyan). The color-coded stereotyped trending patterns on the brain surface are illustrated in Fig. 3.

For the diffusion function (Fig. 3, left), which models the spread of tau protein, the results reveal distinct regional characteristics: (1) Frontal and temporal lobes show a clear increasing trend (highlighted in red), aligning with clinical observations that tau pathology often begins in these regions during the early stages of AD. (2) Parietal lobes exhibit more diverse behaviors, with some areas displaying oscillatory patterns (highlighted in yellow) and others showing

[2] The complex pattern suggests that the underlying function's shape may be composed of multiple simpler patterns.

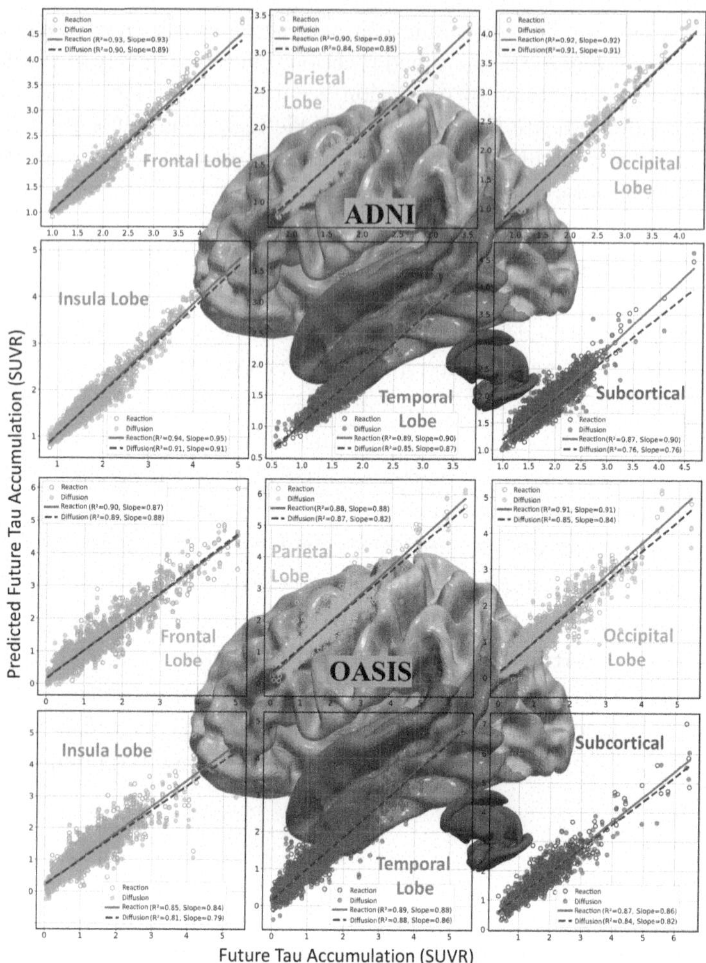

Fig. 2. The linear fitting results between the ground truth and the predicted regional tau accumulation by the learned analytic expressions of $\hat{g}$ (dashed lines) and $\hat{h}$ (solid lines), on ADNI (top two rows) and OASIS data cohorts (bottom two rows).

complex diffusion behaviors (highlighted in blue). These findings may indicate regional variability in tau dynamics, possibly influenced by differences in connectivity or vulnerability to pathology. (3) Few regions maintain stable level of tau accumulation (in orange), suggesting that tau diffusion is an ongoing process in most areas, with varying rates and complexities. (4) None of the brain regions manifests the trending of reduced tau burden, as no region is marked in blue.

The reaction function (Fig. 3, right), which models the interaction between tau and amyloid, reveals more pronounced variability: (1) Oscillatory trends (highlighted in yellow) across brain regions suggest cyclic interactions, potentially reflecting the dynamics of amyloid-triggered tau aggregation and clearance mechanisms. (2) Complex patterns in the temporal lobe (highlighted in blue) might indicate the influence of additional factors that contribute to tau-amyloid interactions. Note, while we cannot guarantee the uniqueness of the mathematical expression across different datasets, we emphasize that in this work, the model's predictive performance and interpretability are firmly grounded in real data. The derived expression reflects meaningful patterns that align with established domain knowledge.

Furthermore, we display the representative diffusion functions and reaction functions at the frontal lobe, parietal lobe, temporal lobe, occipital lobe, and insula lobe, where the color map of curves is shown in Fig. 3 right. Overall, these findings underscore the complicated (not a simple linear process) and region-specific nature of tau spreading along with tau-amyloid interactions.

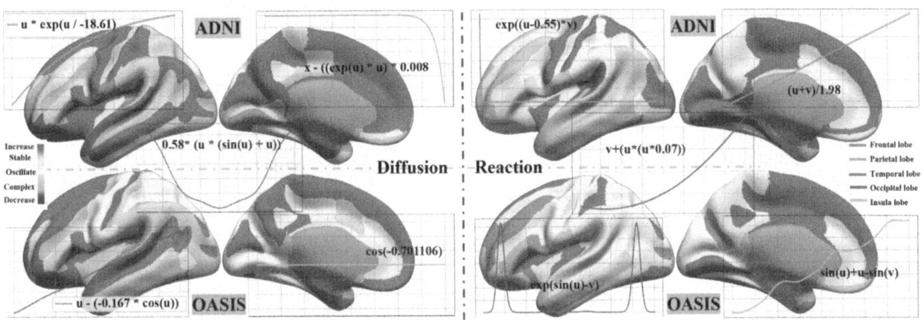

Fig. 3. The brain mapping of characteristic behavior associated with the analytic expressions of the identified diffusion g (left) and reaction h (right) function on ADNI (top) and OASIS (bottom), where red for 'increasing', orange for 'stable', yellow for 'oscillating', and cyan for 'complex'. Furthermore, we overlay the representative g and h in the temporal lobe (blue), parietal lobe (avocado), temporal lobe (blue), occipital lobe (pink), and insula lobe (green). (Color figure online)

New Insight of Aβ-tau Interaction in AD. We switch the gear to explore the underpinning of expressions for Aβ-tau interaction. Recall that the interaction $h(u, v)$ represents the bilateral relationship describing every possible influence from region z_j on the tau development at region z_i. Thus, we consider the amyloid burden at region z_j contributes to the tau aggregation at region z_i by examining whether the amyloid burden v_j occurrence presents in the analytic expression $h_j(u, v)$. By repeating this process for all brain regions, we obtain a brain mapping of amyloid activity map, as shown in Fig. 4, where dark color indicates the cascade of amyloid at the underlying brain region is proactively involved in the tau development of other areas. It is clear that regions with intense Aβ-tau interaction are concentrated in the frontal lobe, temporal lobe, and subcortical regions, particularly at the entorhinal region, indicating that these regions serve as critical hubs for tau propagation. The prominent activity in these regions aligns with established neuropathological observations [26], where tau and amyloid interactions are key drivers of neurodegenerative processes. In addition, the active Aβ-tau interaction in the subcortical regions (such as hippocampus and amygdala) supports the evidence that deep gray matter structures facilitate tau spread, indicating the importance of these regions in early disease progression. Our results also show that amyloid deposits occurred at temporal lobe (blue circles) is responsible for whole-brain tau propagation, as indicated by the high degree of involvement in the tau development. Furthermore, our results suggest that Aβ deposit primarily promote the local tau development, although a few regions play an active role in regulating the production of tau pathology on other regions through the lens of white matter fibers.

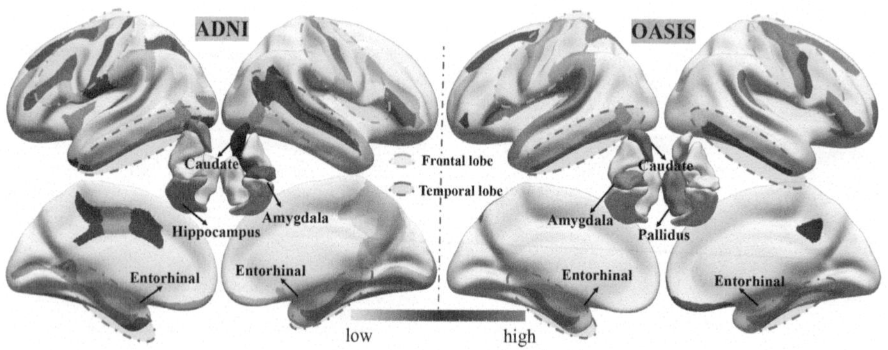

Fig. 4. The brain surface mapping of Aβ-tau interaction. Dark color indicates active involvement of amyloid cascade in the tau propagation.

4 Discussion and Conclusion

An in-depth understanding of the pathophysiological mechanism is vital for developing effective therapeutic treatments for AD and AD-related dementias. In this work, we present a novel physics-informed deep model to uncover the novel mechanism of tau propagation from *in-vivo* longitudinal pathology images, by combining the insight of computational neuroscience and the power of machine learning. The backbone of our deep model is neural symbolic regression for discovering latent analytic expression of diffusion and reaction functions, which allows us to explicitly explore the kinetic characteristics of local tau spreading as well as the interaction between amyloid and tau. Due to the well principled model design, we not only achieve enhanced prediction accuracy for future tau accumulation but also provide a new window to elucidate the novel biological mechanisms using data-driven approaches.

We summarize the answers to key neuroscience questions as follows: (1) Our findings highlight the complex, non-linear, and region-specific nature of tau spreading. (2) Our results suggest that amyloid deposits promote the tau aggregation in a hybrid manner which involves both local and remote interactions. Future work includes cross-validation on other AD and AD-related data cohorts, investigating the interaction between tau and other AD biomarkers, and disease simulation using the elucidated analytic expressions.

Acknowledgments. The National Institutes of Health AG070701, AG073927, AG068399, and Foundation of Hope supported this work.

Disclosure of Interests. The authors declare no competing interests.

References

1. Balaji, V., Song, T.A., Yang, F., Jacobs, H., Johnson, K., Dutta, J.: A graph neural network model for the prediction of longitudinal tau aggregation (2022)
2. Cai, H., Dan, T., Huang, Z., Wu, G.: OSR-net: Ordinary differential equation-based brain state recognition neural network. In: 2023 IEEE 20th International Symposium on Biomedical Imaging (ISBI), pp. 1–5. IEEE (2023)
3. Chamberlain, B., Rowbottom, J., Gorinova, M.I., Bronstein, M., Webb, S., Rossi, E.: Grand: Graph neural diffusion. In: International Conference on Machine Learning, pp. 1407–1418. PMLR (2021)
4. Chen, M., Wei, Z., Huang, Z., Ding, B., Li, Y.: Simple and deep graph convolutional networks. In: International Conference on Machine Learning, pp. 1725–1735. PMLR (2020)
5. Chen, R.T., Rubanova, Y., Bettencourt, J., Duvenaud, D.K.: Neural ordinary differential equations. Adv. Neural Inform. Process. Syst. **31** (2018)
6. Cranmer, M., et al.: Discovering symbolic models from deep learning with inductive biases. Adv. Neural. Inf. Process. Syst. **33**, 17429–17442 (2020)
7. Dan, T., Dere, M., Kim, W.H., Kim, M., Wu, G.: Tauflownet: revealing latent propagation mechanism of tau aggregates using deep neural transport equations. Med. Image Anal. **95**, 103210 (2024)
8. Dan, T., et al.: Re-think and re-design graph neural networks in spaces of continuous graph diffusion functionals. Adv. Neural. Inf. Process. Syst. **36**, 59375–59387 (2023)
9. Dan, T., Kim, M., Kim, W.H., Wu, G.: Tauflownet: Uncovering propagation mechanism of tau aggregates by neural transport equation. In: International Conference on Medical Image Computing and Computer-Assisted Intervention, pp. 77–86. Springer (2023)
10. Destrieux, C., Fischl, B., Dale, A., Halgren, E.: Automatic parcellation of human cortical gyri and sulci using standard anatomical nomenclature. Neuroimage **53**(1), 1–15 (2010)
11. Garbarino, S., Lorenzi, M.: Modeling and inference of spatio-temporal protein dynamics across brain networks. In: Chung, A., Gee, J.C., Yushkevich, P.A., Bao, S. (eds.) IPMI 2019. LNCS, vol. 11492, pp. 57–69. Springer, Cham (2019). https://doi.org/10.1007/978-3-030-20351-1_5
12. Garbarino, S., Lorenzi, M., Initiative, A., et al.: Investigating hypotheses of neurodegeneration by learning dynamical systems of protein propagation in the brain. Neuroimage **235**, 117980 (2021)
13. Hasani, R., Lechner, M., Amini, A., Rus, D., Grosu, R.: Liquid time-constant networks. In: Proceedings of the AAAI Conference on Artificial Intelligence, vol. 35, pp. 7657–7666 (2021)

14. Hasegawa, M., Nonaka, T., Masuda-Suzukake, M.: Prion-like mechanisms and potential therapeutic targets in neurodegenerative disorders. Pharmacol. Therap. **172**, 22–33 (2017)
15. Kingma, D.P.: Adam: A method for stochastic optimization. arXiv preprint arXiv:1412.6980 (2014)
16. Leuzy, A., Chiotis, K., Lemoine, L., Gillberg, P.G., Almkvist, O., Rodriguez-Vieitez, E., Nordberg, A.: Tau pet imaging in neurodegenerative tauopathies–still a challenge. Mol. Psychiatry **24**(8), 1112–1134 (2019)
17. Li, G., Muller, M., Thabet, A., Ghanem, B.: Deepgcns: Can GCNs go as deep as CNNs? In: Proceedings of the IEEE/CVF International Conference on Computer Vision, pp. 9267–9276 (2019)
18. Marcus, D.S., Wang, T.H., Parker, J., Csernansky, J.G., Morris, J.C., Buckner, R.L.: Open access series of imaging studies (OASIS): cross-sectional MRI data in young, middle aged, nondemented, and demented older adults. J. Cogn. Neurosci. **19**(9), 1498–1507 (2007)
19. Petersen, B.K., Landajuela, M., Mundhenk, T.N., Santiago, C.P., Kim, S.K., Kim, J.T.: Deep symbolic regression: Recovering mathematical expressions from data via risk-seeking policy gradients. arXiv preprint arXiv:1912.04871 (2019)
20. Petersen, B.K., Larma, M.L., Mundhenk, T.N., Santiago, C.P., Kim, S.K., Kim, J.T.: Deep symbolic regression: Recovering mathematical expressions from data via risk-seeking policy gradients. In: International Conference on Learning Representations (2020)
21. Petersen, R.C., et al.: Alzheimer's disease neuroimaging initiative (ADNI). Neurology **74**(3), 201–209 (2010)
22. Raj, A., Kuceyeski, A., Weiner, M.: A network diffusion model of disease progression in dementia. Neuron **73**(6), 1204–1215 (2012)
23. Rathore, S., et al.: Predicting regional tau accumulation with machine learning-based tau-pet and advanced radiomics. Alzheimer's Dementia: Transl. Res. Clin. Intervent. **10**(4), e70005 (2024)
24. Velickovic, P., Cucurull, G., Casanova, A., Romero, A., Lio, P., Bengio, Y., et al.: Graph attention networks. Stat **1050**(20), 10–48550 (2017)
25. Vogel, J.W., et al.: Connectome-based modelling of neurodegenerative diseases: towards precision medicine and mechanistic insight. Nat. Rev. Neurosci. **24**(10), 620–639 (2023)
26. Vogel, J.W., et al.: Spread of pathological tau proteins through communicating neurons in human alzheimer's disease. Nat. Commun. **11**(1), 2612 (2020)
27. Vogels, T., et al.: Propagation of tau pathology: integrating insights from post-mortem and in vivo studies. Biol. Psychiat. **87**(9), 808–818 (2020)
28. Weickenmeier, J., Jucker, M., Goriely, A., Kuhl, E.: A physics-based model explains the prion-like features of neurodegeneration in Alzheimer's disease, Parkinson's disease, and amyotrophic lateral sclerosis. J. Mech. Phys. Solids **124**, 264–281 (2019)
29. Zhang, J., et al.: Uncovering the system vulnerability and criticality of human brain under dynamical neuropathological events in alzheimer's disease. J. Alzheimers Dis. **95**(3), 1201–1219 (2023)
30. Zhang, Z., Zou, Z., Kuhl, E., Karniadakis, G.E.: Discovering a reaction-diffusion model for Alzheimer's disease by combining PINNs with symbolic regression. Comput. Methods Appl. Mech. Eng. **419**, 116647 (2024)

A Multi-layer Neural Transport Model for Characterizing Pathology Propagation in Neurodegenerative Diseases

Haifeng Huang[1], Yi Wang[1], Tingting Dan[2], Yang Yang[1], and Guorong Wu[2,3](✉)

[1] School of Information Science and Technology, Yunnan Normal University, Kunming, China

[2] Department of Psychiatry, University of North Carolina at Chapel Hill, Chapel Hill, NC, USA

grwu@med.unc.edu

[3] Department of Computer Science, University of North Carolina at Chapel Hill, Chapel Hill, NC, USA

Abstract. There is a growing consensus in neuroscience that pathological proteins accumulate and spread along specific large-scale brain networks, indicating the mechanistic role of connectome architecture in the progression of neurodegenerative diseases. Although mounting evidence shows that pathology spreading is a dynamic biological process shaped by the complex interplay between the wiring mechanism of neuronal fibers and the self-organized synchronization of functional fluctuations, current computational methods model the propagation of pathology burden through either structural connectivity (SC) or functional connectivity (FC). To address this limitation, we present a multi-layer transport model to capture the SC/FC-specific propagation of neuropathological burdens and their interactions from longitudinal imaging data. Furthermore, we propose to parameterize the spreading pathways using a physics-informed neural network, enabling the prediction of the progression of pathological events at the baseline. We have evaluated the prediction accuracy of tau aggregates in Alzheimer's disease (AD), where our method achieves a significantly higher accuracy compared to existing approaches. In addition, the physics principle in our deep model allows us to explore the biological underpinning of how SC-FC interaction contributes to pathology propagation in AD. Taken together, enhanced prediction accuracy and model interpretability suggest the great potential of our deep model in uncovering the pathophysiological mechanism in neurodegenerative diseases through data-driven approaches.

Keywords: Pathology Propagation Mechanism · Optimal Transport · Brain Networks · Positron Emission Tomography · Systems Biology

H. Huang, Y. Wang, and T. Dan—Contributed equally to this work.

I. Oguz et al. (Eds.): IPMI 2025, LNCS 15830, pp. 51–64, 2026.
https://doi.org/10.1007/978-3-031-96625-5_4

1 Introduction

Many neurodegenerative diseases such as Alzheimer's disease (AD) are characterized by a long preclinical period and progressive course, where the presence of pathology burdens eventually leads to cognitive decline and dementia [13]. Although older adults with an increased magnitude of pathological burden often show a greater cognitive decline over time, the pathophysiological mechanisms by which neuropathology spreads in the brain and how this determines the associated pattern of cognitive decline are still largely elusive.

A growing body of evidence shows that abnormal protein aggregates are often co-localized in brain regions with strong structural connectivity (SC) [2,21]. These findings lead to the development of the "prion-like" spreading hypothesis, suggesting that pathogenic proteins initially accumulate in seed regions and then propagate along neuronal pathways within large-scale brain networks [15]. Since functional connectivity (FC) represents dynamic interactions between brain regions that collaborate during cognitive and physiological processes, pathogenic proteins, such as tau in Alzheimer's disease, can hijack these synchronized interactions to propagate more efficiently [4]. Many neuroimaging studies show that pathology tends to accumulate in regions with a high degree of FC, supporting the notion that functional networks also play a critical role in determining the spreading of neuropathology [14].

Along with the hypothesis of connectome-based neurodegeneration, a number of computational methods have been proposed to model the pathology evolution on top of the network topology, which can be generally classified into two categories: (1) network diffusion models and (2) epidemic models. The network diffusion model [22,25] has been used to predict longitudinal patterns of atrophy and metabolism in AD, characterizing the evolution of pathology burdens as a graph heat process. Although the solution is a linear representation of eigenvectors associated with the graph Laplacian matrix, this pioneering work has shown the potential of capturing microscopic properties of disease progression from a systems biology perspective [23]. Since the propagation of pathology burdens is constrained to the law of mass effect with first-order kinetics, network diffusion models have limited explanatory power. Recently, systems biology approaches have become popular in modeling neurodegeneration and the interaction between neuropathologies using a set of pre-defined partial differential equations (PDEs). For example, the reaction-diffusion model is used to model the tau protein misfolding and spreading, where the diffusion process can be constrained by the geometry of cortical surface [28] and brain connectomes [26]. On the other hand, epidemic models have the advantage of incorporating additional emergent properties of a system as the epidemic spreads [12,29]. For example, a brain region may lose its ability to propagate the disease once it is severely affected. Although epidemic models offer great flexibility in modeling the production and degradation of neuropathology at each brain region, the application is often limited by parameter sensitivity to the extent that small errors in these parameters can result in large deviations in predictions.

Although striking successes have been achieved by current computational approaches in predicting future pathology burdens, there are several limitations. *First*, none of the current methods characterize the pathology propagation by taking both SC and FC into account, not to mention the interaction between SC and FC. Since mounting evidence shows that SC and FC both contribute to the spreading of pathological events, it is vital to understand the mechanistic role of SC-FC interaction in the progression of neurodegenerative diseases. *Second*, the majority of current methods use conventional optimization techniques to fit the model with the observed neuroimaging data. However, the absence of feature selection in model fitting makes current methods sensitive to potential noise and the high dimensionality of neuroimaging data. *Third*, machine learning technique has attracted growing interest in recent studies [7,8,24,28]. Although the prediction accuracy is promising, current deep models lack the power to uncover the biological mechanism by which SC and FC simultaneously contribute to pathology propagation and eventually lead to cognitive decline.

To address these limitations, we conceptualize that the propagation of tau pathology forms a dynamical system where the evolution of tau aggregates is expressed in terms of fluxes along the SC/FC topology and concentration changes on each brain region. Specifically, the tau propagation system consists of two biological pathways. The first pathway underlines the topology of structural brain networks, which restricts the diffusion of tau protein along the white matter fibers [22]. Alternatively, the second pathway considers tau spreading along large-scale functional networks, as FC may facilitate the prion-like trans-synaptic transmission of misfolded tau aggregates [6]. We hypothesize that (1) SC and FC work together to drive tau propagation in AD and (2) the spatial pattern of SC-FC interactions may shift as AD progresses.

To that end, we characterize the evolution of tau aggregations via a multi-layer mass transport model, where the dynamic behavior of tau propagation is jointly shaped by SC and FC. Furthermore, we formulate the SC-FC interaction in the notion of optimal control where the efforts of SC and FC on whole-brain tau propagation reach a balanced dynamic of mutually regulating each other. Together, we present a physics-informed neural network, principled by the multi-layer neural transport model, to predict future tau accumulation for unseen subjects. Enhanced prediction accuracy has been achieved by our multi-layer deep model, compared to the existing methods using either SC or FC. Furthermore, the principled model interpretability allows us to investigate the novel tau propagation mechanism in the context of SC-FC interaction, demonstrating the great applicability of data-driven approaches in AD studies.

2 Methods

Problem Formulation. Without loss of generality, we represent either structural and functional network in a graph structure, $\mathcal{G} = (V, W)$, where $V = \{v_i | i = 1, ..., N\}$ is the set of N graph nodes and $W = [w_{ij}]_{i,j=1}^N$ is the weighted adjacency matrix with each element w_{ij} measures the connectivity degree between region v_i and v_j. Meanwhile, we suppose the observed tau SUVR (standardized uptake value ratio) is associated with each graph node, denoted by $x = [x_i]_{i=1}^N$. In this context, the inputs of our model involve: (1) baseline tau SUVR measurements $x^0 = [x_i^0]_{i=1}^N$ as the graph embedding, (2) SC and FC as the adjacency matrices. The driving force is to predict future tau SUVR x^1 using the baseline observation x^0, where we use superscript '0' and '1' to denote baseline and follow-up, respectively.

2.1 Mass Transport System on Graphs

Suppose $\rho = [\rho_i]_{i=1}^N$ indicates the potential energy at each node v_i, which is obtained by projecting the tau SUVR x through a mapping function ϕ, i.e., $\rho = \phi(x)$. We assume the evolution of tau spreading is formulated as a conservative system of mass transport, governed by the continuity equation: $\frac{\partial \rho}{\partial t} + div(q) = 0$, where $div(\cdot)$ denotes the graph divergence operator, and q represents the flux field responsible for propagating tau-specific potential energy u across time. Drawing an analogy to how a gravitational field drives water flow, the transport equation reflects that variations in energy density ρ influence the propagation of tau aggregates throughout the graph $\mathcal{G}$. According to the Fourier's law of heat conduction, the heat flux $q = -h\nabla\rho$ is proportional to the graph gradient $\nabla\rho = w_{ij}(\rho_i - \rho_j)$, where h is the diffusivity describing the thermal conductance property on $\mathcal{G}$. In the homogeneous case, i.e., $g = c$ is a constant scalar, the mass transport equation on graph $\mathcal{G}$ is expressed as:

$$\frac{\partial \rho}{\partial t} = -div(c\nabla\rho) = -c\Delta\rho, \quad \rho = \phi(x), \tag{1}$$

where Δ is a graph Laplacian operator. Equation 1 encapsulates the interplay between tau aggregates x, potential energy field u, and graph topology Δ, allowing for the learnable modeling of tau spreading dynamics throughout brain networks.

2.2 Multi-Layer Neural Transport Equations with Optimal Control

In this work, we extend the mass transport equation for a single graph to a multi-layer graph topology. Suppose Δ_s and Δ_f represent the graph Laplacian formed by the structural network and the functional network, respectively. First, we model the portion of tau-related potential energy u_s driven by SC using the mass transport system in Eq. 1, i.e., $\frac{\partial u_s}{\partial t} = -c\Delta_s u_s$. Similarly, the spreading of FC-specific potential energy u_f can be modeled as $\frac{\partial u_f}{\partial t} = -c\Delta_f u_f$.

Multi-layer Transport System. To couple tau spreading pathways emerging from SC and FC, we formulate the FC-specific energy u_f as the feedback control on the underlying SC-specific transport system $\frac{\partial u_s}{\partial t}$, and vice versa. The propagation of tau on structural and functional networks is integrated through SC-specific and FC-specific interaction matrices, denoted by M_s and M_f, where the dimension of each matrix is $N \times N$. In this context, we propose a multi-layer neural transport equation, which can be expressed as follows:

$$\begin{cases} \frac{\partial u_s}{\partial t} = -\Delta_s u_s + \lambda_s M_s u_f, \\ \frac{\partial u_f}{\partial t} = -\Delta_f u_f + \lambda_f M_f u_s, \end{cases} \quad \text{subject to:} \quad \phi^{-1}(u_s + u_f) = x, \tag{2}$$

where λ_s and λ_f are used to balance the contribution of feedback control. It is clear that the coupled equations form a closed-loop feedback system such that the output u_s and u_f are fed back to the dynamic system of the counterpart network conduit. For clarity, we further rewrite Eq. 2 as:

$$\frac{\partial}{\partial t}\begin{bmatrix} u_s \\ u_f \end{bmatrix} = \begin{bmatrix} -\Delta_s u_s + \lambda_s M_s u_f \\ -\Delta_f u_f + \lambda_f M_f u_s \end{bmatrix} = \begin{bmatrix} -\Delta_s, \lambda_s M_s \\ \lambda_f M_f, -\Delta_f \end{bmatrix} \begin{bmatrix} u_s \\ u_f \end{bmatrix} = \mathcal{A} \begin{bmatrix} u_s \\ u_f \end{bmatrix}, \tag{3}$$

where $\mathcal{A}$ can be interpreted as a multi-layer network that bridges structural and functional transport. Assume that the multi-layer neural transport system in Eq. 2 is controllable. In this context, it becomes possible to influence the eigenvalues of the closed-loop system $\frac{\partial}{\partial t}u(t)$ in Eq. 3 through an appropriate choice of a full-state feedback control law [27] as $\frac{\partial}{\partial t}u(t) = \Delta u(t) - MKu(t)$, where $M = [M_s, M_f]$ and $K = [K_s, K_f]^\intercal$. Specifically, this is achieved by reformulating u_f and u_s by $u_f = -K_s u_s$ and $u_s = -K_f u_f$, where K_s and K_f are the gain matrices.

Optimal Control. In our multi-layer formulation, the SC-specific flow u_s and FC-specific flow u_f are considered as the feedback control on each others' propagation pathway. By following the notion of optimal control, we introduce the widely used Linear Quadratic Regulator (LQR) to constrain the SC-FC interaction, which is expressed in a quadratic cost function $\mathcal{L}$:

$$\mathcal{L} = \frac{1}{2}u_s^\intercal Q u_s + \frac{1}{2}u_f^\intercal R u_f, \tag{4}$$

where matrices Q and R represent the weights of states u_s and u_f, respectively. To reduce computational complexity, both matrices are diagonal, with the values on the diagonal representing the contribution of each node to tau propagation. We seek for Q and R towards a balance between system performance (i.e., the terminal state $u^1 = u_s^1 + u_f^1$ exhibit the largest correlation with the outcome of the underlying learning task) and control energy consumption (i.e., using the minimal cost of u_s and u_f combined), thereby obtaining the optimal control strategy. The LQR constraint is crucial in this context, as it provides a systematic approach to derive the optimal control strategy by solving the associated Riccati equation, ensuring an efficient trade-off between performance and energy expenditure.

The solution to Eq. 3 with the LQR constraint is fundamentally derived through a combination of the calculus of variations and the method of Lagrange multipliers, which form the basis of *Pontryagin's Minimum Principle* [20]. To do so, we introduce a value function $\mathcal{J}(u)$ as a quadratic form of the state vector $u = \begin{bmatrix} u_s \\ u_f \end{bmatrix}$, given by: $\mathcal{J}(u) = \frac{1}{2}u^\intercal P u$, where $P \in \mathbb{R}^{2N \times 2N}$ is a symmetric positive definite matrix, where $P = \begin{bmatrix} P_{ss} & P_{sf} \\ P_{sf}^{\intercal} & P_{ff} \end{bmatrix}$. P_{ss} and P_{ff} represent the contributions of u_s and u_f to the overall cost, respectively. P_{sf} encodes the coupling between u_s and u_f, reflecting the interactions between the SC-specific and FC-specific tau spreading pathways. The matrix P is fundamental in shaping the optimal weighting matrices Q and R, which are key to the derivation of the control law.

Given P, the gain matrix K has a closed-form solution as: $K = -R^{-1}MP$. We compute P by solving the Riccati equation [1]: $P\mathcal{A} + \mathcal{A}^{\intercal}P + \begin{bmatrix} Q & 0 \\ 0 & R \end{bmatrix} = 0$, where the block structure of P and $\mathcal{A}$ allows the equation to be decomposed into three coupled sub-equations:

$$\begin{cases} P_{ss}(-\Delta_s) + (-\Delta_s)^{\intercal}P_{ss} + P_{sf}\lambda_f M_f + (\lambda_f M_f)^{\top}P_{sf}^{\top} + Q = 0, \\ P_{ff}(-\Delta_f) + (-\Delta_f)^{\intercal}P_{ff} + P_{sf}^{\top}\lambda_s M_s + (\lambda_s M_s)^{\top}P_{sf} + R = 0, \\ P_{sf}(-\Delta_f) + (-\Delta_s)^{\intercal}P_{sf} + P_{ss}\lambda_s M_s + (\lambda_f M_f)^{\top}P_{ff} = 0. \end{cases} \tag{5}$$

Based on the above derivation, we summarize our optimization strategy of optimal control by alternating the following three steps.

- *Solve u_s and u_f.* The optimization is performed separately for SC-specific and FC-specific propagations. For the SC neural transport equation, Q serves as the weighting matrix, and u_s is optimized by solving the Riccati equation to compute P_{ss} and derive the feedback gain K_s. Similarly, for the FC neural transport equation, R is the weighting matrix, and u_f is optimized by calculating P_{ff} and solving for K_f.
- *Estimate SC-FC interaction term P_{sf}.* The interaction between the layers is captured by optimizing the coupling matrix P_{sf}, which minimizes the overall cost function and improves coordination and stability between the SC and FC equations.
- *Estimate M_s and M_f.* We regard M_s and M_f as learnable parameters in the deep model, where we fine-tune the diagonal elements in M_s and M_f through gradient descent.

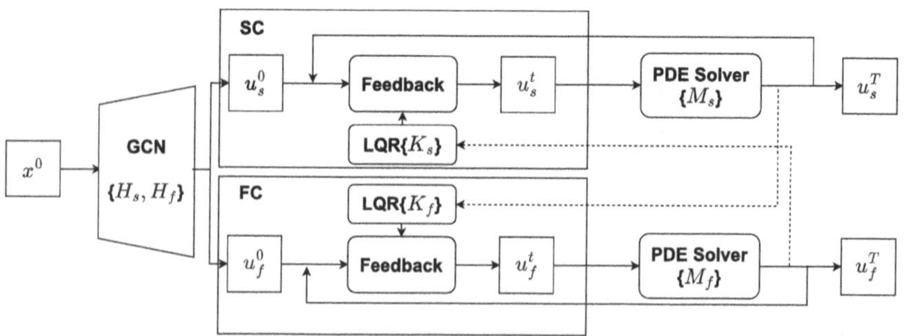

Fig. 1. The overall network architecture of closed-loop feedback multilayer neural transport model with the integrated supervised optimal control. The model input is the baseline regional tau SUVR vector x^0. The output is the disentangled portion of SC-specific and FC-specific tau propagation u_s and u_f.

Design of an Explainable Deep Model for Characterizing Pathology Propagation in AD. Building on the learning components introduced earlier, we propose an explainable deep model that enables the characterization of pathology propagation from spatiotemporal data. The model's network hyperparameters, denoted by $\Gamma = \{M_s, M_f, H_s, H_f, K_s, K_f\}$, represent the network structure and hyperparameter settings. These include the interaction matrices M_s and M_f, which capture the interactions between the SC-specific and FC-specific tau propagation, as well as the GNN hyperparameters H_s and H_f for projecting the observed regional tau SUVR to the potential energy u_s and u_f respectively. The gain matrices K_s and K_f are employed to implement feedback control and regulate the dynamic behavior of the system. The behavior of the entire system is characterized by solving the PDEs, and the optimal control strategy is determined using multi-layer optimal control theory.

The overall network design is illustrated in Fig. 1. First, a graph convolutional network (GCN) is employed for diffusion over the graph topology Δ. The hidden layer incorporates a recurrent component, where a PDE solver is used to characterize the system behavior based on the neural transport equation with the optimal control law. Subsequently, a fully connected layer maps the state space data to the observation space, i.e., $\hat{x}^1$. Since the time-evolving feedback u_s and u_f depend on the prediction error $\|x^1 - \hat{x}^1\|_F^2$, our deep model fine-tunes the network parameters Γ in a supervised manner. This approach ensures that the learned multi-layer system aligns with the mechanical principles, while preserving interpretability in accordance with neuroscience insights.

3 Experiments

Data Description. In our experiments, we use the Open Access Series of Imaging Studies-3 (OASIS-3) dataset [18] to evaluate our method. We select 66 subjects, each with longitudinal Tau-PET scans, T1-weighted magnetic resonance imaging (MRI), diffusion-weighted image, and resting-state fMRI. For each subject, we apply the following image processing steps. To construct the structural network, we employ Destrieux atlas [10] which includes 148 cortical (frontal, insula, temporal, occipital, parietal, and limbic lobe) and 12 subcortical (left and right hippocampus, caudate nucleus, thalamus, amygdala, pallidum, and preumentum) regions. In accordance with the methodology outlined in [9], the pipeline generates SC and FC matrices, each with dimensions of 160×160. For each tau-PET scan, we first apply rigid registration to align the PET scan to the space of its T1-weighted MRI. Then, we calculated the SUVR for each brain region, where brain stem is used as the reference region. We concatenate SUVRs from 160 regions into a data array as the whole-brain tau pathology burden.

Experimental Setup. Our model integrates deep learning to optimize LQR parameters. For evaluation, we extracted data at both initial and final time points, using the initial time point data as input. The model output was compared to the final time point data by calculating the mean absolute error (MAE). Predicted results were further classified to assess their diagnostic capabilities. The model was trained using the Adam [16] optimizer with a learning rate of

0.001 over 100 gradient backpropagation iterations. All results are reported based on 5-fold cross-validation. We compared our method with several state-of-the-art deep learning models, including: (1) Graph-based model: vanilla graph convolutional networks (GCN) [17], GCNII [5], graph attention networks (GAT), and GATv2 [3]. For these models, the baseline x^0 is used as the graph embedding with SC and/or FC serving as the adjacency matrix. (2) A classic sequential model, recurrent neural networks (RNN) [19]. (3) Partial differential equation-based model, liquid time constant networks (LTCNet) [11]. For models other than the graph-based ones, the baseline tau SUVR is used as input.

Ablation experiments have been performed to disentangle the individual and joint effects of SC and FC on the dynamics of the tau aggregate, providing an in-depth insight into the interplay between SC and FC in AD. Furthermore, we uncover the impact of SC and FC on prediction outcomes through their total contribution to tau pathology evolution, which allows us to explore the following neuroscience questions in AD.

Table 1. Prediction of tau accumulation and AD risk (shaded) by different deep models in the context of SC only, FC only, and SC-FC interaction.

Methods	SC	FC	SC-FC	SC	FC	SC-FC
GCN	$0.482_{\pm 0.058}$	$0.501_{\pm 0.040}$	$0.479_{\pm 0.032}$	$0.846_{\pm 0.038}$	$0.815_{\pm 0.069}$	$0.862_{\pm 0.046}$
GCNII	$0.533_{\pm 0.023}$	$0.575_{\pm 0.023}$	$0.577_{\pm 0.038}$	$0.899_{\pm 0.043}$	$0.882_{\pm 0.022}$	$0.901_{\pm 0.025}$
GAT	$0.526_{\pm 0.027}$	$0.570_{\pm 0.016}$	$0.517_{\pm 0.029}$	$0.833_{\pm 0.028}$	$0.803_{\pm 0.030}$	$0.848_{\pm 0.046}$
GATv2	$0.471_{\pm 0.033}$	$0.551_{\pm 0.024}$	$0.495_{\pm 0.026}$	$0.869_{\pm 0.032}$	$0.832_{\pm 0.027}$	$0.881_{\pm 0.033}$
RNN	$0.519_{\pm 0.045}$	$0.552_{\pm 0.039}$	$0.522_{\pm 0.027}$	$0.862_{\pm 0.035}$	$0.843_{\pm 0.023}$	$0.880_{\pm 0.036}$
LTC-Net	$0.457_{\pm 0.048}$	$0.469_{\pm 0.023}$	$0.442_{\pm 0.029}$	$0.877_{\pm 0.040}$	$0.846_{\pm 0.048}$	$0.862_{\pm 0.028}$
OURS	*$0.392_{\pm 0.028}$*	*$0.415_{\pm 0.013}$*	**$0.356_{\pm 0.035}$**	*$0.908_{\pm 0.030}$*	*$0.892_{\pm 0.046}$*	**$0.954_{\pm 0.015}$**

3.1 Quantitative Evaluation on Prediction Accuracy

In Table 1, we show the prediction accuracy of tau pathology by all deep models under comparison, under multiple settings where the propagation of tau aggregates is constrained by SC only, FC only, and SC-FC interaction[1]. It is clear that our multi-layer neural transport model achieves significantly higher prediction over all counterpart deep models (at a significance level $p < 0.05$). Furthermore, we connect the predicted $\hat{u}^1$ with a fully connected layer to predict the risk of AD onset at the follow-up scan. Again, our model outperforms all other methods, achieving the highest disease diagnosis accuracy (shaded in Table 1) of 95.4% by leveraging both SC and FC pathways.

[1] Since conventional deep models are not designed for the setting of multilayer tau propagation, we parameterize the tau spreading on SC and FC separately and then combine them using a fully-connected layer.

3.2 Ablation Study

There is a consistent pattern in Table 1 that utilizing both SC and FC information leads to increased learning performance. Here, we conduct an ablation study to evaluate the contribution of SC and FC to tau propagation in our multi-layer model. As shown in Fig. 2 top, we apply a linear regression model to examine the correlation between ground truth (from the follow-up tau-PET scan) and the predicted degree of regional SUVR by using SC only (left), FC only (middle), and SC-FC interaction (right), where the color represents the cerebral lobe of the underlying brain region. The slope coefficients are 0.620 by SC only ($R^2 = 0.895$), 0.588 by FC only ($R^2 = 0.893$), and 1.006 by modeling SC-FC interaction ($R^2 = 0.944$), suggesting that characterizing the interaction between SC-specific and FC-specific tau propagation plays a critical role in enhancing the prediction accuracy. Furthermore, we evaluate the regional prediction accuracy at the bottom of Fig. 2, where red and blue indicate the large and small MAE of prediction, respectively. At the scale of the cerebral lobe (shown in Fig. 3), we show the mean and standard deviation of MAE in the frontal lobe, insula lobe, temporal lobe, occipital lobe, parietal lobe, limbic lobe, and subcortical regions by SC-only (in red), FC only (in orange), and SC-FC interaction (in blue). Taken together, converging evidence underscores the importance of accounting for SC-FC interactions in the multi-layer tau propagation model.

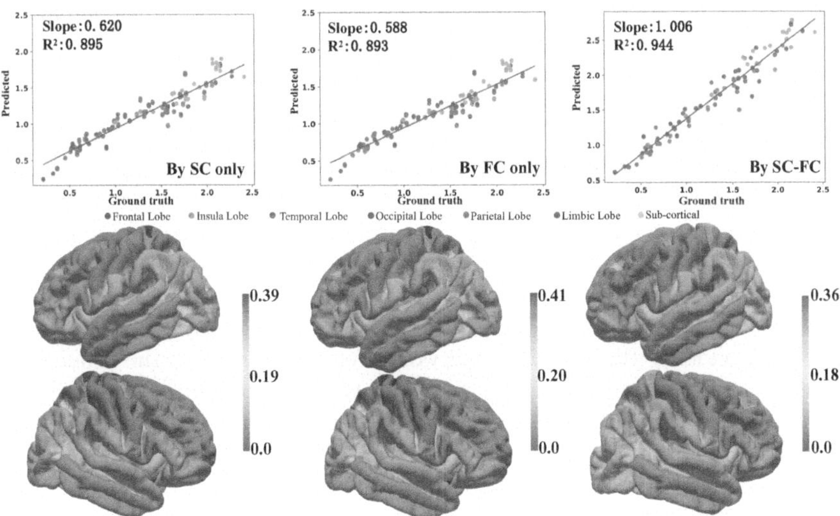

Fig. 2. *Top*: MAE between ground truth (follow-up imaging scans) and predicted tau aggregates using SC only (left), FC only (middle), and SC-FC interaction (right), with colors representing the cerebral lobes. *Bottom*: The brain mapping of average MAE associated with each brain region by SC only (left), FC only (middle), and SC-FC interaction (right), where red and blue indicate high and low MAE, respectively. (Color figure online)

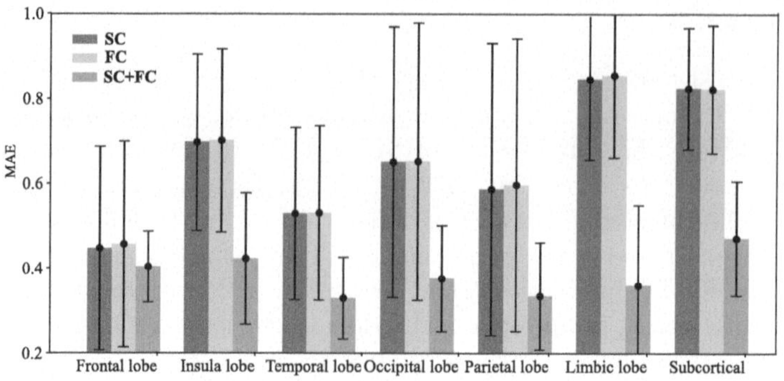

Fig. 3. Mean and standard deviation of MAE in predicting future tau pathology by SC only (red), FC only (orange), and SC-FC interaction (blue). (Color figure online)

3.3 Explore Novel Mechanism in Connectome-Based Tau Propagation Hypothesis

Since our multilayer neural transport model characterizes the dynamic propagation of tau pathology underlying the regional structure-function coupling, we seek to initiate the exploration of novel understandings of tau propagation by examining the hypotheses: (1) Is tau propagation driven by single or multiple connectome conduits? (2) Does the spatial pattern of SC-FC interaction differ across various regions of the brain? (3) What is the relationship between multilayer tau propagation and the disease progression in AD?

To answer the first question, the results in Sect. 3.1 and 3.2 provide strong evidence that both SC and FC contribute to the tau pathology progression since taking the SC-FC interaction into account allows us to significantly enhance the model performance in forecasting the future accumulation of tau aggregates. Furthermore, we examine the proportion of u_s^1 and u_f^1 predicted by our multilayer neural tau transport model. As shown in Fig. 4 top, we plot the mean and standard deviation of u_s^1 (blue) and u_f^1 (red) in the frontal lobe, insula lobe, temporal lobe, occipital lobe, parietal lobe, limbic lobe, and sub-cortical regions. It is evident that no single network conduit dominates the other in tau propagation, supporting the multi-layer tau propagation hypothesis.

Regarding the second question, the bar plot in Fig. 4 top indicates the spatial patterns of SC-FC interaction vary across cerebral lobes. Although the portion of SC-specific tau spreading u_s^1 exceeds u_f^1 in most of the brain regions, FC-specific tau propagation associated with the frontal lobe is significantly greater than the SC-specific counterpart. Furthermore, we examine the proportion of predicted u_s^1 and u_f^1 at each brain region. As shown in Fig. 4 bottom, we use blue color to represent the regional SC-specific tau propagation outnumbers the FC-specific counterpart. Otherwise, we use red color. The node size corresponds to the absolute degree of the difference $|u_x^1 - u_f^1|$. Collectively, our multi-layer neural transport model suggests that the spatial pattern of SC-FC interaction varies across different regions.

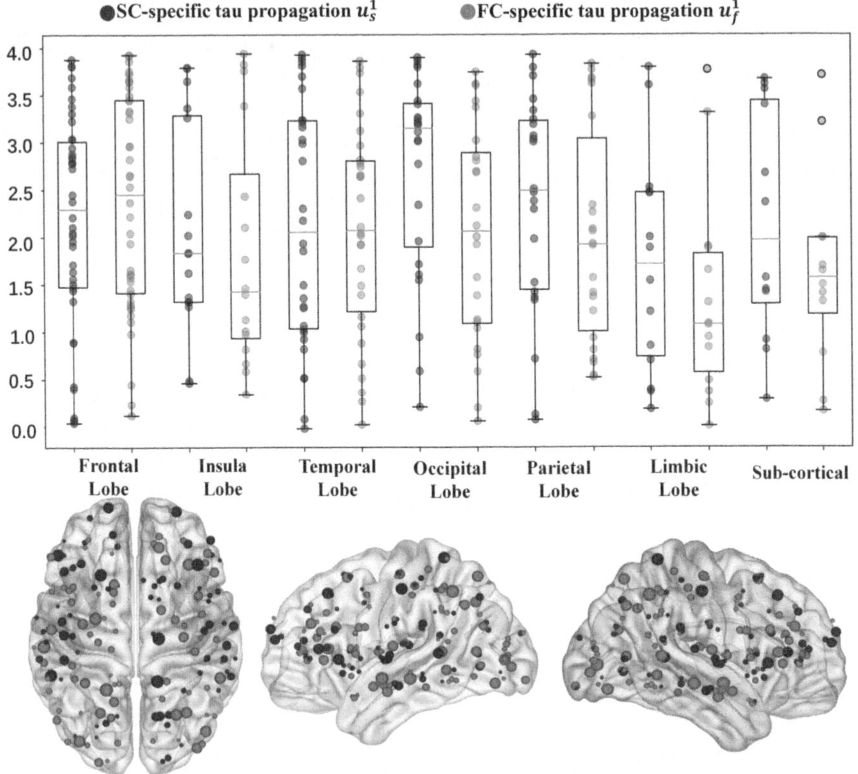

Fig. 4. *Top*: The proportion tau propagation associated with SC u_s^1 (blue) and FC u_f^1 (red). *Bottom*: The competition result between u_s^1 and u_f^1. Blue indicates regions where SC-specific tau propagation (u_x^1) is greater than the FC-specific counterpart (u_f^1), while red represents the opposite. Node size is in proportion to the magnitude of the difference, i.e., $|u_s^1 - u_f^1|$. (Color figure online)

Lastly, we seek to investigate whether the spatial patterns might change during the disease progression by stratifying the regional and lobe-wise analysis in answering Question #2 for CN (cognitive normal) and AD cohorts. As highlighted by the dash box in Fig. 5 right, the SC vs FC competition results has been flipped in the insula lobe, temporal lobe, occipital lobe, parietal lobe, and limbic lobe. Specifically, SC-specific tau spreading predominates over its FC-specific counterpart prior to the onset of AD. However, as the disease progresses, FC conduits increasingly take on a leading role in tau propagation. Therefore the answer to Question #3 is that connectome conduits are deeply involved in the tau propagation along the disease progression while an in-depth understanding of the interaction between SC and FC might be the gateway to translate tau hypothesis to the clinical applications.

As a *proof-of-concept* approach, we assess the clinical utility of our multi-layer neural transport model by evaluating the diagnostic power of using the u_s^1/u_f^1 ratio as the AD biomarker. As a reference, we apply a linear regression model to express the diagnostic label with the regional tau SUVR (i.e., x^1), where age and gender are used as confounders. At a significance level of $p < 0.05$, 33 brain regions manifest a strong association between tau SUVR and clinical label. Similarly, applying the same statistical analysis to investigate the relationship between u_s^1/u_f^1 ratio and clinical outcomes reveals that 42 brain regions show significant results. These results suggest the great potential of our deep model in early diagnosis of AD.

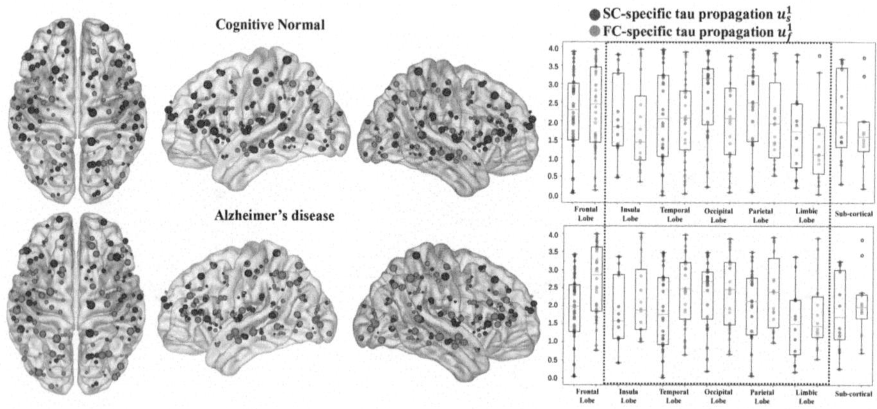

Fig. 5. The spatial patterns of SC-FC interaction in tau propagation at cognitive normal (top) and AD (bottom), where the SC vs FC competition results at region level and cerebral lobe scale are shown in the left and right panels respectively. The definitions of color and node size are the same as in Fig. 4.

4 Conclusion

In this study, we proposed a novel multi-layer neural transport model to investigate the dynamic nature of tau propagation in AD. Our multi-layer framework not only enhances the prediction accuracy but also sheds light on the tau spreading mechanism in the context of both SC and FC conduits. Moving forward, we aim to extend the application of this model to other neurological disorders characterized by network dysfunction, with the goal of uncovering key mechanisms underlying disease propagation.

Acknowledgments. The National Institutes of Health AG070701, AG073927, AG068399, and Foundation of Hope supported this work.

Disclosure of Interests. The authors declare no competing interests.

References

1. Abou-Kandil, H., Freiling, G., Ionescu, V., Jank, G.: Matrix Riccati equations in control and systems theory. Birkhäuser (2012)
2. Braak, H., Braak, E.: Neuropathological stageing of alzheimer-related changes. Acta Neuropathol. **82**(4), 239–259 (1991)
3. Brody, S., Alon, U., Yahav, E.: How attentive are graph attention networks? In: International Conference on Learning Representations (2022). https://openreview.net/forum?id=F72ximsx7C1
4. Buckner, R.L., et al.: Network analysis of the human connectome in Alzheimer's disease. Neuron **63**(6), 748–760 (2009)
5. Chen, M., Wei, Z., Huang, Z., Ding, B., Li, Y.: Simple and deep graph convolutional networks. In: International Conference on Machine Learning, pp. 1725–1735. PMLR (2020)
6. Cope, T.E., et al.: Functional connectivity predicts tau accumulation in Alzheimer's disease. Brain **146**(10), 4040–4055 (2022). https://doi.org/10.1093/brain/awac301
7. Dan, T., Dere, M., Kim, W.H., Kim, M., Wu, G.: Tauflownet: revealing latent propagation mechanism of tau aggregates using deep neural transport equations. Med. Image Anal. **95**, 103210 (2024)
8. Dan, T., Kim, M., Kim, W.H., Wu, G.: Developing explainable deep model for discovering novel control mechanism of neuro-dynamics. IEEE Trans. Med. Imaging **43**(1), 427–438 (2023)
9. Dan, T., Wei, Z., Kim, W.H., Wu, G.: Exploring the enigma of neural dynamics through a scattering-transform mixer landscape for Riemannian manifold. In: Forty-first International Conference on Machine Learning (2024)
10. Destrieux, C., Fischl, B., Dale, A., Halgren, E.: Automatic parcellation of human cortical gyri and sulci using standard anatomical nomenclature. Neuroimage **53**(1), 1–15 (2010)
11. Hasani, R., Lechner, M., Amini, A., Rus, D., Grosu, R.: Liquid time-constant networks. In: Proceedings of the AAAI Conference on Artificial Intelligence. vol. 35, pp. 7657–7666 (2021)
12. Iturria-Medina, Y., Sotero, R.C., Toussaint, P.J., Evans, A.C., Initiative, A.: Epidemic spreading model to characterize misfolded proteins propagation in aging and associated neurodegenerative disorders. PLoS Comput. Biol. **10**(11), e1003956 (2014)
13. Jack, C.R., et al.: Hypothetical model of dynamic biomarkers of the Alzheimer's pathological cascade. Lancet Neurol. **9**(1), 119–128 (2010)
14. Jacobs, H.I., Van Boxtel, M.P., Gronenschild, E.H., Verhey, F.R., Uylings, H.B.: Tau and functional connectivity in Alzheimer's disease: a neuroimaging study. Neuroimage **83**, 471–479 (2013)
15. Jucker, M., Walker, L.C.: Propagation and spread of pathogenic protein assemblies in neurodegenerative diseases. Nat. Neurosci. **21**(10), 1341–1349 (2018)
16. Kingma, D.P., Ba, J.: Adam: A method for stochastic optimization. arXiv preprint arXiv:1412.6980 (2014)
17. Kipf, T.N., Welling, M.: Semi-supervised classification with graph convolutional networks. arXiv preprint arXiv:1609.02907 (2016)
18. Marcus, D.S., Wang, T.H., Parker, J., Csernansky, J.G., Morris, J.C., Buckner, R.L.: Open access series of imaging studies (oasis): cross-sectional MRI data in young, middle aged, nondemented, and demented older adults. J. Cogn. Neurosci. **19**(9), 1498–1507 (2007)

19. Nguyen, M., et al.: Predicting Alzheimer's disease progression using deep recurrent neural networks. Neuroimage **222**, 117203 (2020)
20. Onori, S., Serrao, L., Rizzoni, G., Onori, S., Serrao, L., Rizzoni, G.: Pontryagin's minimum principle. Hybrid Electric Vehicles: Energy Management Strategies, pp. 51–63 (2016)
21. Pearson, R.C., Esiri, M., Hiorns, R., Wilcock, G., Powell, T.: Anatomical correlates of the distribution of the pathological changes in the neocortex in Alzheimer disease. Proc. Natl. Acad. Sci. **82**(13), 4531–4534 (1985)
22. Raj, A., Kuceyeski, A., Weiner, M.: A network diffusion model of disease progression in dementia. Neuron **73**(6), 1204–1215 (2012)
23. Raj, A., LoCastro, E., Kuceyeski, A., Tosun, D., Relkin, N., Weiner, M.: Network diffusion model of progression predicts longitudinal patterns of atrophy and metabolism in Alzheimer's disease. Cell Rep. **10**(3), 359–369 (2015)
24. Rathore, S., et al.: Predicting regional tau accumulation with machine learning-based tau-pet and advanced radiomics. Alzheimer's Dementia: Transl. Res. Clin. Intervent. **10**(4), e70005 (2024)
25. Schäfer, A., Mormino, E.C., Kuhl, E.: Network diffusion modeling explains longitudinal tau pet data. Front. Neurosci. **14**, 566876 (2020)
26. Schäfer, A., Peirlinck, M., Linka, K., Kuhl, E., (ADNI), A.D.N.I.: Bayesian physics-based modeling of tau propagation in alzheimer's disease. Front. Physiol. **12**, 702975 (2021)
27. Smith, J.D., Johnson, E.R.: Full-state feedback control design with eigenvalue placement. Int. J. Contr. Syst. **89**(4), 123–135 (2023)
28. Zhang, Z., Zou, Z., Kuhl, E., Karniadakis, G.E.: Discovering a reaction-diffusion model for Alzheimer's disease by combining PINNs with symbolic regression. Comput. Methods Appl. Mech. Eng. **419**, 116647 (2024)
29. Zheng, Y.Q., et al.: Local vulnerability and global connectivity jointly shape neurodegenerative disease propagation. PLoS Biol. **17**(11), e3000495 (2019)

Enhancing Alzheimer's Diagnosis: Leveraging Anatomical Landmarks in Graph Convolutional Neural Networks on Tetrahedral Meshes

Yanxi Chen[1], Mohammad Farazi[1], Zhangsihao Yang[1], Yonghui Fan[2], Nicholas Ashton[3], Eric M. Reiman[3], Yi Su[3], and Yalin Wang[1](✉)

[1] School of Computing and Augmented Intelligence, Arizona State University, Tempe, AZ, USA
ylwang@asu.edu
[2] Amazon AGI, Redmond, WA, USA
[3] Banner Alzheimer's Institute, Phoenix, AZ, USA

Abstract. Alzheimer's disease (AD) is a major neurodegenerative condition that affects millions around the world. As one of the main biomarkers in the AD diagnosis procedure, brain amyloid positivity is typically identified by positron emission tomography (PET), which is costly and invasive. Brain structural magnetic resonance imaging (sMRI) may provide a safer and more convenient solution for the AD diagnosis. Recent advances in geometric deep learning have facilitated sMRI analysis and early diagnosis of AD. However, determining AD pathology, such as brain amyloid deposition, in preclinical stage remains challenging, as less significant morphological changes can be observed. As a result, few AD classification models are generalizable to the brain amyloid positivity classification task. Blood-based biomarkers (BBBMs), on the other hand, have recently achieved remarkable success in predicting brain amyloid positivity and identifying individuals with high risk of being brain amyloid positive. However, individuals in medium risk group still require gold standard tests such as Amyloid PET for further evaluation. Inspired by the recent success of transformer architectures, we propose a geometric deep learning model based on transformer that is both scalable and robust to variations in input volumetric mesh size. Our work introduced a novel tokenization scheme for tetrahedral meshes, incorporating anatomical landmarks generated by a pre-trained Gaussian process model. Our model achieved superior classification performance in AD classification task. In addition, we showed that the model was also generalizable to the brain amyloid positivity prediction with individuals in the medium risk class, where BM alone cannot achieve a clear classification. Our work may enrich geometric deep learning research and improve AD diagnosis accuracy without using expensive and invasive PET scans.

Keywords: Geometric deep learning · Anatomical landmarking · Tetrahedral meshes · Cortex · Alzheimer's disease

Y. Chen and M. Farazi—These authors contributed equally to this work.

I. Oguz et al. (Eds.): IPMI 2025, LNCS 15830, pp. 65–78, 2026.
https://doi.org/10.1007/978-3-031-96625-5_5

1 Introduction

Alzheimer's disease (AD) has been recognized as a major health problem since the past century, affecting patient cognition and causing dementia [9,28]. Although there is still no cure for AD, several treatments have been developed to alleviate clinical symptoms and slow the disease progression, necessitating the development of early AD diagnosis techniques [4,22]. Among brain imaging techniques, magnetic resonance imaging (MRI) has emerged as a leading and cost-effective method for detecting brain structural changes associated with AD [7]. Compared to more specialized techniques like positron emission tomography (PET), MRI is less invasive, less costly, and more accessible [3,22].

While MRI images can be represented by 3-D voxels, the limited grid resolution prevents detailed study of arbitrarily small regions [12]. Volumetric mesh data structures, such as tetrahedral meshes, offer comprehensive insights by modeling both surface and interior aspects of substances, and have been widely used in scientific modeling such as finite element analysis and physical simulations [5]. Graph neural networks (GNNs) present a natural choice when analyzing graphs, which lack irregular grid-like structures as images. While algorithms based on graph convolution have developed for more commonly used surface meshes [17], convolutions on surface meshes suffer from the over-squashing nature of the message-passing neural networks, potentially aggregating unnecessary or redundant information from intermediate nodes when interacting with long-range nodes. This drawback can naturally be resolved by using volumetric mesh representation, as more detours are available in pathfinding. One notable method for tetrahedral mesh analysis is TetCNN, which was proposed as an improved graph neural networks for tetrahedral mesh data analysis [11] by replacing graph Laplacian with volumetric Laplace Beltrami operator (volumetric LBO) [32,33].

Attention mechanism and transformer architectures are being used ubiquitously in deep learning applications [8,30,31]. Fundamental to the attention mechanism is a set of vectors associated with the input data, namely query vectors, key vectors, and value vectors. The outputs of each attention layer consist of the weighted sum of the value vectors, where the weights are determined by attention scores representing the similarity between the query and key vectors. Transformer architectures have proved their superiority over traditional CNNs in various tasks, such as machine translation, text generation, image classification and graph classification. However, these methods are not directly suitable for tetrahedral meshes with varying sizes and topologies, necessitating a novel transformer-based framework tailored to this structure with strong scalability.

To address the challenge of applying a transformer to tetrahedral meshes with raw 3D coordinate inputs, we propose generating tokens based on learned local features. Unlike ViT for images, generating patches in tetrahedral meshes is not straightforward. To overcome this, we identify key landmarks (super nodes) using a pre-trained Gaussian Process (GP) model. These landmarks serve as the centers of patches, and an efficient methodology assigns nearest neighbors to each super node, ultimately forming the final patch structure for tokenization. The more important reason we create these patches is to learn features and use a pooling strategy to represent the final token feature, unlike the ViT style models concatenating the raw input without any learning strategy beforehand. This is essential to make sure all patches have the same input dimension.

Recent advancements in blood-based biomarkers (BBBMs), such as plasma Amyloid-β 42/40 ratios and phosphorylated tau (pTau) isoforms, offer a promising alternative for detecting AD pathology [2,25]. Among these, plasma pTau-217 stands out as the most effective biomarker for classifying brain amyloid positivity and distinguishing AD from other neurodegenerative diseases [15,26,29]. By integrating MRI with BBBMs as auxiliary inputs into our transformer model, we aim to enhance diagnostic performance, particularly for cases where structural imaging alone may be insufficient.

In this study, we proposed landmark-enhanced TetCNN (LETetCNN) for AD diagnosis and brain amyloid positivity classification [10]. Our primary contributions are summarized as follows:

1. We propose, for the first time, a novel tetrahedral mesh autoionization and transformer-based model designed to handle large and varying-sized meshes.
2. To address the computational challenges of full self-attention, we adopt a sparse local attention strategy, significantly reducing computational cost.
3. A node-level feature learning module is introduced to improve performance and local contextual awareness before feeding geometry into the transformer.
4. Our model is rigorously evaluated on subsets of the ADNI dataset for multiple tasks, consistently achieving improvements over state-of-the-art methods.
5. Integrating BBBMs with learned features significantly boosts performance, highlighting the complementary role of sMRI analysis in the AD studies.

2 Methods

2.1 Notation

In this paper, we represent the tetrahedral mesh as $M = (V, T)$, in which V and T represent vertices and tetrahedra set of the volumetric mesh equipped with its corresponding Laplace Beltrami Operator (LBO) L. The mesh consists of $N = |V|$ nodes and $E = |E|$ edges and $|P|$ distinct patches after partitioning the mesh. Node-level attributes are denoted by their associated patch numbers as subscripts and is small characters. Accordingly, patch level attributes are stated with capital letters. Layer-level attributes are indicated with superscripts $()^{layer}_{node,patch}$, like $x^l_{i,p}$ is the node feature of node i in patch p at layer l.

Definition. The volumetric Laplace-Beltrami Operator (LBO), Δ_{tet}, on a tetrahedral mesh M with vertices $\{v_i\}_{i=1}^n$ satisfies the eigen-system $\Delta_{tet} f = -\lambda f$ for a real-valued function $f \in C^2$. Its discrete approximation [32] is:

$$\Delta f(v_i) = \frac{1}{d_i} \sum_{j \in N(i)} k_{i,j}(f(v_i) - f(v_j)), \tag{1}$$

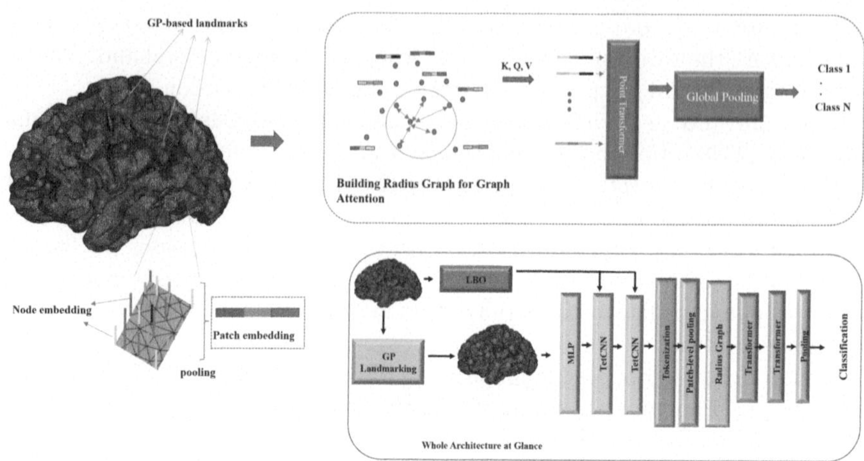

Fig. 1. The architecture of our model begins with identifying patch centers using Gaussian Process (GP)-based landmarks. Local-aware features are then learned through TetCNN, followed by constructing a radius graph to compute sparse attention over the previously learned tokens. Additionally, we pre-compute the Laplace-Beltrami Operator (LBO) and landmarks to streamline the process.

where $N(i)$ are vertices adjacent to v_i, d_i is the volume of adjacent tetrahedra, and $k_{i,j}$ depends on edge lengths and dihedral angles. The stiffness matrix $A = W - K$, with $W = \mathrm{diag}(w_1, \ldots, w_n)$ and $w_i = \sum_{j \in N(i)} k_{i,j}$. The lumped discrete LBO is $L_{tet} = D^{-1}A$ where $D = \mathrm{diag}(d_1, \ldots, d_n)$.

2.2 Gaussian Process-Based Anatomical Landmarking

Anatomical landmarking was initially developed for dimensionality reduction, or saliency detection [23]. A recent study proposed a Gaussian process-based method for automatically generating anatomical landmarks [10]. To be specific, a multi-frequency multi-phase periodic diffusion kernel (mmPDK) was designed to encode high-quality geometric features into the prior knowledge of a Gaussian process (GP) model, resulting in a geometry-aware landmark generation framework for generating an arbitrary number of landmarks. In this study, we leveraged the proposed landmark generation algorithm to predefine a group of landmarks for each tetrahedral mesh. These generated landmarks were then incorporated into our GNN-based model as additional inputs. The overall architecture of the model is also illustrated in Fig. 1.

2.3 Tokenization

Based on our landmarks (super nodes), we use the K-nearest Neighbor algorithm to assign each node of the mesh to these super nodes to create our tokens. We assign the labels to each node based on their proximity to their immediate close

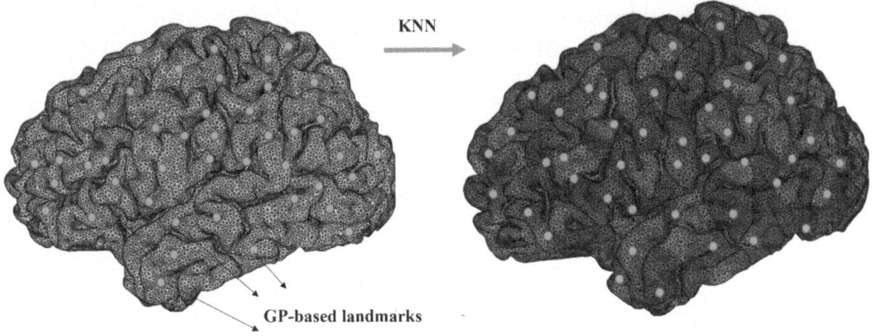

Fig. 2. Visualization of tokenization of the tetrahedral mesh. We assigned each node of the mesh the the nearest super node to create our tokens, resulting in patchwise visualization of the mesh. The supernodes are located in the centers of each patch. Each color represent unique patch (token).

super nodes. Unlike in images with a uniform grid structure, here on volumetric meshes, patches generated in any fashion always have different sizes in terms of nodes. Figure 2 better illustrates the way how tokens are generated. After creating tokens, the center of each patch is also saved for transformer processing.

2.4 Node and Patch-Level Embeddings

Unlike traditional ViT-style transformers that rely on fixed-size patches with a consistent number of nodes per patch, we propose leveraging patch embeddings without relying on padding, accommodating inputs with highly variable node counts. For patch embedding, we first propose learning primary local feature representation using TetCNN layers [11]. TetCNN is tailored for graph convolution on tetrahedral meshes by substituting the graph Laplacian operator with a customized volumetric Laplacian Beltrami Operator (volumetric LBO). The computational overhead of the graph convolution is substantially reduced by approximating the operation with spectral filtering of mesh signals, accomplished by Chebyshev polynomial approximation [11] This helps to learn further features not limited to raw 3D coordinates. In order to represent one patch, we use pooling to average the node features of that patch. We can define node-level embeddings and patch-level embeddings based on the following:

$$x_{i,0} = f_\theta([x_{in}]) \tag{2}$$

$$x_{i,\ell+1} = \sum_{m=0}^{K} \theta_m T_m(L_{tet}) x_{i,\ell}, \tag{3}$$

For the first layer, we employ a point-wise MLP to project the raw node features—specifically the 3D coordinates—into a higher-dimensional space using an MLP layer followed by an activation function, i.e. f_θ. Here, K defines the maximum polynomial order, θ_m are learnable weights for each order, and $T_m(L_{tet})$ represents the m-th Chebyshev polynomial of the tetrahedral Laplacian L_{tet}.

Next, we use a simple pooling strategy per patch to generate each token embedding with uniform embedding size across all $|P|$ patches.

$$X^p = \frac{1}{|p|} \sum_{i \in p} x_i \tag{4}$$

In this equation p is an arbitrary patch on the mesh, and $|p|$ shows the number of nodes in the patch. X shows the patch embedding. With these embeddings, the tokens are now prepared. One approach is to add positional encodings and input them into a conventional transformer. However, due to the computational cost of full self-attention and the opportunity to leverage geometric information, a graph-based approach for calculating attention is more suitable.

2.5 Super Node Representation

A super node is represented as a point in 3D space equipped with an embedding in a higher-dimensional feature space. Each super node is denoted as $s_i = (p_i, e_i)$, where $p_i \in \mathbb{R}^3$ corresponds to the spatial coordinates (x, y, z) of the super node, and $e_i \in \mathbb{R}^d$ is its feature embedding in a d-dimensional space. The set of all super nodes is defined as:

$$\mathcal{S} = \{s_1, s_2, \ldots, s_N\}, \quad s_i = (p_i, e_i), \quad p_i \in \mathbb{R}^3,\ e_i \in \mathbb{R}^d.$$

Radius Graph Construction. A radius graph $G = (\mathcal{V}, \mathcal{E})$ is constructed by defining edges between super nodes that are within a fixed radius r. Specifically, for each super node s_i, an edge $(i, j) \in \mathcal{E}$ exists if the Euclidean distance between their spatial coordinates p_i and p_j satisfies:

$$\|p_i - p_j\|_2 \leq r,$$

where $\|\cdot\|_2$ denotes the ℓ_2-norm. The graph is defined as:

$$G = (\mathcal{V}, \mathcal{E}), \quad \mathcal{E} = \{(i, j) \mid \|p_i - p_j\|_2 \leq r\}.$$

Relative Positional Encoding. For each edge (i, j), the relative positional encoding [27,34,35] Δp_{ij} is computed as:

$$\Delta p_{ij} = p_j - p_i,$$

which captures the spatial relationship between neighboring super nodes. This positional encoding is incorporated into the attention mechanism.

2.6 Attention Mechanism

The proposed transformer-based framework leverages an attention mechanism designed for tetrahedral meshes. The attention mechanism integrates both feature similarity and geometric relationships by computing attention scores for each edge (i, j) in the graph. The attention score is given by:

$$\text{Attention}(q_i, k_j, \Delta p_{ij}) = \sigma\big(\phi(q_i, k_j) + \delta(\Delta p_{ij})\big), \quad (5)$$

where:

- q_i and k_j are the query and key vectors of super nodes i and j, respectively.
- ϕ: A learnable function capturing feature similarity between q_i and k_j.
- δ: A learnable function encoding relative positional information Δp_{ij}.
- σ: A normalization function, such as softmax, ensuring valid attention scores.

In this formulation, Δp_{ij} encodes geometric and topological relationships between super nodes i and j, allowing the model to capture both local and global structure in tetrahedral meshes. Unlike prior works focused on point clouds or fixed-sized patches, our method adapts to varying mesh topologies by incorporating anatomical landmarks as super nodes and constructing task-specific graphs for attention computation.

2.7 Integrating BBBMs

After extracting the graph features, we integrated blood-based biomarkers by concatenating the BBBM values with the feature vector in the first fully connected layer, resulting in $d + 1$ dimension inputs.

2.8 Computational Efficiency

By using a radius graph, the Point Transformer focuses only on local neighborhoods, significantly reducing computational complexity compared to full self-attention over all points. The sparsity of the graph ensures that the attention mechanism is applied efficiently.

3 Experimental Design

3.1 Datasets

Data used in this study was downloaded the Alzheimer's Disease Neuroimaging Initiative (ADNI) database [18]. We evaluated our model on two subsets. For A total of 947 samples from 906 individuals were contained in the dataset used for AD classification experiments, including 313 with Alzheimer's disease (AD), 402 with mild cognitive impairment (MCI), and 229 being cognitively normal (CN). In addition, the dataset for brain amyloid positivity classification contained 909 data samples with matching T1-weighted brain MRI, PET scan and pTau-217 measured by PrecivityAD2 released by C2N diagnostics. (Table 1). All samples within the same dataset were processed using the same protocol.

Table 1. Dataset used for AD classification task and brain amyloid positivity classification task. Note that most Aβ+ samples have high pTau-217 levels, and most Aβ- samples have low pTau-217 levels. We aimed at distinguishing Aβ positivity for the medium risk group besides the overall classification performance.

AD Classification			Aβ+			Aβ-		
N(AD)	N(MCI)	N(CN)	Low	Medium	High	Low	Medium	High
313	402	229	19	70	265	335	96	23

3.2 Data Preprocessing

The collected MRI images were first processed by Freesurfer to reconstruct the pial and white surfaces. We then generated the tetrahedral meshes using the TetGen pipeline [16], resulting in tetrahedral meshes containing around 70k-100k vertices. Finally, 1000 anatomical landmarks were extracted from each tetrahedral mesh using a GP-based landmarking generation pipeline. The ground truth class labels were identified by computing the Centiloid values from PET images. Specifically, the PET images were processed by PET Unified Pipeline (PUP), and the positivity threshold was set to $Centiloid > 20$, as proposed and validated in previous studies [21]. In addition, we applied a two-threshold approach to classify individuals into low, medium and high risk of being amyloid positive [14]. The two thresholds were 1.53 and 2.602, determined by 95% sensitivity and 95% specificity, respectively. Since the pTau-217 alone has already achieved significant success as a sole predictor of brain amyloid positivity (AUC=0.94 (95% CI 0.92–0.97)), but not in the medium group (as shown in our results), we focus our effort on the medium group in our study [1].

3.3 Model Training

Prior to feeding data into our model, we followed the approach proposed in [11] to pre-compute the mass matrix and cotangent matrix for each tetrahedral mesh. We initially applied two layers of TetCNN for feature learning. To construct the graph for self-attention computation, we generated a radius graph with $r = 0.5$ using the PyG library [13]. Subsequently, we employed two Point Transformer layers [13,35] on the generated graph, with a hidden dimension size of 128 for all layers. Finally, a global pooling operation was applied for both final classification and patch embeddings. The model was trained using binary cross-entropy loss, the ADAM optimizer with a learning rate of 10^{-4}, weight decay of 10^{-4}, and a total of 500 epochs. Due to the small batch size of 2, gradient accumulation was employed to improve training efficiency.

3.4 Evaluation Protocol

For AD classification, we evaluated our model performance on classification tasks between all pairs of diagnosis groups. We compared our LETetCNN model with

Table 2. Classification results among clinical diagnosis groups. Our LETetCNN model achieved the best performance across all classification tasks. In general, AD vs MCI and MCI vs CN were less separable by all classifiers.

	AD vs CN			AD vs MCI			MCI vs CN		
Model	ACC	SEN	SPE	ACC	SEN	SPE	ACC	SEN	SPE
ChebyNet	0.870	0.881	0.850	0.703	**0.790**	0.616	0.735	0.778	0.667
GAT	0.858	0.873	0.836	0.727	0.630	0.773	0.722	0.763	0.660
TetCNN	0.876	0.886	0.859	0.709	0.660	0.769	0.730	0.761	0.700
LETetCNN	**0.917**	**0.915**	**0.920**	**0.755**	0.615	**0.835**	**0.794**	**0.778**	**0.822**

baselines including ChebyNet [6] and GAT [31]. For brain amyloid positivity classification, we compared our LETetCNN model and the incorporated LETetCNN + pTau-217 model with three additional baseline models besides the aforementioned GNN-based models, including logistic regression models using hippocampal volume, pTau-217 alone, and hippocampal volume + pTau-217 as predictors. For each classification task, we calculated the accuracy, sensitivity and specificity and compared the metrics among all models.

4 Results

4.1 Alzheimer's Disease Classification

As shown in Table 2, our model consistently outperformed existing methods across all clinical diagnosis group pairs, with varied degrees of improvement observed across groups. Specifically, we achieved 91.7% accuracy in AD vs CN classification, and 75.5% accuracy in AD vs MCI classification, and 79.4% in MCI vs CN classification. While the individuals in AD diagnosis group typically exhibit a higher stage of dementia, leading to a significant geometric distinction between the AD and MCI/CN groups, discriminating between MCI and CN subgroups based on cortical structural change poses more challenge. Therefore, our model's superior classification performance, particularly in MCI vs CN comparisons, demonstrates the effectiveness and robustness of our method. Moreover, integrating anatomical landmarks further enhanced the model's performance, highlighting the strength of our approach.

4.2 Brain Amyloid Positivity Classification

We generated a total of 909 brain cortical tetrahedral meshes from structural MR images. All samples were from ADNI dataset with matching PET scans and pTau-217 measures. We followed the widely used protocol [21] to determine brain amyloid positivity by calculating Centiloid value for each PET image, and

Table 3. Amyloid positivity prediction results in the medium risk groups. An integration of LETetCNN and pTau-217 showed enhanced accuracy, outperforming all other baseline classifiers.

Model	ACC	SEN	SPE
Hippocampal Volume	0.450	0.529	0.391
pTau-217	0.675	0.750	0.625
Hippocampal Volume + pTau-217	0.675	0.750	0.625
ChebyNet	0.677	0.563	0.800
GAT	0.677	0.611	0.769
TetCNN	0.690	0.684	0.694
LETetCNN	0.758	0.785	0.733
LETetCNN+pTau-217	**0.798**	**0.785**	**0.811**

labeled any PET image with $Centiloid > 20$ as amyloid positive. We employed the aforementioned two-threshold method to classify samples into low, medium and high groups, and primarily focus on the medium group, where pTau-217 alone has demonstrated limited classification power. The subset of the medium group contained 70 positive and 96 negative samples. Hippocampal volume was also included as a baseline classifier, given its well-documented correlation with AD-related dementia. However, as shown in Table 3, hippocampal volume yielded results close to random classification. In comparison, pTau-217 alone achieved 0.675 classification accuracy, with similar performance observed from other GNN-based methods. In contrast, our LETetCNN model outperformaned all baseline models by achieving an accuracy of 0.758. Notably, the performance was further enhanced after integrating pTau-217, reaching a classification accuracy of 0.798 with a balanced prediction (SEN = 0.785, SPE = 0.811). While further validations are needed, our current results suggest that our work could complement BBBM research, enhancing its accuracy through a cost-effective and less invasive approach.

4.3 Ablation Studies

We first investigated the impact of learned features extracted by a few TetCNN layers before they were passed to the transformer. To assess this, we removed these layers and directly feed the raw features into the transformer. In the second ablation stude, we evaluated whether integrating pTau-217 with LETetCNN significantly enhanced performance. This analysis was crucial, as pTau-217 itself was a highly predictive biomarker, and it was important to determine whether the features learned by our network provided additional complementary information. The results of both studies were summarized in Table 4. It demonstrated TetCNN learned important geometric features and the proposed framework further enhanced the performance of BBBMs for preclinical AD research.

Table 4. The ablation study results for amyloid positivity prediction in the medium-risk group indicate that the transformer-only architecture (LE) did not yield competitive performance compared to using geometric features learned from TetCNN as input. Additionally, our proposed approach provided complementary information to enhance the performance of BBBMs.

Model	ACC	SEN	SPE
LE	0.725	0.674	0.750
LETetCNN	0.765	0.738	0.790
pTau-217	0.675	0.750	0.625
LETetCNN+pTau-217	**0.798**	**0.785**	**0.811**

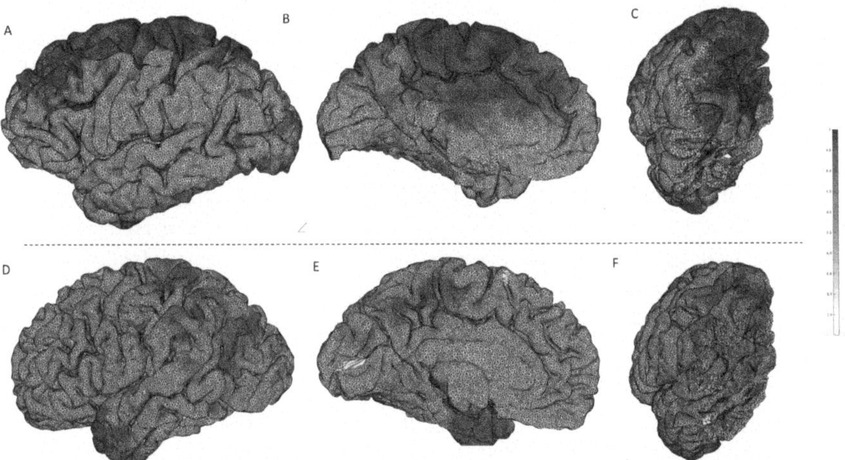

Fig. 3. Grad-CAM visualization highlighting regions of interest that align with Alzheimer's Disease (AD) pathology. Both results from AD classification task (A-C) and brain amyloid positivity prediction task (D-F) exhibited strong activations in the temporal gyrus and parts of the inferior temporal lobe. Mild activation was also observed in superior frontal cortex, which was related to later-stage disease progression. These results closely align with neuroimaging studies identifying key biomarkers for AD pathology.

4.4 Grad-CAM Visualization

We visualized the activation map of our model prediction for the AD classification task (Fig. 3). The highlighted areas were spreading around posterior cingulate and medial temporal lobe, with some less significant regions located in frontal lobe. Medial temporal lobe has been widely reported as the first region affected by AD, and posterior cingulate was also shown to be functionally important in impairment of learning and memory [19,24]. In addition, frontal lobe was also shown to be correlated in late stage AD [20]. Therefore, the Grad-CAM heatmap demonstrated high correspondence of pathologically significant regions identified by our model with those from existing literature.

5 Conclusion and Future Work

In this work, we proposed a novel transformer-based framework tailored specifically for tetrahedral meshes with varying sizes and topologies. To the best of our knowledge, this is the first approach to extend transformer architectures to tetrahedral meshes, leveraging relative positional encodings and task-specific graph tokenization for efficient and geometry-aware representation learning. Unlike prior methods that rely on fixed-size re-meshing or hierarchical pooling, our approach dynamically adapts to the inherent variability of tetrahedral mesh structures. We evaluated our model on two distinct tasks: clinical AD classification and brain amyloid positivity prediction. We compared our model with baseline classifiers, including ChebyNet and GAT in AD classification, and hippocampal volume and pTau-217 in amyloid positivity prediction. In both experiments, our model outperformed all baseline methods, indicating both effectiveness and generalizability. In particular, our LETetCNN model can be empowered by leveraging the pTau-217 information. With the rapid advancement of BBBM studies in AD research [15,26,29], our work indicates that traditional brain imaging research will still be useful and relevant.

Compared to TetCNN, which utilizes localized filters defined on the volumetric LBO our method introduces several key advantages. First, the attention mechanism in our framework enables dynamic modeling of global relationships across the mesh, overcoming the locality limitation of TetCNN. While TetCNN layers effectively capture local features, their hierarchical pooling may result in loss of critical geometric details, particularly for meshes with irregular or anisotropic topologies. In contrast, our method retains finer global and local information through attention mechanisms that explicitly consider both feature similarity and geometric relationships.

Second, the incorporation of relative positional encodings allows our framework to encode spatial relationships between nodes and patches directly in the feature space. This improves the model's ability to represent both local geometric features and global structural context. Additionally, our use of anatomical landmarks as super nodes ensures robust and semantically meaningful tokenization, which is particularly beneficial for handling large volumetric datasets.

Empirical results validate the effectiveness of our approach. Our framework achieves superior performance in clinical group classification, with an average of 5% increase in accuracy compared to TetCNN. Visualizations of attention maps further highlight the ability of our model to focus on critical regions, demonstrating richer feature representations. Furthermore, ablation studies show the significant contributions of relative positional encodings and landmark-based tokenization to overall performance.

In addition to its representational advantages, our framework maintains computational efficiency by leveraging sparse graph structures and localized attention. This scalability makes it well-suited for large-scale tetrahedral meshes, addressing a key limitation of TetCNN and other traditional methods.

References

1. Arranz, J., et al.: Diagnostic performance of plasma ptau217, ptau181, aβ1-42 and aβ1-40 in the lumipulse automated platform for the detection of Alzheimer disease. Alzheimer's Res. Ther. **16**(1), 139 (2024)
2. Ashton, N.J., et al.: Plasma and CSF biomarkers in a memory clinic: head-to-head comparison of phosphorylated tau immunoassays. Alzheimer's Dementia **19**(5), 1913–1924 (2023)
3. Atri, A.: The Alzheimer's disease clinical spectrum: diagnosis and management. Med. Clin. **103**(2), 263–293 (2019)
4. Atri, A.: Current and future treatments in Alzheimer's disease. In: Seminars in neurology, vol. 39, pp. 227–240. Thieme Medical Publishers (2019)
5. Chentanez, N., Feldman, B.E., Labelle, F., O'Brien, J.F., Shewchuk, J.R.: Liquid simulation on lattice-based tetrahedral meshes (2007)
6. Defferrard, M., Bresson, X., Vandergheynst, P.: Convolutional neural networks on graphs with fast localized spectral filtering. In: Advances in Neural Information Processing Systems, vol. 29 (2016)
7. Dickerson, B.C., et al.: The cortical signature of Alzheimer's disease: regionally specific cortical thinning relates to symptom severity in very mild to mild ad dementia and is detectable in asymptomatic amyloid-positive individuals. Cereb. Cortex **19**(3), 497–510 (2009)
8. Dosovitskiy, A., et al.: An image is worth 16x16 words: transformers for image recognition at scale. In: Proceedings of the International Conference on Machine Learning (ICML) (2021)
9. El-Hayek, Y.H., et al.: Tip of the iceberg: assessing the global socioeconomic costs of Alzheimer's disease and related dementias and strategic implications for stakeholders. J. Alzheimers Dis. **70**(2), 323–341 (2019)
10. Fan, Y., Wang, Y.: Convolutional Bayesian models for anatomical landmarking on multi-dimensional shapes. In: Martel, A.L., et al. (eds.) MICCAI 2020. LNCS, vol. 12264, pp. 786–796. Springer, Cham (2020). https://doi.org/10.1007/978-3-030-59719-1_76
11. Farazi, M., Yang, Z., Zhu, W., Qiu, P., Wang, Y.: TetCNN: convolutional neural networks on tetrahedral meshes. In: Frangi, A., de Bruijne, M., Wassermann, D., Navab, N. (eds.) International Conference on Information Processing in Medical Imaging, pp. 303–315. Springer (2023). https://doi.org/10.1007/978-3-031-34048-2_24
12. Fawaz, A., et al.: Benchmarking geometric deep learning for cortical segmentation and neurodevelopmental phenotype prediction. bioRxiv (2021)
13. Fey, M., Lenssen, J.E.: Fast graph representation learning with PyTorch geometric. arXiv preprint arXiv:1903.02428 (2019)
14. Figdore, D.J., et al.: Optimizing cutpoints for clinical interpretation of brain amyloid status using plasma P-TAU217 immunoassays. Alzheimers Dement. **20**(9), 6506–6516 (2024)
15. Groot, C., et al.: Diagnostic and prognostic performance to detect Alzheimer's disease and clinical progression of a novel assay for plasma p-tau217. Alzheimer's Res. Therapy **14**(1), 67 (2022)

16. Hang, S.: TetGEN, a delaunay-based quality tetrahedral mesh generator. ACM Trans. Math. Softw. **41**(2), 11 (2015)
17. Hanocka, R., Hertz, A., Fish, N., Giryes, R., Fleishman, S., Cohen-Or, D.: MeshCNN: a network with an edge. ACM Trans. Graph. (ToG) **38**(4), 1–12 (2019)
18. Jack Jr, C.R., et al.: The Alzheimer's disease neuroimaging initiative (ADNI): MRI methods. J. Magn. Reson. Imag. **27**(4), 685–691 (2008)
19. Jack, C.R., et al.: Rate of medial temporal lobe atrophy in typical aging and Alzheimer's disease. Neurology **51**(4), 993–999 (1998)
20. Johnson, J.K., Head, E., Kim, R., Starr, A., Cotman, C.W.: Clinical and pathological evidence for a frontal variant of Alzheimer disease. Arch. Neurol. **56**(10), 1233–1239 (1999)
21. Klunk, W.E., et al.: The centiloid project: standardizing quantitative amyloid plaque estimation by pet. Alzheimer's Dementia **11**(1), 1–15 (2015)
22. Liss, J., et al.: Practical recommendations for timely, accurate diagnosis of symptomatic Alzheimer's disease (MCI and dementia) in primary care: a review and synthesis. J. Intern. Med. **290**(2), 310–334 (2021)
23. Liu, X., Liu, L., Song, W., Liu, Y., Ma, L.: Shape context based mesh saliency detection and its applications: a survey. Comput. Graph. **57**, 12–30 (2016)
24. Minoshima, S., Giordani, B., Berent, S., Frey, K.A., Foster, N.L., Kuhl, D.E.: Metabolic reduction in the posterior cingulate cortex in very early Alzheimer's disease. Ann. Neurol. Off. J. Am. Neurol. Assoc. Child Neurol. Soc. **42**(1), 85–94 (1997)
25. Nakamura, A., et al.: High performance plasma amyloid-β biomarkers for Alzheimer's disease. Nature **554**(7691), 249–254 (2018)
26. Palmqvist, S., et al.: Discriminative accuracy of plasma phospho-tau217 for Alzheimer disease vs other neurodegenerative disorders. JAMA **324**(8), 772–781 (2020)
27. Park, W., Chang, W., Lee, D., Kim, J., Hwang, S.: GRPE: relative positional encoding for graph transformer. arXiv preprint arXiv:2201.12787 (2022)
28. Qiu, C., Kivipelto, M., Von Strauss, E.: Epidemiology of Alzheimer's disease: occurrence, determinants, and strategies toward intervention. Dialogues Clin. Neurosci. **11**(2), 111–128 (2009)
29. Therriault, J., et al.: Equivalence of plasma p-tau217 with cerebrospinal fluid in the diagnosis of Alzheimer's disease. Alzheimer's Dementia **19**(11), 4967–4977 (2023)
30. Vaswani, A., et al.: Attention is all you need. In: Advances in Neural Information Processing Systems (NeurIPS) (2017)
31. Veličković, P., Cucurull, G., Casanova, A., Romero, A., Lio, P., Bengio, Y.: Graph attention networks. arXiv preprint arXiv:1710.10903 (2017)
32. Wang, G., Wang, Y., Initiative, A., et al.: Towards a holistic cortical thickness descriptor: heat kernel-based grey matter morphology signatures. Neuroimage **147**, 360–380 (2017)
33. Wang, Y., Gu, X., Chan, T.F., Thompson, P.M., Yau, S.T.: Volumetric harmonic brain mapping. In: 2004 2nd IEEE International Symposium on Biomedical Imaging: Nano to Macro (IEEE Cat No. 04EX821), pp. 1275–1278. IEEE (2004)
34. Wu, X., Lao, Y., Jiang, L., Liu, X., Zhao, H.: Point transformer V2: grouped vector attention and partition-based pooling. Adv. Neural. Inf. Process. Syst. **35**, 33330–33342 (2022)
35. Zhao, H., et al.: Point transformer. In: Proceedings of the IEEE/CVF Conference on Computer Vision and Pattern Recognition (CVPR) (2021)

Hierarchical Variable Importance with Statistical Control for Medical Data-Based Prediction

Joseph Paillard[1,2(✉)], Antoine Collas[2], Denis A. Engemann[1], and Bertrand Thirion[2]

[1] Roche Pharma Research and Early Development, F. Hoffmann-La Roche Ltd., Basel, Switzerland
[2] Université Paris-Saclay, Inria, CEA, Paris, Palaiseau, France
joseph.paillard@roche.com

Abstract. Recent advances in machine learning have greatly expanded the repertoire of predictive methods for medical imaging. However, the interpretability of complex models remains a challenge, which limits their utility in medical applications. Recently, model-agnostic methods have been proposed to measure conditional variable importance and accommodate complex non-linear models. However, they often lack power when dealing with highly correlated data, a common problem in medical imaging. We introduce Hierarchical-CPI, a model-agnostic variable importance measure that frames the inference problem as the discovery of groups of variables that are jointly predictive of the outcome. By exploring subgroups along a hierarchical tree, it remains computationally tractable, yet also enjoys explicit family-wise error rate control. Moreover, we address the issue of vanishing conditional importance under high correlation with a tree-based importance allocation mechanism. We benchmarked Hierarchical-CPI against state-of-the-art variable importance methods. Its effectiveness is demonstrated in two neuroimaging datasets: classifying dementia diagnoses from MRI data (ADNI dataset) and analyzing the Berger effect on EEG data (TDBRAIN dataset), identifying biologically plausible variables.

Keywords: Statistics · neuroimaging · interpretable machine learning

1 Introduction

Within the field of medcial imaging, machine learning holds great promise to facilitate prediction of clinical outcomes, see e.g. [1–6]. However, these advances have also opened major interpretability challenges. A key issue is how to infer the importance of features from prediction models going beyond ordinary least squares to accommodate a large number of predictors and represent non-linear associations between features and outcomes. Therefore, developing methods to measure variable importance in a model-agnostic manner is critical in order to

I. Oguz et al. (Eds.): IPMI 2025, LNCS 15830, pp. 79–93, 2026.
https://doi.org/10.1007/978-3-031-96625-5_6

obtain clinical insights and develop biomarkers, for instance, using brain images for the diagnosis of Alzheimer Disease (AD) based on existing cohorts, or data from clinical trials [2,6–8]. However, to develop trustworthy methods, it is essential to understand their theoretical guarantees, particularly concerning the risk of making false discoveries, which can be captured by the Family-Wise Error Rate (FWER) (see e.g. [9,10]). Only few variable importance methods give access to such guarantees. Moreover, we focus here on *conditional importance*, meaning the importance measure whether a variable is *directly* predictive of the outcome, without being explained away by other variables [11,12]. Such conditional importance is needed to establish that a marker carries independent information about the outcome, rather than merely reflecting distributed factors that are also present in other variables. Conditional importance analysis is particularly difficult in datasets that exhibit strong correlation structures such as image- or genomics-based biomarkers, or health data that reflect common latent factors [13].

We assess the face validity of the approach with two tasks that have been extensively studied in the literature—the effect of AD on structural MRI and the Berger effect on electroencephalography (EEG)—to allow for a form of confirmation, addressing the challenge posed by the absence of ground truth in variable importance methods.

1.1 Related Work

This work focuses *global* variable importance, as opposed to local variable importance methods such as *LIME* [14] or *SHAP* [15]. Global variable importance is estimated using methods such as global sensitivity analysis [11] or the popular Leave One Covariate Out (LOCO) approach [16,17]. These methods can accommodate different types of learners, taking advantage of advances in machine learning to measure importance in complex and nonlinear models [17,18]. Similarly, conditional permutation approaches have been shown to estimate a quantity equivalent to LOCO at a cheaper computational cost and with a faster convergence rate [19]. These methods have in common to provide a good control of the type-1 error rate, that is considering a null variable (or group) as important. However, all approaches suffer from an inherent limitation: conditional importance decreases as correlation increases. For instance, considering two random normal variables X_1, X_2 with correlation ρ in a simple linear model $y = \beta_1 X_1 + \beta_2 X_2$ the importance of X_1 decreases proportionally to $(1 - \rho^2)$.

To mitigate this issue, methods based on variables grouping have been proposed to identify groups of highly-correlated, hence indistinguishable variables that predict the outcome [12,13]. Variable grouping can be performed based on prior knowledge about the data or by using clustering techniques. While this effectively increases the statistical power by averaging correlated variables, it also reduces the precision in the sense that error control only holds at the group level. When performing the grouping in a data-driven way, choosing the clustering scheme and parameters has a critical impact yet has no obvious solution. A line of work relying on linear models and agglomerative clustering has been proposed in that direction with applications to genomics data [9,10,20]. Agglomerative clustering offers a compelling solution as it naturally explores different groupings at various resolutions along the hierarchical tree learned from the data. However, this approach relied on Lasso regression and would consequently limit the user in the choice of the model used to predict the outcome of interest from the variables.

Another popular model explainability approach is Shapley Additive Global importancE (*SAGE*) [21]. Based on Shapley values, this approach estimates an importance score for a given variable by conditioning on all subgroups that include this variable. This procedure provides a more nuanced view than strict conditional importance, because it decomposes additively the variance explained by the model into variable importance. However, it suffers from two main limitations. First, as an aggregated statistic, it obcures the role of variables in the prediction function [18], and is unable to distinguish between a predictive variable and another, non-predictive variable yet correlated with a predictive one. Second, the exploration of all submodels comes with an exponential explosion of computation cost, making this approach intractable. While implementations rely on Monte-Carlo sampling instead of exhausting the full combinatorial sum, the number of sampling steps needed to obtain accurate estimates still leads to intractable computation costs [18]. This two limitations are clearly visible in our experiments in Fig. 2.

Our contributions are *i)* to introduce Hierarchical-CPI, a model-agnostic variable importance measure that improves FWER control; it explores subgroups in a tree-guided manner, using agglomerative clustering to provide more information than variable-level importance while remaining tractable; *ii)* to present an approach that enforces importance conservation through downstream importance allocation strategy, addressing the issue of vanishing importance under high correlation.

2 Methods

Notations: We denote X as the variables, y as the outcome, and μ as the predictive model. X_G represents the set of variables belonging to a group G, and X_{-G} denotes the set of variables in the complement of G. The importance of a group is denoted as ψ_G. We use S^* to denote the support (or set of active variables) and S_0 for the set of null variables. In the hierarchical tree defined by the clustering, P refers to a parent node, and L and R refer to its left and right child nodes, respectively.

2.1 Hierarchical-CPI

We present a method for measuring variable importance while conditioning on others, with conditioning sets taken in a tree-organized hieracrchical representation of the variables. It balances *precision* and *statistical power*; precision refers to extracting the information located in groups of variables that are as small

Algorithm 1. Hierarchical CPI

Input: K: number of folds, μ: predictive model, ν: imputation model, (X, y): data

tree $\leftarrow$ Fit hierarchical clustering on X
for k in $[1, \cdots, K]$ **do**
  $\hat{\mu}_k \leftarrow$ Fit using (X_{train}, y_{train}) // fit the full model
  for node in tree **do**
    G $\leftarrow$ traversal(node) // search variables belonging to the node
    $\hat{\nu}_G^k \leftarrow \mathbb{E}[X_G^{train} | X_{-G}^{train}]$ // estimate the conditional distribution
    $\widetilde{X}_G^{test} \sim \hat{\nu}_G^k(X_{-G}^{test})$ // sample from the conditional distribution
    $\hat{\psi}_G^k \leftarrow \mathcal{L}\left(y, \mu(\widetilde{X}_G^{test})\right) - \mathcal{L}\left(y, \mu(X^{test})\right)$ // compute variable importance
  end for
end for
$p_G \leftarrow$ t-test$(\hat{\psi}_G^1, \cdots, \hat{\psi}_G^K)$ // compute p-value over folds
$p_G^h \leftarrow \max_{G \subseteq D} p_D^h$ // hierarchical adjustment
return $p_{G_i}^h$ for $i = 1, \cdots, 2p-1$

as possible; statistical power is achieved by considering condition sets different from the set of all variables. The proposed method builds on top of Conditional Permutation Importance (CPI) [12] which, given a group of variables G, a model μ and loss $\mathcal{L}$ estimates the conditional importance

$$\psi_G = \mathcal{L}\left(y, \mu(\tilde{X}_G)\right) - \mathcal{L}\left(y, \mu(X)\right), \tag{1}$$

where $\tilde{X}_G$ is obtained by substituting into the group X_G variables sampled from the conditional distribution $(X_G | X_{-G})$ and leaving the X_{-G} variables unchanged. In brief, CPI quantifies the loss increase when conditioning on all other variables than those in G. This approach estimates the well known total Sobol index [11]. In addition, hierarchical-CPI leverages Ward's minimum variance method for agglomerative clustering to learn the hierarchical group structure [22]. The proposed method is presented in Algorithm 1, for a problem with p variables, it consists in estimating the conditional permutation importance of each group of variables within the hierarchical structure. Empirical importance values are obtained in a K-fold cross-validation scheme, yielding K estimates per group, $(\psi_G^1, \cdots, \psi_G^K)$. A p-value p_G is then derived based on a one-sample t-test. Finally, the node-level p-values p_G are hierarchically adjusted by,

$$p_G^h = \max_{G \subseteq D} p_D, \tag{2}$$

to enforce that the p-value of a node is larger than the p-value of its parent.

2.2 Hierarchical CPI Achieves FWER Control

In this section, we demonstrate that the hierarchical-CPI approach controls the FWER under assumptions of estimator optimality and regularity. The assumptions (A.1, A.2, B.1, B.2) from [17] stipulate that the estimator μ must be optimal and exhibit sufficient regularity. These assumptions have been validated by independent work and are considered not too restrictive [17–19]. We refer to a tree cut as a set of non-overlapping nodes within a hierarchical tree. Let S_0 denote the set of groups that only contain null variables, and let $\hat{S}_\alpha = \{G \mid p_G \leq \alpha\}$ be the estimated set of active variables for a given significance level $\alpha \in [0, 1]$. The following result holds.

Theorem 1. *Under the assumption that the conditions (A.1, A.2, B.1, B.2) stated in [17] on μ, for any significance level $\alpha \in [0,1]$ the multiplicity corrected p-values $\tilde{p}^h_G = \min(1, C \cdot p^h_G)$, with $C = p$ control the family-wise error rate at level α, i.e. $\mathbb{P}(S_0 \cap \hat{S}_\alpha \neq \emptyset) \leq \alpha$. Where G is a node of a tree cut.*

Proof.

$$\mathbb{P}(S_0 \cap \hat{S}_\alpha \neq \emptyset) = \mathbb{P}\left(\min_{G \subseteq S_0} \tilde{p}^h_G \leq \alpha\right) = \mathbb{P}\left(\bigcup_{G \subseteq S_0} p^h_G \leq \frac{\alpha}{C}\right)$$

Then, given Boole's inequality,

$$\begin{aligned} \mathbb{P}(S_0 \cap \hat{S}_\alpha \neq \emptyset) &\leq \sum_{G \subset S_0} \mathbb{P}\left(p^h_G \leq \frac{\alpha}{C}\right) \\ &\leq C \cdot \mathbb{P}\left(p^h_G \leq \frac{\alpha}{C}\right) \end{aligned}$$

Since a tree cut contains less than p nodes, $\mathrm{card}(\{G_i \mid G_i \subseteq S_0\}_{i \in [1,C]}) \leq C$. Furthermore, given that $p^h_G = \max_{G \subseteq D}\{p_D\}$, where the maximum is taken over ancestor nodes, it comes that $\mathbb{P}\left(p^h_G \leq \alpha\right) \leq \mathbb{P}\left(p_G \leq \alpha\right)$. Finally, under assumptions (A.1, A.2, B.1, B.2), it has been shown in [19] that, $\forall G \subset S_0$, $\mathbb{P}\left(p_G \leq \alpha\right) \leq \alpha$. We then have, $\mathbb{P}(S_0 \cap \hat{S}_\alpha \neq \emptyset) \& \leq \alpha$ which completes the proof.

While this result holds when considering inference at the variable level, it is more general and applies to any node of the tree. Hierarchical CPI allows to learn a tree structure from the data and make inference at different levels.

2.3 Importance Conservation to Prevent Importance Vanishing

A common pitfall of conditional importance is that it vanishes under high correlation: For a parent node P with two strongly correlated children nodes L and R, then we have that $\psi^k_L + \psi^k_R \ll \psi^k_P$ for all k in $[1..K]$. This effect is illustrated in Fig. 1a and c. Hierarchical CPI can infer that P is important, but will give little to no importance to L and R (and all downstream groups). To obtain a Shapley-like additive decomposition of the model fit, inspired by variance partitioning

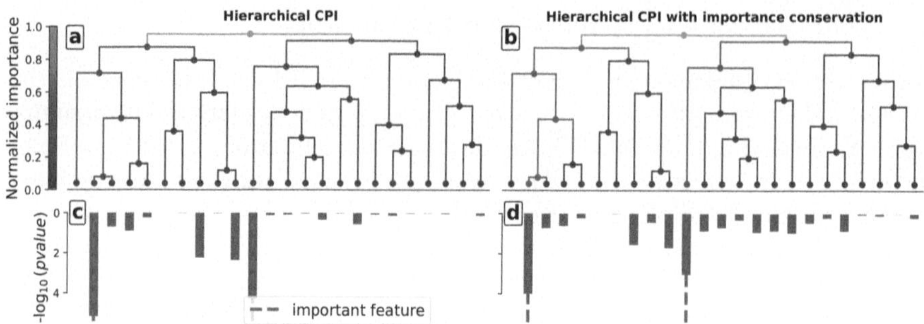

Fig. 1. Importance conservation prevents the importance from vanishing as the group size decreases in a high-correlation setting. Example using simulated data with $n = 300$ samples and $d = 24$ variables. The data is generated by blocks, each corresponding to an AR(1) with autocorrelation parameter $\rho_{\max} = 0.95$. **a** and **b** show the same dendrogram obtained through Ward's clustering. Each node's color encodes the conditional importance of the variables it contains. Without importance conservation, importance quickly vanishes down the tree. **c** and **d** present the p-value distributions, demonstrating that both methods accurately rank important variables.

ideas [23], we introduce a transfer mechanism that ensures additivity at the node level, meaning that the importance of a parent node is split into the sum of the importance of its children. This is meant to allow a more refined allocation of the importance budget compared to traditional clustering methods. For a node R, let $\tilde{\psi}_R^k$ be the corrected value for ψ_R^k that ensures importance conservation through the hierarchical structure. Then, $\mathbb{1}_R(\epsilon)$ is the indicator function equal to one when $\psi_R/\hat{\sigma}_R \geq \epsilon > 0$, where $\hat{\sigma}_R$ is the standard deviation of the right node importance estimated over the k-folds and zero otherwise. Importance conservation aims at satisfying the equation $\tilde{\psi}_P^k = \tilde{\psi}_L^k + \tilde{\psi}_R^k$. This condition ensures that the importance of the top node, which measures the full model's importance, is allocated to its children nodes. The allocation mechanism for the child node L with sibling R and parent P proceeds as follows:

$$\tilde{\psi}_L^k = \begin{cases} \psi_L^k + \mathbb{1}_R(\epsilon)\frac{\tilde{\psi}_P^k - \psi_L^k - \psi_R^k}{2} & \text{if } \mathbb{1}_L(\epsilon) = 1 \\ \tilde{\psi}_P^k \frac{\psi_L^k}{\psi_R^k + \psi_L^k}(1 - \mathbb{1}_R(\epsilon)) + \mathbb{1}_R(\epsilon)(\tilde{\psi}_P^k - \psi_R^k) & \text{if } \mathbb{1}_L(\epsilon) = 0 \end{cases} \tag{3}$$

This equation distributes the parent's importance proportionally to the children's importance when both nodes' importance values are greater than a threshold ϵ. If the importance ψ_L of node L is smaller than ϵ, it remains unchanged, avoiding false positives (not all children of an important parent are important). When both children have sub-threshold importance, indicating that their correlation leads to mutual importance cancellation, the importance of the parent is allocated equally between the two nodes.

3 Results

3.1 Control of Family-Wise Error Rate on Simulated Data

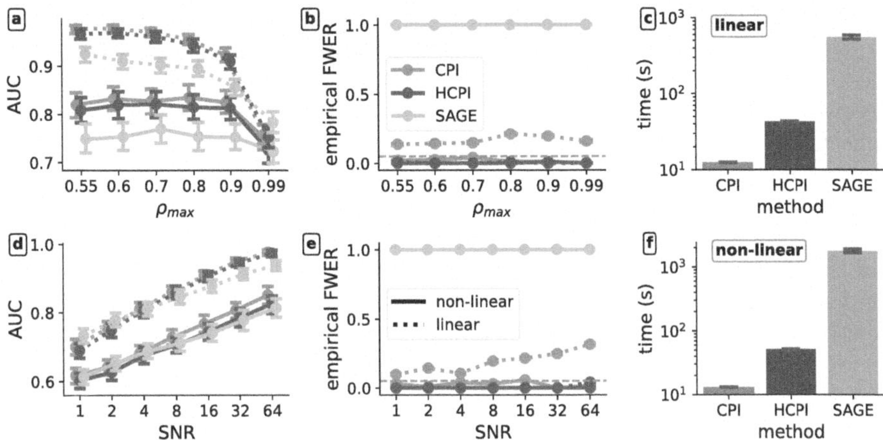

Fig. 2. Hierarchical CPI empirically controls the FWER with high statistical power. Results from simulated data with $n = 400$ samples and $p = 124$ variables sampled from a normal distribution with block correlations described in Subsect. 3.1. The results present a summary of 100 repetitions of the simulation. **a** and **d** present the AUC for important/non-important variable classification as a function of correlation and SNR. Error bars represent 95% confidence interval. **b** and **c** show the evolution of the FWER at level $\alpha = 0.05$ with Bonferroni correction. The top row explores varying ρ_{max} values at a fixed SNR=2, while the bottom row examines varying SNRs at a fixed $\rho_{max} = 0.9$. **c** and **f** shows the average computation time taken by each method over the simulations for the linear and non-linear scenario respectively.

This experiment benchmarks the hierarchical-CPI approach described in Algorithm 1 with other state of the art variable importance methods on simulated data. The data is generated by blocks, each corresponding to an AR(1) with autocorrelation parameter $\rho_{\max}$. The outcome is modeled using two different scenarios. The first is a linear model with additive noise $y = X\beta + \sigma_N \varepsilon$, represented with dotted lines in Fig. 2. The support $S^* = \{j; |; \beta_j \neq 0\}$ is kept sparse, with $|S^*|$ set to either 5 or 10, and coefficients values sampled from the set $\beta_j \in \{-2, -1, 1, 2\}$ with uniform probability. The second is a non linear scenario presented in [19]: $y = X_{j_1} + 2\log(1 + 2X_{j_2}^2 + (X_{j_3} + 1)^2) + X_{j_4}X_{j_5} + \sigma_N \epsilon$. The support $S^* = \{j_1, \cdots, j_5\} \in [\![0, p]\!]^5$ is randomly sampled at each simulation run. In both cases, additive noise $\varepsilon \sim \mathcal{N}(0, I_n)$, controls the signal-to-noise ratio (SNR), defined as $SNR = ||y^*||_2^2/\sigma_N^2||\epsilon||_2^2$, where y^* is the noiseless outcome. The SNR is a simulation parameter. In the experiments shown in Fig. 2, we used $p = 124$ variables grouped into five blocks of correlated features with respective sizes 4, 8, 16, 32, and 64. The number of samples was fixed at $n = 400$ to match the dimensionality commonly found in medical imaging applications. To solve the non-linear regression task, a multilayer perceptron (MLP) with 100 hidden units was used. It was trained using *Adam* optimizer for 400 epochs with early stopping (patience 10 epochs). For the linear scenario, a Ridge-regularized model was used. Its regularization parameter was learned via nested cross-validation. For CPI and HCPI, the loss used in Eq. 1 is the root mean squared error.

The top row (**a, b, c**) explores the influence of ρ_{max} while the bottom row the influence of the SNR. We considered two metrics. First, the Area Under the Receiver Operating Characteristic Curve (AUC): This compares the predicted importance to the true importance. It aims to assess each method's ability to recover the true support. Second, the FWER: This measures the probability of making at least one false discovery, estimated over 100 simulation repetitions. We compare three methods: CPI, Hierarchical-CPI (HCPI) and SAGE. For CPI and HCPI, the predicted importance correspond to $1-p$-value and the estimated support is $\hat{S}_\alpha = \{j \mid p_j \leq \alpha\}$, we considered a level $\alpha = 0.05$. Regarding SAGE, we used a publicly available implementation[1] which provides estimated standard deviation from which 95% confidence interval were derived. The estimated support for SAGE consisted of variables for which the confidence interval did not include 0.

As shown in Fig. 2, the hierarchical approach effectively controls the FWER even in challenging simulation settings, e.g. with very high correlation or low SNR. This can be attributed to the hierarchical adjustment, described in Eq. 2, that bounds a node's p-value below using the p-value of its parent. As shown in subpannels **a** and **d** of the Fig. 2, this additional control does not decrease the power of the method when compared to CPI. While the exploration of the nodes entails an additional computation cost, it remains of the same order as CPI, which is a fast method. Indeed, for a hierarchical clustering of p variables, the total number of nodes is $2p - 1$ which makes the computation scale linearly with the dimension instead of the exponential explosion inherent to SAGE [21]. This fact is illustrated on the panels **c** and **f** of Fig. 2, where the logarithmic axis illustrates the untractable computation time of SAGE. In the non-linear case where a neural network is used, this trend ibecomes even more pronounced.

3.2 Hierarchical CPI Identifies Characteristic Markers of AD

This study explores image-based diagnosis using the ADNI dataset [7]. Cohort selection was based on the availability of T1-weighted images, similarly to [8]. A total of 1616 patients were included: 760 controls (CN), 529 diagnosed with Mild Cognitive Impairment (MCI), and 327 with AD. Gray Matter (GM) density maps were computed using the *sMRIPrep* pipeline [24], which is part of the widely used *fMRIPrep* pipeline [25]. The mean GM densities were extracted from 116 Regions of Interest (ROIs) defined by the Automated Anatomical Labeling Atlas 3 [26] using *Nilearn* [27] and used as features for classification tasks. The high correlation between ROIs (Pearson correlation ranging from 0.32 to 0.94) and

[1] https://github.com/iancovert/sage.

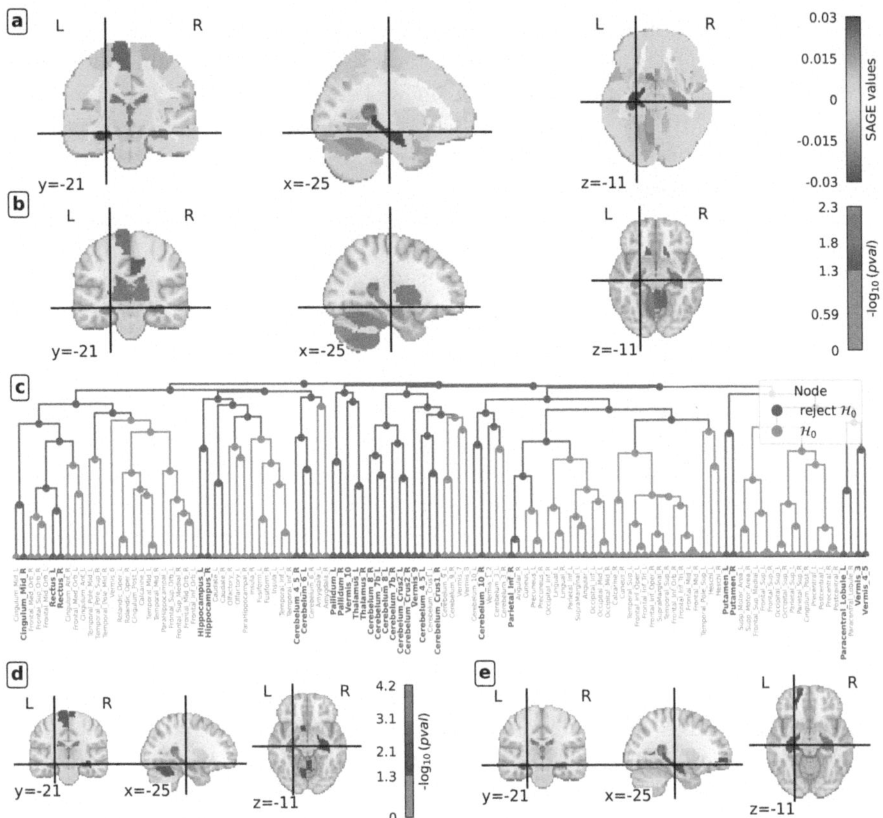

Fig. 3. Hierarchical CPI discovers groups of characteristic markers of AD progression. Importance obtained for classifying AD and MCI subjects from the ADNI dataset using a support vector classifier and grey matter densities in the 116 AAL regions. **a** Signed importance values obtained using the SAGE method. Only values for which the 95% CI did not overlap with 0 are reported. **b** Important regions identified by hierarchical CPI at the $\alpha = 0.05$ level. **c** Dendrogram derived from the hierarchical CPI approach. The important nodes ($\alpha = 0.05$) of the tree learned by hierarchical clustering are colored in red. Regions identified as important are labeled in red. **d** Important regions identified by HCPI for classifying patients with AD versus controls. **e** Important regions for classifying controls versus patients with MCI. (Color figure online)

the interest in locating the pathology's impact precisely motivated the proposed approach. The methodology for data processing, model optimization, and hyper-parameter tuning followed [8], to ensure result reproducibility. We used a Support Vector Classifier (SVC) as implemented in *libsvm* on GM densities in the 116 regions of the AAL atlas. For the HCPI method, importance was measured by the hinge loss difference in Eq. 1. All results are reported using 10-fold cross-validation with stratification. Hyper-parameter tuning was performed using a nested cross-validation loop to avoid information leakage from the test set [28]. Using a linear or *rbf* kernel led to similar predictive performance and importance scores. The results were reported for a linear kernel. Three classification tasks were considered: MCI vs. AD, AD vs. CN, and MCI vs. CN. The average AUC on the test set over 10 folds were 0.78, 0.93, and 0.74, respectively, which is consistent with existing literature [8].

Figure 3 presents the importance maps at the individual feature resolution, computed using the SAGE method (**a**) and the proposed hierarchical-CPI (**b**) for classifying patients with AD and MCI. Similar to the previous section, for both methods, results are reported at a significance threshold of $\alpha = 0.05$. While SAGE identifies more regions as important, it is likely that many of these are only marginally, not conditionally, associated with the outcome. By contrast, HCPI identifies fewer regions that summarize the specific markers of the disease. Importantly, all regions identified by HCPI are also identified by SAGE, revealing a form of consistency. These regions include the hippocampi, and the orbitofrontal cortex (rectus in the AAL) which have been extensively described in the literature as areas where atrophy is substantial [29,30]. The putamen and thalamus were also identified as predictive, consistently with published work documenting the association between AD and decreased global GM in these regions [31]. Moreover, HCPI allows to learn clusters of predictive variables with varying resolutions, This is depicted in **c**, which presents the dendrogram learned by agglomerative clustering, with nodes having a p-value below the significance threshold highlighted in red. This information complements panel **b** by highlighting the importance of selected subgroups. For instance, the node including (`Caudate_L, Caudate_R`) is important whereas individual variables are not, due to the high correlation between these two regions, (Pearson correlation of 0.84) leading to a cancellation of their conditional importance. In addition to these results Fig. 3 presents the importance map for two additional tasks: AD vs CN (**d**) and MCI vs CN (**e**). Similar to Fig. 3, the HCPI approach identifies hallmarks of AD pathology, such as the gray matter density in the hippocampi, which has been extensively described in the literature [29].

3.3 Importance Conservation Enables Inference on Highly Correlated EEG Data

The importance conservation approach was then applied to the EEG data from the TDBRAIN dataset to characterize the known Berger effect [32]. EEG data is known to exhibit high correlation due to latent sources spreading across the scalp as a result of field spreads. Resting state EEG were acquired from 1234 healthy subjects who were asked to open and close there eyes during labeled periods. The dataset was preprocessed using the pipeline presented in [33] in order to remove artifacts generated by non-brain sources. Specifically, independant component analysis was applied in order to remove eye-movement artifacts which would make the task trivial. The power at each of the 26 electrodes was computed across 17 logarithmically spaced frequency bands, ranging from 1 to

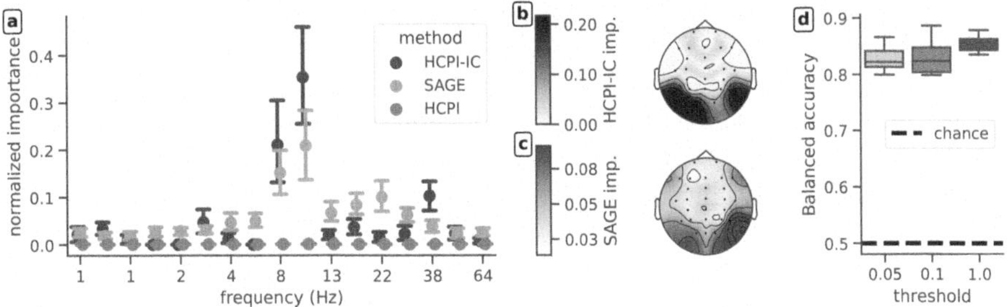

Fig. 4. Importance conservation enables variable importance inference in high-dimensional settings with very high correlation. Comparison of the variable importance obtained using hierarchical CPI with importance conservation (HCPI-IC), without (HCPI) and using SAGE. Absolute SAGE values are represented for readability. **a** The distribution of importance over frequencies for significant variables at the $\alpha = 0.05$ threshold. Each point represents the sum of important channels at a given frequency. **b, c** The distribution of importance over scalp topography for significant variables at the $\alpha = 0.05$ threshold. At a given channel, the sum is taken over all important frequencies. **d** Presents the performance of sub-models that use only a fraction of variables identified as significant at a level α, with $\alpha = 1$ corresponding to the full model. The boxes represent the distribution over 10-fold cross-validation. The dotted line represents the chance level.

64 Hz, using Morlet wavelets. The 442 resulting features present a very high correlation structure, with minimum Pearson correlation above 0.9. We considered the task of classifying the eyes status (closed vs open) using a pipeline consisting of a logarithm computation followed by logistic regression. Similarly to the previous section, 10-fold nested cross validation was used. The loss used to measure the conditional importance in Eq. 1 is the cross-entropy loss. The transmission threshold ϵ was set to the 95% quantile of the normal distribution.

As illustrated by Fig. 4a, the high correlation in the data causes the conditional importance to vanish, resulting in no significant discoveries at a threshold of $\alpha = 0.05$ (HCPI in red). To mitigate this issue and increase statistical power, the importance conservation mechanism (HCPI-IC, in blue) introduced in Sect. 2.3 is applied. The importance however remains more focal than with SAGE (orange), which spreads the importance over a wide range of frequencies. Panels **a**, **b** and **c** show the distribution of importance in the frequency and sensor spaces among significant variables at a threshold of $\alpha = 0.05$. The pattern observed, with most of the importance located around 10 Hz at occipital electrodes, corresponds to the well-studied Berger effect, characterized by increased occipital activity in the alpha-band [34]. The pattern is precisely identified by HCPI-IC, while SAGE distributes more broadly the importance over electrodes and frequencies. Finally, panel **d** demonstrates the performance of submodels using only significant variables at thresholds $\alpha < 0.05$, $\alpha < 0.1$, and 1 (all variables). It reveals that at the strictest threshold ($\alpha = 0.05$), the procedure selects 55 variables out of 442, recovering 96% of the full model's performance.

4 Discussion and Conclusion

The HCPI approach was motivated by the challenge of making inference on high-dimensional and highly correlated neuroimaging data. To achieve this, it frames the inference problem as the discovery of groups of variables that are jointly predictive of the outcome. It can recover statistical control in high correlations regimes where standard methods lose consistency. Statistical guarantees were empirically validated on simulated data, and the method was applied to two neuroimaging modalities using publicly available datasets. Its effectiveness was demonstrated on both classification and regression tasks. By successfully testing different tasks, models and losses, we proved the practical utility of this model-agnostic approach. HCPI flexibility exceeds that of existing methods relying on linear models or Lasso-based knockoffs [9,13,20].

By exploring subgroups within a learned hierarchical tree, HCPI balances precision and statistical power, allowing the identification of groups that are important, even if none of the individual variables is significant. It thus identifies the highest resolution at which importance can be narrowed down, without needing to optimize clustering parameters [12,13,20]. This information can easily be visualized using a dendrogram. Unlike additive methods like SAGE, which exhaustively explore all subgroups including a variable, eliminating many costly and useless evaluations. It provides a FWER control, thus contrasting with SAGE's known lack of type-1 error control [18]. Moreover, it remains tractable, requiring only $2p - 1$ importance evaluations. Finally, the importance conservation mechanism introduced in Eq. 3 mitigates power loss due to vanishing importance, as shown with highly correlated EEG data features.

When tested on MRI and EEG data, our method identified biologically well-studied features consistent with existing literature such as hallmarks of AD in MRI and the Berger effect in EEG. Lastly, while we used Conditional Permutation Importance, because known to be more stable and efficient than LOCO, the latter could also be used as a drop-in replacement for estimating importance.

Limitations: We have explored scenarios where a single agglomerative clustering is performed, demonstrating that it can yield insightful learnings about data structure, with clusters of variables being predictive even if individual variables are not. However, this step can introduce randomness. For applications not requiring hierarchical tree learning, like voxel-level or applications to raw images–it may be beneficial to repeat the procedure and leverage p-value aggregation strategies or e-values [35,36]. Future work could involve repeated agglomerative clustering on random data subsets, followed by aggregation to improve robustness. Another limitation concerns the theoretical guarantees of the importance conservation approach. While we empirically observed a type-1 error rate much lower than SAGE, a formal result remains to be established. The transmission threshold ϵ is critical in this context: it defines a threshold below which the importance of a parent node becomes indivisible because the contributions of its children nodes cancel each other out. We conjecture that it is possible to obtain guarantees for type-1 error control outside a neighborhood (which size depends on ϵ) around the support.

The algorithm builds on open-source software available on Github[2].

Acknowledgments. This research has received funding from the H2020 Research Infrastructures Grant EBRAIN-Health 101058516 and the VITE ANR-23-CE23-0016 and PEPR Santé numérique, Brain health Trajectories ANR-22-PESN-0012 projects.

References

1. Wen, J., et al.: Convolutional neural networks for classification of Alzheimer's disease: overview and reproducible evaluation. Med. Image Anal. **63**, 101694 (2020)
2. Tosun, D., Alzheimer's Disease Neuroimaging Initiative.: Identifying individuals with non-Alzheimer's disease co-pathologies: a precision medicine approach to clinical trials in sporadic Alzheimer's disease. Alzheimer's Dementia **20**(1), 421–436 (2024)
3. Gemein, L.A., et al.: Machine-learning-based diagnostics of EEG pathology. NeuroImage **220**, 117021 (2020)
4. Cuingnet, R., Alzheimer's Disease Neuroimaging Initiative, et al.: Automatic classification of patients with Alzheimer's disease from structural MRI: a comparison of ten methods using the ADNI database. Neuroimage **56**(2), 66–781 (2011)
5. Abraham, A., et al.: Machine learning for neuroimaging with Scikit-learn. Front. Neuroinform. **8**, 71792 (2014)
6. Tousignant, A., Lemaître, P., Precup, D., Arnold, D.L., Arbel, T.: Prediction of disease progression in multiple sclerosis patients using deep learning analysis of MRI data. In: International Conference on Medical Imaging with Deep Learning, pp. 483–492. PMLR (2019)
7. Petersen, R.C., et al. Alzheimer's disease neuroimaging initiative (ADNI) Clin. Charact. Neurology, **74**(3), 201–209 (2010)
8. Samper-González, J., et al.: Reproducible evaluation of classification methods in Alzheimer's disease: framework and application to MRI and pet data. Neuroimage **183**, 504–521 (2018)
9. Mandozzi, J., Bühlmann, P.: Hierarchical testing in the high-dimensional setting with correlated variables. J. Am. Stat. Assoc. **111**(513), (2016). ISSN 0162-1459. https://doi.org/10.1080/01621459.2015.1007209
10. Chevalier, J.-A., Salmon, J., Thirion, B.: Statistical inference with ensemble of clustered desparsified lasso. In: Frangi, A.F., Schnabel, J.A., Davatzikos, C., Alberola-López, C., Fichtinger, G. (eds.) MICCAI 2018. LNCS, vol. 11070, pp. 638–646. Springer, Cham (2018). https://doi.org/10.1007/978-3-030-00928-1_72
11. Sobol, I.M.: Global sensitivity indices for nonlinear mathematical models and their Monte Carlo estimates. Math. Comput. Simul. **55**(1–3), 271–280 (2001). https://doi.org/10.1016/s0378-4754(00)00270-6
12. Chamma, A., Thirion, B., Engemann, D.: Variable importance in high-dimensional settings requires grouping. In: Proceedings of the AAAI Conference on Artificial Intelligence, vol. 38, no. 10 (2024). https://doi.org/10.1609/aaai.v38i10.28997

[2] https://github.com/mind-inria/hidimstat.

13. Chevalier, J.-A., Nguyen, T.-B., Salmon, J., Varoquaux, G., Thirion, B.: Decoding with confidence: statistical control on decoder maps. Neuroimage **234**, 117921 (2021). https://doi.org/10.1016/j.neuroimage.2021.117921
14. Ribeiro, M.T., Singh, S., Guestrin, C.: Why should i trust you?: explaining the predictions of any classifier. In: Proceedings of the 22nd ACM SIGKDD International Conference on Knowledge Discovery and Data Mining (2016). https://doi.org/10.48550/arxiv.1602.04938
15. Štrumbelj, E., Kononenko, I.: Explaining prediction models and individual predictions with feature contributions. Knowl. Inf. Syst. **41**(3), 647–665 (2013). https://doi.org/10.1007/s10115-013-0679-x
16. Homma, T., Saltelli, A.: Importance measures in global sensitivity analysis of nonlinear models. Reliab. Eng. Syst. Saf. **52** (1996). https://doi.org/10.1016/0951-8320(96)00002-6
17. Williamson, B.D., Gilbert, P.B., Simon, N.R., Carone, M.: A general framework for inference on algorithm-agnostic variable importance. J. Am. Stat. Assoc. **118** (2023). https://doi.org/10.1080/01621459.2021.2003200
18. Verdinelli, I., Wasserman, L.: Feature importance: a closer look at shapley values and loco. Stat. Sci. **39**(4), 623–636 (2024)
19. Chamma, A., Engemann, D.A., Thirion, B.: Statistically valid variable importance assessment through conditional permutations. In: Advances in Neural Information Processing Systems, vol. 36 (2024)
20. Chung, A., Gee, J.C., Yushkevich, P.A., Bao, S. (eds.): IPMI 2019. LNCS, vol. 11492. Springer, Cham (2019). https://doi.org/10.1007/978-3-030-20351-1
21. Covert, I., Lundberg, S.M., Lee, S.I.: Understanding global feature contributions with additive importance measures. In: Advances in Neural Information Processing Systems, vol. 32 (2020)
22. Ward, J.H.: Hierarchical grouping to optimize an objective function. J. Am. Stat. Assoc. **58**(301), 236–244 (1963)
23. Lescroart, M.D., Stansbury, D.E., Gallant, J.L.: Fourier power, subjective distance, and object categories all provide plausible models of bold responses in scene-selective visual areas. Front. Comput. Neurosci. **9**(November), 135 (2015)
24. Esteban, O., Markiewicz, C.J., Blair, R., Poldrack, R.A., Gorgolewski, K.J.: sMRIPrep: structural MRI preprocessing workflows (2021)
25. Esteban, O., et al.: fMRIPrep: a robust preprocessing pipeline for functional MRI. Nat. Methods **16**(1), 111–116 (2019)
26. Rolls, E.T., Huang, C.C., Lin, C.P., Feng, J., Joliot, M.: Automated anatomical labelling atlas 3. Neuroimage **206**, 116189 (2020)
27. Nilearn contributors: Nilearn. https://github.com/nilearn/nilearn
28. Pedregosa, F., et al.: Scikit-learn: machine learning in Python. J. Mach. Learn. Res. **12**, 2825–2830 (2011)
29. Knopman, D.S., et al.: Alzheimer disease. Nat. Rev. Disease Prim. **7**(1), 33 (2021)
30. Van Hoesen, G.W., Parvizi, J., Chu, C.C.: Orbitofrontal cortex pathology in Alzheimer's disease. Cerebral Cortex **10**(3), 243–251 (2000)
31. de Jong, L.W., et al.: Strongly reduced volumes of putamen and thalamus in Alzheimer's disease: an MRI study. Brain **131**(12), 3277–3285 (2008)
32. Van Dijk, H., Van Wingen, G., Denys, D., Olbrich, S., Van Ruth, R., Arns, M.: The two decades brainclinics research archive for insights in neurophysiology (TDBRAIN) database. Sci. Data **9**(1), 333 (2022)
33. Bomatter, P., Paillard, J., Garces, P., Hipp, J., Engemann, D.A.: Machine learning of brain-specific biomarkers from EEG. biorxiv (2023)

34. Hohaia, W., Saurels, B.W., Johnston, A., Yarrow, K., Arnold, D.H.: Occipital alpha-band brain waves when the eyes are closed are shaped by ongoing visual processes. Sci. Rep. **12**(1), 1194 (2022)
35. Meinshausen, N., Meier, L., Bühlmann, P.: P-values for high-dimensional regression. J. Am. Stat. Assoc. **104**(488), 1671–1681 (2009)
36. Vovk, V., Wang, R.: E-values: calibration, combination and applications. Ann. Stat. **49**(3), 1736–1754 (2021)

Disentangle Disease-Relevant Patterns from Irrelevant Patterns in fMRI Analysis Using Equivariant and Contrastive Learning

Xin Shen[1], Shengjie Zhang[2], Wenbin Liu[3](✉), and Yuan Zhou[4](✉)

[1] School of Mathematical Sciences, Beijing Normal University, Beijing, China
sd228xins@gmail.com

[2] Institute of Science and Technology for Brain-Inspired Intelligence, Fudan University, Shanghai, China
zsjxll@gmail.com

[3] Research Center for Mathematics, Beijing Normal University, Zhuhai, China
wbliu@uic.edu.cn

[4] School of Data Science, Fudan University, Shanghai, China
yuanzhou@fudan.edu.cn

Abstract. Functional magnetic resonance imaging (fMRI) holds great potential for diagnosing and understanding brain disorders. However, the complexity and subtlety of disease-relevant variations in fMRI present significant challenges. To address this issue, we propose a framework that combines equivariant learning and contrastive learning (ECL) to disentangle disease-relevant patterns from irrelevant patterns in fMRI. The framework uses a personalized mask to separate the functional connectivity network from fMRI into a disease-relevant subgraph and an irrelevant subgraph. The disease-relevant subgraph undergoes an equivariant learning pipeline to align the orbit of the encoded features with the orbit of the augmented views of the inputs. The disease-irrelevant subgraph undergoes a contrastive learning pipeline that pulls the encoded features to be close from augmented views of the same input. By combining these 2 learning processes, the learned encoder can be invariant to perturbations to disease-irrelevant patterns while equivariant to disease-relevant variations. The proposed approach achieved state-of-the-art classification performance across 3 benchmark datasets: ABIDE I, ABIDE II, and ADHD-200, with significant improvements in accuracy (improved by up to 5%). Interpretability experiments identified disease-related regions of interest (ROIs) of clinical relevance. These results establish our framework as a promising tool for analyzing brain networks in fMRI. The code is available at https://github.com/CXshen468/ecl.

Keywords: Equivariant learning · Contrastive learning · Functional connectivity network · fMRI · Graph neural network

I. Oguz et al. (Eds.): IPMI 2025, LNCS 15830, pp. 94–108, 2026.
https://doi.org/10.1007/978-3-031-96625-5_7

1 Introduction

Functional magnetic resonance imaging (fMRI) can be used to study functional connectivity networks (FCN) in the brain [1]. The process usually parcellates an fMRI by an atlas into several regions of interest (ROIs) and constructs a graph with nodes as brain regions and edges as functional connectivity. The functional connectivity can be measured by Pearson's correlation coefficient of blood oxygen level-dependent (BOLD) signals between 2 ROIs, which reflects their temporal activity correlation. Traditional methods for analyzing FCN assess network properties like degree, clustering coefficient, average path length, etc. [2]. These metrics struggle to capture complex nonlinear interactions. With the recent breakthroughs of graph neural networks (GNNs) in deep learning, there is growing interest in applying GNNs to FCNs from fMRI [3]. GNNs [17–20] use an encoder to extract features from a graph via a message-passing and aggregation mechanism [4]. Recent studies show that the GNN-based methods can outperform traditional methods by capturing the rich underlying structure in a graph, offering new insights into disease classification, multimodal data integration, and biomarker identification [3].

Despite the great potential of GNNs in analyzing FCNs, challenges such as data scarcity, subtle inter-class differences, and diverse noise interference arise [5]. A leading approach to mitigating such issues is to use graph augmentation techniques [7] in the self-supervised learning (SSL) framework, particularly in the form of graph contrastive learning (GCL). GCL defines a graph augmentation strategy and learns graph representations by requiring that a graph remains close to its augmented views while being distinct from other graphs in the latent space. Early examples include InfoGraph [9] which maximizes the mutual information between node-level and graph-level features, and GraphCL [8] which proposes 4 augmentation techniques over graphs. MVGRL [10] uses the degree matrix-normalized adjacency matrix as the augmented view. LaGraph [11] integrates a decoder to reconstruct the node features of the graphs. In fMRI analysis, GCL models have also been developed to enhance diagnostic accuracy and uncover insights into brain functional networks [18,20]. For example, both GATE and CFCG-Net [18,20] introduce temporal augmentations to the BOLD signals through overlapping time windows.

The graph augmentation methods mentioned above are random and may significantly distort disease-relevant patterns. To address this issue, recent studies propose learnable augmentation strategies [12,15]. For example, BraGCL [19] uses importance scores to guide edge and node feature removal, and A-GCL [17] employs adversarial contrastive learning to train a Bernoulli mask for edge removal. Although these approaches generate controllable augmentations, they typically treat all variations in the graph equally. Consequently, subtle yet disease-relevant changes may become indistinguishable from disease-irrelevant changes in learned representations. Therefore, a question arises: Can we design a method that handles disease-relevant and irrelevant patterns separately?

Inspired by Chen et al. [26], we propose a group-theoretical framework to solve this question. In this framework, graph augmentations are treated as group actions and the augmented graphs form an orbit. This leads to two distinct maps for feature extraction: an invariant map and an equivariant map. The invariant map requires the features to remain invariant under graph perturbations and can be learned via contrastive learning (see Fig. 1a). The equivariant map asks that the learned representations vary according to augmentations in the graph space by using a *predictor* that maps the parameters of the augmentation to transformations in the latent space, which can be learned via equivariant learning (see Fig. 1b).

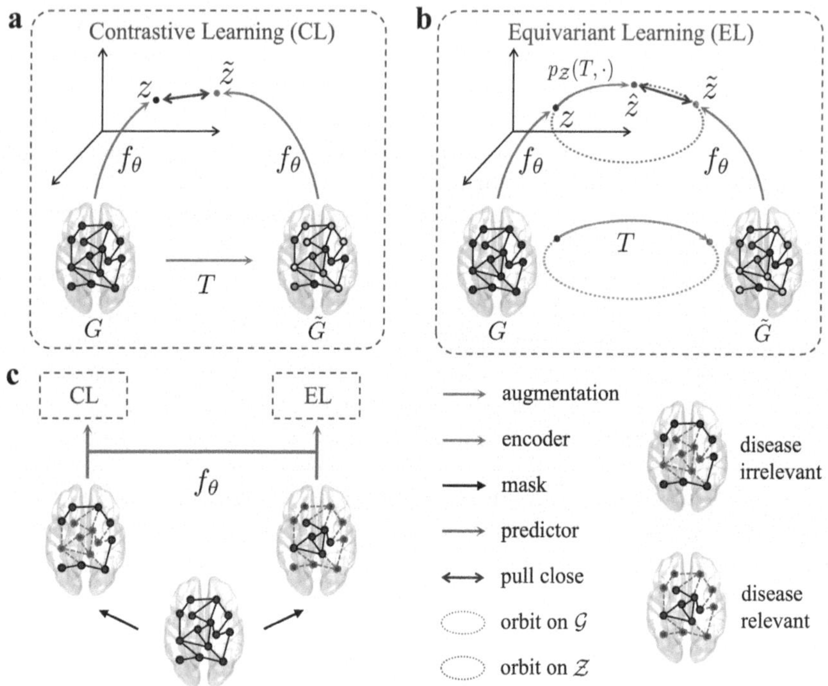

Fig. 1. Motivation of disentangling disease-relevant patterns from irrelevant patterns by combining equivariant and contrastive learning. (a) The contrastive learning (CL) framework learns an encoder f_θ invariant to input perturbations by maximizing the similarity of the latent embeddings (z and $\tilde{z}$) of the original graph G and its augmented view $\tilde{G}$. (b) The equivariant learning (EL) framework learns an encoder f_θ and a predictor $p_{\mathcal{Z}}(T, \cdot)$ by treating augmentations as a group and augmented views as an orbit and pulling close the predicted embedding $\hat{z}$ and augmented embedding $\tilde{z}$. (c) By splitting the graph into a disease-relevant subgraph and a disease-irrelevant subgraph and applying EL and CL on them respectively, our model learns an encoder f_θ invariant to perturbations to disease-irrelevant patterns and equivariant to disease-relevant variations.

Building upon these theoretical insights, our framework is also based on clinical observations that brain disorders are often associated with specific networks composed of sparsely distributed regions (e.g., the default mode network in autism spectrum disorder (ASD) and the cognitive control network in attention deficit hyperactivity disorder (ADHD) [36,39]). Hence, the brain graph can be decomposed into a disease-relevant subgraph (containing the disease-associated networks) and a disease-irrelevant subgraph. We then apply equivariant learning on the disease-relevant subgraph to capture disease-associated variations, and apply contrastive learning on the disease-irrelevant subgraph to filter out irrelevant noise (see Fig. 1c).

Our contributions are three-fold. First, we propose a novel framework, Equivariant and Contrastive Learning (ECL) to disentangle disease-relevant patterns while preserving their inherent variability. Second, we formulate the graph augmentation based on group theory, treating different augmented graphs and their embeddings as orbits to explore disease-induced variations. This leads to a novel principle for predictor design in equivariant learning. Third, our framework with a mask to separate disease-relevant and -irrelevant regions is inherently interpretable. Interpretability results reveal key regions in ASD and ADHD which align well with clinical research.

2 Methodology

In functional brain network analysis, each individual's brain is modeled as a graph $G = (V, X, A)$, where $V = \{v : v = 1, \ldots, N\}$ is the set of nodes (ROIs), $A = [a_{uw}] \in [-1, 1]^{N\times N}$ is the adjacency matrix, $X = \{x_v \in \mathbb{R}^m : v \in V\}$ are the node features. Each off-diagonal element in the adjacency matrix is computed using Pearson's correlation coefficient between the BOLD signals in 2 ROIs, reflecting their functional connectivity. The node features are derived from the amplitude of low-frequency fluctuations (ALFFs) of the BOLD signal [32] in m frequency bands.

We start with explaining group theory in the graph space.

2.1 Preliminaries

Group. A group is a set $\mathcal{T}$ equipped with a binary operation $\cdot : \mathcal{T} \times \mathcal{T} \to \mathcal{T}$ that satisfies 4 properties: closure, associativity, identity, and invertibility. In the context of our framework, we use augmentation on the node features and treat the augmentation parameters as a group $\mathcal{T}$. Specifically, we focus on the additive group $(\mathbb{R}^{N\times m}, +)$, where each element is added to the original node features to create an augmented graph and the augmentation composition is achieved through element-wise addition $+$.

Group Action. A group action is a formal way in which a group $\mathcal{T}$ acts on a set $\mathcal{G}$, mapping elements of $\mathcal{T}$ to transformations on $\mathcal{G}$. The action of $T \in \mathcal{T}$ on $G \in \mathcal{G}$ is defined as a function $p_\mathcal{G} : \mathcal{T} \times \mathcal{G} \to \mathcal{G}$ that satisfies two properties:

- Identity: $p_\mathcal{G}(e, G) = G, \quad \forall G \in \mathcal{G}$
- Associativity: $p_\mathcal{G}(T_1 \cdot T_2, G) = p_\mathcal{G}(T_1, p_\mathcal{G}(T_2, G)), \quad \forall T_1, T_2 \in \mathcal{T}, G \in \mathcal{G}$

where e is the identity element in the group.

Data augmentation is modeled as a group action, where each transformation $T \in \mathcal{T}$ acts on a graph $G \in \mathcal{G}$ through $p_{\mathcal{G}}(T, G)$. Similarly, we can also define a group action $p_{\mathcal{Z}}(T, \cdot)$ on the latent space $\mathcal{Z}$. For a graph $G = (V, X, A)$ constructed from an fMRI image, where node features X are the ALFFs, augmentations applied to X correspond to perturbations in the spectra of the signal.

Group Orbit. Given a group $\mathcal{T}$ and its action on a set $\mathcal{G}$, the group orbit of an element $G \in \mathcal{G}$ is the set of all possible transformations of G by elements of $\mathcal{T}$. Formally, the orbit of G under $\mathcal{T}$ is denoted as $p_{\mathcal{G}}(\mathcal{T}) \cdot G = \{p_{\mathcal{G}}(T, G) \mid T \in \mathcal{T}\}$. The orbit captures all transformations (augmented views) of G under the group action.

Equivariant Map. A function $f : \mathcal{G} \to \mathcal{Z}$ is said to be equivariant with respect to a group $\mathcal{T}$ iff

$$f(p_{\mathcal{G}}(T, G)) = p_{\mathcal{Z}}(T, f(G)), \quad \forall T \in \mathcal{T}, G \in \mathcal{G} \tag{1}$$

In this framework, $p_{\mathcal{G}}$ and $p_{\mathcal{Z}}$ represent the group actions on the input space $\mathcal{G}$ and the latent space $\mathcal{Z}$, respectively. Equivariant maps maintain the relationship between augmented graphs and their corresponding latent representations, preserving structural consistency across transformations. The feature extractor f is designed to be an equivariant map on disease-relevant variations.

2.2 General Architecture

Figure 2 shows an overview of our architecture. The input graph G is separated into a disease-relevant subgraph G^r and a disease-irrelevant subgraph G^{ir} by a mask. The disease-relevant subgraph undergoes one augmentation to generate its augmented view $\tilde{G}^r$. The disease-irrelevant subgraph undergoes 2 augmentations to generate its augmented views $\tilde{G}^{1,ir}$, $\tilde{G}^{2,ir}$. An encoder $f_\theta : \mathcal{G} \to \mathbb{R}^d$, parameterized by θ, produces latent embeddings of all these views and the original graph. For the latent embeddings of the irrelevant subgraph $\tilde{z}^{1,ir}, \tilde{z}^{2,ir}$, a loss function from contrastive learning is used to enforce these embeddings to be similar. For the embeddings of the relevant subgraph z^r, $\tilde{z}^r$, a predictor $p_{\mathcal{Z}}(T^r, \cdot)$ transforms z^r to approximate $\tilde{z}^r$. In this way, we can learn an encoder f_θ that captures disease-relevant features independent of augmentations applied to irrelevant patterns. The details are given below.

Decomposing the Graph into Disease-Relevant and Irrelevant Subgraphs. First, we introduce a mask M to separate disease-relevant regions and irrelevant ones. The mask partitions the graph into the disease-related subgraph $G^r = M \odot G$ and the irrelevant subgraph $G^{ir} = (1 - M) \odot G$, where $\odot$ retains the corresponding node features and extracts the subgraph by keeping only the edges that connect the retained nodes.

For the irrelevant subgraph, the augmented views are generated with the relevant subgraph preserved:

$$\tilde{G}^{k,ir} = p_{\mathcal{G}}(T^{k,ir}, G) = G^r \oplus p_{\mathcal{G}}(T^{k,ir}, G^{ir}),\ k = 1, 2$$

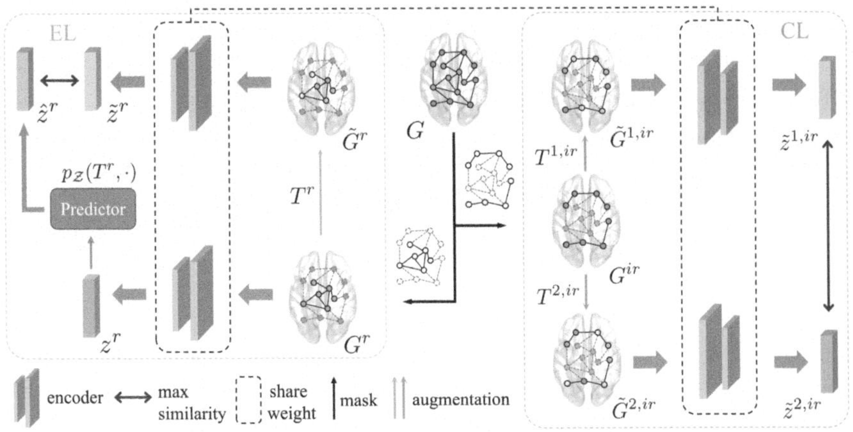

Fig. 2. The general architecture of ECL. First, we generate a personalized mask for each individual, which identifies the disease-relevant subgraph G^r and the irrelevant subgraph G^{ir}. These subgraphs are then processed through separate branches. In the contrastive learning branch (green), the disease-irrelevant subgraph is perturbed to generate two views $(\tilde{G}^{1,ir}, \tilde{G}^{2,ir})$ whose encoder outputs are enforced to be similar. In the equivariant learning branch (yellow), the disease-relevant subgraph G^r is perturbed to generate an augmented view $\tilde{G}^r$. A learnable predictor is used to transform the embedding of the original input z^r to approximate the embedding of the augmented view $\tilde{z}^r$.

where $\oplus$ represents element-wise addition and $T^{k,ir}$ indicates the augmentation parameters of the k-th view on G^{ir} while keeping G^r unchanged. Conversely, for the relevant subgraph, one view is generated:

$$\tilde{G}^r = p_{\mathcal{G}}(T^r, G) = G^{ir} \oplus p_{\mathcal{G}}(T^r, G^r),$$

where T^r affects only G^r while leaving G^{ir} unchanged. Here, both augmentation parameters are drawn from the additive group $\mathcal{T} = (\mathbb{R}^{N \times m}, +)$.

Contrastive Learning for Disease-Irrelevant Subgraphs. For the irrelevant subgraph G^{ir}, we want the extracted latent representations to be similar for different augmentations of this subgraph. This is achieved by using the SimCLR framework [13]. Specifically, given a mini-batch of graphs $\{G_i\}_{i=1}^B$ where B is the number of graphs in the batch, we compute the InfoNCE loss [13] for the 2 latent embeddings of the augmented views $\tilde{z}_i^{1,ir} = f_\theta(\tilde{G}_i^{1,ir})$ and $\tilde{z}_i^{2,ir} = f_\theta(\tilde{G}_i^{2,ir})$ where $\tilde{G}_i^{k,ir} = p_{\mathcal{G}}(T_i^{k,ir}, G_i), k = 1, 2$:

$$\mathcal{L}_{CL}(\theta) = -\frac{1}{B}\sum_{i=1}^{B} \log \frac{\exp\left(\text{sim}\left(\tilde{z}_i^{1,ir}, \tilde{z}_i^{2,ir}\right)/\tau\right)}{\sum_{j=1}^{B} \exp\left(\text{sim}\left(\tilde{z}_i^{1,ir}, \tilde{z}_j^{2,ir}\right)/\tau\right)},$$

where $\text{sim}(z, z') = z^\top z' / \|z\| \|z'\|$ is the cosine similarity function, τ is a hyperparameter. By applying contrastive learning only on the disease-irrelevant regions, the method ensures that disease-related features remain consistent across different views of the disease-irrelevant subgraph.

Equivariant Learning for Disease-Relevant Subgraphs. For the disease-relevant subgraph G^r, we use equivariant learning to enforce that transformations applied to G^r are mirrored in the latent space, preserving meaningful variations in disease-relevant features. To achieve this, we introduce a learnable predictor h_ψ that approximates the group action $p_\mathcal{Z}$, ensuring consistency between transformations in the input and latent spaces:

$$f_\theta(p_\mathcal{G}(T^r, G)) = h_\psi(T^r, f_\theta(G)). \tag{2}$$

Denoting $\hat{z}_i = h_\psi(T_i^r, f_\theta(G_i))$ and $\tilde{z}_i^r = f(\tilde{G}_i^r)$ in the context of mini-batches, the prediction task becomes minimizing the mean squared error (MSE) loss between the true augmented representation $\tilde{z}_i^r$ and the predicted representation $\hat{z}_i$:

$$\mathcal{L}_{EL}(\theta, \psi) = \frac{1}{B}\sum_{i=1}^{B} \|\tilde{z}_i^r - \hat{z}_i\|^2.$$

This setup enforces that the disease-relevant patterns exhibit equivariance with respect to augmentations.

Transformation Predictor. In equivariant learning, the key is to design the predictor h_ψ which approximates the group action $p_\mathcal{Z}$ and should satisfy the identity property and the associativity property (Sect. 2.1), i.e., $h_\psi(\mathbf{0}, z) = z$ and $h_\psi(T_1 \cdot T_2, z) = h_\psi(T_1, h_\psi(T_2, z))$ for all $T_1, T_2 \in \mathcal{T}$ and $z \in \mathcal{Z}$. These properties are intuitive considering the equivariant equation (2). The identity property is equivalent to $f_\theta(G) = h_\psi(\mathbf{0}, f_\theta(G))$, i.e., when the augmentation leaves G unchanged, the predictor should also leave the embedding unchanged. The associativity property is equivalent to $f_\theta(p_\mathcal{G}(T_1 \cdot T_2, G)) = h_\psi(T_1, h_\psi(T_2, f_\theta(G))$, i.e., when a sequence of augmentations is applied to G in the graph space, the embedding is equal to a sequence of transformations in the latent space.

Given these properties, we propose a candidate predictor design for the additive group that satisfies these properties:

$$h_\psi(T, \boldsymbol{z}) = \boldsymbol{z} + u_\psi(T). \tag{3}$$

Here, $u_\psi(T)$ can take a linear form $u_\psi(T) = v^\top TW$, or a graph-based form $u_\psi(T) = v^\top m_\psi(A)TW$, where $W \in \mathbb{R}^{m \times d}, v \in \mathbb{R}^N$ are learnable parameters and $m_\psi(\cdot)$ is a learnable spectral graph filter on the adjacency matrix [21]. In this formulation, we avoid bias terms and non-linear activations, which accidentally aligns with empirical findings that bias can lead to model collapse in equivariant learning [27]. Through this design, the predictor guides latent features along their group orbits, allowing the model to capture subtle variations in the graph.

2.3 Loss Function and Optimization

Combining the contrastive learning and the equivariant learning, our model optimizes the objective function

$$\mathcal{L}(\theta, \psi) = \mathcal{L}_{CL}(\theta) + \mathcal{L}_{EL}(\theta, \psi).$$

We propose an alternate optimization scheme that updates each mask M_i and model parameters θ, ψ alternately in each epoch. The mask M_i is updated for each training instance using graph class activation mapping (GCAM) [12] to compute node-level gradients to identify nodes that contribute most to the contrastive loss as disease-relevant. Intuitively, the rationale of this practice can be understood in this way: Suppose the encoder can extract features by focusing only on disease-relevant patterns. For a random augmentation applied to the graph, the contrastive loss can be decreased most efficiently by changing the disease-relevant pattern in the original graph to match the disease-relevant pattern in the augmented view, since then the encoder extracts identical features from the original graph and its augmented view, optimizing the contrastive loss. Hence, the gradient of the contrastive loss should have large values in disease-relevant patterns in the original graph and can serve as the mask.

In our GCAM, the above gradient (saliency map) is refined in 3 iterations. At each iteration, it begins with randomly augmented graphs to compute the InfoNCE loss between the original graph representation z and its augmented representation $\tilde{z}$. Then, we calculate the partial derivative of the InfoNCE loss with respect to each node-level feature z_v, and normalize these contributions into saliency scores from 0 to 1 to indicate each node's relevance to the disease state. Finally, the saliency maps from the 3 iterations are aggregated to produce a final saliency map $\tilde{M}_i \in [0,1]^N$ which is binarized by a threshold to produce the final mask M_i. Once the mask M_i is updated, θ, ψ are updated through gradient descent via backpropagation.

3 Results

3.1 Experimental Setup

Dataset and Preprocessing. We utilized 3 publicly available fMRI datasets: The Autism Brain Imaging Data Exchange (ABIDE) I, ABIDE II, and ADHD-200. The fMRI images were preprocessed using the fMRIPrep pipeline [31], which includes reference image estimation, head motion correction, slice timing correction, and susceptibility distortion correction. After aligning the volumes to MNI152 space, we regressed out confounders such as framewise displacement, global signals, and mean tissue signals. Quality control measures from [17] enabled retention of 456 ASD and 504 normal control (NC) subjects for ABIDE I, 236 ASD and 276 NC for ABIDE II, and 275 ADHD and 205 NC for ADHD-200.

Then, we employed the AAL1 atlas [30] to divide the brain into 116 regions. Mean time series (BOLD signals) were calculated for each region. The adjacency matrix was calculated by Pearson's correlation coefficient between the time series of 2 regions. The node features were calculated via the Fourier transform of the mean time series, encompassing 3 frequency bands of the ALFFs (Slow-5: 0.01–0.027 Hz, Slow-4: 0.027–0.073 Hz, Classical: 0.01–0.08 Hz) and representing the total power within the low-frequency range.

Implementation Details. Our model is implemented in PyTorch to run on a single GPU. Experiments were accelerated using two servers equipped with 8 NVIDIA V100 GPUs and 2 NVIDIA A6000 GPUs. Our encoder backbone consists of two graph isomorphism network (GIN) [25] layers, a global sum pooling layer, and a multi-layer perceptron (MLP). The MLP has 1 hidden layer and the same number of nodes d in the input, hidden, and output layer. The spectral graph filter in the predictor is implemented using a Chebyshev kernel [21]. The implementation parameters include a learning rate of 0.001, a mask threshold of 0.5, an embedding dimension $d = 32$, and a batch size $B = 32$.

Competing Methods. We compare our proposed models with competing self-supervised GCL methods that are designed for general purposes and fMRI. The general purpose GCL methods include InfoGraph [9], JOAO [15], LaGraph [11], MVGRL [10], GraphCL [8], and CCA [14]. The fMRI methods are BraGCL [19], CFCGL [20], GATE [18], and AGCL [17]. The graph representations learned by all models are evaluated using a linear SVM classifier to assess classification accuracy, following the 5-fold cross-validation scheme in ADGCL [28] which trains on training/validation/test sets and selects the final test score from the best validation epoch using SVM.

Table 1. Classification results (mean±std) on 3 datasets using 5-fold cross-validation. Best performances are highlighted in bold, with the top competing method underlined.

Type	Method	ABIDE I		ABIDE II		ADHD	
		Accuracy	AUC	Accuracy	AUC	Accuracy	AUC
SSL GCL	InfoGraph	73.35±2.00	73.41±2.08	75.19±3.45	74.61±3.59	63.41±4.39	**64.11±4.53**
	JOAO	77.00±3.42	77.60±3.55	79.31±3.23	79.23±3.16	62.04±3.88	62.15±3.79
	LaGraph	76.78±2.36	77.11±2.66	80.64±3.37	80.34±3.22	64.05±2.71	63.95±2.56
	MVGRL	56.97±3.71	52.85±2.18	61.88±4.37	55.67±2.35	63.41±3.46	63.34±3.43
	GraphCL	73.85±2.13	74.01±2.32	81.38±2.44	81.17±2.41	63.00±4.90	63.24±4.49
	CCA	68.28±3.07	67.65±3.17	80.26±4.32	80.17±4.49	62.99±5.12	63.10±5.13
SSL fMRI	BraGCL	71.02±2.91	71.21±2.86	84.06±4.04	83.76±4.05	62.07±3.63	61.48±3.73
	CFCGL	75.68±2.12	75.80±2.71	79.51±3.19	78.93±3.22	61.28±3.24	61.86±3.31
	GATE	74.56±3.00	74.91±3.07	79.70±2.73	79.10±2.72	62.27±3.98	62.72±3.99
	AGCL	76.59±3.63	76.53±3.49	79.33±3.70	78.36±4.02	63.24±4.26	62.57±4.37
Ours	ECL-L	81.37±2.75	81.11±2.61	**86.67±4.04**	**86.90±4.13**	64.62±3.57	63.69±3.13
	ECL-S	**82.47±2.39**	**83.03±2.44**	84.97±4.06	84.63±4.22	**65.02±4.12**	62.36±4.19

3.2 Classification Results and Ablation Studies

Accuracy Comparison. Table 1 presents the classification performance of our method, which includes two variants, linear transformation (ECL-L) and spectral graph filter (ECL-S), corresponding to the linear form and the graph-based form in Eq. (3) respectively. Both variants achieve state-of-the-art accuracy and area under the curve (AUC), with improvements up to 5% over competing methods. On the ABIDE I dataset, ECL-S outperforms all methods, achieving the highest accuracy (82.47%) and AUC (83.03%). Meanwhile, ECL-L also demonstrates competitive performance with an accuracy of 81.37% and an AUC of 81.11%, surpassing the best competing method, JOAO. On the ABIDE II dataset, ECL-L delivers the highest accuracy (86.67%) and AUC (86.90%), while ECL-S achieves strong results with an accuracy of 84.97% and AUC of 84.63%. On the ADHD dataset, ECL-S achieves the best accuracy (65.02%). Though ECL-L has slightly lower performance with an accuracy of 64.62%, it still outperforms most competing methods. These results illustrate the robustness of both variants across diverse datasets.

Impact of Equivariant Learning and Mask. For ablation studies, we first validated the contributions of two critical components in our model: the equivariant learning and the mask to separate disease-relevant and -irrelevant regions. First, to remove the equivariant learning structure, we only trained the disease-irrelevant subgraphs using contrastive learning with the mask to differentiate between disease-relevant and irrelevant subgraphs. The results show a significant drop in classification accuracy across all datasets (Fig. 3a). This underscores the critical role of equivariant learning in enhancing disease-relevant features. Second, we excluded the mask and applied the equivariant learning to the entire graph. Again, this led to notable accuracy declines, particularly on ABIDE II (from 86.67% to 80.46%) (see Fig. 3a). Third, we removed both the equivariant learning and the mask, leaving only the standard contrastive learning on the entire graph [13]. This resulted in the worst performance on 2 datasets. These results collectively suggest the importance of equivariant learning and the mask in our architecture.

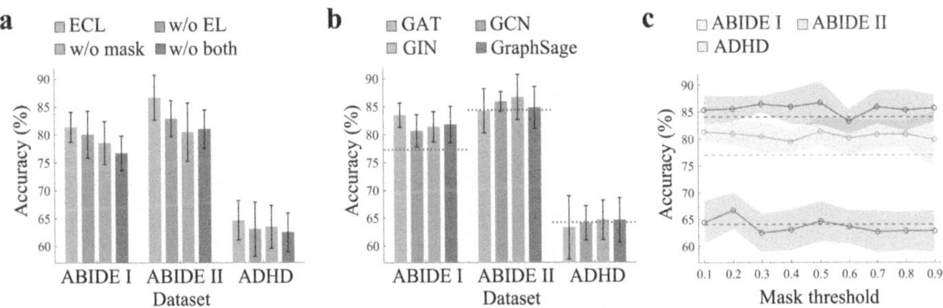

Fig. 3. Ablation studies and sensitivity analysis. (a) Ablation studies illustrating the impact of the equivariant learning structure, the mask, and their combination on fMRI classification. (b,c) Impact of (b) different graph kernels and (c) mask thresholds on the performance. The dashed lines indicate the best results from the competing methods.

Impact of Graph Kernel. To analyze the impact of different graph kernels on classification performance, we evaluated our framework using 4 commonly used graph kernels: graph convolutional network (GCN) [22], GIN [25], GraphSAGE [24], and graph attention network (GAT) [23] (see Fig. 3b). Across all datasets, GIN generally performed the best, with the exception on the ABIDE I dataset, where GAT achieved the highest accuracy (83.48%).

Overall, the architecture is robust to change of graph kernels, with nearly all kernels performing better than existing methods across the 3 datasets.

Impact of Mask Threshold. We further explored the choice of mask threshold on fMRI classification by binarizing the saliency map using thresholds ranging from 0.1 to 0.9 with an increment of 0.1 (see Fig. 3c). The optimal thresholds varied across datasets, with the optimal one for ASD classification being 0.5 and the optimal one for ADHD being 0.2. Notably, almost all thresholds yielded classification accuracies exceeding the highest performance from the competing methods in ASD classification, reinforcing the robustness of our model.

3.3 Interpretability Analysis

The disease-associated brain regions can be identified by analyzing the masks. Therefore, for each patient, we generate a personalized mask using the model. To find the common disease-associated regions, we calculate how frequently each ROI appears across all these masks and select the top 10 most frequently retained ROIs. Results, visualized in Fig. 4 via BrainNet Viewer [33], show strong concordance with regions reported in clinical studies [34–39]. The significance of these identified regions is explained below.

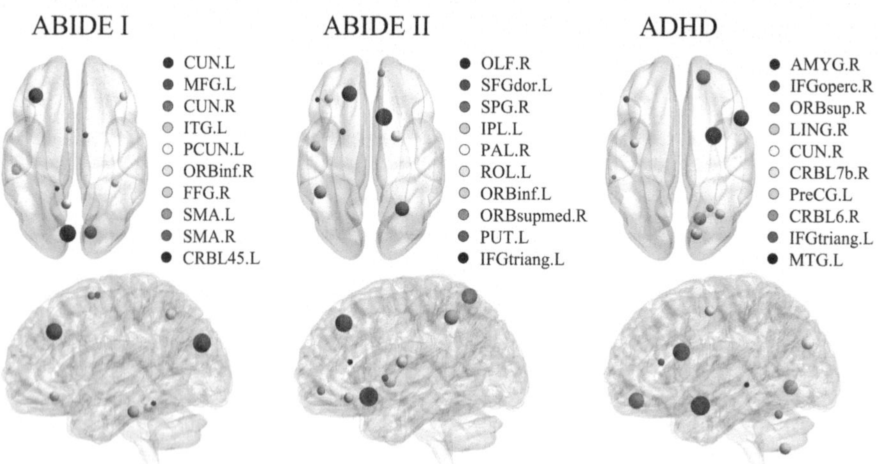

Fig. 4. Top 10 most frequently retained ROIs selected as common disease-related ROIs across the 3 datasets. A larger node size indicates that the region has a higher frequency.

ROI-Level Analysis. In the ABIDE I dataset, the middle frontal gyrus (MFG.L) is central to executive functions, reflecting difficulties with sustained attention and cognitive tasks. The cuneus (CUN.L) points to deficits in visual-spatial processing, while the fusiform gyrus (FFG.R) [36] underscores impaired social perception. In the ABIDE II dataset, the superior frontal gyrus (SFG-dor.L), orbital inferior frontal gyrus (ORBinf.L), and inferior parietal lobule (IPL.L) further highlight deficits in cognitive control, emotion regulation, and action planning. Overall, both datasets point to the frontal lobe central to ASD pathology.

For ADHD, identified regions include the amygdala (AMYG.R) for emotional dysregulation, inferior frontal gyrus (IFGoperc.R) for inhibitory control, precentral gyrus (PreCG.L) for hyperactivity, and cerebellum (CRBL7b.R, CRBL6.R) for motor coordination and attention regulation, aligning with ADHD symptoms.

Neural System-Level Analysis. For ASD, the ABIDE I dataset suggests disruptions in the visual network (VN), default mode network (DMN), and somatomotor network (SMN). The cuneus, middle frontal gyrus, and fusiform gyrus are within the VN which supports visual processing and higher-order visual functions. The precuneus and orbital inferior frontal gyrus are within the DMN which is essential for self-referential thinking and social cognition. The supplementary motor area (SMA) and cerebellum are within the SMN which could be responsible for repetitive behaviors in ASD. In the ABIDE II dataset, the SMN continues to be implicated through regions such as the putamen and rolandic operculum. The analysis further identifies disruptions in the cognitive control network (CCN) which includes the superior frontal gyrus and inferior parietal lobule and governs attention and executive functions. Collectively, findings from both datasets point to the CCN, DMN, and SMN as core neural systems disrupted in ASD.

For ADHD, the primarily affected neural systems include the CCN, bilateral limbic network (BLN), and SMN. The CCN, including the inferior frontal gyrus, precentral gyrus, and orbital superior gyrus, is crucial for attention regulation, inhibitory control. The BLN, which includes the amygdala, plays a central role in emotional regulation and impulsivity and highlights the emotional dysregulation characteristic of ADHD. The SMN, including the cerebellum and precentral gyrus, is responsible for motor coordination and control, reflecting the hyperactivity commonly observed in the disorder. Overall, ADHD is marked by disruptions in the CCN, SMN, and BLN, underscoring impairments in attention, emotion regulation, and motor control.

4 Conclusion

In this paper, we introduced a novel framework for disentangling disease-relevant patterns in fMRI analysis by combining equivariant learning and contrastive learning. The proposed method incorporates a group-theoretic perspective on graph augmentation. By distinguishing disease-relevant and irrelevant subgraphs through a mask, the framework effectively suppresses noise and amplifies disease-related features. Our model achieved state-of-the-art classification performance across 3 datasets, ABIDE I, ABIDE II, and ADHD-200. Ablation studies validate the critical roles of the mask and the equivariant learning structure in isolating and preserving disease-relevant patterns. Interpretability experiments identified disease-related brain regions that align with established findings in clinical research.

References

1. Smitha, K.A., Akhil Raja, K., Arun, K.M., et al.: Resting state fMRI: a review on methods in resting state connectivity analysis and resting state networks. Neuroradiol. J. **30**(4), 305–317 (2017)
2. Bassett, D., Sporns, O.: Network neuroscience. Nat. Neurosci. **20**, 353–364 (2017)
3. Bessadok, A., Mahjoub, M.A., Rekik, I.: Graph neural networks in network neuroscience. IEEE Trans. Pattern Anal. Mach. Intell. **45**(5), 5833–5848 (2022)
4. Canario, E., Chen, D., Biswal, B.: A review of resting-state fMRI and its use to examine psychiatric disorders. Psychoradiology **1**(1), 42–53 (2021)
5. Song, Y., Zhang, X., Wang, X.: Large margin local estimate with applications to medical image classification. IEEE Trans. Med. Imaging **34**(6), 1362–1377 (2015)
6. Xie, Y., Xu, Z., Zhang, J., Wang, Z., Ji, S.: Self-supervised learning of graph neural networks: a unified review. IEEE Trans. Pattern Anal. Mach. Intell. **45**(2), 2412–2429 (2022)
7. Ding, K., Xu, Z., Tong, H., Liu, H.: Data augmentation for deep graph learning: a survey. SIGKDD Explor. Newsl. **24**(2), 61–77 (2022)
8. You, Y., Chen, T., Sui, Y., Chen, T., Wang, Z., Shen, Y.: Graph contrastive learning with augmentations. Adv. Neural. Inf. Process. Syst. **33**, 5812–5823 (2020)
9. Sun, F.-Y., Hoffman, J., Verma, V., Tang, J.: InfoGraph: unsupervised and semi-supervised graph-level representation learning via mutual information maximization. In: Proceedings of the 8th International Conference on Learning Representations (ICLR). Addis Ababa, Ethiopia (2020)
10. Hassani, K., Khasahmadi, A.H.: Contrastive multi-view representation learning on graphs. In: Proceedings of the 37th International Conference on Machine Learning (ICML). Proceedings of Machine Learning Research, vol. 119, pp. 4116–4126. PMLR (2020)
11. Xie, Y., Xu, Z., Ji, S.: Self-supervised representation learning via latent graph prediction. In: Proceedings of the 39th International Conference on Machine Learning (ICML). Proceedings of Machine Learning Research, vol. 162, pp. 24460–24477. PMLR (2022)
12. Wei, C., Wang, Y., Bai, B., Ni, K., Brady, D., Fang, L.: Boosting graph contrastive learning via graph contrastive saliency. In: Proceedings of the 40th International Conference on Machine Learning (ICML). Proceedings of Machine Learning Research, vol. 202, pp. 36839–36855. PMLR (2023)
13. Chen, T., Kornblith, S., Norouzi, M., Hinton, G.: A simple framework for contrastive learning of visual representations. In: Proceedings of the 37th International Conference on Machine Learning (ICML). Proceedings of Machine Learning Research, vol. 119, pp. 1597–1607. PMLR (2020)

14. Zhang, H., Wu, Q., Yan, J., Wipf, D., Yu, P.S.: From canonical correlation analysis to self-supervised graph neural networks. Adv. Neural. Inf. Process. Syst. **34**, 76–89 (2021)
15. You, Y., Chen, T., Shen, Y., Wang, Z.: Graph contrastive learning automated. In: Proceedings of the 38th International Conference on Machine Learning (ICML). In: Proceedings of Machine Learning Research, vol. 139, pp. 12121–12132. PMLR (2021)
16. Wen, G., Cao, P., Liu, L., et al.: Graph self-supervised learning with application to brain networks analysis. IEEE J. Biomed. Health Inform. **27**(8), 4154–4165 (2023)
17. Zhang, S., Chen, X., Shen, X., et al.: A-GCL: adversarial graph contrastive learning for fMRI analysis to diagnose neurodevelopmental disorders. Med. Image Anal. **90**, 102932 (2023)
18. Peng, L., Wang, N., Xu, J., Zhu, X., Li, X.: GATE: graph CCA for temporal self-supervised learning for label-efficient fMRI analysis. IEEE Trans. Med. Imag. **42**(2), 391–402 (2022)
19. Luo, X., Dong, G., Wu, J., Beheshti, A., Yang, J., Xue, S.: An interpretable brain graph contrastive learning framework for brain disorder analysis. In: Proceedings of the 17th ACM International Conference on Web Search and Data Mining, pp. 1074–1077. Association for Computing Machinery, New York, NY, USA (2024)
20. Liu, S., Wang, S., Liang, B., Li, B., Xu, J.: Diagnosis of autism spectrum disorder based on contrastive functional connectivity graph learning network. In: ICASSP 2024 - 2024 IEEE International Conference on Acoustics, Speech and Signal Processing, pp. 12991-12995. IEEE (2024)
21. Defferrard, M., Bresson, X., Vandergheynst, P.: Convolutional neural networks on graphs with fast localized spectral filtering. Adv. Neural. Inf. Process. Syst. **29**, 3844–3852 (2016)
22. Kipf, T. N., Welling, M.: Semi-supervised classification with graph convolutional networks. arXiv preprint arXiv:1609.02907 (2016)
23. Veličković, P., Cucurull, G., Casanova, A., et al.: Graph attention networks. arXiv preprint arXiv:1710.10903 (2017)
24. Hamilton, W., Ying, Z., Leskovec, J.: Inductive representation learning on large graphs. Adv. Neural. Inf. Process. Syst. **30**, 1025–1035 (2017)
25. Xu, K., Hu, W., Leskovec, J., et al.: How powerful are graph neural networks? arXiv preprint arXiv:1810.00826 (2018)
26. Chen, S., Dobriban, E., Lee, J.H.: A group-theoretic framework for data augmentation. J. Mach. Learn. Res. **21**(245), 1–71 (2020)
27. Garrido, Q., Najman, L., Lecun, Y.: Self-supervised learning of split invariant equivariant representations. In: Krause, A., Brunskill, E., Cho, K., Engelhardt, B., Sabato, S., Scarlett, J. (eds.) Proceedings of the 40th International Conference on Machine Learning, vol. 202, pp. 10975-10996. PMLR (2023)
28. Suresh, S., Li, P., Hao, C., et al.: Adversarial graph augmentation to improve graph contrastive learning. Adv. Neural. Inf. Process. Syst. **34**, 15920–15933 (2021)
29. Fan, R.E., Chang, K.W., Hsieh, C.J., et al.: LIBLINEAR: a library for large linear classification. J. Mach. Learn. Res. **9**, 1871–1874 (2008)
30. Tzourio-Mazoyer, N., Landeau, B., Papathanassiou, D., et al.: Automated anatomical labeling of activations in SPM using a macroscopic anatomical parcellation of the MNI MRI single-subject brain. Neuroimage **15**(1), 273–289 (2002)
31. Esteban, O., Markiewicz, C.J., Blair, R.W., et al.: fMRIPrep: a robust preprocessing pipeline for functional MRI. Nat. Methods **16**(1), 111–116 (2019)

32. Yang, H., Long, X.Y., Yang, Y., et al.: Amplitude of low frequency fluctuation within visual areas revealed by resting-state functional MRI. Neuroimage **36**, 144–152 (2007)
33. Xia, M., Wang, J., He, Y.: BrainNet viewer: a network visualization tool for human brain connectomics. PLoS ONE **8**(7), e68910 (2013)
34. Lord, C., Brugha, T.S., Charman, T., et al.: Autism spectrum disorder. Nat. Rev. Dis. Primers. **6**(1), 1–23 (2020)
35. Schultz, R.T.: Developmental deficits in social perception in autism: the role of the amygdala and fusiform face area. Int. J. Dev. Neurosci. **23**, 125–141 (2005)
36. Padmanabhan, A., Lynch, C.J., Schaer, M., Menon, V.: The default mode network in autism. Biol. Psychiatry Cognit. Neurosci. Neuroimag. **2**(6), 476–486 (2017)
37. Seidman, L.J., Valera, E.M., Makris, N.: Structural brain imaging of attention-deficit/hyperactivity disorder. Biol. Psychiat. **57**(11), 1263–1272 (2005)
38. Swanson, J.M., Sergeant, J.A., Taylor, E., Sonuga-Barke, E., Jensen, P.S., Cantwell, D.P.: Attention-deficit hyperactivity disorder and hyperkinetic disorder. Lancet **351**(9100), 429–433 (1998)
39. Cai, W., Griffiths, K., Korgaonkar, M.S., et al.: Inhibition-related modulation of salience and frontoparietal networks predicts cognitive control ability and inattention symptoms in children with ADHD. Mol. Psychiatry **26**, 4016–4025 (2021)

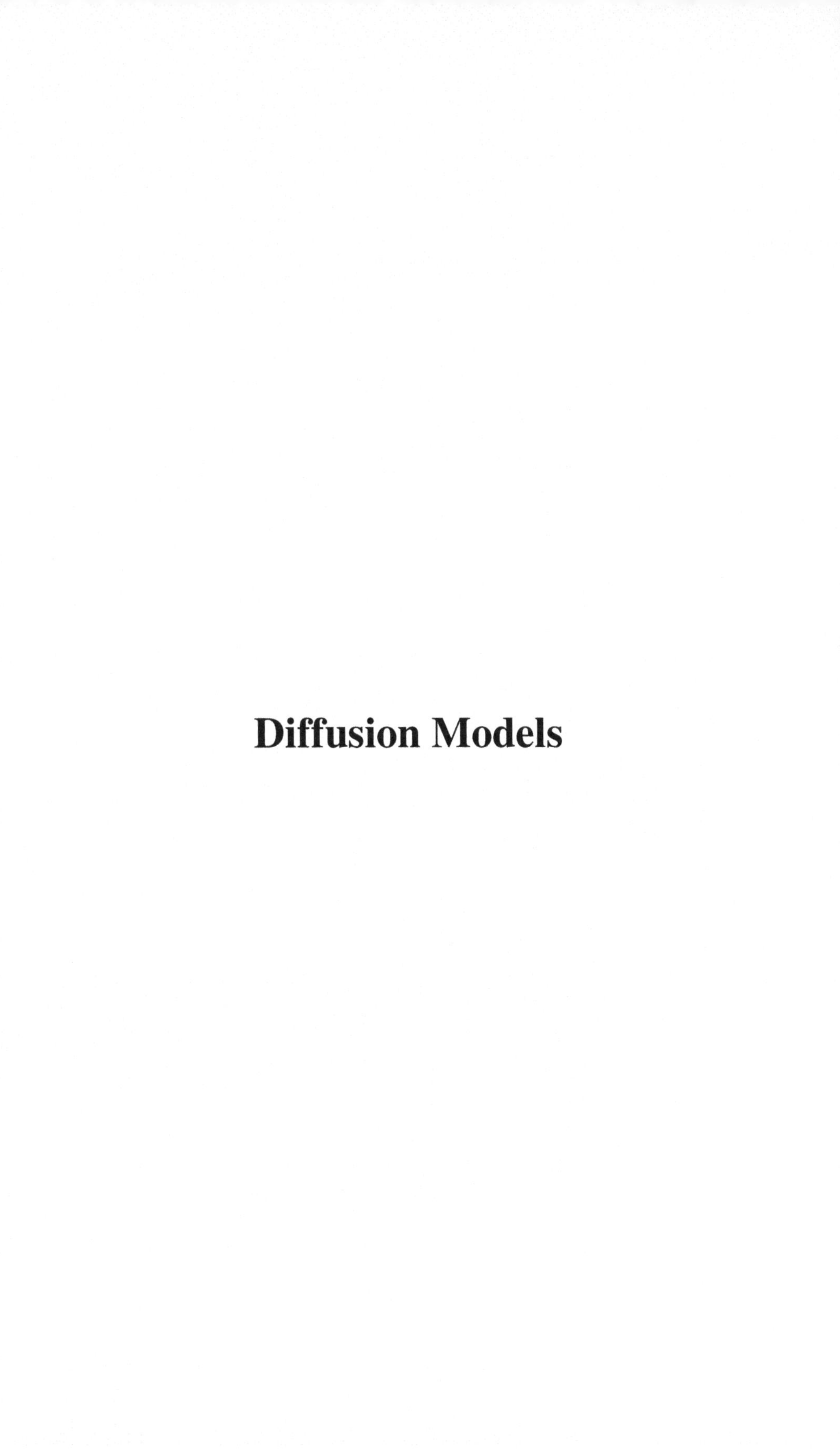

Diffusion Models

Continuous Diffusion Model for Self-supervised Denoising and Super-Resolution on Fluorescence Microscopy Images

Colin S. C. Tsang(✉) and Albert C. S. Chung

Department of Computer Science and Engineering,The Hong Kong University of Science and Technology, Clear Water Bay, Hong Kong
{sctsangab,achung}@cse.ust.hk

Abstract. Recent studies have shown that Joint Denoising and Super-Resolution (JDSR) approaches can produce high-quality medical images. However, obtaining ground truth images for training is often infeasible in fluorescence microscopy (FM). Moreover, current JDSR methods can be impractical as they only rely on a fixed scale for resolution. Existing methods aim to learn a direct mapping between low-quality input and high-quality output in pixel space. In this paper, we introduce the Continuous Diffusion Model. This novel self-supervised method iteratively retrieves a noise-free image in continuous space, enabling us to generate clean images at any arbitrary resolutions. Our quantitative analysis confirms the effectiveness of our proposed method. We outperformed both supervised and self-supervised methods by an average of 30.5%/7.4% in terms of RMSE/SSIM. Our output preserves fine details and structures better than other state-of-the-art methods, as demonstrated by the qualitative analysis. The source code is available at https://github.com/colinsctsang/ContinuousDiffusionModel.

Keywords: Super-resolution · Denoising · Self-supervised

1 Introduction

FM is a critical tool in the fields of biology and chemistry, enabling the visualization of biochemical processes at the molecular level. Despite its success in the field, FM faces several significant issues. First, the mechanism of FM unavoidably induces noise and restricted resolution [1], resulting in unsatisfactory image quality. Researchers have employed techniques such as Structure Illumination Microscopy (SIM) to obtain high-quality images [8]. However, SIM can be time-consuming and require high illumination doses that may damage or even kill cells [2]. Furthermore, the state-of-the-art methods for Super-Resolution (SR) typically have a fixed scale [21], which is very inconvenient for practical use, as cell analysis requires zooming in and out at different scales.

I. Oguz et al. (Eds.): IPMI 2025, LNCS 15830, pp. 111–124, 2026.
https://doi.org/10.1007/978-3-031-96625-5_8

In this paper, we introduce the Continuous Diffusion Model, a new approach for JDSR that eliminates the need for training data and scale limitations. Our contributions are three-fold: (1) Since our model does not require ground truth training data, our approach can avoid any risk of damaging the cell samples. (2) Our model operates in a continuous space, enabling JDSR at arbitrary scales and reducing computational costs. (3) We demonstrate superior performance compared to other state-of-the-art methods on two publicly available datasets.

2 Related Work

JDSR is a promising approach to producing high-quality FM images. Zhou *et al.* utilized the supervised approach for FM images using SIM ground truth [23]. Tsang *et al.* employed multiple noisy captures to perform weakly-supervised JDSR on FM images [17]. However, obtaining multiple captures or SIM images could kill the cells. Also, these methods aim to learn a mapping with a fixed scale, which is a significant drawback. Conversely, our model learns to retrieve a high-quality image in continuous space iteratively.

Another avenue of research is implicit neural representation. Chen *et al.* proposed an implicit function as a latent representation of an input image [4]. Rombach *et al.* adopt this approach to develop a Latent Diffusion Model (LDM) [14]. However, there is no upscale component in the LDM. In contrast, our continuous approach specifically caters to the arbitrary scale super-resolution task, making it one of our major contributions. The ability to achieve flexible upscale is particularly valuable in FM applications.

Also, Gao *et al.* introduced the Implicit Diffusion Model (IDM) [7]. However, unlike our self-supervised approach, IDM is trained in a supervised manner for natural images. This is important since ground truth training images are often unavailable in FM. Hence, IDM cannot be applied to our FM application. Moreover, our model's entire structure differs from IDM. While IDM alternates between implicit representation and denoising model, our method solely applies the denoising model in the continuous space.

Recently, the diffusion model has shown impressive effectiveness in image generation, prompting some researchers to adapt it as a denoiser. One simple approach is to utilize the backward sampling process directly as a denoiser [10]. On the other hand, Xiang *et al.* developed a diffusion model for MRI denoising [22]. This method estimates the noise level in an input image using a blind-spot model and maps this noise level to a pre-defined noise schedule for denoising. However, neither of these methods includes an upscaling module, making them unsuitable for our JDSR task. Moreover, all of these approaches are performed in the pixel space. Hence, the computational cost is prohibitive [5].

3 The Diffusion Model

We follow the basic ideas mentioned in [6,9,16] to derive a diffusion model.

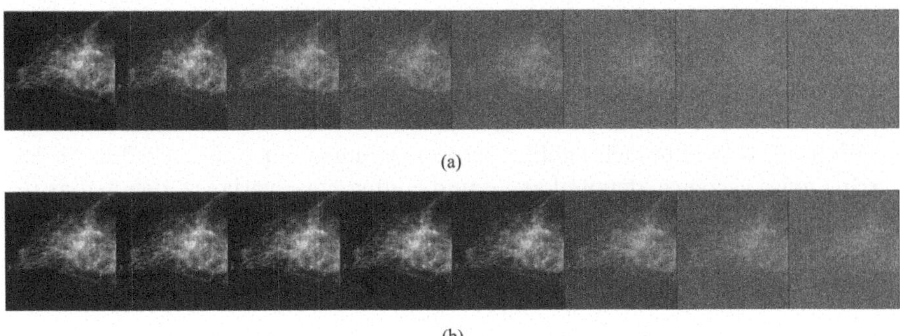

Fig. 1. Comparison of (a) the linear noise schedule and (b) our slow-growth noise schedule. The total time step is $T = 1000$ for illustration purposes. We have $t = 1, 150, 300, 450, 600, 750, 900, 1000$ from left to right. For (b) the slow-growth noise schedule, the noise level remains at the starting level until $t = 500$ and then increases linearly.

3.1 Forward Diffusion Process

Suppose x_0 is our original input image. Let x_t indicate the image at the time step t, where $t \in \{1, 2, .., T\}$. In the forward diffusion process, we define

$$q(x_t|x_{t-1}) = \mathcal{N}(x_t; x_{t-1}\sqrt{1-\beta_t}, \beta_t I), \tag{1}$$

where $\{\beta_1, ..., \beta_T\}$ is a noise schedule. In practice, it does not need to add noise for t time steps to obtain the image x_t. Let $\alpha_t = 1 - \beta_t$ and $\bar{\alpha}_t = \prod_{s=1}^{n} \alpha_s$. In this case, Eq. 1 can be rewritten as

$$q(x_t|x_{t-1}) = x_0\sqrt{\bar{\alpha}_t} + \epsilon\sqrt{1-\bar{\alpha}_t}, \tag{2}$$

where $\epsilon \sim \mathcal{N}(0, I)$. Therefore, with a predefined noise schedule $\{\beta_1, ..., \beta_T\}$, we can obtain x_t directly from x_0 using Eq. 2.

3.2 Backward Sampling Process

Following the derivation in [9], we want to predict the noise at time step t during the backward sampling process. Thus, we have a loss function in pixel space for some model parameters θ:

$$\mathcal{L}_{px} = ||\epsilon - \epsilon_\theta(x_t, t)||^2. \tag{3}$$

3.3 Slow-Growth Noise Schedule

Typical diffusion models utilize noise schedules $\{\beta_1, ..., \beta_T\}$ that are designed for image generation tasks. However, the noise levels in FM images are mild compared to those used in image generation tasks. To address this issue, we propose a slow-growth noise schedule. This schedule allows the noise variance parameter, denoted as β_t, to remain constant for the first half of the total time steps and then increase linearly to the final time step.

Adopting this approach enables us to customize our noise schedule to any starting noise variance, denoted as β_1, and ending noise variance, denoted as β_T, resulting in a slow-growth noise schedule that is well-suited for FM images. By using this noise schedule, we can effectively balance the need to remove noise from the images with the need to preserve the underlying details of the cells in the images, resulting in high-quality, super-resolved images. The slow-growth noise schedule is given by

$$\beta_t = \begin{cases} \beta_1, & t = 1, 2, ..., \frac{T}{2} \\ \beta_1 + \frac{\beta_T - \beta_1}{T/2}(t - \frac{T}{2}), & t = \frac{T}{2} + 1, ..., T. \end{cases} \quad (4)$$

Figure 1 illustrates the difference between our slow-growth noise schedule and the typical linear noise schedule.

4 Continuous Diffusion Model

The operation of diffusion models in pixel space, as mentioned in Sect. 3, is known to be computationally expensive [5]. Also, this approach is not suitable for our JDSR task, as the diffusion model does not include an upscaling module. Note that adding an upscaling layer at the decoder would only enable the model to perform super-resolution in a fixed scale, which is not ideal in FM images.

In this section, we suggest a novel approach that makes use of continuous space, allowing us to produce high-resolution images at any scale. This approach not only reduces computational costs but also provides greater flexibility to the model, making it well-suited for the JDSR task in FM applications.

We delve into the loss function within a continuous space in Sect. 4.1. Following this, we elaborate on the encoder E_φ and the decoder D_ω in Sect. 4.2. Subsequently, the self-supervised training procedure of our model is outlined in Sect. 4.3.

4.1 The Loss Function in Continuous Space

We propose to execute the forward diffusion process and the backward sampling process, mentioned in Sect. 3.1 and Sect. 3.2 in a continuous space. In this way, $z_t = E_\varphi(x_t)$ is used to represent the image in the continuous space, where E_φ is an encoder with parameters φ. After both processes, we use D_ω to output a pixel-based image with an arbitrary scale, where D_ω is a decoder with parameters ω. Therefore, Eq. 3 can be rewritten as a loss function in the continuous space:

$$\mathcal{L}_{cont} = ||E_\varphi(\epsilon) - \epsilon_\theta(z_t, t)||^2. \quad (5)$$

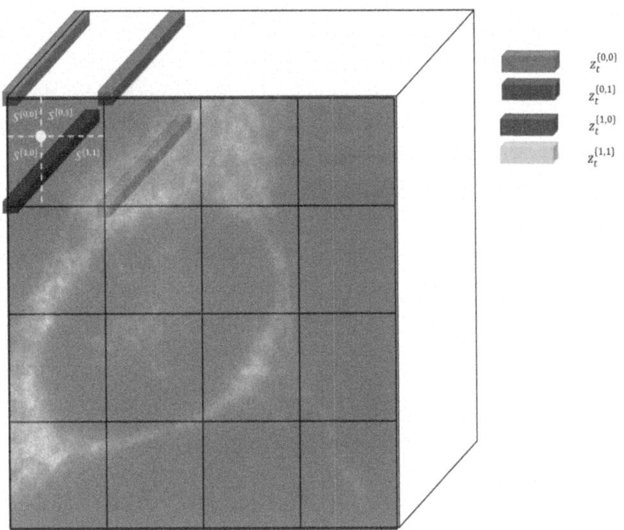

Fig. 2. This figure demonstrates a tensor z_t with dimensions $5 \times 5 \times \hat{C}$. The red dot indicates the input pixel coordinates (u_i, v_j). The cubes represent four feature vectors with different colours. The predicted pixel value is generated by an MLP M_ω and weighted by the area ratio, as shown in Eq. 7.

4.2 Encoder E_φ and Decoder D_ω

As discussed in Sect. 4.1, we use an encoder E_φ and a decoder D_ω to translate the image between pixel space and continuous space. Suppose we have an image $x_t \in \mathbb{R}^{H \times W \times C}$, where H, W, and C denote the height, width, and number of color channels in the image, respectively.

We construct the continuous space representation $z_t \in \mathbb{R}^{\hat{H} \times \hat{W} \times \hat{C}}$ by applying an encoder E_φ to the image x_t. Here, $\hat{H}$, $\hat{W}$, and $\hat{C}$ denote the height, width, and number of channels (or depths) in the encoded tensor, respectively. We have

$$z_t = E_\varphi(x_t), \quad \text{where} \quad z_t \in \mathbb{R}^{\hat{H} \times \hat{W} \times \hat{C}}. \tag{6}$$

Therefore, the output 3D tensor z_t contains $\hat{H} \times \hat{W}$ feature vectors in $\mathbb{R}^{\hat{C}}$. We consider this tensor z_t to be the continuous space representation of the original image x_t. The deep latent code stored in the corresponding feature vectors should provide rich information for recovering pixel values in a continuous manner.

To generate a pixel value from z_t, we have:

$$\rho = \sum_{k \in \{00,01,10,11\}} \frac{S^{\{k\}}}{S} M_\omega\Big((u_i, v_j), z_t^{\{k\}}\Big), \tag{7}$$

where (u_i, v_j) denotes the input coordinate. Note that z_t^k refers to the four feature vectors closest to (u_i, v_j) in Euclidean distance, which are located at the top-left, top-right, bottom-left, and bottom-right corners of the grid cell. The function M_ω is an MLP that outputs a predicted value. The predicted pixel value ρ at (u_i, v_j) is weighted by the area in the grid cell. Specifically, S^k denotes the four areas from (u_i, v_j) to the four closest feature vectors, and S represents the total area of the grid cell. Figure 2 illustrates the process of generating a pixel value from the continuous space representation z_t.

In traditional interpolation methods, we map the surrounding pixels to predict an output pixel value. By assuming that neighboring pixels exhibit some level of correlation or similarity, a reliable prediction can be made even in the absence of ground truth information. Similarly, Eq. 7 can be viewed as a function that maps the surrounding feature vectors to a pixel value. These feature vectors are expected to encapsulate sufficient information to facilitate accurate prediction of pixel values in the upscaled image. Therefore, it is possible to reconstruct an output image in the desired scale by inputting all of the coordinates (u_i, v_j) and using Eq. 7 to generate the corresponding pixel values.

In this paper, the coordinates are normalised to $(u, v) \in [-1, 1] \times [-1, 1]$. Therefore, it is easy to create a coordinate grid according to the desired resolution. For any target resolution $H \times W$, the corresponding coordinate grid is given by the following formula:

$$u_i = -1 + \frac{2}{H-1} i, \quad \text{and} \quad v_j = -1 + \frac{2}{W-1} j, \tag{8}$$

where $i = 0, 1, ..., H-1$ and $j = 0, 1, ..., W-1$. For simplicity, we denote the process of generating an output image by Eq. 7 as a decoder D_ω:

$$\tilde{x}_t = D_\omega\Big((u, v), z_t\Big), \tag{9}$$

where $\tilde{x}_t$ is an output image and (u, v) is a coordinate grid that contains all (u_i, v_j).

4.3 Self-supervised Training of the Continuous Diffusion Model

The general framework of our proposed training process is shown in Fig. 3. In the training process, the input image x_0 is divided into two branches. The upper branch is just the forward diffusion process to obtain $\{x_0, ..., x_T\}$ with Eq. 2. In the lower branch, we obtain a low-resolution image $\hat{x}_0$ by applying a random downsampling with scale s. By Eq. 2 once again, we can obtain $\{z_0, ..., z_T\}$. The continuous representation is given by $z_t = E_\varphi(\hat{x}_t)$. Next, we sample an integer $t \in [1, T]$ uniformly. A U-Net model U_θ [15] is used to perform the backward sampling process, that is, to predict noise at the time step t and recover $\tilde{z}_t$. The predicted $\tilde{z}_t$ is compared in two loss functions to train our model. First, we have the continuous space loss function in Eq. 5 to compare $\tilde{z}_t$ and z_t. Second, we reconstruct $\tilde{x}_t$ by Eq. 9. In this case, the input coordinates (u, v) is the coordinate

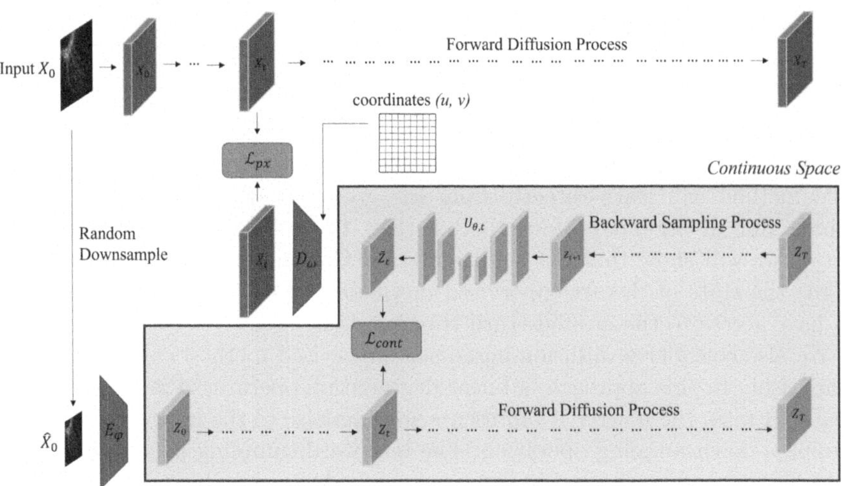

Fig. 3. The training process of the Continuous Diffusion Model for JDSR.

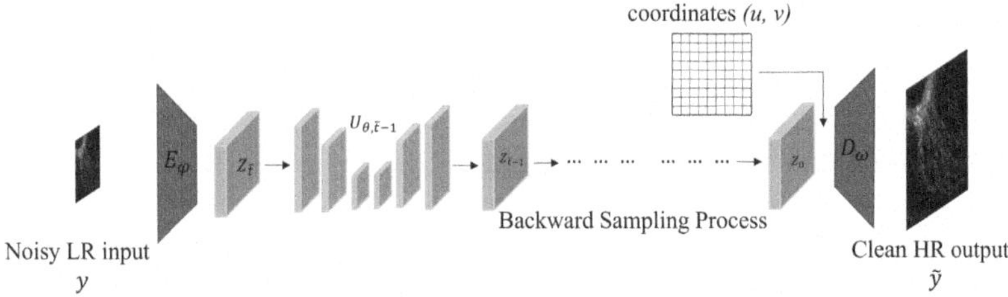

Fig. 4. The sampling process to generate a high-resolution clean image by our Continuous Diffusion Model.

grid of the input image x_0. Hence, the output $\tilde{x}_t$ is a prediction of x_t. They are compared using the loss function described in Eq. 3.

Figure 4 demonstrated the sampling process of our proposed method. For the sampling process, suppose that we have a low-resolution noisy input image y. We select a time step $\bar{t}$ as the starting state of the backward sampling process and generate the continuous representation $z_{\bar{t}} = E_{\varphi}(y)$. Then, we can obtain $\{z_{\bar{t}}, ..., z_0\}$ iteratively using our model U_θ. Lastly, we obtain a clean high resolution output $\tilde{y}$ with an arbitrary resolution by Eq. 9.

Note that both the training and sampling processes only have access to the input image x_0. As a self-supervised method, we do not have access to the ground truth SIM image.

5 Experiments

5.1 State-of-the-Art Methods

For the JDSR task in medical images, supervised methods such as GAN-based and RCAN-based architectures have shown the best results [23]. As the supervised method can learn directly from the ground-truth SIM image, one may assume that it should perform better than the self-supervised approach. In Sect. 6, we will show that our self-supervised Continuous Diffusion Model outweighs the state-of-the-art supervised methods, even though our method does not have access to the ground-truth training data.

We also consider a diffusion-based self-supervised method called DDNM for comparison. In this approach, a linear degradation operator A is pre-defined. In the JDSR task, the matrix A represents adding noise to the image and then performing a downsampling operation. The backward sampling process is updated with the information from A in the pixel space. More details about this method can be found in [20].

5.2 Dataset

We have evaluated our method using a publicly available dataset W2S [23] consisting of 360 FM image sets. All image sets contain human cell images captured by widefield imaging. We randomly selected 240 sets and 120 sets as training and test datasets. In each set, there were 400 noisy captures of the same scene with a resolution of 512×512 pixels. We generated a single capture x_0^i, where $i = 1, ..., 360$, by randomly selecting η noisy captures and taking the average.

Since the average of multiple noisy captures can provide a high-quality image with a lower noise level, different noise levels can be generated with different η. Thus, a small η corresponds to an image with a high noisy level. We assume that an image is noise-free if $\eta = 400$. On the other hand, a high-resolution clean image with a resolution of 1024×1024 is captured by SIM for each set. Our task is to recover a high-resolution SIM image from a low-resolution, noisy, widefield image. Note that the noise-free images and the SIM images are used only for evaluation purposes. Our self-supervised method does not have access to these ground-truth images.

We have also evaluated the transferability of our method by applying it to a different publicly available dataset, FNC [13]. This dataset consists of 283 fluorescent microscopy images of brain cells in mice. The image resolution is 1600×1200 for all images. We center-cropped every image to 1024×1024 for simplicity. The FNC dataset was originally developed for the cell counting task. In this paper, we downsampled the image by interpolation with a scale of $s = 4$ and added Gaussian noise to generate a low-resolution noisy input image. Our task is to recover the high-resolution clean image from it.

Table 1. Quantitative comparison of different methods. We present the average RMSE/SSIM values. A smaller η corresponds to a higher noise level. The upscale factor is $s = 2$. The t-tests performed in Table 2 indicate that our findings are statistically significant.

Approach	Method	W2S dataset				FNC dataset
		$\eta = 1$	$\eta = 2$	$\eta = 4$	$\eta = 8$	
Supervised	W2S-RCAN	0.343/0.737	0.331/0.769	0.313/0.798	0.298/0.819	0.313/0.691
	W2S-GAN	0.346/0.769	0.296/0.776	0.306/0.783	0.301/0.788	0.102/0.862
Self-supervised	DDNM	0.381/0.740	0.331/0.757	0.293/0.774	0.266/0.787	0.102/0.869
	Ours	**0.290/0.811**	**0.251/0.823**	**0.224/0.839**	**0.207/0.855**	**0.098/0.897**

Table 2. The t-tests of RMSE/SSIM of our model against other methods. The t-tests were performed on a combination of all noise levels (i.e., $\eta = 1, 2, 4, 8$). The size of the dataset was 120 for each noise level, resulting in a total degree of freedom of 958 for all t-tests.

Approach	Method	Mean	S.D.	t-value	p-value
Supervised	W2S-RCAN	0.322/0.781	0.090/0.156	15.8/5.87	$<10^{-4}$
	W2S-GAN	0.312/0.779	0.082/0.122	14.6/7.10	$< 10^{-4}$
Self-supervised	DDNM	0.318/0.765	0.084/0.125	15.6/8.93	$<10^{-4}$
	Ours	0.243/0.832	0.063/0.109	-	-

5.3 Implementation Details

We use standard metrics such as Root Mean Squared Error (RMSE) and Structural Similarity Index Measure (SSIM) [19]. These metrics provide a quantitative analysis for fair comparisons, as used in previous studies [23].

For a fair comparison, we followed [20] to use a standard version of U-Net [15] with a depth of 2 and transformer sinusoidal position embedding [18] as the model architecture of U_θ. In the forward diffusion process, we chose $\beta_1 = 50^{-5}$ and $\beta_T = 10^{-2}$. The total time step was $T = 500$, and the starting time step in the sampling process was $\bar{t} = 200$. The choice of these parameters was based on the simple trial-and-error and could be further optimized in future studies. We used the EDSR-baseline [12] and the local implicit image function [4] as our encoder E_φ. The decoder D_ω was a 5-layer MLP with 256 hidden dimensions and ReLU activation. All models were trained with 1000 epochs with the exponential moving average in PyTorch. For each training batch, the input image was randomly cropped into a 256×256 image patch. We performed both horizontal and vertical random flips with a probability of 0.5. The low-resolution noisy input image was generated by bicubic downsampling with a scale s, where s was randomly selected from $\{2, 4, 8\}$ uniformly for each training batch. We used Adam optimiser [11] with learning rate 10^{-4} and weight decay 10^{-2}. The loss functions in Eq. 3 and 5 were added together with equal weight. All experiments in this paper were carried out on GPU: NVIDIA GeForce RTX 4080 with 16 GB memory and CPU: Intel Core i7-12700F.

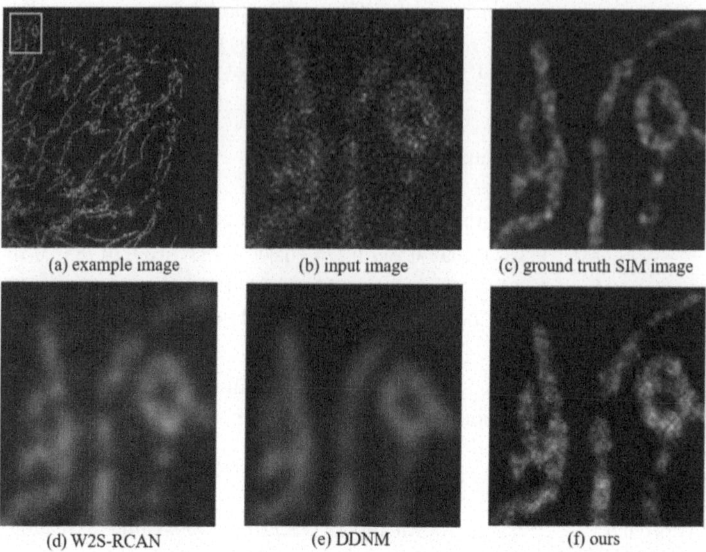

Fig. 5. The qualitative results of different methods in the W2S dataset with a scale of $s = 2$ and a noise level of $\eta = 1$. The example image is shown in (a), with a zoomed-in area indicated by the red box for subsequent images (b)-(f). W2S-GAN is omitted to save space as it is similar to W2S-RCAN. (Color figure online)

6 Results

6.1 Quantitative Analysis

Table 1 compares our model to the state-of-the-art methods. Our Continuous Diffusion Model outperforms the other methods consistently at all noise levels. Our model achieves an average RMSE of 0.243 and SSIM of 0.832, with improvements of 30.5% in RMSE and 7.4% in SSIM. T-tests in Table 2 confirm the statistical significance of our results ($p < 10^{-4}$).

The DDNM method attempts to sample an output image in pixel space while using the degradation matrix as guidance. In contrast, our approach iteratively recovers a noise-free image in the continuous space. Our method surpasses the DDNM method across all noise levels, with an average improvement of 30.7% in RMSE and 8.8% in SSIM. These results demonstrate the unique advantage of working in continuous space rather than in pixel space.

For the FNC dataset, our method achieves the best quantitative results without re-training for different upscaling factors, demonstrating robustness across various FM images and modalities. We outperform other methods with improvements of 75.9% in RMSE and 11.1% in SSIM. While our method achieves the best performance, the other methods exhibit similar performance, with the exception of the W2S-RCAN approach. This finding suggests that supervised methods

Table 3. Comparison of RMSE/SSIM of our method and the pixel space version in the W2S dataset. The experiment is done on the medium noise level $\eta = 4$.

Method	RMSE/SSIM	GPU memory
Ours	0.224/0.839	~12 GB
Pixel space + RCAN	0.449/0.726	~30 GB
Pixel space + GAN	0.430/0.803	~30 GB

could be unreliable when applied to a different dataset, as they learn a direct mapping between the input and ground truth, which makes them less adaptable to new data.

6.2 Qualitative Analysis

An example of low-resolution noisy input and ground truth SIM image is shown in Image (b) and Image (c) of Fig. 5, respectively. We compare the performance of our Continuous Diffusion Model with that of W2S-RCAN and DDNM.

As shown in Image (d) of Fig. 5, the W2S-RCAN method produces a denoised image that is visibly smoother than the input image. However, this improvement in denoising comes at the cost of significant blurring. The output of DDNM, shown in Image (e), provides a similar but even smoother result compared to the W2S-RCAN method. Both methods produce smooth output images but at the expense of lost structural details. In contrast, our Continuous Diffusion Model denoises the image in continuous space, which allows for the preservation of fine details and structures. This continuous-space approach not only speeds up the diffusion process but also results in a higher-quality image that is closer to the SIM ground truth, as shown in Image (f) of Fig. 5.

Furthermore, it is worth noting that our approach is capable of producing results at arbitrary scales and does not rely on ground truth data during training. This makes it a highly flexible and practical solution for real-world applications.

6.3 Ablation Study

In this section, we have performed an ablation study on different parts of our model. First, we provide justification for the use of continuous space. To do so, we run the training and sampling processes (i.e., Figs. 3 and 4) in the pixel space by removing the encoder E_φ and decoder D_ω. Note that our model cannot perform super-resolution if the continuous space component is removed. Therefore, we applied some state-of-the-art upscale methods as post-processing steps to obtain a JDSR result. We used RCAN and GAN again for a fair comparison. On the other hand, training our model in pixel space would require a GPU with a memory of around 30 GB. Our continuous space approach can reduce the memory requirement to around 12 GB. We need to use one more GPU with parallel computing to perform the training in pixel space in these experiments.

Table 4. Performance of different noise schedule on the W2S dataset. The result is the medium noise level $\eta = 4$ with an upscale factor of $s = 2$.

Noise Schedule	RMSE/SSIM
Slow-growth	0.224/0.839
linear	0.260/0.810
quadratic	0.280/0.802
cosine	0.267/0.815

In Table 3, we compare the performance of our proposed model with the pixel space version of our model at $s = 2$. Table 3 demonstrates that our continuous approach not only produces output at arbitrary scales but also yields higher quality results.

Second, we have validated the use of the slow-growth noise schedule in our proposed approach empirically. To compare its performance with other common noise schedules such as linear, quadratic, and cosine [3], we have conducted the experiments and reported the results. Our empirical results demonstrate that while all of these noise schedules can provide a satisfactory result, our proposed noise schedule can yield the best overall performance. On average, our proposed scheme offers a 20.1%/3.7% performance boost in terms of RMSE/SSIM compared to the other standard noise schedules. These results suggest that the slow-growth noise schedule is well-suited for the JDSR task in FM applications and can effectively balance the need to remove noise from the images with the need to preserve the underlying details of the cells in the images. Moreover, we notice that the result with other standard noise schedules is still better than all state-of-the-art methods. It further validates the robustness of our proposed Continuous Diffusion Model (Table 4).

7 Conclusion

We have introduced a novel Continuous Diffusion Model for the JDSR task, which operates in the continuous space rather than the pixel space. By using this approach, it becomes possible to perform diffusion-based JDSR with a much lower GPU memory requirement. In addition, this method provides the ability to generate images at any desired resolution. As a self-supervised method, it provides researchers with access to high-quality FM images without the need for SIM or multiple exposures, thereby avoiding any potential harm to sample cells. Our method has demonstrated robust performance across different noise levels and datasets, and we expect it to be easily adopted in FM, where acquiring ground truth data is often infeasible. The simplicity and general applicability of our approach make it a promising tool for advancing research in this field.

References

1. Belthangady, C., Royer, L.A.: Applications, promises, and pitfalls of deep learning for fluorescence image reconstruction. Nat. Methods **16**(12), 1215–1225 (2019)
2. Chakrova, N., Canton, A.S., Danelon, C., Stallinga, S., Rieger, B.: Adaptive illumination reduces photo-bleaching in structured illumination microscopy. Biomed. Opt. Express **7**(10), 4263–4274 (2016)
3. Chen, T.: On the importance of noise scheduling for diffusion models. In: arXiv reprint. p. arXiv:2301.10972 (2023)
4. Chen, Y., Liu, S., Wang, X.: Learning continuous image representation with local implicit image function. In: Proceedings of the of IEEE/CVF Conference on Computer Vision and Pattern Recognition, pp. 8628–8638 (2021)
5. Croitoru, F.A., Hondru, V., Ionescu, R.T., Shah, M.: Diffusion models in vision: a survey. In: IEEE Trans. Pattern Anal. Mach. Intell. **45**, 10850–15869 (2023)
6. Dhariwal, P., Nichol, A.: Diffusion models beat GANs on image synthesis. Adv. Neural Inf. Process. Syst. **34**, 8780–8794 (2021)
7. Gao, S., et al.: Implicit diffusion models for continuous super-resolution. In: Proceedings of IEEE/CVF Conference on Computer Vision and Pattern Recognition, pp. 10021–10030 (2023)
8. Gustafsson, M.G.: Surpassing the lateral resolution limit by a factor of two using structured illumination microscopy. J. Microsc. **198**(2), 82–87 (2000)
9. Ho, J., Jain, A., Abbeel, P.: Denoising diffusion probabilistic models. Adv. Neural Inf. Process. Syst. **33**, 6840–6851 (2020)
10. Hu, D., Tao, Y.K., Oguz, I.: Unsupervised denoising of retinal oct with diffusion probabilistic model. Med. Imag. Image Process. **12032**, 25–34 (2022)
11. Kingma, D.P., Ba, J.: Adam: a method for stochastic optimization. In: International Conference on Learning Representations (2015)
12. Lim, B., Son, S., Kim, H., Nah, S., Mu Lee, K.: Enhanced deep residual networks for single image super-resolution. In: Proceedings of the of IEEE/CVF Conference on Computer Vision and Pattern Recognition, pp. 136–144 (2017)
13. Morelli, R., et al.: Automating cell counting in fluorescent microscopy through deep learning with C-RESUNET **11**(1), 22920 (2021)
14. Rombach, R., Blattmann, A., Lorenz, D., Esser, P., Ommer, B.: High-resolution image synthesis with latent diffusion models. In: Proceedings of the of the IEEE/CVF International Conference on Computer Vision (2022)
15. Ronneberger, O., Fischer, P., Brox, T.: U-Net: convolutional networks for biomedical image segmentation. In: International Conference on Medical Image Computing and Computer-Assisted Intervention, pp. 234–241 (2015)
16. Sohl-Dickstein, J., Weiss, E., Maheswaranathan, N., Ganguli, S.: Deep unsupervised learning using nonequilibrium thermodynamics. In: International Conference on Machine Learning, pp. 2256–2265 (2015)
17. Tsang, C.S., Mok, T.C., Chung, A.C.: Joint denoising and super-resolution for fluorescence microscopy using weakly-supervised deep learning. In: Medical Optical Imaging and Virtual Microscopy Image Analysis: First International Workshop, pp. 32–41 (2022)
18. Vaswani, A., et al.: Attention is all you need. In: Advances in Neural Information Processing System, pp. 5998–6008 (2017)
19. Wang, X., Bovik, A., Sheikh, H., Simoncelli, E.: Image quality assessment: from error visibility to structural similarity. IEEE Trans. Image Process. **13**, 600–612 (2004)

20. Wang, Y., Yu, J., Zhang, J.: Zero-shot image restoration using denoising diffusion null-space model. arXiv:2212.00490 (2022)
21. Wang, Z., Chen, J., Hoi, S.: Deep learning for image super-resolution: a survey. IEEE Trans. Pattern Anal. Mach. Intell. **43**, 3365–3387 (2020)
22. Xiang, T., Yurt, M., Syed, A.B., Setsompop, K., Chaudhari, A.: DDM ^{2}: self-supervised diffusion MRI denoising with generative diffusion models. In: The International Conference on Learning Representations (2023)
23. Zhou, R., El Helou, M., Sage, D., Laroche, T., Seitz, A., Süsstrunk, S.: W2s: microscopy data with joint denoising and super-resolution for wide-field to sim mapping. In: European Conference on Computer Vision, pp. 474–491 (2020)

Self-supervised Denoising of Diffusion MRI Data with Efficient Collaborative Diffusion Model

Xiaoyu Bai, Haotian Jiang, and Geng Chen(✉)

National Engineering Laboratory for Integrated Aero-Space-Ground-Ocean Big Data Application Technology, School of Computer Science and Engineering, Northwestern Polytechnical University, Xi'an, Shaanxi, China
geng.chen@ieee.org

Abstract. Diffusion MRI (dMRI) suffers from heavy noise, which undermines the accuracy and reliability of the subsequent quantitative analysis. Traditional deep learning denoising methods typically depend on training with paired noisy and clean data, which are unavailable in practice. Self-supervised techniques, such as DDM^2, overcomes this limitation with the diffusion model. However, DDM^2 is plagued by high computational cost and unsatisfactory performance when dealing with heavy noise. To tackle these challenges, we propose a novel self-supervised dMRI denoising model, called Efficient Collaborative Diffusion Model (ECDM). Specifically, we first employ a Noise2Noise-like method to obtain coarse denoised dMRI data. Subsequently, we use a latent encoder to compress the coarse data into a highly compact latent space. A diffusion model is then trained within this latent space to generate prior features. These features are passed to the denoising network through a hierarchical architecture and a cross-attention component for collaborative fine noise reduction. Our method not only achieves effective noise reduction with a collaborative coarse-to-fine framework but also enhances the efficiency of the diffusion model by utilizing the compact latent representation. Extensive experiments on both simulated and real datasets demonstrate that ECDM surpasses existing dMRI denoising methods remarkably.

Keywords: Diffusion MRI · Denoising · Self-supervised Learning · Latent Space · Diffusion Model

1 Introduction

Diffusion MRI (dMRI) is a powerful imaging technique capable of probing the diffusion pattern of water molecules in vivo [20]. This capability is essential for

---

X. Bai and H. Jiang—Contributed equally to this work.

This work was supported in part by the National Natural Science Foundation of China (No. 61540047) and the Practice and Innovation Funds for Graduate Students of Northwestern Polytechnical University (No. PF2024012).

I. Oguz et al. (Eds.): IPMI 2025, LNCS 15830, pp. 125–138, 2026.
https://doi.org/10.1007/978-3-031-96625-5_9

understanding brain white matter structure, providing valuable information for the diagnosis of brain diseases. However, dMRI data are corrupted with heavy noise, resulting in a lower Signal-to-Noise Ratio (SNR) that hinders subsequent quantitative analysis. Therefore, it is imperative to enhance the SNR of dMRI data through effective denoising techniques.

Before the advent of deep learning, a variety of traditional denoising methods were proposed. These methods are based on different techniques, such as total variation minimization [15], cosine transform filtering [17], Principal Component Analysis (PCA) [27], non-local means [2,3], etc. With the rise of deep learning, traditional methods began to be outperformed due to the superior ability to learn feature representations directly from data. As a result, deep learning techniques have been increasingly applied to dMRI denoising. Initially, supervised denoising methods [5] were developed. However, the utility of supervised methods is limited by the availability of such paired data, which is not always accessible in many real-world scenarios, especially in the medical field. To address this challenge, researchers have proposed numerous unsupervised and self-supervised methods [7,25], which significantly reduce the necessity for paired data.

Recently, Diffusion Models (DMs) have emerged as powerful probabilistic generative models, surpassing the state-of-the-art capabilities of Generative Adversarial Networks (GANs) [9]. This has been especially evident with the introduction of the Denoising Diffusion Probabilistic Model (DDPM) [11] and its derivatives [12,24], which have achieved impressive results across various fields, including image super-resolution [23] and beyond [1,14,16,19]. They also demonstrate significant potential in restoration and denoising tasks. Xiang *et al.* were the first to propose a three-stage self-supervised diffusion model, termed DDM^2 [28], specifically for denoising dMRI data. This method begins with the first stage, where a denoising model is trained to generate an approximation of clean data. In the second stage, it estimates the noise distribution of the input noisy data by leveraging the outcomes from the first stage. Finally, in the third stage, DDM^2 utilizes the knowledge acquired from the previous two stages and trains a diffusion model to obtain the denoised data.

Although DDM^2 shows great potential in dMRI denoising, it encounters several limitations. *(i) Over-Smoothing*: We observed an over-smoothing issue in the denoised data provided by the first stage. This may hinder the accurate estimation of diffusion steps, eventually leading to unsatisfactory performance. *(ii) Noneffective Information Integration*: DDM^2 is unable to effectively integrate information from both the diffusion model and regression-based self-supervised denoising methods, simultaneously. This separation restricts the model's comprehensive advantages during processing. *(iii) Heavy Computational Cost*: The training and testing of the diffusion model typically demand significant computational time. This inherent challenge not only escalates computing costs but may also constrain the model's real-time performance and efficiency in practical applications.

To address these limitations, we propose a novel self-supervised denoising model, called Efficient Collaborative Diffusion Model (ECDM). Specifically,

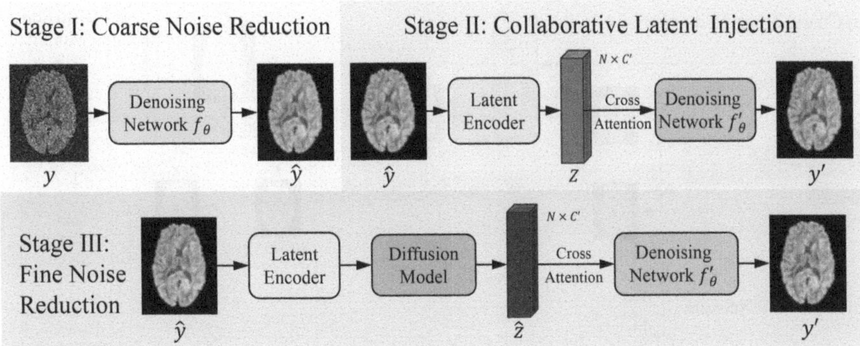

Fig. 1. Overview of ECDM. *Stage I Coarse Noise Reduction:* We use a Noise2Noise-like method for self-supervised denoising. *Stage II Collaborative Latent Injection:* We utilize estimated $\hat{y}$ to create latent prior features z in a highly compact latent space. *Stage III Fine Noise Reduction:* The diffusion model is used to generate prior features $\hat{z}$ and hierarchically conduct the denoising network cross-attention component. *Denoising Network:* The difference between f_θ and $f_\theta^{'}$ is that f_θ without cross-attention component.

the ECDM first employs a Noise2Noise-like denoising model to obtain coarse denoised dMRI data. Next, we design a Latent Encoder (LE) to map this coarse denoised data into latent space features. Finally, we implement a diffusion model in the latent space to conduct the denoising network via cross-attention and hierarchical structure, providing fine denoised dMRI data. The main contributions are summarized as follows:

- We propose a novel self-supervised denoising method, ECDM, which utilizes a coarse-to-fine strategy along with a diffusion model to effectively remove the noise.
- We exploit the latent encoder to generate efficient and compact features from the coarse denoised data, which provides vital guidance for fine denoising and significantly reduces the computational cost.
- We design a collaborative latent injector, which leverages latent features to collaboratively train a denoising model for fine noise reduction.
- Qualitative and quantitative experimental results demonstrate that ECDM outperforms DDM^2 and other competitive methods. It provides the denoised dMRI data and the associated microstructure indices with the best quality. Furthermore, compared with DDM^2, ECDM achieves training and inference times that are 7 times and 219 times shorter, respectively.

2 Methods

The overview framework of ECDM is shown in Fig. 1, with detailed components shown in Fig. 2. The ECDM is divided into three stages. *Stage I:* Feeding noisy

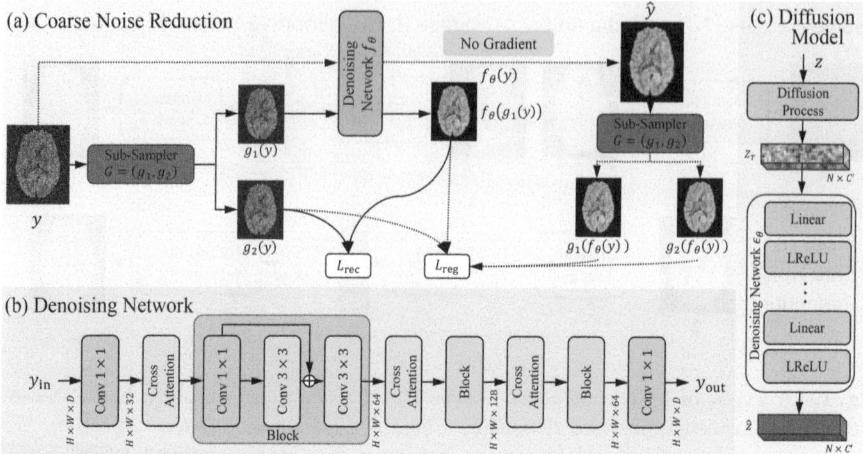

Fig. 2. Details of the model implementation. (a) Specific denoising process and loss computation in the first stage. (b) Architecture of the denoising network. The cross-attention component is introduced into the network starting from the second stage. (c) The architecture of the diffusion model.

data y into a Noise2Noise-like denoising model to obtain coarse dMRI data $\hat{y}$; *Stage II:* Learning the latent features of $\hat{y}$ to achieve more compact and efficient representations; *Stage III:* Utilizing a latent diffusion model to further generate prior features z, which hierarchically guide the denoising network to collaboratively use both generation and regression information, achieving fine denoised dMRI data. In this section, we will elaborate on the three stages.

2.1 Stage I: Coarse Noise Reduction

Inspired by [13], the method is carried out in two main steps. First, to create the data pairs needed for self-supervised training, the ECDM employs a random neighbor sub-sampler $G = \{g_1(\cdot), g_2(\cdot)\}$ to generate a pair of similar but not identical sub-sampled data pairs, $g_1(y)$ and $g_2(y)$. To ensure that the two generated noisy images $g_1(y)$ and $g_2(y)$ are very similar at the voxel level, the method begins by dividing the original noisy image into $k \times k$ cells, and we set $k = 2$ here. Within each cell, two voxels are randomly selected: one voxel is assigned to the noisy image $g_1(y)$, and the other is assigned to the noisy image $g_2(y)$, thus constructing a noise pair in each cell.

Next, even if the two sub-sampled data are very similar at the voxel level, the network may still struggle to accurately learn the subtle differences between them. To address this issue, a regularization term is introduced to the loss function to account for the non-zero true gap between the sub-sampled data pairs, as shown in Fig. 1 (Stage I). The overall loss function is defined as follows:

$$\begin{aligned} L &= L_{\text{rec}} + \gamma L_{\text{reg}} \\ &= ||f_\theta(g_1(y)) - g_2(y)||_2^2 \\ &+ \gamma||f_\theta(g_1(y)) - g_2(y) - (g_1(f_\theta(y))) - g_2(f_\theta(y)))||_2^2, \end{aligned} \tag{1}$$

where L_{reg} is the regularization term that imposes an additional constraint on image differences during training. The regularization coefficient γ controls the influence of this constraint on the overall loss, and $f_\theta(\cdot)$ is the denoising network.

By adding this regularization, the model focuses more on learning the subtle features and differences between the sub-sampled noisy images, rather than solely on voxel-level similarity. This effectively mitigates the issue of overly smooth denoising images, leading to more stable denoising performance and helping us achieve coarse denoising results $\hat{y}$.

2.2 Stage II: Collaborative Latent Injection

To address the substantial computational demands inherent in training diffusion models, we compress the high-dimensional input data into a low-dimensional latent space. In this way, not only can the most crucial information in the image be extracted to form an efficient representation, but it can also significantly reduce computational costs. The process is as follows:

Latent Encoder. Given the coarse clean image $\hat{y} \in \mathbb{R}^{(H \times W \times D)}$ obtained from the first stage, it is fed into the LE to generate the prior latent feature $z \in \mathbb{R}^{(N \times C')}$. Here, H, W, and D represent the height, width, and gradient directions of the dMRI data, while N and C' represent the number of tokens and the channel dimension of z, respectively. Since N is much smaller than $H \times W$, utilizing this latent space can effectively reduce the computational burden of the subsequent latent diffusion model.

Collaborative Latent Injector. To fully leverage the learned latent features, we introduce the cross-attention component [26]. By incorporating cross-attention before each convolutional block, the prior features extracted by LE are hierarchically injected into the CNN-based denoising model, enabling the effective fusion of input and latent features, as shown in Fig. 1 (Stage II).

Specifically, the input features X_{in} are reshaped into $z_{\text{r}} \in \mathbb{R}^{(\hat{H}\hat{W} \times \hat{D})}$, where $\hat{H}\hat{W}$ represents the spatial resolution and $\hat{D}$ denotes the gradient directions of dMRI. Subsequently, the reshaped features z_{r} are linearly projected into a query matrix $Q \in \mathbb{R}^{(\hat{H}\hat{W} \times \hat{D})}$. The latent features $z_i \in \mathbb{R}^{(N \times C')}$ are then projected as a key $K \in \mathbb{R}^{(\hat{N} \times \hat{C})}$ and a value $V \in \mathbb{R}^{(\hat{N} \times \hat{C})}$. The cross-attention calculation is defined as follows:

$$\begin{aligned} \text{Attention}(Q, K, V) &= \text{softmax}(\frac{QK^\top}{\sqrt{\hat{C}}} \cdot v), \\ Q = W_Q z_r, \quad K &= W_K z_i, \quad V = W_V z_i, \end{aligned} \tag{2}$$

where $W_Q \in \mathbb{R}^{(\hat{C}\times\hat{C})}$, $W_K \in \mathbb{R}^{(C'\times\hat{C})}$, and $W_V \in \mathbb{R}^{(C'\times\hat{C})}$ are learnable projection matrices. Finally, we reshape and project the output of the cross-attention component and add it with X_{in}, resulting in the output features $z_{\text{out}} \in \mathbb{R}^{(\hat{H}\times\hat{W}\times\hat{C})}$.

During training, the LE and the denoising model, which has already been trained in Stage I, undergo further joint optimization, collaborating to enhance each other's learning.

2.3 Stage III: Fine Noise Reduction

In the third stage, we trained a Latent Diffusion Model (LDM) to generate prior features in the latent space [4,22]. It is important to note that the information extracted by traditional regression methods differs fundamentally from what is learned by generative models. Generative models focus on modeling the data distribution by recovering real data from noise.

LDM consists of two steps: the forward process simulates the diffusion process where data is progressively corrupted by random noise, and the reverse process gradually recovers the noisy data back to the original data. In the forward process, the distribution q is defined as follows:

$$q\left(z_t \mid z_{t-1}\right) = \mathcal{N}\left(z_t; \sqrt{1-\beta_t} z_{t-1}, \beta_t I\right), \tag{3}$$

where z_t represents the noisy latent features obtained through an iterative process of noise addition; $\mathcal{N}(\cdot)$ represents a Gaussian distribution with mean 0 and covariance matrix I. The amount of noise added at each step in the sequence is controlled by a variance schedule $\beta_1, ..., \beta_T$, where $t \in \{0, ..., T\}$ denotes a particular moment during the noise addition process.

Here, we use the LE trained in the second stage to generate the corresponding prior latent features $z \in \mathbb{R}^{(N\times C')}$, and use z as the starting point of the forward process to gradually add Gaussian noise in T iterations. The noisy features at time t can be expressed as:

$$z_t = \sqrt{\overline{\alpha}_t} z_0 + \sqrt{1-\overline{\alpha}_t} \epsilon_t, \tag{4}$$

with $\alpha_t = 1 - \beta_t$ and $\overline{\alpha}_t = \prod_{j=1}^{t} \alpha_j$, z_0 being the original latent feature, and ϵ_t being Gaussian noise added at step t.

The reverse process starts from a prior distribution $p(z_T)$ and progressively reduces noise through a series of steps, ultimately approaching the observed data z_0. The distribution can be written as:

$$p_\theta\left(z_{t-1} \mid z_t\right) := \mathcal{N}\left(z_{t-1}; \mu_\theta\left(z_t, t\right), \Sigma_\theta\left(z_t, t\right)\right). \tag{5}$$

Our reverse process employs eight steps, as the latent space distribution is significantly simpler than that of images, enabling prior feature generation with fewer iterations. The denoising network ϵ_θ relied on in the reverse process consists of linear layers, which optimize the parameters θ by predicting the noise from z_t to z_{t-1}. This process can be expressed as:

$$z_{t-1} = \frac{1}{\sqrt{a_t}}(z_t - \frac{1-\alpha_t}{\sqrt{1-\overline{a}_t}}\epsilon_\theta) + \sqrt{1-\alpha_t}\epsilon_t. \quad (6)$$

By performing eight iterations of sampling on z_t using Eq. (6), we generated the prior features.

During training, these prior features guide the network from the second stage through the hierarchical structure to obtain the final fine-denoised image, as illustrated in Fig. 1 (Stage III), where $z_1 = z_2 = z_3$. To reduce the deviation between the generated prior features and the previous priors and improve the denoising performance, we jointly train the LDM and the network.

Table 1. Quantitative evaluation using RMSE, PSNR, and SSIM for dMRI data from DW-POSSUM at SNRs of 20 and 15. The best result is **bold**.

Method	SNR 20			SNR 15		
	PSNR ↑	SSIM ↑	RMSE ↓	PSNR ↑	SSIM ↑	RMSE ↓
Noisy	29.742	0.594	2.965	24.907	0.368	5.172
Patch2Self [7]	32.590	0.904	2.136	28.669	0.848	3.354
DDM^2 [28]	32.130	0.814	2.250	27.106	0.678	4.015
NLM [2]	34.075	0.902	1.800	28.957	0.823	3.245
Zero-Shot [18]	35.112	0.872	1.598	29.814	0.736	2.940
MPPCA [27]	37.905	0.939	1.158	31.879	0.886	2.318
ECDM	**37.974**	**0.942**	**1.149**	**32.432**	**0.901**	**2.175**

3 Experiment

3.1 Datasets and Implementation Details

Datasets. To evaluate the model's generalization ability, we conducted experiments on two datasets. The first dataset is a simulated dataset generated using DW-POSSUM [6,10]. After generating the Ground Truth (GT) data, noise was added to create noisy datasets with SNRs of 15 and 20, respectively. The second dataset is the publicly available real brain dMRI dataset, the Stanford HARDI dataset (HARDI) [21], with $b = 2000$ s/mm^2.

Implementation Details. The model is implemented using PyTorch and runs on an RTX 3060 GPU with 12GB of RAM. We utilized the ADAM optimizer with an initial learning rate of $1e^{-4}$. For model training, we empirically set the batch size to 32 and the number of iterations to 10,000 for each stage. The λ in stage I is set to eight. Our code will be made publicly available at https://github.com/Nwtbbetter/ECDM.

To simulate clinical data, this study used dMRI data with the first 31 volumes, including one b_0 volume in all experiments. The choice of 31 volumes was primarily because clinical scans typically have a smaller number of gradient directions. This selection makes the results more representative of actual clinical data.

3.2 Quantitative Results

We compared ECDM with four state-of-the-art methods: Patch2Self [7], NLM [2], MPPCA [27] and Zero-Shot Noise2Noise (Zero-Shot) [18], as well as the baseline model DDM^2 [28]. The first three methods are from the dipy package [8], while the experimental setups for the other two methods follow their original configurations. To comprehensively evaluate the performance of these methods, we perform experimental comparisons on the simulated DW-POSSUM dMRI dataset with an SNR of 20 and 15 and the real HARDI dataset, respectively.

In the DW-POSSUM dataset, the experimental results are evaluated using Root Mean Square Error (RMSE), Peak Signal-to-Noise Ratio (PSNR), and Structural Similarity Index (SSIM), as shown in Table 1. The results show that

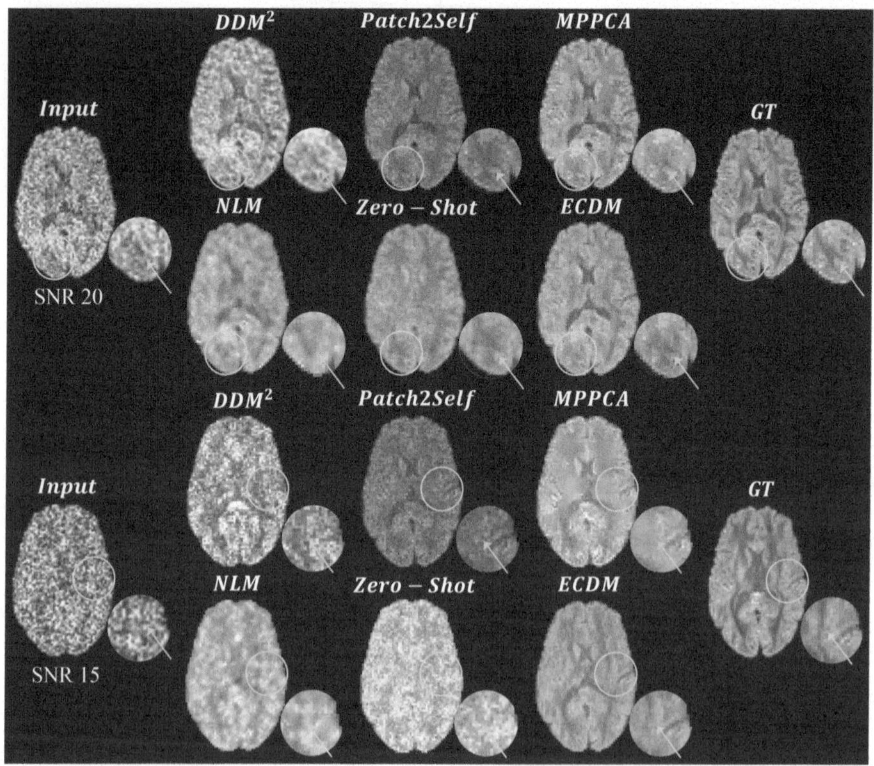

Fig. 3. Visual comparison of the DW-POSSUM dataset with SNR=15 and SNR=20.

our method outperforms other competing methods in terms of PSNR and RMSE under different SNR conditions, fully demonstrating its robustness. The advantage of our method is particularly pronounced at lower SNRs. Compared with the second-best method, our method improves the RMSE metrics by 0.8% and 6% at SNRs of 20 and 15. It shows different degrees of improvement in the SSIM and PSNR metrics. These results indicate that our model can better retain the structural information of the image after denoising and restoring the data more accurately, making the processed image closer to the real image.

In the HARDI dataset, we use SNR and Contrast-to-Noise Ratio (CNR) as evaluation metrics since GT images are unavailable. Both metrics require the segmentation of regions of interest. We employed the method by [8] to segment the corpus callosum, which allowed us to delineate foreground and background signals. Then we obtained the corresponding scores by subtracting the input noise image from the denoised image. We show the box plot of relative SNR and CNR on the right side of Fig. 4. Although the MPPCA, DDM^2 and Patch2Self methods perform better on the worst results, their improvement in signal and contrast is not significant enough, and the results have certain volatility. In contrast, our method shows significant improvements in SNR and CNR.

To validate the acceleration effects of ECDM on diffusion models during training and inference, we compared the training and inference times across simulated and real datasets, as shown in Table 2. The results show that during training, obtaining high-quality results using DDM^2 requires five times longer compared to our method. In the inference process, our method achieves up to 20 times speedup in experiments with the HARDI dataset. Additionally, in data with higher noise levels, DDM^2 typically requires more steps to simulate the initial noise conditions of the data. This results in more than 1000 times longer inference time than ours, as evidenced by the DW-POSSUM dataset results.

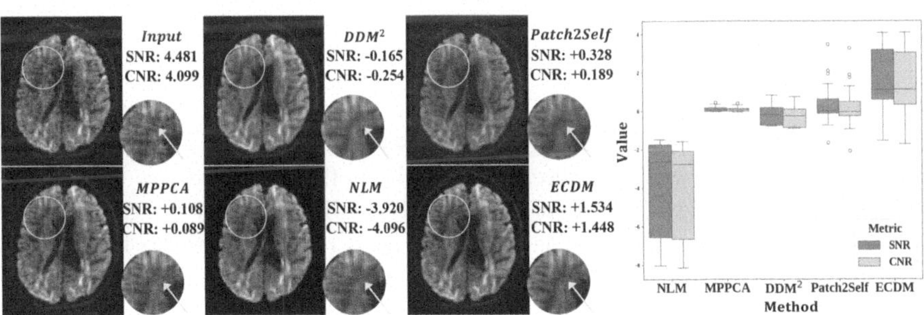

Fig. 4. Visual comparison on the HARDI dataset. Box plot of SNR and CNR by method.

Table 2. Comparison of training and inference time on DW-POSSUM and HARDI datasets.

Time (min)	Stage I	Stage II	Stage III	Training	Inference
ECDM (HARDI)	22.43	47.13	48.64	118.20	0.14
DDM^2 (HARDI)	53.96	1.06	648.77	703.79	22.08
Difference	**31.53**	**-46.07**	**600.13**	**585.59**	**21.94**
ECDM (DW-POSSUM)	17.79	37.35	49.16	104.30	0.09
DDM^2 (DW-POSSUM)	34.87	1.29	479.81	515.97	186.68
Difference	**17.08**	**-36.06**	**430.65**	**411.67**	**186.59**

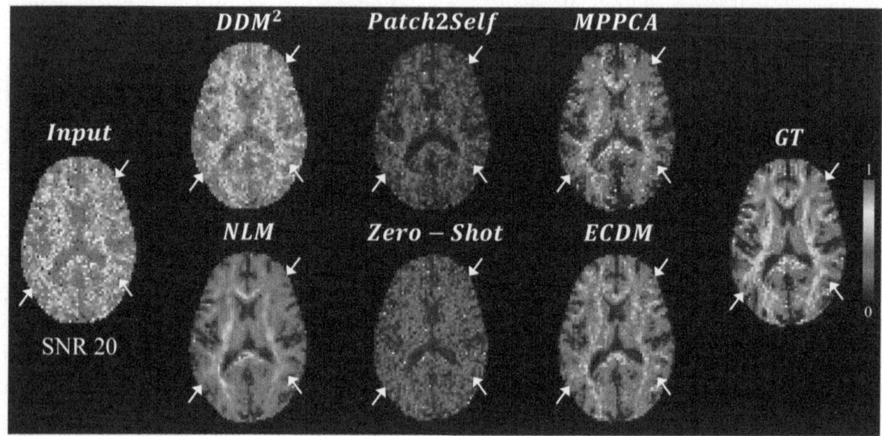

Fig. 5. FA map comparisons (major differences marked with arrows).

3.3 Visualization of Results

Results on Simulated Data. The visual comparison results of the DW-POSSUM are shown in Fig. 3. It can be observed that other methods exhibit issues with over-smoothing, whereas our method effectively reduces noise while preserving clear edges, making it closer to the GT, as shown in the close-up views. When the noise level in the dataset is higher, other methods face significant challenges during the denoising process. At the same time, ECDM can still remove noise and reveal clear structures. These visual comparisons are consistent with the quantitative results, further demonstrating the superiority of our method.

Results on Real Data. To evaluate our method more comprehensively, we conducted experiments on the HARDI dataset, which is relatively large and less noisy. As shown in the left side of Fig. 4, these methods generally perform well, but our method still performs better in reconstructing details and structural

Table 3. Quantitative evaluation of FA maps on the DW-POSSUM with SNR of 20 using RMSE, PSNR, and SSIM. The best result is **bold**.

Method	PSNR ↑	SSIM ↑	RMSE ↓
Noisy	18.395	0.840	0.120
Patch2Self [7]	17.637	0.785	0.131
Zero-Shot [18]	18.073	0.797	0.125
DDM^2 [28]	20.351	0.850	0.096
MPPCA [27]	21.482	0.878	0.084
NLM [2]	22.692	0.882	0.073
ECDM	**23.149**	**0.906**	**0.070**

information. Specifically, our method can observe clear structures after removing noise, thereby more effectively improving the quality of dMRI data.

Fractional Anisotropy Comparison. To comprehensively evaluate the denoising performance, we also analyzed the Fractional Anisotropy (FA) maps in the DW-POSSUM with an SNR of 20. To ensure the accuracy and comprehensiveness of the results, we selected the volume data of the first 31 volumes to calculate the FA value. As shown in Fig. 5, ECDM demonstrated superior denoising performance compared with other methods, providing a clearer depiction of fiber tract directions and higher consistency with the GT. We further validated these results by calculating several quantitative metrics, with the specific outcomes presented in Table 3. These analyses fully demonstrate the

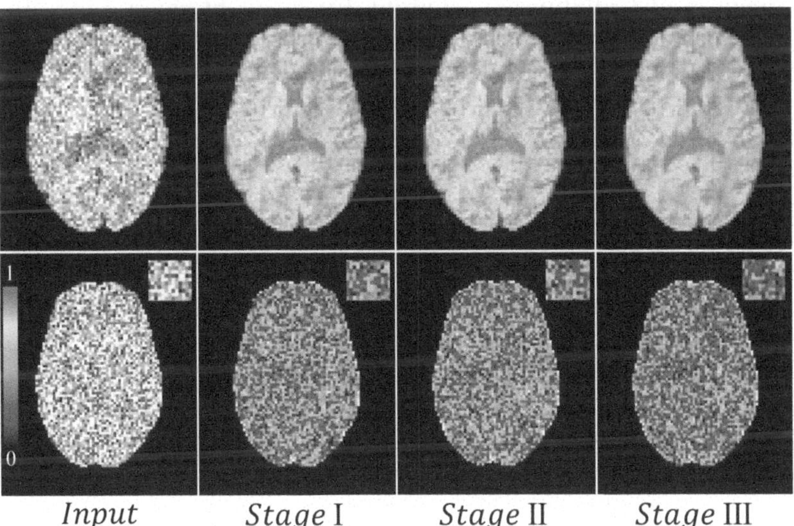

Fig. 6. Denoising results and residual maps at different stages.

Table 4. Quantitative results for ablation study. The best results are in **bold**.

Version	CNRS	CLIS	FNRS	PSNR↑	SSIM↑	RMSE↓
(A)				32.130	0.814	2.252
(B)	✓			37.181	0.930	1.259
(C)	✓	✓		37.404	0.930	1.227
(D)	✓	✓	✓	**37.974**	**0.942**	**1.149**

advantages of ECDM in noise suppression and the preservation of fiber tract structure fidelity.

3.4 Ablation Study

To study the effects of different designs of our proposed method, we conducted ablation experiments using simulated DW-POSSUM dMRI data with an SNR of 20, as shown in Table 4.

Effectiveness of Coarse Noise Reduction Stage (CNRS). In this ablation experiment, we introduced an ablated version (A), where the model was trained using the modules from DDM^2. To verify the effectiveness of CNRS for denoising, we designed an ablated version (B). As shown in Table 4, training with this method resulted in a significant PSNR increase of 5.051 dB. This substantial improvement indicates that the Noise2Noise-like model enhancement effectively boosts the model's initial capabilities and significantly improves image reconstruction quality.

Effectiveness of Collaborative Latent Injection Stage (CLIS). To validate the compactness and efficiency of the latent space representation, we designed an ablated version (C). The experimental results demonstrate improvements in both PSNR and RMSE, while SSIM remains unchanged. This suggests that the optimization of the latent space representation primarily focuses on noise reduction and detail recovery, providing a more efficient feature representation.

Effectiveness of Fine Noise Reduction Stage (FNRS). Finally, we compare the full version of ECDM (D) with version (C) to verify the effectiveness of using the prior features generated by the latent diffusion model to guide denoising. It can be seen that all metrics are better than (C), which shows that the simultaneous use of regression information and generation information can more accurately restore the details and structures in the image, and has obvious advantages in denoising tasks.

We also present the visual results of each stage, as shown in Fig. 6. It can be seen that as the training progresses, the noise in the image gradually decreases, and the detailed structure is presented more and more clearly. In addition, we show the residual map of dMRI data in the second row. As shown in the close-up view, after three stages of training, the difference between the image prediction value and the true value becomes smaller and smaller, which effectively improves the quality of dMRI data. This further verifies the effectiveness of the proposed method and demonstrates the potential and advantages of ECDM in image-denoising tasks.

4 Conclusion

In this paper, we proposed ECDM, an efficient collaborative diffusion model for self-supervised denoising of dMRI data. Our ECDM involves three stages. In Stage I, the self-supervised denoising model is trained using a Noise2Noise-like method to obtain the coarse clean image from the noisy input data. In Stage II, the coarse denoised image is compressed into latent feature representations by LE, and these features are further optimized via joint training. This process makes the features more compact and significantly improves computational efficiency, laying a solid foundation for fine denoising in subsequent stages. In Stage III, a latent diffusion model is trained to generate prior features that guide the denoising process hierarchically, enabling the collaborative use of detailed information from both regression and generative methods to achieve fine denoising results. We conducted extensive experiments on both simulated and real datasets. The experimental results demonstrate that ECDM achieves superior performance compared with state-of-the-art methods.

References

1. Bansal, A., et al.: Cold diffusion: inverting arbitrary image transforms without noise. In: Advances in Neural Information Processing Systems, vol. 36 (2024)
2. Buades, A., Coll, B., Morel, J.M.: Non-local means denoising. Image Process. Line **1**, 208–212 (2011)
3. Chen, G., Wu, Y., Shen, D., Yap, P.T.: XQ-NLM: denoising diffusion MRI data via x-q space non-local patch matching. In: International Conference on Medical Image Computing and Computer-Assisted Intervention, pp. 587–595 (2016)
4. Chen, Z., et al.: Hierarchical integration diffusion model for realistic image deblurring. In: Advances in Neural Information Processing Systems, vol. 36 (2024)
5. Cheng, H., et al.: Denoising diffusion weighted imaging data using convolutional neural networks. PLoS ONE **17**(9), e0274396 (2022)
6. Drobnjak, I., Gavaghan, D., Süli, E., Pitt-Francis, J., Jenkinson, M.: Development of a functional magnetic resonance imaging simulator for modeling realistic rigid-body motion artifacts. Magn. Resonance Med. Off. J. Int. Soc. Magn. Resonance Med. **56**(2), 364–380 (2006)
7. Fadnavis, S., Batson, J., Garyfallidis, E.: Patch2Self: denoising diffusion MRI with self-supervised learning. Adv. Neural. Inf. Process. Syst. **33**, 16293–16303 (2020)
8. Garyfallidis, E., et al.: Dipy, a library for the analysis of diffusion MRI data. Front. Neuroinform. **8**, 8 (2014)
9. Goodfellow, I., et al.: Generative adversarial nets. In: Advances in Neural Information Processing Systems, vol. 27 (2014)

10. Graham, M.S., Drobnjak, I., Zhang, H.: Realistic simulation of artefacts in diffusion MRI for validating post-processing correction techniques. Neuroimage **125**, 1079–1094 (2016)
11. Ho, J., Jain, A., Abbeel, P.: Denoising diffusion probabilistic models. Adv. Neural. Inf. Process. Syst. **33**, 6840–6851 (2020)
12. Ho, J., Salimans, T.: Classifier-free diffusion guidance. arXiv preprint arXiv:2207.12598 (2022)
13. Huang, T., Li, S., Jia, X., Lu, H., Liu, J.: Neighbor2Neighbor: self-supervised denoising from single noisy images. In: Proceedings of the IEEE/CVF Conference on Computer Vision and Pattern Recognition, pp. 14781–14790 (2021)
14. Kingma, D., Salimans, T., Poole, B., Ho, J.: Variational diffusion models. Adv. Neural. Inf. Process. Syst. **34**, 21696–21707 (2021)
15. Knoll, F., Bredies, K., Pock, T., Stollberger, R.: Second order total generalized variation (TGV) for MRI. Magn. Reson. Med. **65**(2), 480–491 (2011)
16. Kong, Z., Ping, W., Huang, J., Zhao, K., Catanzaro, B.: DiffWave: a versatile diffusion model for audio synthesis. arXiv preprint arXiv:2009.09761 (2020)
17. Manjón, J.V., Coupé, P., Buades, A., Collins, D.L., Robles, M.: New methods for MRI denoising based on sparseness and self-similarity. Med. Image Anal. **16**(1), 18–27 (2012)
18. Mansour, Y., Heckel, R.: Zero-shot noise2noise: efficient image denoising without any data. In: Proceedings of the IEEE/CVF Conference on Computer Vision and Pattern Recognition, pp. 14018–14027 (2023)
19. Mittal, G., Engel, J., Hawthorne, C., Simon, I.: Symbolic music generation with diffusion models. arXiv preprint arXiv:2103.16091 (2021)
20. Novikov, D.S., Fieremans, E., Jespersen, S.N., Kiselev, V.G.: Quantifying brain microstructure with diffusion MRI: theory and parameter estimation. NMR Biomed. **32**(4), e3998 (2019)
21. Rokem, A.: Stanford HARDI surfaces (2016)
22. Rombach, R., Blattmann, A., Lorenz, D., Esser, P., Ommer, B.: High-resolution image synthesis with latent diffusion models. In: Proceedings of the IEEE/CVF Conference on Computer Vision and Pattern Recognition, pp. 10684–10695 (2022)
23. Saharia, C., Ho, J., Chan, W., Salimans, T., Fleet, D.J., Norouzi, M.: Image super-resolution via iterative refinement. IEEE Trans. Pattern Anal. Mach. Intell. **45**(4), 4713–4726 (2022)
24. Song, J., Meng, C., Ermon, S.: Denoising diffusion implicit models. arXiv preprint arXiv:2010.02502 (2020)
25. Tian, Q., et al.: SDnDTI: self-supervised deep learning-based denoising for diffusion tensor MRI. Neuroimage **253**, 119033 (2022)
26. Vaswani, A., Shazeer, N., Parmar, N., Uszkoreit, J., Jones, L., Gomez, A.N., Kaiser, Ł., Polosukhin, I.: Attention is all you need. Advances in Neural Information Processing Systems **30** (2017)
27. Veraart, J., Novikov, D.S., Christiaens, D., Ades-Aron, B., Sijbers, J., Fieremans, E.: Denoising of diffusion MRI using random matrix theory. Neuroimage **142**, 394–406 (2016)
28. Xiang, T., Yurt, M., Syed, A.B., Setsompop, K., Chaudhari, A.: DDM^2: self-supervised diffusion MRI denoising with generative diffusion models. arXiv preprint arXiv:2302.03018 (2023)

MAD-AD: Masked Diffusion for Unsupervised Brain Anomaly Detection

Farzad Beizaee[1,2(✉)], Gregory Lodygensky[2,3], Christian Desrosiers[1], and Jose Dolz[1]

[1] ÉTS Montreal, Montreal, Canada
[2] CHU Sainte-Justine Hospital, Montreal, Canada
farzad.beizaee.1@ens.etsmtl.ca
[3] University of Montreal, Montreal, Canada

Abstract. Unsupervised anomaly detection in brain images is crucial for identifying injuries and pathologies without access to labels. However, the accurate localization of anomalies in medical images remains challenging due to the inherent complexity and variability of brain structures and the scarcity of annotated abnormal data. To address this challenge, we propose a novel approach that incorporates masking within diffusion models, leveraging their generative capabilities to learn robust representations of normal brain anatomy. During training, our model processes only normal brain MRI scans and performs a forward diffusion process in the latent space that adds noise to the features of randomly-selected patches. Following a dual objective, the model learns to identify which patches are noisy and recover their original features. This strategy ensures that the model captures intricate patterns of normal brain structures while isolating potential anomalies as noise in the latent space. At inference, the model identifies noisy patches corresponding to anomalies and generates a normal counterpart for these patches by applying a reverse diffusion process. Our method surpasses existing unsupervised anomaly detection techniques, demonstrating superior performance in generating accurate normal counterparts and localizing anomalies. The code is available at hhttps://github.com/farzad-bz/MAD-AD.

Keywords: Unsupervised Anomaly Detection · Brain MRI · Diffusion

1 Introduction

The accurate detection and localization of brain anomalies in medical images, particularly in Magnetic Resonance Imaging (MRI) data, is paramount to diagnosing and understanding neurological injuries and pathologies. However, the complexity of brain structures and the scarcity of labeled abnormal data present

I. Oguz et al. (Eds.): IPMI 2025, LNCS 15830, pp. 139–153, 2026.
https://doi.org/10.1007/978-3-031-96625-5_10

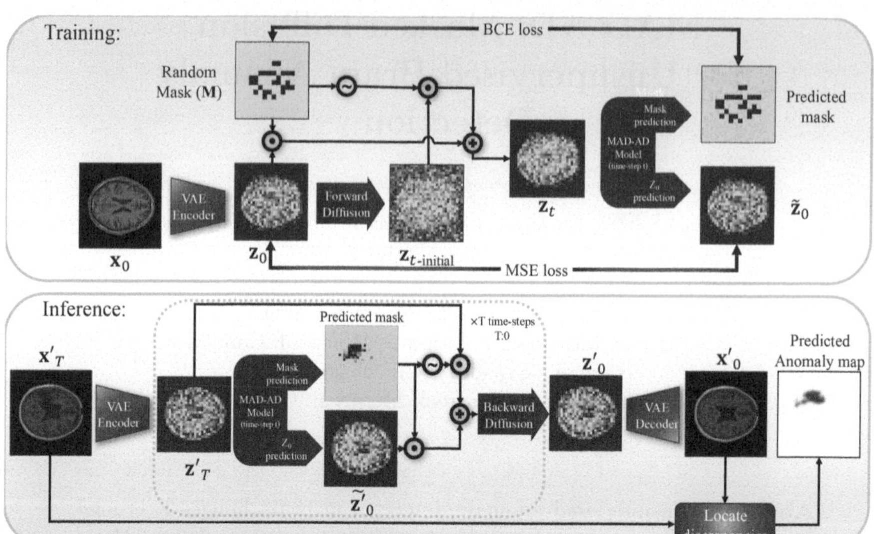

Fig. 1. Overview of the proposed method. During *training*, normal samples are encoded into latent space. A binary mask and a time-step t are applied, and non-masked regions undergo forward diffusion to produce z_t. The model is then trained to predict z_0 and the incorporated mask for forward diffusion. At *inference*, the model undergoes a selective reverse process using the predicted mask at each step.

significant challenges in developing robust and generalizable solutions. Traditionally, brain anomaly detection has been framed as a supervised learning task, which aims at identifying well-defined pathologies such as brain tumor [15,18,36], atrophy [29] or white matter hyper-intensities [20,22], among many others. Nevertheless, casting anomaly detection as a supervised problem introduces an inherent bias towards the targeted lesions, limiting the scope of detectable pathologies. Moreover, collecting large amounts of annotated samples encompassing the entire spectrum of potential brain abnormalities is expensive and impractical for novel structures or rare abnormal patterns.

Unsupervised anomaly detection (UAD), which involves modeling the distribution of normal data and identifying deviations as anomalies, has gained attention as a promising alternative [6,10,35,42]. Conventional unsupervised methods, such as autoencoders [32] and generative adversarial networks (GANs) [13], attempt to reconstruct normal anatomical structures and flag areas with high reconstruction errors as anomalies. Despite their potential, these approaches suffer from notable limitations. Autoencoders often fail to capture the fine-grained details of normal anatomy, whereas GANs are prone to mode collapse and instability during training. Moreover, these models frequently reconstruct anomalies as part of normal structures, reducing their reliability in clinical applications.

Recent advances in diffusion models [16,21,37,38] have opened new avenues in generative modeling. Such models [37] leverage a stochastic process to gradually corrupt data and learn to reverse this process, enabling them to model complex data distributions with remarkable precision. Their success in generating high-quality images and their ability to capture intricate patterns in the data have prompted researchers to explore their use for anomaly detection [14,23,26,28,40]. While these methods have improved the accuracy of anomaly detection, their application to brain images introduces several challenges. Firstly, the forward diffusion process can cause a loss of distinctive features across brain regions, especially when the number of steps is large. This loss may compromise the model's ability to differentiate between normal and anomalous brain regions. Also, reducing the number of forward diffusion steps introduces the risk of an *"identity shortcut"* problem. In this problem, the model can easily recover the fine details of the input image, resulting in anomalous regions being preserved in the reconstruction. This is a significant concern in brain anomaly detection, where subtle but critical deviations such as tumors or lesions may be overlooked due to this shortcut behavior. Another issue arises from the indiscriminate application of forward and reverse diffusion across the entire brain image. This approach can hinder the model's ability to effectively reconstruct normal brain patterns.

To address these limitations, we propose MAD-AD, a Masked Diffusion for brain anomaly detection with the following key contributions. First, we leverage latent diffusion models to treat anomalies as partial noise in the latent space, enabling their effective restoration through the denoising process. Our method also removes the reliance on forward diffusion steps during inference, thereby preventing the loss of critical visual details and enabling highly accurate reconstructions of the underlying normal appearance. This is accomplished by masking the forward diffusion process and training the model to reverse it effectively. Furthermore, we incorporate a mask-prediction module into the diffusion framework, allowing the prediction of the incorporated mask in the diffusion process. This approach ensures the selective correction of anomalous regions while preserving normal regions intact, ultimately delivering more precise and reliable anomaly detection results. The overview of our method is depicted in Fig. 1.

2 Related Works

Recent approaches for unsupervised anomaly detection (UAD) in brain MRI can mainly be divided in three categories: methods based on different variants of autoencoders (AEs), those using generative adversarial networks (GANs) and the ones based on diffusion models.

AE-Based Methods. Approaches in this category train an autoencoder on normal data to accurately reconstruct input images. At inference, the reconstruction error measured at each pixel is used to localize anomalies. Different networks have been explored for the reconstruction, including standard autoencoders (AE) [2,4], variational autoencoders (VAEs) [4,35,42] and denoising autoencoders (DAEs) [19]. A common issue with these methods is their propensity to overfit the training data, leading to a poor generalization on unseen data. Furthermore, they are prone to blurry reconstructions, struggling to accurately distinguish subtle anomalies from normal variations, especially when relying solely on reconstruction error as a measure of abnormality.

GAN-Based Methods. These approaches employ an adversarial learning strategy where a generator and a discriminator are jointly trained on healthy subject images to learn a latent representation of normal variability. AnoGAN [34] measures anomaly scores based on a combination of reconstruction error and distance in the latent space. f-AnoGAN [33] improves upon this work by incorporating an additional feature-level reconstruction strategy, yielding a more precise localization of anomalies. The work in [5] uses a style transfer method based on CycleGANs to map real MR images of healthy brains to synthetic ones, and vice versa. Anomalies are then detected by comparing input images to their reconstruction. While the ability of GANs to generate high-quality images can translate in a more detailed delineation of anomalies, they are also prone to training instability and are often sensitive to hyperparameter choices.

Diffusion-Based Method. Diffusion models have gained significant attention in computer vision for their ability to generate high-fidelity images [12]. Recently, these models have also shown promise in various medical image analysis tasks including UAD [7–9,17,24,27,28,39]. A prominent diffusion-based method for UAD in medical images, AnoDDPM [39] utilizes a partial diffusion strategy, adding noise to an image up to a specific timestep and then recovering the original image with a reverse diffusion process. This method has shown success in detecting anomalies in brain MRI and other domains. PDDPM [7] instead applies the diffusion process in a patch-wise manner, aiming to improve the understanding of local image context and achieve better anatomical coherence in the reconstruction. This method divides the image into overlapping patches and reconstructs each patch while considering its unperturbed surroundings. CDDPM [8] generates multiple reconstructions via the reverse diffusion process and pinpoints anomalies by examining the distribution of these reconstructions with the Mahalanobis distance, subsequently labeling outliers as anomalies. MDDPM [17] incorporates masking-based regularization, applied on both image patches and in the frequency domain, to enhance unsupervised anomaly detection. AutoDDPM [11] incorporates automatic masking, stitching, and resampling techniques within the DDPM framework to enhance its robustness and accuracy in anomaly detection. This approach also addresses the challenge of selecting an appropriate noise level for detecting lesions of various sizes. However, the diffusion-based UAD models mentioned above rely heavily on a forward diffusion process that inherently results in information loss. Consequently, these methods often fail to accurately reconstruct the original healthy brain structures, leading to false-positive detections where normal regions are incorrectly identified as anomalous. This issue is particularly prominent in brain anomaly detection tasks, as brain structures, especially cortical regions, vary uniquely across individuals, thereby increasing the difficulty of accurately recovering normal anatomical variations.

A recently proposed method, DISYRE [27,28], uses a diffusion-like pipeline to train a model to restore images that have been corrupted with synthetic anomalies. Anomalies in a new image are detected based on the model's ability to restore the image to a healthy state. A key limitation of this method is that the synthetic anomalies may not encompass all types of real-world anomalies, limiting its generalization ability. THOR [9] integrates implicit guidance into the DDPM's denoising process using intermediate masks to preserve the integrity of healthy tissue details. It aims to ensure a faithful reconstruction of the original image in areas unaffected by pathology, minimizing false positives. However, since these intermediate masks are determined based on the perceptual differences between input images and their reconstruction at each step, the model may struggle to detect subtle or small anomalies, as they might be masked out due to their minimal differences with the input image. Additionally, reconstruction errors may occur due to the loss of details during the forward process, with normal regions not getting masked due to their high perceptual differences. Inspired by diffusion-based models, IterMask^2 [24] incorporates an iterative spatial mask refinement process and frequency masking to enhance UAD performance. This strategy minimizes information loss in normal areas by iteratively shrinking a spatial mask, starting from the whole brain towards the anomaly. Although the model performs well in detecting hypo- or hyper-intense areas, it can fail to localize structural anomalies such as atrophy or enlarged ventricles as their reconstruction is conditioned on structural information from high-frequency image components which can be recovered by the model.

3 Method

3.1 Modeling the Normal Feature Space

We resort to diffusion models for learning the space of normal data and reconstructing the normal counterpart of anomalous regions. Denoising Diffusion Probabilistic Models (DDPMs) [16] learn a data distribution by gradually adding noise to the data (i.e., forward process) and then training a model to reverse this process. While DDPMs are highly effective at generating high-quality images, there are certain limitations when using them directly for detecting anomalous regions. Firstly, the number of steps in the forward diffusion process can have a considerable impact on the performance. If this number is too large, semantic information of the brain structure can be lost, resulting in an uncorrelated brain reconstruction and the incorrect detection of normal regions as abnormal. On the other hand, if not enough steps are used, the model can too easily recover the fine details in the image. As a result, abnormal regions will incorrectly be detected as normal. Moreover, as normal patches are also affected by noise, they cannot be fully exploited to reconstruct abnormal regions.

To overcome the aforementioned limitations, we propose to incorporate a random masking strategy in the diffusion model and modify the reverse process so that the diffusion model can selectively alter anomalous parts of an image, while keeping the normal regions untouched. Following [30,31], we employ a diffusion model operating in the *latent* space. This has two important advantages. First, whereas adding Gaussian noise directly on the image yields corruptions that have no meaningful structure, injecting this noise on latent features and then reconstructing these noisy features results in more complex corruptions that better represent real anomalies in brain MRI. Moreover, this also mitigates the "identity shortcut" problem, enhances computational efficiency, and improves stability, particularly with limited training data.

Let $\mathcal{X} = \{\boldsymbol{x}^{(i)}\}_{i=1}^{N}$ be the training set consisting exclusively of normal images $\boldsymbol{x}^{(i)} \in \mathbb{R}^{H\times W\times C}$, where H, W, and C correspond to the image height, width, and number of channels, respectively. We employ a pre-trained variational autoencoder [31], which is adapted and fine-tuned for medical images. This model can encode high-dimensional image data into a compact latent representation and reconstruct this data from the latent space while preserving essential structural and semantic information. Denoting the encoder network as $\mathrm{V}_{E,\phi}$, an input image $\boldsymbol{x}^{(i)}$ is mapped to its latent space representation $\boldsymbol{z}^{(i)} = \mathrm{V}_{E,\phi}(\boldsymbol{x}^{(i)})$, where $\boldsymbol{z}^{(i)} \in \mathbb{R}^{H'\times W'\times C'}$.

Random Masking. To incorporate random masking into the forward diffusion process, given the latent features of an input normal sample $\boldsymbol{z}_0 \sim p(\boldsymbol{z}_0)$, we spatially divide $\boldsymbol{z}_0$ into non-overlapping patches defined by a random mask $M \in [0,1]^{H\times W}$. The forward Markov diffusion process to generate samples $\boldsymbol{z}_t$ gradually applies noise to the non-masked patches of sample $\boldsymbol{z}_0$ for t time steps, where $t \in [1,T]$. Following [16], the forward noising process in the latent space with masking can be characterized as:

$$\boldsymbol{z}_t = \left(\sqrt{1-\beta_t}\boldsymbol{z}_{t-1} + \sqrt{\beta_t}\boldsymbol{\epsilon}\right) \odot M + \boldsymbol{z}_0 \odot (1-M), \tag{1}$$

where $\boldsymbol{z}_t$ is the partially diffused image at step t, $\epsilon \sim \mathcal{N}(0,I)$ is the sampled Gaussian noise and β_t is the noise schedule at step t, which controls the amount of noise added at each step. Using the reparameterization trick, $\boldsymbol{z}_t$ can be obtained implicitly using the following equation:

$$\boldsymbol{z}_t = \left(\sqrt{\bar{\alpha}_t}\boldsymbol{z}_0 + \sqrt{1-\bar{\alpha}_t}\boldsymbol{\epsilon}\right) \odot M + \boldsymbol{z}_0 \odot (1-M), \tag{2}$$

with $\alpha_t = 1-\beta_t$ and $\bar{\alpha}_t = \prod_{i=1}^{T}\alpha_i$. The reverse process aims to recover the original data $\boldsymbol{z}_0$ by gradually removing the noise. This process is modeled as a learned distribution that reverses the forward noising steps. Given the masked sample $\boldsymbol{z}_t$ at step t and mask M at spatial location k, the reverse process can be modeled as follows:

$$p(\boldsymbol{z}_{t-1}^k|\boldsymbol{z}_t^k) = \begin{cases} \mathcal{N}(\boldsymbol{z}_{t-1}^k; \mu_\theta(\boldsymbol{z}_t^k,t),\ \beta_t\mathbf{I}), & \text{if } M^k = 1 \\ \boldsymbol{z}_t^k, & \text{otherwise;} \end{cases} \tag{3}$$

In this equation, $\mu_\theta(\boldsymbol{z}_t,t)$ is a trainable function, which can be reparameterized as a predicted noise ϵ or a predicted clean image $\boldsymbol{z}_0$. Due to the incorporated random masking strategy, we prefer the latter one for simplicity. Therefore, $\mu_\theta(\boldsymbol{z}_t,t)$ can be formally expressed as:

$$\mu_\theta(\boldsymbol{z}_t,t) = \frac{\sqrt{\bar{\alpha}_{t-1}}\beta_t}{1-\bar{\alpha}_t} f_{\theta,\boldsymbol{z}_0}(\boldsymbol{z}_t,t) + \frac{\sqrt{\alpha_t}\,(1-\bar{\alpha}_{t-1})}{1-\bar{\alpha}_t}\boldsymbol{z}_t, \tag{4}$$

where $f_{\theta,z_0}(z_t, t)$ is a function that predicts $\tilde{z}_0$ at time step t, given z_t.

Mask Prediction. By parameterizing f_θ as a neural network, the model can be trained using a simple mean-square error loss between z_0 and the predicted clean image. Moreover, in Eq. (3), we assumed that the mask M is available in the reverse process. However, this assumption is unrealistic since the mask used in diffusing the image, which contains the location of anomalous regions, is not accessible at inference. Therefore, we include an additional head $f_{\theta,M}$ to the diffusion model that predicts the mask used in the forward diffusion. This can be achieved by applying a binary cross-entropy ($\mathcal{L}_{BCE}$) loss between the predicted mask from this head and a randomly sampled mask used during partial diffusion in training. The final training objective of our model is defined by:

$$\min_{\theta} \ \mathbb{E}_{z_0 \sim q(z_0),\epsilon,t,M} \left[\| z_0 - f_{\theta,z_0}(z_t, t) \|_2^2 + \lambda \mathcal{L}_{BCE}\big(M,\, f_{\theta,M}(z_t, t)\big) \right], \quad (5)$$

where λ is a hyper-parameter that balances the contributions of the two terms.

3.2 Recovering Normal Images

During inference, the goal is to recover a normal version of an abnormal brain image, where anomalous regions are replaced with their normal counterpart while normal areas remain unchanged. As previously discussed, a pre-trained VAE, $V(\cdot)$, is employed to project the image into a latent space where the data follows a normal distribution. In this space, abnormal brain regions can be interpreted as normal noise, as they fall outside the learned normal distribution of the model. These abnormal areas can also be considered as non-masked regions through the forward diffusion process using a mask that points out anomalous regions. Consequently, the proposed method incorporates all the necessary components to first predict the location of anomalies using the mask prediction head and then progressively denoise these regions to reconstruct their normal counterpart. Finally, by comparing the input image with its corrected version, anomalies can be accurately localized. The following section provides a detailed explanation of the sampling process in the MAD-AD model during inference.

Let $\mathcal{X}' = \{x'^{(i)}\}_{i=1}^{N'}$ denote the test set at inference time, which consists of samples with potential anomalies. We first map these images into the latent space using $\mathrm{V}_{E,\phi}$. As explained before, we treat the latent space of an anomalous image as step T of the masked forward diffusion process applied on its normal counterpart, i.e., $z'_T = \mathrm{V}_{E,\phi}(x'_T)$. By predicting the mask that corresponds to the anomaly location and the reconstructed $\tilde{z}'_0$ at each time-step t, using Eq. (3), we can progressively correct the anomaly regions and obtain the normal counterpart ($z'_T \rightarrow z'_0$) while preserving fine details of the normal regions.

Nevertheless, one drawback of sampling with DDPM is that it requires many reverse sampling steps to obtain the normal version. Therefore, we instead opted for DDIM [38] which, by reducing the stochasticity of DDPM, makes the reverse process more deterministic and requires fewer sampling steps. Consequently, we modify the reverse process of DDIM for the MAD-AD model as:

$$z'_{t-1} = \underbrace{B\Big(f_{\theta,M}(z'_t)\Big)}_{\text{"predicted mask"}} \Big(\sqrt{\bar{\alpha}_{t-1}}\ \underbrace{f_{\theta,z_0}(z'_t)}_{\text{"predicted } \tilde{z}'_0\text{"}} + \underbrace{\sqrt{1-\bar{\alpha}_{t-1}}\tilde{\epsilon}_t(z'_t)}_{\text{"direction pointing to } z'_t\text{"}} + \sigma_t \epsilon'_t\Big) + \Big(1 - B\big(f_{\theta,M}(z'_t)\big)\Big) \cdot z'_t \tag{6}$$

where $B(.)$ is a binarization function, ϵ'_t is random normal noise, σ_t is a hyper-parameter that controls the stochasticity of reverse process, and $\tilde{\epsilon}$ is the predicted noise calculated based on the predicted $\tilde{z}_0$ and z_t as follows:

$$\tilde{\epsilon}_t = \frac{z'_t \cdot f_{\theta,z'_0}(z'_t)}{\sqrt{1-\bar{\alpha}_t}} \tag{7}$$

As mentioned above, σ_t controls the noise level and stochasticity of the sampling process in DDIMs. Specifically, $\sigma_t = 0$ makes the model deterministic, while $\sigma_t > 0$ introduces stochasticity. For $\sigma_t = 1$, the model behaves like a DDPM, where the sampling process involves full stochasticity with noise added at each step. While having a fully deterministic model can be desirable for UAD applications, introducing a bit of noise to the non-masked (anomalous) regions helps bring the distribution closer to normal. This makes it easier for the model to recover the normal variation of the input. Therefore, we propose to use an in-between value of $\sigma_t = 0.5$. A qualitative example of the reverse process in MAD-AD is depicted in Fig. 2.

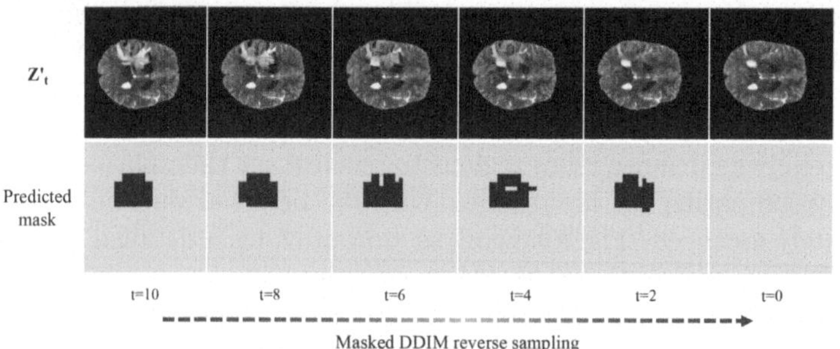

Fig. 2. Visual example of the reverse process. Both the predicted mask and the decoded latent representation of the intermediate reverse step (z'_t) at multiple time steps are depicted to highlight the masked reverse sampling in MAD-AD.

3.3 Anomaly Localization

Equation 6 enables a correcting trajectory from z'_T to z'_0, resulting in generating high-quality normal variation of the anomalous image in fewer steps. To accurately localize anomalies, we used the discrepancy between the input image and its reconstructed normal counterpart. More concretely, using the "normal" latent embedding $\tilde{z}'_0$, we generated a reconstructed normal sample in the image-space as $\tilde{x}'_0 = \mathcal{V}_{D,\phi}(\tilde{z}'_0)$, where $\mathcal{V}_{D,\phi}$ is the pre-trained VAE decoder. The predicted anomaly map is then given by:

$$\boldsymbol{a} = G * \min(\|\tilde{\boldsymbol{x}}'_0 - \boldsymbol{x}'_0\|, \gamma)/\gamma, \tag{8}$$

where G is a Gaussian kernel to smooth the predicted mask, $*$ is the convolution operator, and γ is a threshold designed to prevent assigning excessive weight to patches with significant deviations.

4 Experiments

4.1 Experimental Setting

Datasets. We employ three datasets to asses the performance of UAD methods. **IXI Dataset** [1]: a publicly available resource with brain MRI scans from approximately 600 healthy subjects. **ATLAS 2.0** [25] includes 655 T1-weighted MRI scans accompanied by expert-segmented lesion masks. As a pre-processing, all brain scans of both IXI and ATLAS 2.0 datasets were registered to MNI152 1mm templates and normalized to the 98th percentile. Then, mid-axial slices were extracted and padded to the resolution of 256 × 256 pixels. **BraTS'21** [3]: following the experimental setup of IterMask[2] [24], we also employ this dataset, which comprises 1251 brain scans across four modalities: T1-weighted, contrast-enhanced T1-weighted (T1CE), T2-weighted, and T2 Fluid Attenuated Inversion Recovery (FLAIR). For each scan, 20 middle axial slices of the skull-stripped brain are extracted, which are padded to the resolution of 256 × 256 pixels.

Training/Testing Protocol. We found that the training and testing protocols considerably differ in the UAD literature. For a fair comparison with prior methods, we evaluated our approach in two widely-adopted settings, comparing against the methods that were originally evaluated in each of these settings. **Setting-1 (*S1*)** [9]: training is performed on the middle slices of IXI subjects, whereas only middle slices of ATLAS 2.0 are used for testing. **Setting-2 (*S2*)** [24]: in this setting, only normal slices from a given modality are used for training, while the abnormal slice of that modality with the largest pathology is employed for inference. The BRATS'21 dataset is used in this case, which is split into training (80%), validation (10%) and testing (10%).

Evaluation Metrics. To evaluate the performance of our brain anomaly detection model, we use the Maximum Dice score, which reports the highest value obtained for thresholds ranging from 0 to 1. Following [9], we employ the global Maximum Dice score in setting *S1*, which first flattens and concatenates all segmentations and predictions before calculating the maximum Dice score. For setting *S2*, we instead consider the regular Maximum Dice score.

Implementation Details. To project the data into the latent space, we employed a pre-trained perceptual compression VAE model [31]. This model

Table 1. Performance in setting *S1*: results across different lesion sizes, where bold highlights the best method and improvements of our approach compared to the best baseline are indicated in green.

Method	Pathology (Global Max Dice)↑			
	Average	Small	Medium	Large
DDPM [16]$_{NeurIPS'20}$	8.1	1.4	9.5	25.7
AnoDDPM [39]$_{CVPRw'22}$	18.1	4.8	23.5	46.7
AutoDDPM [11]$_{ICMLW'23}$	17.0	4.5	22.1	43.5
pDDPM [7]$_{MIDL'24}$	22.3	8.0	30.2	47.7
THOR [9]$_{MICCAI'24}$	29.7	11.5	39.2	63.6
MAD-AD *(Ours)*	**51.6**$_{+21.9}$	**15.5**$_{+4.0}$	**50.1**$_{+10.9}$	**64.1**$_{+0.5}$

leverages an autoencoder trained using a combination of perceptual loss [41] and a patch-based adversarial objective, allowing for effectively reducing the spatial dimension by a factor of 8 ($256 \rightarrow 32$). As this model was originally trained for RGB images, we further adapted it and fine-tuned it for single-channel brain MRI data. Then, the VAE remained frozen throughout training the diffusion model (we used a UNet with attention as the diffusion model). The number of training and inference time-steps (T) is set to 10. To form the random mask at each iteration, the masking ratio is drawn from a uniform distribution $U[0, 0.4]$, and the patch sizes of the mask along the X and Y axes are sampled independently from the following set: $\{1, 2, 4, 8\}$. The random mask is then multiplied by the brain mask to prevent noise in non-brain (i.e., background) patches. The model was trained for 300 epochs using a batch size of 96 and AdamW optimizer with a learning rate of 5×10^{-4}.

4.2 Results

Quantitative Results. We empirically validated our method against a set of relevant state-of-the-art brain unsupervised anomaly detection methods in the two settings described in Sect. 4.1. Table 1 reports the results under the first setting, which uses middle slices of the IXI dataset for training, and middle slices of ATLAS 2.0 for evaluation. We can observe that the proposed approach substantially outperforms existing diffusion-based methods, particularly on small- and medium-sized lesions. More concretely, our approach improves the best baseline (the recent THOR method [9]) by 4.0% and 10.9% in small and medium lesions, respectively, and by 21.9% when using the whole dataset (referred to as *"Average"*, as in [9]). The performance gap further increases if we consider the second best baseline (i.e., pDDPM), where average differences are equal to nearly 30%. Note that even though our model yields superior performance for small pathologies, it still struggles to accurately locate these type of small abnormalities, similarly to existing approaches. In MAD-AD, this low performance may be due to the use of a diffusion model on a compressed latent space, which can lead to overlooking very small pathologies.

Table 2. Performance in setting *S2*: results across different modalities, where bold highlights the best method and performance improvements (*resp.* decrease) of our approach compared to the best baseline are indicated in **green** (*resp.* **red**).

Method	Modality (Max Dice)↑				
	FLAIR	T1CE	T2-w	T1-w	Avg
AE [4]$_{MedIA'21}$	33.4	32.3	30.2	28.5	31.1
DDPM [16]$_{Neurips'20}$	60.7	37.9	36.4	29.4	41.1
AutoDDPM [11]$_{ICMLW'23}$	55.5	36.9	29.7	33.5	38.9
Cycl.UNet [23]$_{MICCAI'23}$	65.0	42.6	49.5	37.0	48.5
DAE [19][0, ∞]$_{MIDL'22}$	79.7	36.7	69.6	29.5	53.9
IterMask^2 [24]$_{MICCAI'24}$	**80.2**	61.7	71.2	58.5	67.9
MAD-AD *(Ours)*	76.2$_{-4.0}$	**68.5**$_{+6.8}$	**73.2**$_{+2.0}$	**63.4**$_{+4.9}$	**70.3**$_{+2.4}$

Under the second setting (*S2*), the proposed approach yields the best scores in three out of four modalities, leading to the highest average score (Table 2). While the differences with respect to the best baseline are smaller in this setting, improvements over the second best baseline are still considerably high, with an overall boost near to 16%. Thus, quantitative results under two common settings in the UAD literature demonstrate the superior performance of our approach for this task, highlighting its potential as a powerful alternative to existing methods.

Ablation on Using Different Sources for the Anomaly Score. In this section, we investigate the impact of using different strategies to form the anomaly map: pixel-level discrepancies $(\boldsymbol{x}'_0, \boldsymbol{x}'_T)$, latent-space discrepancies $(\boldsymbol{z}'_0, \boldsymbol{z}'_T)$, and the average of the predicted mask at reverse diffusion steps $(\frac{1}{T}\sum_{t=1}^{T} f_{M,\theta}(\boldsymbol{z}'_t))$. These results, which are reported in Table 3, showcase the better performance of resorting to the image-level difference, motivating our design choice.

Impact of Hyper-parameters. Next, we evaluate the influence of key hyper-parameters on the performance of the proposed method, whose results on the BRATS dataset are depicted in Table 4. From these results, we can observe that the choices made for the hyper-parameters lead to the best results overall.

Qualitative Results. To further highlight the effectiveness of our unsupervised anomaly detection method, we present qualitative results obtained on the ATLAS 2.0 dataset (*S1*) and across all modalities of the BraTS dataset (*S2*). Figure 3. Figure 3 showcases representative examples of anomalous instances, their normal counterpart reconstructions, segmentation, and anomaly map by MAD-AD. These qualitative results underscore the ability of our approach to accurately localize anomalous regions without relying on supervised labels.

Table 3. Effect of different sources for the anomaly score in MAD-AD (BRATS'21).

Anomaly source	Modality (Max Dice)↑				
	T1-w	T1CE	T2-w	FLAIR	Avg
Average predicted mask	60.4	62.3	65.6	66.1	63.6
Latent-level diff	63.0	66.2	69.6	75.5	68.6
Image-level diff	**63.4**	**68.5**	**73.2**	**76.2**	**70.3**

Table 4. Ablation study on two key hyper-parameters of MAD-AD.

Hyper- parameter	Value	Modality (Max Dice)↑				
		T1-w	T1CE	T2-w	FLAIR	Avg
#DDIM steps	2	62.3	**70.1**	68.5	75.3	69.0
	5	63.1	69.4	71.1	74.0	69.4
	10	**63.4**	68.5	**73.2**	**76.2**	**70.3**
γ	0.2	**63.4**	**68.5**	73.2	**76.2**	70.3
	0.4	63.3	68.3	**73.8**	76.0	**70.3**
	✗	62.0	67.9	72.6	74.9	69.3

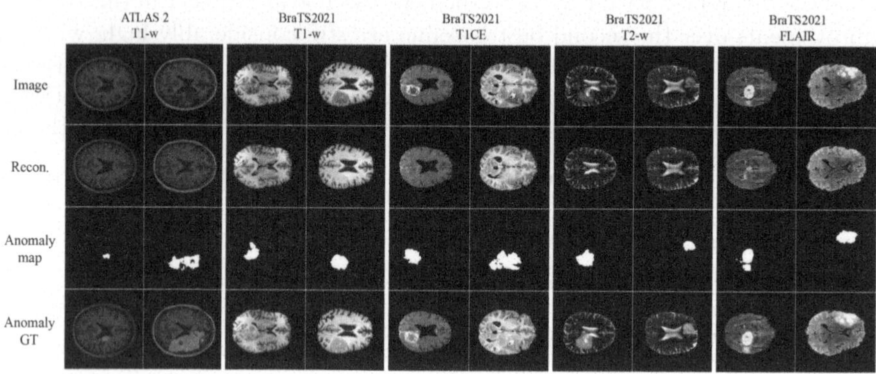

Fig. 3. Qualitative results. Anomaly segmentation performance obtained by our approach (i.e., "Anomaly map") in brain MRI for different modalities and datasets.

5 Conclusion

This paper introduces a novel unsupervised anomaly detection method for brain MRI using a latent diffusion model with a random masking strategy. The approach leverages latent space, as brain anomalies in the latent space could be considered as noise and therefore be removed during the denoising process of diffusion models. Furthermore, by using a mask prediction module in the diffusion model, the model can selectively modify anomalous regions while preserving normal areas, enabling accurate identification of anomalous regions. Experiments on two datasets and two common brain UAD experimental settings demonstrate the superiority of our approach, validating its effectiveness in detecting and localizing brain anomalies without requiring labeled data, and showcasing its promising potential as an alternative to existing methods.

References

1. Ixi dataset. https://brain-development.org/ixi-dataset/, Accessed 15 Feb 2023
2. Atlason, H.E., Love, A., Sigurdsson, S., Gudnason, V., Ellingsen, L.M.: Unsupervised brain lesion segmentation from MRI using a convolutional autoencoder. In: Medical Imaging 2019: Image Processing, vol. 10949, pp. 372–378. SPIE (2019)
3. Bakas, S., et al.: Advancing the cancer genome atlas glioma mri collections with expert segmentation labels and radiomic features. Sci. Data **4**(1), 1–13 (2017)
4. Baur, C., Denner, S., Wiestler, B., Navab, N., Albarqouni, S.: Autoencoders for unsupervised anomaly segmentation in brain MR images: a comparative study. Med. Image Anal. **69**, 101952 (2021)
5. Baur, C., Graf, R., Wiestler, B., Albarqouni, S., Navab, N.: Steganomaly: inhibiting cyclegan steganography for unsupervised anomaly detection in brain mri. In: International Conference on Medical Image Computing and Computer-Assisted Intervention, pp. 718–727. Springer (2020)
6. Behrendt, F., Bengs, M., Rogge, F., Krüger, J., Opfer, R., Schlaefer, A.: Unsupervised anomaly detection in 3D brain MRI using deep learning with impured training data. 2022 IEEE 19th International Symposium on Biomedical Imaging (ISBI), pp. 1–4 (2022)
7. Behrendt, F., Bhattacharya, D., Krüger, J., Opfer, R., Schlaefer, A.: Patched diffusion models for unsupervised anomaly detection in brain MRI. In: Medical Imaging with Deep Learning, pp. 1019–1032. PMLR (2024)
8. Behrendt, F., et al.: Leveraging the Mahalanobis distance to enhance unsupervised brain MRI anomaly detection. In: International Conference on Medical Image Computing and Computer-Assisted Intervention, pp. 394–404. Springer (2024)
9. Bercea, C.I., Wiestler, B., Rueckert, D., Schnabel, J.A.: Diffusion models with implicit guidance for medical anomaly detection. In: International Conference on Medical Image Computing and Computer-Assisted Intervention, pp. 211–220. Springer (2024)
10. Bercea, C.I., Wiestler, B., Rueckert, D., Schnabel, J.A.: Generalizing unsupervised anomaly detection: towards unbiased pathology screening. In: Medical Imaging with Deep Learning, pp. 39–52. PMLR (2024)
11. Bercea, C.I., Neumayr, M., Rueckert, D., Schnabel, J.A.: Mask, stitch, and re-sample: enhancing robustness and generalizability in anomaly detection through automatic diffusion models. In: ICML 3rd Workshop on Interpretable Machine Learning in Healthcare (IMLH)
12. Croitoru, F.A., Hondru, V., Ionescu, R.T., Shah, M.: Diffusion models in vision: a survey. IEEE Trans. Pattern Anal. Mach. Intell. **45**(9), 10850–10869 (2023)
13. Goodfellow, I., et al.: Generative adversarial networks. Commun. ACM **63**(11), 139–144 (2020)

14. Graham, M.S., Pinaya, W.H., Tudosiu, P.D., Nachev, P., Ourselin, S., Cardoso, J.: Denoising diffusion models for out-of-distribution detection. In: Proceedings of the IEEE/CVF Conference on Computer Vision and Pattern Recognition, pp. 2948–2957 (2023)
15. Havaei, M., et al.: Brain tumor segmentation with deep neural networks. Med. Image Anal. **35**, 18–31 (2017)
16. Ho, J., Jain, A., Abbeel, P.: Denoising diffusion probabilistic models. Adv. Neural. Inf. Process. Syst. **33**, 6840–6851 (2020)
17. Iqbal, H., Khalid, U., Chen, C., Hua, J.: Unsupervised anomaly detection in medical images using masked diffusion model. In: International Workshop on Machine Learning in Medical Imaging, pp. 372–381. Springer (2023)
18. Kamnitsas, K., et al.: Ensembles of multiple models and architectures for robust brain tumour segmentation. In: Brainlesion: Glioma, Multiple Sclerosis, Stroke and Traumatic Brain Injuries: Third International Workshop, BrainLes 2017, Held in Conjunction with MICCAI 2017, Quebec City, QC, Canada, 14 September 2017, Revised Selected Papers 3, pp. 450–462 (2018)
19. Kascenas, A., Pugeault, N., O'Neil, A.Q.: Denoising autoencoders for unsupervised anomaly detection in brain MRI. In: International Conference on Medical Imaging with Deep Learning (pp. 653-664). PMLR (2022)
20. Kervadec, H., Bouchtiba, J., Desrosiers, C., Granger, E., Dolz, J., Ben, I.: Boundary loss for highly unbalanced segmentation. Med. Image Anal. **67**(10185), 101851 (2021)
21. Kingma, D., Salimans, T., Poole, B., Ho, J.: Variational diffusion models. Adv. Neural. Inf. Process. Syst. **34**, 21696–21707 (2021)
22. Kuijf, H.J., et al.: Standardized assessment of automatic segmentation of white matter hyperintensities and results of the WMH segmentation challenge. IEEE Trans. Med. Imaging **38**(11), 2556–2568 (2019)
23. Liang, Z., Anthony, H., Wagner, F., Kamnitsas, K.: Modality cycles with masked conditional diffusion for unsupervised anomaly segmentation in MRI. In: International Conference on Medical Image Computing and Computer-Assisted Intervention, pp. 168–181 (2023)
24. Liang, Z., Guo, X., Noble, J.A., Kamnitsas, K.: Itermask 2: iterative unsupervised anomaly segmentation via spatial and frequency masking for brain lesions in mri. In: International Conference on Medical Image Computing and Computer-Assisted Intervention, pp. 339–348. Springer (2024)
25. Liew, S.L., et al.: A large, curated, open-source stroke neuroimaging dataset to improve lesion segmentation algorithms. Sci. Data **9**(1), 320 (2022)
26. Lu, F., Yao, X., Fu, C.W., Jia, J.: Removing anomalies as noises for industrial defect localization. In: Proceedings of the IEEE/CVF International Conference on Computer Vision, pp. 16166–16175 (2023)
27. Naval Marimont, S., Baugh, M., Siomos, V., Tzelepis, C., Kainz, B., Tarroni, G.: DISYRE: diffusion-inspired synthetic restoration for unsupervised anomaly detection. In: Proceedings/IEEE International Symposium on Biomedical Imaging (ISBI), IEEE (2024)
28. Naval Marimont, S., Siomos, V., Baugh, M., Tzelepis, C., Kainz, B., Tarroni, G.: Ensembled cold-diffusion restorations for unsupervised anomaly detection. In: International Conference on Medical Image Computing and Computer-Assisted Intervention, pp. 243–253 (2024)
29. Pagnozzi, A.M., Fripp, J., Rose, S.E.: Quantifying deep grey matter atrophy using automated segmentation approaches: a systematic review of structural mri studies. Neuroimage **201**, 116018 (2019)

30. Peebles, W., Xie, S.: Scalable diffusion models with transformers. In: Proceedings of the IEEE/CVF International Conference on Computer Vision, pp. 4195–4205 (2023)
31. Rombach, R., Blattmann, A., Lorenz, D., Esser, P., Ommer, B.: High-resolution image synthesis with latent diffusion models. In: Proceedings of the IEEE/CVF Conference on Computer Vision and Pattern Recognition, pp. 10684–10695 (2022)
32. Rumelhart, D.E., Hinton, G.E., Williams, R.J.: Learning internal representations by error propagation. In: Rumelhart, D.E., McClelland, J.L., the PDP Research Group (eds.) Parallel Distributed Processing: Explorations in the Microstructure of Cognition, vol. 1, pp. 318–362. MIT Press, Cambridge, MA, USA (1986)
33. Schlegl, T., Seeböck, P., Waldstein, S.M., Langs, G., Schmidt-Erfurth, U.: f-AnoGAN: fast unsupervised anomaly detection with generative adversarial networks. Med. Image Anal. **54**, 30–44 (2019)
34. Schlegl, T., Seeböck, P., Waldstein, S.M., Schmidt-Erfurth, U., Langs, G.: Unsupervised anomaly detection with generative adversarial networks to guide marker discovery. In: International Conference on Information Processing in Medical Imaging, pp. 146–157. Springer (2017)
35. Silva-Rodríguez, J., Naranjo, V., Dolz, J.: Constrained unsupervised anomaly segmentation. Med. Image Anal. **80**, 102526 (2022)
36. Sinha, A., Dolz, J.: Multi-scale self-guided attention for medical image segmentation. IEEE J. Biomed. Health Inform. **25**(1), 121–130 (2020)
37. Sohl-Dickstein, J., Weiss, E., Maheswaranathan, N., Ganguli, S.: Deep unsupervised learning using nonequilibrium thermodynamics. In: International Conference on Machine Learning, pp. 2256–2265. PMLR (2015)
38. Song, J., Meng, C., Ermon, S.: Denoising diffusion implicit models. In: International Conference on Learning Representations (2021)
39. Wyatt, J., Leach, A., Schmon, S.M., Willcocks, C.G.: ANODDPM: anomaly detection with denoising diffusion probabilistic models using simplex noise. In: Proceedings of the IEEE/CVF Conference on Computer Vision and Pattern Recognition Workshops, pp. 650–656 (2022)
40. Yan, C., Zhang, S., Liu, Y., Pang, G., Wang, W.: Feature prediction diffusion model for video anomaly detection. In: Proceedings of the IEEE/CVF International Conference on Computer Vision, pp. 5527–5537 (2023)
41. Zhang, R., Isola, P., Efros, A.A., Shechtman, E., Wang, O.: The unreasonable effectiveness of deep features as a perceptual metric. In: Proceedings of the IEEE Conference on Computer Vision and Pattern Recognition, pp. 586–595 (2018)
42. Zimmerer, D., Isensee, F., Petersen, J., Kohl, S., Maier-Hein, K.: Unsupervised anomaly localization using variational auto-encoders. In: Shen, D., et al. (eds.) MICCAI 2019. LNCS, vol. 11767, pp. 289–297. Springer, Cham (2019). https://doi.org/10.1007/978-3-030-32251-9_32

Self-supervised Learning

Taming Masked Image Modeling for Chest X-ray Diagnosis by Incorporating Clinical Visual Priors

Zihao Zhao[1], Mei Wang[1,2], Zhiming Cui[1], Sheng Wang[1], Qian Zhou[3], Li Fan[3], Qian Wang[1,5], and Dinggang Shen[1,4,5](✉)

[1] School of Biomedical Engineering and State Key Laboratory of Advanced Medical Materials and Devices, ShanghaiTech University, Shanghai 201210, China
dgshen@shanghaitech.edu.cn

[2] School of Biomedical Engineering, Southern Medical University, Guangzhou 510515, China

[3] Department of Radiology, Second Affiliated Hospital of Naval Medical University, Shanghai, China

[4] Shanghai United Imaging Intelligence Co., Ltd., Shanghai 200230, China

[5] Shanghai Clinical Research and Trial Center, Shanghai 201210, China

Abstract. Accurate chest X-ray diagnosis typically requires a nuanced understanding of image semantics. Hence, masked image modeling (MIM) has gained attention as a promising approach for learning local image semantics by reconstructing masked patches from their surrounding context. However, existing MIM methods have two key limitations when applied to chest X-ray images: (1) the use of semantic-unaware masking strategies which may corrupt clinically significant visual features, and (2) insufficient multi-scale supervision which fails to consider abnormalities of varying sizes. To address these challenges, we introduce CXR-MIM, a novel MIM framework that incorporates clinical visual priors to guide the image modeling process. Specifically, we leverage these priors to indicate key aspects of a visual feature, including its **semantics, location, size, and shape**. For semantic and location awareness, CXR-MIM utilizes radiologists' gaze data to differentiate between clinically significant and insignificant regions in the masking phase, implementing a controlled masking strategy that moderately masks significant diagnostic features. To capture abnormalities of different sizes and shapes, we develop a pyramid adaptive reconstruction module to provide supervision across multiple scales in the reconstruction phase, which is further enhanced by semantic-aware recalibrated gaze heatmaps. Experiments on both two publicly available datasets and one private dataset demonstrate the superior performance of CXR-MIM compared to existing MIM methods. Further evaluation involving pre-training on an additional large-scale dataset indicates promising scalability with increasing data size, underscoring its potential in the age of foundation models. Our code will be available at Github.

Keywords: Masked Image Modeling · Chest X-ray · Clinical Visual Priors · Human Visual Attention

I. Oguz et al. (Eds.): IPMI 2025, LNCS 15830, pp. 157–171, 2026.
https://doi.org/10.1007/978-3-031-96625-5_11

1 Introduction

Accurate diagnosis of chest X-rays necessitates a fine-grained understanding of local image semantics, as certain regions may contain subtle and complex features that are critical for identifying potential thoracic abnormalities compared to normal regions [2]. Therefore, advanced techniques that extend beyond simple image-level analysis are required. Masked image modeling (MIM) [3,12] emerges as a promising solution, as it learns representations at the patch level by predicting missing parts from the surrounding context.

However, existing MIM methods struggle to capture clinically significant features in chest X-rays (as will be shown in Sect. 4.2) due to design choices that ignore clinical visual priors, limiting their applicability in medical imaging. This conflict primarily involves two aspects. First, traditional MIMs employ semantic-unaware masking, disregarding the clinical prior knowledge that certain regions are more important than others. Second, these methods lack sufficient multi-scale supervision to accommodate the diverse sizes and types of abnormalities present in chest X-rays [14,22,25], as illustrated in Fig. 1a. The varying sizes of lesion regions further increase the difficulty for the self-attention mechanism to learn clinical semantics autonomously.

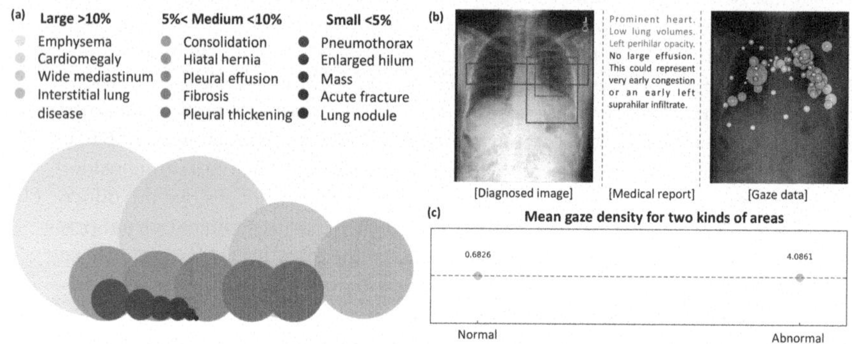

Fig. 1. Illustration of thoracic abnormalities of varying sizes and the alignment of gaze data with these regions. (a) shows the proportion of different abnormalities in relation to the entire image across the REFLACX dataset [4], categorized into three scales: small, medium, and large. (b) qualitatively demonstrates how gaze data aligns closely with lesion regions, indicated by colored bounding boxes. (c) provides quantitative results by computing gaze density for normal and abnormal regions, respectively. The average gaze density within abnormal regions is six times higher than that of normal regions.

To address these identified gaps, we introduce CXR-MIM, a novel approach that integrates clinical visual priors into MIM for chest X-ray analysis. Semantics, location, size, and shape of a certain visual feature are four attributes introduced in this study to indicate clinical visual priors. We expect the MIM method to be aware of these attributes of image regions during image modeling process. Central to our method is the utilization of gaze data from expert radiologists and a multi-scale reconstruction strategy. As qualitatively and quantitatively illustrated in Fig. 1b and c, gaze attention serves as a semantic indicator of clinically relevant areas across diverse scales. By leveraging this insight, CXR-MIM employs an attention-controlled masking strategy that treats clinically significant and insignificant regions separately, strategically masking less critical areas while preserving essential diagnostic features. Furthermore, acknowledging the challenge posed by the varying sizes of thoracic abnormalities, we propose a multi-scale semantic-aware module, termed pyramid adaptive reconstruction, for effective supervision. Augmented with gaze attention, this module provides semantic-aware supervision across multiple scales, dynamically adjusting the model's focus to align with the granularity and scale of the pathologies in the images. These two modules synergistically enhance the effectiveness.

To summarize, the key contributions of our work are as follows:

- We present CXR-MIM, a novel MIM method that integrates clinical visual priors tailored specifically for chest X-rays, composed of attention-controlled masking and pyramid adaptive reconstruction.
- Through our proposed attention-controlled masking, we investigate how to apply masking in the MIM for chest X-ray analysis, i.e., clinically significant regions should be moderately masked to balance the preservation of critical information, avoiding both full masking and full preservation.
- We demonstrate the effectiveness and generalizability of CXR-MIM by evaluating the learned image representations on both two public datasets and one private real-world dataset. **The performance concerning scaling training data** further highlights its potential in the age of foundation models.

2 Related Work

Masked Image Modeling. The concept of masked image modeling derives from [3], which first adapted the masked language modeling [8] to image representation learning by masking and predicting in the latent space. He et al. [12] advanced this research with MAE, demonstrating that directly masking and reconstructing image pixels is sufficiently effective for MIM given certain designs. The simple yet effective design of MAE established it as a common baseline for MIM methods. Given that MIM primarily involves random masking, encoding, and reconstruction, several studies have sought to improve it along these dimensions. AttMask [18] employed a teacher-student architecture to use the teacher's attention map to guide masking. HPM [33] introduced an auxiliary predictor to identify patches that are difficult to reconstruct, thereby deciding where to mask next. Another line of research [27,34,37] aimed to enhance reconstruction effectiveness. Notably, DeepMIM [27] proposed a hierarchical reconstruction strategy, which is methodologically similar to ours but differs in design.

Gaze Data as a Semantic Indicator. Due to its ease of collection and interpretability compared with other annotations [35], an increasing number of studies [16] have incorporated radiologist's gaze data to reflect the semantics of specific regions. For example, an early-stage study [19] proposed a straightforward idea, they developed a multi-task method that simultaneously performs chest X-ray diagnosis and gaze heatmap prediction. Similarly, Wang et al. [35] utilized gaze heatmaps as auxiliary supervision for knee X-ray classification, guiding the network to pay more attention to regions receiving long-lasting human attention. Besides simply adopting gaze data as supervision, more studies investigated the potential of gaze data as an input to the network. Alsharid et al. [1] trained a gaze predictor for ultrasound videos and incorporated it into an image captioning model, improving descriptive accuracy. There are also studies considering gaze data as a kind of valuable guidance. Zhao et al. [41] leveraged gaze data to construct positive pairs for medical contrastive learning, while Kong et al. [20] employed it to reduce false positives for improved detection. Collectively, these studies demonstrate the effectiveness of gaze data as an excellent semantic indicator.

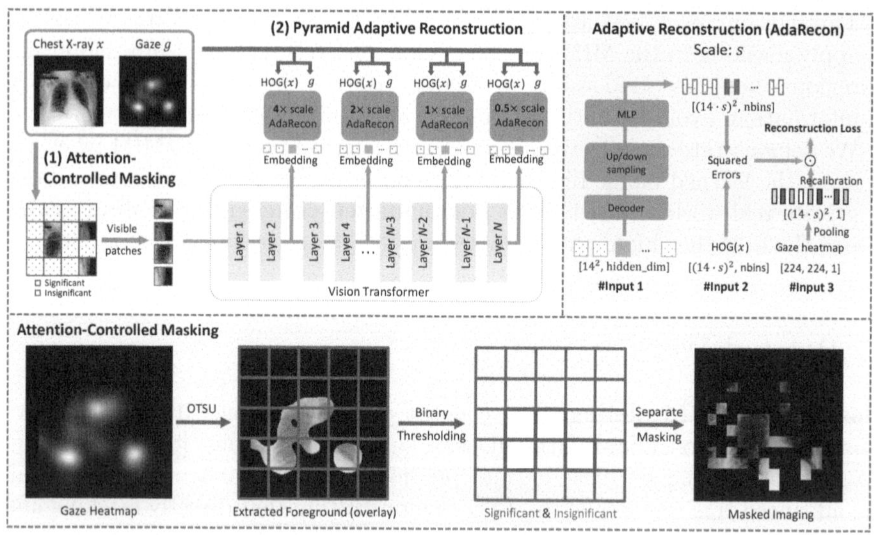

Fig. 2. Overview of our proposed CXR-MIM by using ViT-B/16 as the default vision backbone. CXR-MIM integrates (1) attention-controlled masking to preserve clinically significant features based on radiologist attention, and (2) pyramid adaptive reconstruction to reconstruct multi-scale HOG features using embeddings from selected transformer layers. The reconstruction loss is further adaptively weighted through recalibrated gaze heatmaps, emphasizing important regions during training.

3 Method

This section presents CXR-MIM, a novel MIM approach with semantic-aware capabilities and sensitivity to multi-scale features. We detail its two main components: attention-controlled masking (Sect. 3.2) and pyramid adaptive reconstruction (Sect. 3.3).

3.1 Overview

As illustrated in Fig. 2, CXR-MIM follows the typical MIM framework of masking, encoding, and reconstruction, with significant enhancements in the masking and reconstruction phases.

Given a batch of chest X-rays $\{x_i\}_{i=1}^N$, where each $x_i \in \mathbb{R}^{H\times W}$, CXR-MIM aims to enable Vision Transformers (ViTs) [9] to learn high-quality, meaningful representations specific to medical imaging. Recognizing that **radiologist gaze data** can serve as an effective semantic indicator [20,23,35,41], we incorporate corresponding gaze heatmaps $\{g_i\}_{i=1}^N$, with each $g_i \in \mathbb{R}^{H\times W}$ matching the size of x_i. Since MIM operates on individual samples without inter-sample interactions, we denote a single input image as x, and g for its associated gaze heatmap, in Fig. 2 and subsequent descriptions for simplicity. CXR-MIM begins by generating a mask using our attention-controlled masking method, designed to preserve clinically significant features while reducing redundancy. The masked image $\tilde{x}$ is input into a Vision Transformer (ViT-B/16 by default), producing feature embeddings $z^l \in \mathbb{R}^{(p+1)\times d}$ at each layer $l \in \{1,\dots,L\}$. Here, p denotes the fixed number of image patches (e.g., 14^2 for a 224×224 input in ViT-B/16), and $p+1$ indicates the concatenation of the [CLS] token to image embeddings [9].

Based on empirical validation (see Table 2b), we select $\{z^2, z^4, z^{10}, z^{12}\}$ to capture a hierarchy of features from low-level to high-level representations during reconstruction. The pyramid adaptive reconstruction module decodes histogram of oriented gradients (HOG) features—chosen for their effectiveness in capturing edge and texture information critical in medical images [6,7,10,11,28,30]—at pre-defined scales. It further adaptively weights the reconstruction loss of each masked patch using recalibrated gaze heatmaps, thus emphasizing clinically significant regions and improving diagnostic accuracy.

3.2 Attention-Controlled Masking

Traditional MIM methods apply random masking by uniformly sampling patches, effectively following a uniform distribution. In contrast, our attention-controlled masking differentiates between clinically significant and insignificant regions, resembling sampling from a bimodal distribution [15,21,31].

Significance Discrimination. To identify clinically significant regions, we utilize gaze heatmaps $g \in \mathbb{R}^{H\times W}$ derived from radiologists' visual attention. Assuming that gaze intensities indicate importance but may not strictly follow a bimodal distribution, we employ OTSU's method [24] to automatically segment the heatmap into foreground (significant regions) and background (insignificant regions), yielding a binary mask $\mathbf{B} \in \{0,1\}^{H\times W}$. The binary mask then serves as a reference to discriminate significant regions. Specifically, after partitioning x into p non-overlapping patches, for each patch, we classify it as significant if more than 50% of its pixels belong to the foreground, as illustrated by an indicator

$$S_k = \begin{cases} 1, & \text{if } \sum_{(i,j)\in \text{patch}(k)} \mathbf{B}_{i,j} > 0.5 \cdot n_p \\ 0, & \text{otherwise} \end{cases},$$

where n_p is the number of pixels per patch, and $k = 1, 2, \ldots, p$. Accordingly, $p_s = \sum_{k=1}^{p} S_k$ denotes the number of significant patches, and $p_n = p - p_s$ the number of insignificant patches.

Bimodal Controlled Sampling. Our masking strategy employs two ratios, global masking ratio r, the overall proportion of patches to mask, resulting in $m_g = r \cdot p$ masked patches. Clinical masking ratio r_c, the proportion of significant patches to mask, leading to $m_s = r_c \cdot p_s$ masked significant patches. Hence, the number of insignificant patches to mask is then $m_n = m_g - m_s = r \cdot p - r_c \cdot p_s$. We randomly select m_s significant patches and m_n insignificant patches to mask, resulting in M_g. The masked image $\tilde{x}$ is obtained by:

$$\tilde{x} = \text{patchify}(x) \odot M_g,$$

where $\odot$ denotes element-wise multiplication, and patchify$(\cdot)$ indicates the partitioning of images into patches.

3.3 Pyramid Adaptive Reconstruction

During reconstruction, CXR-MIM employs a pyramid adaptive reconstruction strategy to effectively capture and reconstruct multi-scale HOG features.

Pyramid Strategy. We choose HOG features over direct pixel reconstruction because HOG captures essential local texture information while reducing redundancy inherent in medical images [6,11,30]. Embeddings $\{z^2, z^4, z^{10}, z^{12}\}$ from selected ViT layers are input into scale-specific adaptive reconstruction modules, denoted as $s \times$ AdaRecon (see Fig. 2), where $s \in \{4, 2, 1, 0.5\}$ is the rescaling factor. To encourage the vision backbone to learn meaningful representations, the lightweight AdaRecon modules are made deliberately simple. Each module comprises a single transformer layer, a rescaling layer, and a multi-layer perceptron (MLP). The rescaling layer adjusts the spatial resolution of the embeddings. For example, for $s > 1$, it performs upsampling via transposed convolution, while for $s < 1$, it downsamples using average pooling.

Someone may notice that feature embeddings from a shallower layer tend to have a larger s. This is because the process of encoding images in a ViT is progressively from low-level representations to high-level representations [26]. In ViTs, the encoding process transitions from low-level representations—which capture basic features such as edges and textures—to high-level representations that encode more complex attributes like objects and semantic categories. This encoding process is inherently lossy; as the model progresses through layers, there is a gradual reduction in the informational content of embeddings, with earlier layers retaining more detailed image features. To leverage this characteristic, we assign a larger s to embeddings from the shallower layers. Conversely, we rescale the embeddings from the final layers to half their original size to enhance their capability for high-level information extraction. This scaling approach *not only* enhances the extraction capabilities across different layers *but also* mitigates the risk of overfitting (see Sect. 4.3).

Adaptive Reconstruction. After processing through the MLP, the embeddings are transformed into output features with shape $[(p \times s)^2, n_{\text{bins}}]$, where n_{bins} corresponds to the number of bins in the HOG descriptor. This process enables the reconstruction of HOG features at multiple scales, making the model sensitive to texture features of varying sizes, which is essential for accurately representing complex medical images. To maintain consistency between the reconstruction and masking phases and also ensure semantic awareness throughout, we utilize the gaze heatmap g when calculating the reconstruction loss. Specifically, we perform average pooling on the element-wise product of the gaze heatmap and the binary mask $\mathbf{M}_g$ from the masking phase:

$$g'_s = \text{AvgPool}\left(g \odot \mathbf{M}_g\right),$$

where $g'_s \in \mathbb{R}^{(p \cdot s) \times (p \cdot s)}$, and it is then reshaped to a vector of size $[(p \cdot s)^2, 1]$. To mitigate the impact of noise inherent in eye-tracking data, we recalibrate the gaze heatmap before incorporating it into the loss calculation. We apply softmax normalization to g'_s (with a temperature of 0.1) and add the binary mask $\mathbf{M}_g$:

$$\mathbf{W}_s = \text{softmax}\left(g'_s, \tau\right) + \mathbf{M}_g,$$

where $\mathbf{W}_s$ represents the recalibrated attention weights for each patch, and τ denotes a temperature coefficient. This recalibration sharpens the attention on significant regions and ensures stability during training by preventing noise in the gaze data from adversely affecting the loss computation. The reconstruction loss for the $s\times$ AdaRecon module is computed as the weighted sum of squared errors between the predicted and target HOG features:

$$\mathcal{L}_s = \frac{\mathbf{W}_s \odot \mathbf{M}_g \cdot \left(\hat{\mathbf{H}}_s - \mathbf{H}_s\right)^2}{\sum\left(\mathbf{W}_s \odot \mathbf{M}_g\right)},$$

where $\hat{\mathbf{H}}_s$ and $\mathbf{H}_s$ are the predicted and target HOG features for image x, respectively. The total reconstruction loss $\mathcal{L}_{\text{recon}}$ is then calculated by:

$$\mathcal{L}_{\text{recon}} = \sum_{s \in \{4,2,1,0.5\}} \mathcal{L}_s.$$

By employing this pyramid adaptive reconstruction strategy, CXR-MIM effectively captures multi-scale texture information while emphasizing clinically significant regions, thereby enhancing the quality of learned representations.

4 Experimental Results

In this section, we first introduce the datasets and implementation details. Next, we compare CXR-MIM with various existing MIM methods. A comprehensive ablation study is then implemented to justify the rationality of our design, together with an initial investigation of a proper masking strategy for chest X-rays. Finally, we explore the potential of CXR-MIM when scaling the dataset size.

4.1 Datasets and Implementations

Following prior studies [12,32,41], we assess the effectiveness of different MIM methods by evaluating their linear-probing classification performance on several downstream datasets.

Pre-training Datasets: This study utilizes two pre-training datasets: REFLACX [4] and MIMIC-CXR [17]. REFLACX comprises 2,616 chest X-rays, with some interpreted by multiple radiologists, and includes corresponding 3,032 eye-tracking data. It is used as the pre-training dataset in Sects. 4.2 and 4.3. In contrast, MIMIC-CXR is a large-scale dataset containing 377,100 chest X-rays and is hence employed in Sect. 4.4 to investigate the scaling performance.

Downstream Datasets: We evaluate pre-trained representations on three downstream datasets: RSNA Pneumonia (RSNA) [29], NIH Chest X-ray (NIH-CXR14) [36], and ChestStruct. RSNA Pneumonia contains approximately 29,700 frontal-view chest X-rays annotated for the presence of pneumonia. Following [14], we randomly split the dataset into training, validation, and test sets with a ratio of 70%:15%:15%. While RSNA Pneumonia focuses on a binary classification task, NIH Chest X-ray and ChestStruct involve challenging multi-label classification tasks [5,13]. NIH Chest X-ray consists of 112,120 X-rays covering 14 different thoracic diseases. We randomly split this dataset into training, validation, and test sets with a ratio of 70%:10%:20%. ChestStruct is a private real-world dataset containing 7,497 chest radiographs collected from Changzheng Hospital. It focuses on four common structural abnormalities: pneumothorax, pleural effusion, rib fracture, and hydropneumothorax. We adopt a 70%:10%:20% data split for this dataset. We choose different ratios for different downstream datasets because of their varying size.

Implementations: All pre-training experiments were conducted on a single NVIDIA Tesla V100S GPU with 32 GB memory, using a batch size of 128 and a learning rate (lr) set to $3e^{-3}$. We pre-trained on the REFLACX dataset for 400 epochs but reduced the training to 100 epochs for MIMIC-CXR due to its larger dataset size. Downstream linear probing evaluations were performed over 50 epochs on a single NVIDIA RTX 4090 GPU, with a batch size of 32 and a lr of $1e^{-3}$. Global masking ratio r and clinical masking ratio r_c were set as 0.75 and 0.5, respectively, across all pre-training experiments. Temperature coefficient τ used in pyramid adaptive reconstruction is set as 0.1. ViT-B/16 is the default vision backbone in this study. Unless otherwise specified, all downstream evaluations are repeated three times, and then the mean(std) values are reported.

Table 1. Quantitative comparison between all MIM methods considering the linear probing performance. All methods were pre-trained on the REFLACX dataset. The area under curve (AUC×100) is reported on RSNA, while for the other two datasets with multiple potential abnormalities, the macro-averaged AUC (mAUC)×100 is reported. For each experiment, we repeat 3 times with different random seeds and report mean(std) values.

Method	RSNA [29]			NIH-CXR14 [36]			ChestStruct
	1%	10%	100%	1%	10%	100%	100%
Semantic-unaware							
MAE [12]	79.19(0.06)	83.04(0.01)	84.64(0.01)	56.91(0.59)	63.73(0.56)	68.33(0.34)	73.05(0.03)
MaskFeat [37]	80.99(0.23)	84.29(0.11)	85.67(0.19)	61.18(0.62)	66.19(1.16)	70.21(0.15)	75.34(0.18)
dMAE [38]	81.74(0.40)	84.65(0.10)	85.70(0.03)	58.71(1.53)	65.41(1.73)	69.88(0.42)	74.52(0.08)
MaskAlign [39]	79.64(0.82)	84.56(0.12)	85.26(0.01)	58.37(0.93)	65.10(3.96)	70.52(1.47)	73.83(0.61)
Semantic-aware							
AttMask [18]	82.45(0.04)	84.04(0.09)	84.97(0.07)	60.32(0.26)	65.94(1.12)	70.43(0.39)	73.96(0.25)
HPM [33]	82.01(0.03)	84.30(0.02)	85.46(0.02)	59.93(0.18)	65.57(2.48)	70.87(0.71)	71.08(0.07)
AttnMA [39]	79.15(0.98)	84.06(0.15)	85.09(0.02)	57.55(0.34)	64.81(3.42)	70.01(1.42)	74.20(0.18)
DeepMIM [27]	81.89(0.36)	83.96(0.12)	85.18(0.11)	61.97(0.35)	66.41(1.52)	71.20(0.64)	76.61(0.09)
CXR-MIM	**83.51(0.08)**	**85.49(0.05)**	**86.83(0.01)**	**63.48(0.22)**	**68.53(0.68)**	**72.82(0.02)**	**79.26(0.20)**

4.2 Comparison with Existing Methods

In Table 1, we investigate the quantitative performance of CXR-MIM on all three downstream datasets. Semantic-unaware and -aware methods are both included to ensure a comprehensive comparison. We changed the number of available downstream training data (1%, 10%, and 100%) to further elucidate the effectiveness of MIM methods when downstream data is scarce, except ChestStruct, which has a relatively small data size. Note that in [39], the authors proposed MaskAlign and its semantic-aware version Attentive MaskAlign (AttnMA), which can be a good control group. There are several findings as summarized below:

Superior Performance Over Others. CXR-MIM achieves state-of-the-art results across all three datasets, surpassing both semantic-aware and unaware methods. We observe that while the advantage on RSNA, which involves only a single disease, is smaller, the benefits are more pronounced in tasks involving multiple diseases (NIH-CXR14 and ChestStruct). Specifically, on RSNA with 100% of the training data utilized, CXR-MIM scored an AUC of 86.83(0.01), compared to the lowest-performing MAE, which scored 84.64(0.01). On the NIH-CXR14 and ChestStruct, CXR-MIM achieved significantly better results, improving from 68.33(0.34) to 72.82(0.02) and from 73.05(0.03) to 79.26(0.20), respectively. A similar observation can be found in DeepMIM, a previous study focusing on multi-scale features. This further highlights the necessity of a multi-scale strategy for masked image modeling of chest X-rays.

Table 2. Comprehensive ablation study evaluating different configurations of pyramid adaptive reconstruction, showcasing linear-probing performance across two datasets (100% downstream training data), RSNA and NIH-CXR14. The macro-averaged AUC (mAUC)×100 is reported.

Scales s	RSNA	NIH-CXR14
$\{2, 2, 2, 1\}$	85.59(0.15)	69.56(0.19)
$\{4, 2, 2, 1\}$	86.32(0.06)	70.80(0.09)
$\{4, 2, 1, 1\}$	86.65(0.01)	71.33(0.65)
$\{4, 2, 1, 0.5\}$	**86.83(0.01)**	**72.82(0.02)**

(a) **Rescaling factors** for embeddings.

Source layers	RSNA	NIH-CXR14
$\{12, 12, 12, 12\}$	85.13(0.04)	70.52(0.44)
$\{2, 2, 12, 12\}$	86.19(0.06)	71.22(0.15)
$\{2, 4, 10, 12\}$	**86.83(0.01)**	**72.82(0.02)**
$\{2, 4, 8, 10, 12\}$	84.42(0.09)	68.34(0.88)

(b) **Source layers** of embeddings.

Limitations of Learnable Attention. At present, many studies [33] are prone to adopt learnable Grad-CAM or self-attention maps to reflect image semantics, instead of collecting human visual attention data as suggested in this study. However, such methods may not be effective for chest X-rays. Unlike natural images, where significant visual details may be salient, chest X-ray diagnosis relies more heavily on the radiologist's knowledge and experience, rather than merely identifying high-intensity pixels[1]. In Table 1, AttnMA—the attention variant of MaskAlign—performs masking conditioned on the ViT's self-attention map. Unfortunately, this form of attentive masking yields improvements only on the ChestStruct dataset, with AttnMA performing worse than MaskAlign on both the RSNA and NIH-CXR14 datasets, regardless of the proportion of available training data. Furthermore, when comparing semantic-unaware and semantic-aware methods, we observe that semantic-unaware methods such as MaskFeat and dMAE outperform semantic-aware methods like AttMask and HPM. This indicates the necessity to specifically tailor semantic-aware approaches for effectively handling the unique characteristics of chest X-rays.

4.3 Ablation Studies

Ablation Study of Pyramid Configuration. We examine the specific configuration settings of pyramid adaptive reconstruction in Table 2 from the aspects of rescaling setting and embeddings' source layers. In Table 2a, we fix feature embeddings' source layers as $\{2, 4, 10, 12\}$ and change their corresponding rescaling factor s. Results clarified the impact of rescaling factors and how upsampling embeddings from shallower layers can leverage their richer information,

[1] Youtube: How to Interpret a Chest X-Ray.

enhancing performance. We also find that setting $s = 0.5$ can further slightly improve the linear-probing performance. We assume that this is because down-sampling last-layer embeddings can improve the high-level representations, which is beneficial to downstream classification tasks. Table 2b sets rescaling settings as $\{4, 2, 1, 0.5\}$ and begins by setting source layers as $\{12, 12, 12, 12\}$, meaning that the last-layer embeddings are simultaneously fed into four adaptive reconstruction module with different s. We add shallow and deep layers progressively, until facing a significant performance drop when adding 8_{th} layer. We attribute this to the issue of fragile co-adaptation in neural networks [40] and leave it for future exploration.

Table 3. Ablation studies of key components of ClinicalMIM.

Attention-controlled masking	Pyramid Adaptive Reconstruction			NIH-CXR14	ChestStruct
	HOG	Pyramid	Adaptive		
✓				68.97(0.16)	73.89(0.24)
✓	✓			70.52(0.44)	75.91(0.46)
✓	✓	✓		72.18(0.17)	78.50(0.13)
✓	✓	✓	✓	72.82(0.02)	79.26(0.20)

Ablation Study of Key Components. Table 3 showcases the incremental effectiveness of CXR-MIM's components. The first row corresponds to MAE with attention-controlled masking, the subsequent addition of components reflects the actual evolution of our design. Initial results with attention-controlled masking yield AUCs of 68.97(0.16) for NIH-CXR14 and 73.89(0.24) for ChestStruct, respectively. Replacing pixel values as HOG features and the adoption of the pyramid reconstruction strategy can significantly enhance the performance. Further integration of adaptive reconstruction slightly improves the numerical results and achieves top AUCs of 72.82(0.02) and 79.26(0.20). These increments validate the contribution of each component to the overall performance. These increments validate the contribution of each component to the overall performance.

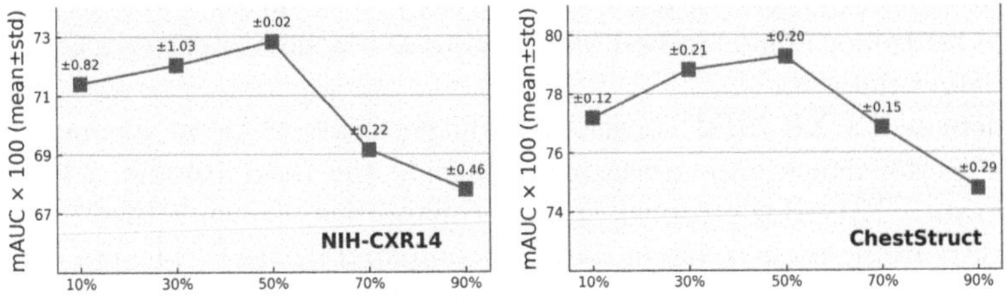

Fig. 3. Ablation study on clinical masking ratio r_c. We change r_c while fixing the global masking ratio $r = 0.75$ throughout pre-training.

Ablation Study on Clinical Masking Ratio. We assess the impact of the clinical masking ratio r_c, with the global masking ratio r maintained at 0.75, as illustrated in Fig. 3, across the NIH-CXR14 and ChestStruct datasets. The performance demonstrates notable variations as r_c is adjusted from 10% to 90%. On both the NIH-CXR14 and ChestStruct datasets, CXR-MIM consistently achieves its best performance at $r_c = 0.5$, suggesting that either excessively high or low masking of clinically significant regions is suboptimal for MIM. Furthermore, we observe that performance significantly decreases when r_c increases beyond 0.5, whereas the decline is more gradual when r_c decreases. This finding underscores the importance of preserving clinically significant regions for effective MIM learning in chest X-ray analysis.

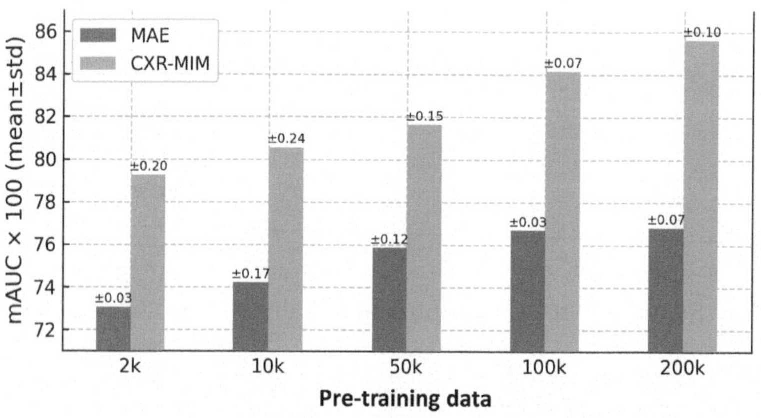

Fig. 4. Linear-probing performance w.r.t. scaling pre-training data.

4.4 CXR-MIM with Scaling Data

This section evaluates the scalability of CXR-MIM and compares it to a traditional MIM method, MAE, by examining performance across various pre-training data sizes, as illustrated in Fig. 4. Due to the absence of radiologist gaze data in MIMIC-CXR, a U-Net trained on the REFLACX dataset was used to predict gaze heatmaps. This method allows us to assess how effectively CXR-MIM utilizes larger datasets.

As depicted, CXR-MIM consistently outperforms MAE at all data scales, with the performance gap increasing alongside the data volume. The results demonstrate CXR-MIM's robust scaling capabilities, showing that it benefits significantly from access to larger datasets compared to MAE. The use of pseudo gaze *not only* improves performance *but also* underscores the adaptability of CXR-MIM to a higher data volume when only a small set of gaze-annotated examples is available. This adaptability highlights CXR-MIM's potential in the era of foundational models.

5 Conclusion

In this study, we propose a novel method named CXR-MIM based on clinical visual prior knowledge. Specifically, the significance of abnormal regions over normal motivates the introduction of gaze data as a semantic indicator, and the varying size of abnormalities encourages us to adopt a pyramid reconstruction strategy. Our analysis of the masking strategy reveals that neither over-masking nor under-masking of **clinically significant regions** yields optimal performance. Instead, we recommend a moderate masking setting (e.g., 0.5). This finding can provide a valuable guideline for the community. The scalability of CXR-MIM further highlights its potential for large-scale pre-training, positioning it as a promising tool for medical imaging in current data-rich environment.

Acknowledgement. This work was supported in part by National Natural Science Foundation of China (grant numbers U23A20295, 82441023, 62131015, 82394432), the China Ministry of Science and Technology (S20240085, STI2030-Major Projects-2022ZD 0209000, STI2030-Major Projects-2022ZD0213100), Shanghai Municipal Central Guided Local Science and Technology Development Fund (No. YDZX2023310000 1001), and HPC Platform of ShanghaiTech University.

References

1. Alsharid, M., Cai, Y., Sharma, H., Drukker, L., Papageorghiou, A.T., Noble, J.A.: Gaze-assisted automatic captioning of fetal ultrasound videos using three-way multi-modal deep neural networks. Med. Image Anal. **82**, 102630 (2022)
2. Bansal, T., Beese, R.: Interpreting a chest x-ray. Br. J. Hosp. Med. **80**(5), C75–C79 (2019)
3. Bao, H., Dong, L., Piao, S., Wei, F.: BEit: BERT pre-training of image transformers. In: International Conference on Learning Representations (2022), https://openreview.net/forum?id=p-BhZSz59o4
4. Bigolin Lanfredi, R., et al.: Reflacx, a dataset of reports and eye-tracking data for localization of abnormalities in chest x-rays. Sci. Data **9**(1), 350 (2022)
5. Chen, B., Li, J., Lu, G., Yu, H., Zhang, D.: Label co-occurrence learning with graph convolutional networks for multi-label chest x-ray image classification. IEEE J. Biomed. Health Inform. **24**(8), 2292–2302 (2020)
6. Chen, S., et al.: Automatic scoring of multiple semantic attributes with multi-task feature leverage: a study on pulmonary nodules in ct images. IEEE Trans. Med. Imaging **36**(3), 802–814 (2016)
7. Cui, Z., et al.: Tsegnet: an efficient and accurate tooth segmentation network on 3d dental model. Med. Image Anal. **69**, 101949 (2021)
8. Devlin, J., Chang, M.W., Lee, K., Toutanova, K.: Bert: pre-training of deep bidirectional transformers for language understanding. In: Proceedings of the 2019 Conference of the North American Chapter of the Association for Computational Linguistics: Human Language Technologies, Volume 1 (Long and Short Papers), pp. 4171–4186 (2019)
9. Dosovitskiy, A., et al.: An image is worth 16x16 words: transformers for image recognition at scale. In: International Conference on Learning Representations (2021), https://openreview.net/forum?id=YicbFdNTTy

10. Fan, J., Cao, X., Wang, Q., Yap, P.T., Shen, D.: Adversarial learning for mono-or multi-modal registration. Med. Image Anal. **58**, 101545 (2019)
11. Ghalati, M.K., Nunes, A., Ferreira, H., Serranho, P., Bernardes, R.: Texture analysis and its applications in biomedical imaging: a survey. IEEE Rev. Biomed. Eng. **15**, 222–246 (2021)
12. He, K., Chen, X., Xie, S., Li, Y., Dollár, P., Girshick, R.: Masked autoencoders are scalable vision learners. In: Proceedings of the IEEE/CVF Conference on Computer Vision and Pattern Recognition, pp. 16000–16009 (2022)
13. Holste, G., et al.: Towards long-tailed, multi-label disease classification from chest x-ray: Overview of the cxr-lt challenge. Med. Image Anal. 103224 (2024)
14. Huang, S.C., Shen, L., Lungren, M.P., Yeung, S.: Gloria: a multimodal global-local representation learning framework for label-efficient medical image recognition. In: Proceedings of the IEEE/CVF International Conference on Computer Vision, pp. 3942–3951 (2021)
15. Huang, X., et al.: Identification of ultrasonic echolucent carotid plaques using discrete fréchet distance between bimodal gamma distributions. IEEE Trans. Biomed. Eng. **65**(5), 949–955 (2017)
16. Ibragimov, B., Mello-Thoms, C.: The use of machine learning in eye tracking studies in medical imaging: a review. IEEE J. Biomed. Health Inf. (2024)
17. Johnson, A.E., et al.: Mimic-cxr, a de-identified publicly available database of chest radiographs with free-text reports. Sci. Data **6**(1), 317 (2019)
18. Kakogeorgiou, I., et al.: What to hide from your students: attention-guided masked image modeling. In: European Conference on Computer Vision, pp. 300–318. Springer (2022)
19. Karargyris, A., et al.: Creation and validation of a chest x-ray dataset with eye-tracking and report dictation for ai development. Sci. Data **8**(1), 92 (2021)
20. Kong, Y., et al.: Gaze-detr: using expert gaze to reduce false positives in vulvovaginal candidiasis screening. In: International Conference on Medical Image Computing and Computer-Assisted Intervention, pp. 133–143. Springer (2024)
21. Li, C., Wang, X., Eberl, S., Fulham, M., Yin, Y., Feng, D.D.: Supervised variational model with statistical inference and its application in medical image segmentation. IEEE Trans. Biomed. Eng. **62**(1), 196–207 (2014)
22. Li, J., et al.: Multiscale attention guided network for covid-19 diagnosis using chest x-ray images. IEEE J. Biomed. Health Inform. **25**(5), 1336–1346 (2021)
23. Men, Q., Teng, C., Drukker, L., Papageorghiou, A.T., Noble, J.A.: Multimodal-guidenet: gaze-probe bidirectional guidance in obstetric ultrasound scanning. In: International Conference on Medical Image Computing and Computer-Assisted Intervention, pp. 94–103. Springer (2022)
24. Otsu, N., et al.: A threshold selection method from gray-level histograms. Automatica **11**(285–296), 23–27 (1975)
25. Ouyang, X., et al.: Learning hierarchical attention for weakly-supervised chest x-ray abnormality localization and diagnosis. IEEE Trans. Med. Imaging **40**(10), 2698–2710 (2020)
26. Prinzi, F., Currieri, T., Gaglio, S., Vitabile, S.: Shallow and deep learning classifiers in medical image analysis. Eur.Radiol. Exp. **8**(1), 26 (2024)
27. Ren, S., Wei, F., Albanie, S., Zhang, Z., Hu, H.: Deepmim: deep supervision for masked image modeling. arXiv preprint arXiv:2303.08817 (2023)
28. Ren, X., et al.: Interleaved 3d-cnn s for joint segmentation of small-volume structures in head and neck ct images. Med. Phys. **45**(5), 2063–2075 (2018)

29. Shih, G., et al.: Augmenting the national institutes of health chest radiograph dataset with expert annotations of possible pneumonia. Radiol. Artif. Intell. **1**(1), e180041 (2019)
30. Shin, H.C., et al.: Deep convolutional neural networks for computer-aided detection: cnn architectures, dataset characteristics and transfer learning. IEEE Trans. Med. Imaging **35**(5), 1285–1298 (2016)
31. Siegel, A.F.: Practical Business Statistics. Academic Press (2016)
32. Wang, F., Zhou, Y., Wang, S., Vardhanabhuti, V., Yu, L.: Multi-granularity cross-modal alignment for generalized medical visual representation learning. Adv. Neural. Inf. Process. Syst. **35**, 33536–33549 (2022)
33. Wang, H., Song, K., Fan, J., Wang, Y., Xie, J., Zhang, Z.: Hard patches mining for masked image modeling. In: Proceedings of the IEEE/CVF Conference on Computer Vision and Pattern Recognition, pp. 10375–10385 (2023)
34. Wang, H., Tang, Y., Wang, Y., Guo, J., Deng, Z.H., Han, K.: Masked image modeling with local multi-scale reconstruction. In: Proceedings of the IEEE/CVF Conference on Computer Vision and Pattern Recognition, pp. 2122–2131 (2023)
35. Wang, S., Ouyang, X., Liu, T., Wang, Q., Shen, D.: Follow my eye: using gaze to supervise computer-aided diagnosis. IEEE Trans. Med. Imaging **41**(7), 1688–1698 (2022)
36. Wang, X., Peng, Y., Lu, L., Lu, Z., Bagheri, M., Summers, R.M.: Chestx-ray: hospital-scale chest x-ray database and benchmarks on weakly supervised classification and localization of common thorax diseases. In: Deep Learning and Convolutional Neural Networks for Medical Imaging and Clinical Informatics, pp. 369–392 (2019)
37. Wei, C., Fan, H., Xie, S., Wu, C.Y., Yuille, A., Feichtenhofer, C.: Masked feature prediction for self-supervised visual pre-training. In: Proceedings of the IEEE/CVF Conference on Computer Vision and Pattern Recognition, pp. 14668–14678 (2022)
38. Wu, Q., Ye, H., Gu, Y., Zhang, H., Wang, L., He, D.: Denoising masked autoencoders help robust classification. In: The Eleventh International Conference on Learning Representations (2023), https://openreview.net/forum?id=zDjtZZBZtqK
39. Xue, H., et al.: Stare at what you see: Masked image modeling without reconstruction. In: Proceedings of the IEEE/CVF Conference on Computer Vision and Pattern Recognition, pp. 22732–22741 (2023)
40. Yosinski, J., Clune, J., Bengio, Y., Lipson, H.: How transferable are features in deep neural networks? In: Ghahramani, Z., Welling, M., Cortes, C., Lawrence, N., Weinberger, K. (eds.) Advances in Neural Information Processing Systems 27 (NIPS '14), pp. 3320–3328. Curran Associates, Inc. (2014)
41. Zhao, Z., Wang, S., Wang, Q., Shen, D.: Mining gaze for contrastive learning toward computer-assisted diagnosis. In: Proceedings of the AAAI Conference on Artificial Intelligence, vol. 38, pp. 7543–7551 (2024)

Diffusion MAE: Paving the Way for Representation Learning of Diffusion MRI

Haotian Jiang and Geng Chen(✉)

National Engineering Laboratory for Integrated Aero-Space-Ground-Ocean Big Data Application Technology, School of Computer Science and Engineering, Northwestern Polytechnical University, Xi'an, Shaanxi, China
geng.chen@ieee.org

Abstract. Deep learning has achieved significant success in diffusion MRI (dMRI) computing. However, most deep learning-based dMRI methods heavily rely on supervised learning, which requires large-scale labeled datasets that are difficult to obtain. Recently, Masked AutoEncoder (MAE) has gained popularity in computer vision for its ability to leverage unlabeled data, yet its potential has not yet been fully explored in dMRI. To fill this gap, we propose the first MAE-based self-supervised learning framework, called DMAE, for representation learning of dMRI data. In DMAE, we first create dMRI patches using a deliberately designed tube q-space masking strategy, which adapts well to the unique q-space characteristics of dMRI data. We then encode the patches into a latent space for pre-training to learn high-level semantic representations. During the fine-tuning stage, we design a task-specific decoder and incorporate the decoder with the pre-trained encoder to achieve superior performance in downstream tasks. Extensive quantitative and qualitative results demonstrate the superiority of our framework over the state-of-the-art self-supervised learning approaches.

Keywords: Diffusion MRI · Self-supervised Learning · Masked AutoEncoder · Latent Space

1 Introduction

Diffusion MRI (dMRI) is a non-invasive imaging technique with significant clinical applications. It not only allows for the mapping of brain structural connectivity but also reveals the microstructural characteristics of the human white matter. As a result, dMRI is widely regarded as a pivotal tool in studying the human brain, playing a crucial role in disease diagnosis [12] and neuroscience [11].

This work was supported by the National Natural Science Foundation of China (No. 61540047).

I. Oguz et al. (Eds.): IPMI 2025, LNCS 15830, pp. 172–185, 2026.
https://doi.org/10.1007/978-3-031-96625-5_12

With the advancement of deep learning, an increasing number of deep learning models have been applied to various tasks in dMRI. For instance, in the popular research area of microstructure estimation, models, such as graph convolutional networks [3], dictionary-based deep networks [21], and Transformer [4], have been utilized to improve the performance of microstructure estimation significantly. Additionally, models, such as UNet, 3D Convolutional Neural Networks (CNNs), and 3D UNet, have demonstrated excellent performance in other key tasks, such as super-resolution [1], fiber bundle segmentation [18,20], etc. All of these achievements demonstrate the remarkable success of deep learning in dMRI research.

Despite these advancements, many deep learning-based dMRI tasks heavily rely on supervised learning that requires large amounts of labeled data, which are often challenging to obtain. In recent years, self-supervised learning has gained significant attention in the field of computer vision due to its ability to effectively leverage unlabeled data to obtain useful visual representations. Within self-supervised learning, contrastive learning [5] and masked image modeling [9] have demonstrated superior feature representation capabilities, with the latter excelling in particular. Among these, Masked AutoEncoder (MAE) [9], which operates by randomly masking patches of the input image and reconstructing the missing patches, exhibits remarkable popularity, especially in medical image segmentation tasks. For example, Zhou et al. [23] pre-trained Vision Transformer (ViT) [7] on the target dataset using MAE, significantly enhancing the performance of medical image classification and segmentation tasks. Additionally, Tang et al. [14] integrated the local feature extraction capability of multi-scale CNNs with the global context modeling ability of Transformers to enhance the effectiveness of medical image pre-training, thereby improving performance in downstream medical image segmentation tasks.

Despite the success of MAE and self-supervised pre-training in both computer vision and medical image analysis, dMRI computing still lacks a pre-trained model capable of effectively capturing the rich semantic information inherent in dMRI data. Such a model can be fine-tuned with limited data to achieve optimal performance in dMRI downstream tasks. However, the application of MAE in dMRI has encountered certain challenges, as outlined below: (1) Due to the unique nature of the q-space in dMRI, current MAE methods have yet to explore or address this characteristic; (2) Designing effective masking strategies is an unresolved challenge. These strategies allow the network to simultaneously reconstruct information from both x-space and q-space; (3) Most dMRI tasks are regression-based, necessitating careful design of the decoder architecture for such tasks in the fine-tuning phase.

To address these challenges, we propose the first MAE-based self-supervised learning framework, called DMAE, for the representation learning of dMRI data. Specifically, we first create dMRI patches with a deliberately designed tube q-space masking strategy. Our strategy effectively facilitates the model's learning of joint x-q space information while preventing information leakage caused by the structural similarity between adjacent gradient directions. Next, we encode

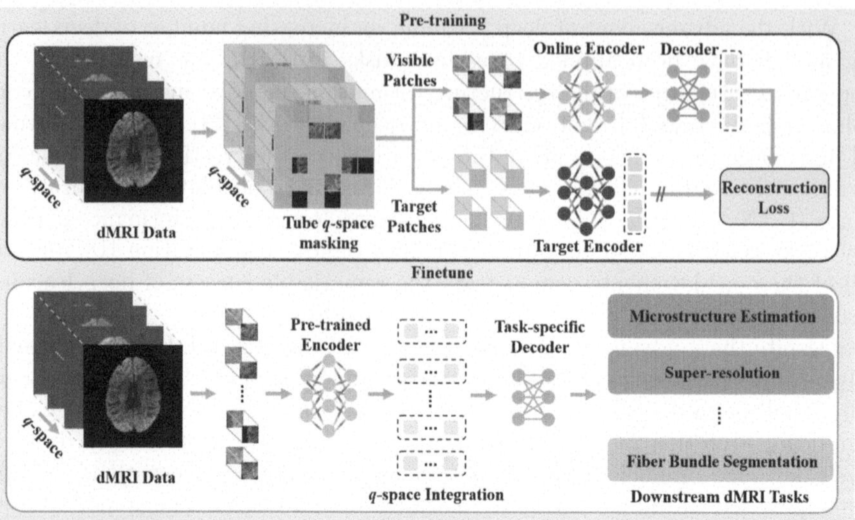

Fig. 1. An overview of DMAE. Our DMAE consists of two stages: pre-training and fine-tuning. In pre-training, dMRI data is cropped into visible and target patches using tube q-space masking strategy. Visible patches are processed through an online encoder and decoder for latent representations, while target patches are processed through a target encoder. Pre-training is done using reconstruction loss. During fine-tuning, DMAE extracts dMRI feature representations using a pre-trained encoder. These representations are then combined with a task-specific decoder to complete the downstream dMRI tasks.

the dMRI patches into a latent space for a masking dMRI data reconstruction task. This latent encoding aims to guide the model to focus on learning high-level semantic features of the data, rather than merely reconstructing low-level raw dMRI data. Finally, during the fine-tuning stage, we aggregate q-space information within the encoder and perform up-sampling to generate multi-scale features, which are then fused with a proposed task-specific decoder. We validate our method by pre-training DMAE on the large-scale Human Connectome Project (HCP) [17] dataset and fine-tuning it on a limited amount of data for downstream tasks. In comparison with previous state-of-the-art methods, the experimental results on dMRI downstream tasks demonstrate that our method demonstrates superior performance.

2 Method

2.1 Overview

Our DMAE is a self-supervised learning framework tailored for dMRI data. As shown in Fig. 1, it consists of two stages: pre-training and fine-tuning. Pre-training involves self-supervised learning on a large amount of unlabeled data to obtain dMRI feature representations. Fine-tuning, on the other hand, involves freezing the pre-trained encoder and training only the decoder for downstream tasks. During the pre-training stage, the input dMRI data $X \in \mathbb{R}^{H \times W \times L \times C}$ is partitioned into patches with a size of k, where H, W, L, and C represent the height, width, length, and number of gradient directions, respectively. This process yields $N = C \cdot \frac{H}{k} \cdot \frac{W}{k} \cdot \frac{L}{k}$ patches, where each patch is represented as x_i with $i \in 1, \ldots, N$. Subsequently, we apply the tube q-space masking strategy to these patches. For finishing the masked image modeling in latent space, we utilize an online encoder $f(\cdot)$, a target encoder $f_{\mathrm{m}}(\cdot)$, and a decoder $d(\cdot)$. During the fine-tuning stage, the data first passes through the pre-trained encoder, and DMAE aggregate q-space information within the encoder via a q-space Integration (QI) operation. These features are then up-sampled to form multi-scale features, which are utilized for downstream tasks. The following sections provide a detailed explanation of each key component.

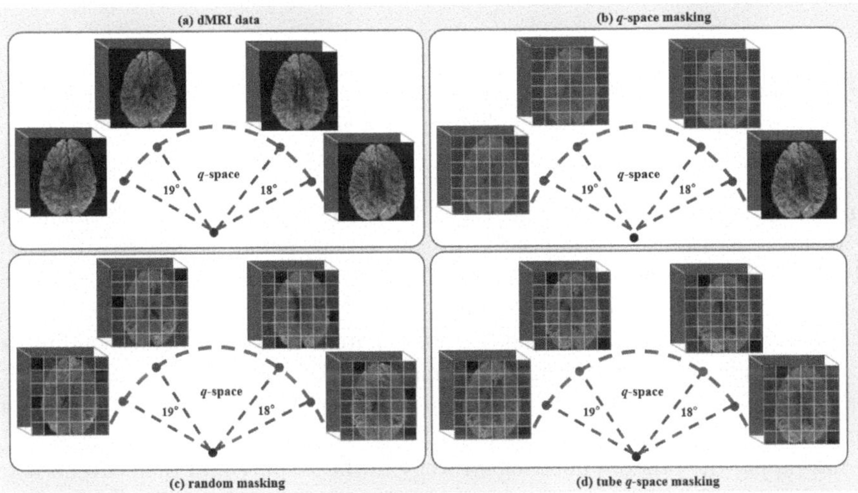

Fig. 2. Examples of masking strategies. (a) illustrates q-space of dMRI. (b), (c), and (d) represent the random masking strategy in q-space, the random masking strategy in x-q space, and the tube masking strategy in q-space, respectively.

2.2 Pre-Training

Tube q-Space Masking. Firstly, dMRI differs from natural images and common 2D-3D medical imaging modalities in its unique q-space structure, which complicates the design of masking strategies. Therefore, it is crucial to consider how to simultaneously apply data masking in both x-space and q-space. As shown in Fig. 2 (a), if the angular difference between two gradient directions in q-space is small, the corresponding data often exhibit similar structures. If random masking is applied solely to the gradient directions in q-space, without

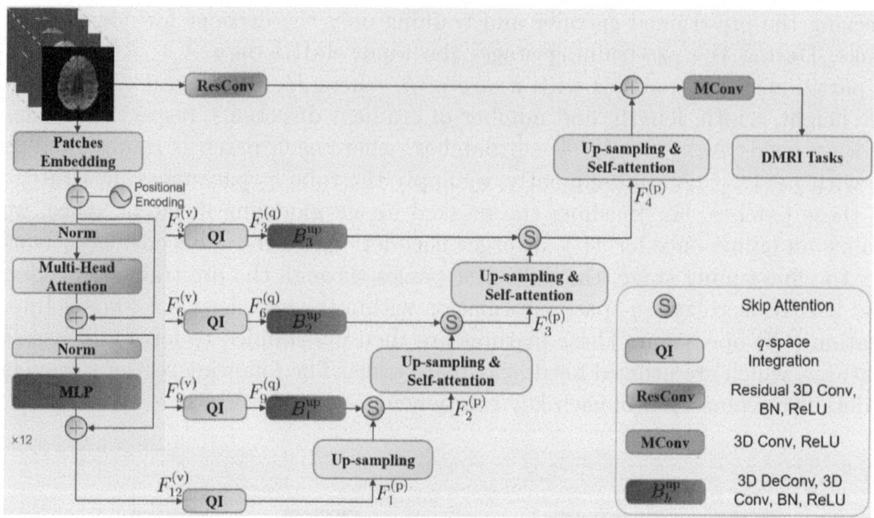

Fig. 3. Overview of DMAE fine-tuning architecture. It leverages a pre-trained Transformer to transfer its features to the decoder via a skip attention mechanism.

considering x-space (as shown in Fig. 2 (b)), the model may overly focus on the q-space information, potentially missing crucial semantic features that are only discernible in the joint x-q space. Furthermore, when a random masking strategy is applied to both q-space and x-space (as shown in Fig. 2 (c)), a patch in x-space may be masked, while a similar patch in q-space corresponding to a nearby gradient direction remains unmasked. This could cause the model to learn the masked patch through the unmasked, similar patch. To address these issues, we propose the tube q-space masking strategy (as shown in Fig. 2 (d)). We first apply the random masking strategy in the x-space and then extend it to all gradient directions in q-space. Specifically, we define the indicator function $\mathbb{I}[x_i \in \Omega]$, where Ω is the set of patches selected for masking. This indicator function follows a Bernoulli distribution, defined as:

$$\mathbb{I}[x_i \in \Omega] \sim \text{Bernoulli}(\rho_{\text{mask}}), \tag{1}$$

where ρ_{mask} is the probability that a patch x_i in the x-space is selected for masking. Specifically, when $\mathbb{I}[x_i \in \Omega] = 1$, the patch x_i is selected for masking; when $\mathbb{I}[x_i \in \Omega] = 0$, the patch is not selected. The mask is then consistently applied to all gradient directions in the q-space. The tube q-space masking strategy is simple yet effective in resolving information leakage issues and has been validated in Sect. 3.7.

Latent Masked dMRI Data Modeling. After processing with the tube q-space masking, the patches are divided into a visible patch set $X_{\mathcal{V}} = \{(x_i, p_i)\}_{i \in \mathcal{V}}$ and a target patch set $X_{\mathcal{T}} = \{(x_i, p_i)\}_{i \in \mathcal{T}}$, where p_i represents the position encoding. $\mathcal{V}$ and $\mathcal{T}$ denote the numbers of visible patches and target patches, respectively. The visible patch set generates latent space representations, while the target patch set serves as the reconstruction target in latent space. Inspired by LatentMAE [19], we utilize InfoNCE loss [10] as reconstruction loss, which is expressed as follows:

$$\mathcal{L}_r = \mathbb{E}\left[\frac{1}{|\mathcal{T}|}\text{InfoNCE}\left(Z_{\mathcal{T}}, \hat{Z}_{\mathcal{T}}\right)\right], \tag{2}$$

where $Z_{\mathcal{T}} = f_{\text{m}}(X_{\mathcal{T}})$ represents the latent space features of the target patches and $\hat{Z}_{\mathcal{T}} = d(f(X_{\mathcal{V}}), P_{\mathcal{T}})$ represents the predicted latent space features of the target patches. Here, $P_{\mathcal{T}}$ denotes the target position encoding. For the target encoder $f_{\text{m}}(\cdot)$, we employ a momentum encoder [15], whose weights $\bar{\theta}$ are updated via the exponential moving average of the weights θ from the encoder $f(\cdot)$. This process is denoted by $\bar{\theta} \leftarrow \alpha\theta + (1-\alpha)\bar{\theta}$, where α is a momentum coefficient, and we set $\alpha = 0.99$. To prevent the features learned in the latent space from becoming overly similar, we introduce an explicit regularization term $\mathcal{R}$ to address this issue. The regularization term is defined as:

$$\mathcal{R} = \left(\gamma - \mathbb{E}_{i,j\in\mathcal{T}}\left[\text{sim}\left(\hat{z}_i, \hat{z}_j\right)\right]\right)^2 + \left(\gamma - \mathbb{E}_{i,j\in\mathcal{V}}\left[\text{sim}\left(z_i, z_j\right)\right]\right)^2, \tag{3}$$

where γ is the target patch similarity threshold, $\text{sim}(\cdot,\cdot)$ computes feature similarity, and z, $\hat{z}$ denote the target and predicted features, respectively. Additionally, to enhance the decoder's focus on finer-grained spatial reasoning, we directly embed visual cues into the decoder's input sequence. Specifically, at each target location $i \in \mathcal{T}$, the input is $m_i = m + p_t + \text{Softmax}\left(P_{\mathcal{T}} P_{\mathcal{V}}^{\top}\right) Z_{\mathcal{V}}$, where m is the mask token, $Z_{\mathcal{V}}$ the visible latents, $P_{\mathcal{V}}$ the visual position encoding, and $t \in \mathcal{T}$. Finally, we introduce a linear projector, consisting of three fully connected layers (multi-layer perceptrons), placed after the decoder.

2.3 Fine-Tuning

q-Space Integration. As shown in Fig. 3, we employ the pre-trained model $f(\cdot)$ to encode the dMRI data X, yielding the feature vector $F_n^{(\text{v})}$, where $n \in \{1,\ldots,M\}$ with M representing the layer number of ViT. We set M to 12 and extract sequence representations $F_n^{(\text{v})}$ with $n \in \{3,6,9,12\}$ to obtain multi-scale features. Additionally, to better leverage the features in the q-space, we perform a QI operation to integrate the q-space information. It is defined as:

$$F_n^{(\text{q})} = \text{Reshape}(\text{MLP}(F_n^{(\text{v})})), \tag{4}$$

where $\text{MLP}(\cdot)$ consists of two linear layers and a ReLU activation function, compressing the features of $F_n^{(\text{v})}$. The $\text{Reshape}(\cdot)$ transforms the dimensionality of $F_n^{(\text{q})}$ from $N \times F_{\text{dim}}$ to $K \times F_{\text{dim}} \cdot C$, where K is the number of patches in x-space, and F_{dim} is the feature dimension of $F_n^{(\text{q})}$. This step integrates q-space information with spatial information. Finally, these integrated features are up-sampled to generate multi-scale features using up-sampling blocks $B_h^{\text{up}}(\cdot)$ via

DeConvolution (DeConv) layers with $h \in \{1, 2, 3\}$. DMAE transforms the $F_n^{(q)}$ into $F_{\text{dim}} \cdot C \times \frac{H}{k} \times \frac{W}{k} \times \frac{L}{k}$ dimensions. The up-sampling blocks first up-sample the $F_3^{(q)}$, $F_6^{(q)}$, and $F_9^{(q)}$ features by 8x, 4x, and 2x, respectively. This yields the up-sampled features $F_4^{(d)}$, $F_3^{(d)}$, and $F_2^{(d)}$.

Task-specific Decoding. We decode the feature $F_{12}^{(q)}$ using a task-specific decoder based on the [22] and fuse it with the previously obtained multi-scale features. The specific steps are as follows:

$$F_{l+1}^{(p)} = \text{SA}\left(\text{SkipAttn}\left(\text{UP}\left(F_l^{(p)}\right), F_{l+1}^{(d)}\right)\right), \quad l \in \{1, 2, 3\}, \tag{5}$$

where $F_{l+1}^{(p)}$ represents the prediction data, and $F_1^{(p)} = F_{12}^{(q)}$. Here, SA$(\cdot)$ and SkipAttn$(\cdot)$ correspond to the self-attention mechanism and skip attention mechanism, respectively. The details of both mechanisms are identical to those presented in [22]. The operation UP$(\cdot)$ denotes up-sampling. Finally, we up-sample $F_4^{(p)}$ and combine it with the features from X as follows:

$$F_{\text{out}}^{(p)} = \text{MConv}\left(\text{UP}\left(F_4^{(p)}\right) + \text{ResConv}\left(X\right)\right), \tag{6}$$

where $F_{\text{out}}^{(p)}$ is the final prediction. The operation MConv$(\cdot)$ consists of multiple 3D Convolution (Conv) layers, ReLU activation functions, and BatchNorm (BN) layers. In contrast, ResConv$(\cdot)$ contains residual 3D Conv layers, the ReLU activation functions, and BN layers.

3 Experiments

3.1 Datasets

We utilize the HCP dataset to validate our method. The HCP data were collected from healthy participants aged 22 to 35 years with each participant undergoing a 55-min scan. The scanning protocol includes three b-values (1000, 2000, 3000 s/mm^2), with 90 diffusion directions measured for each b-value, as well as 18 b=0 s/mm^2 scans. The isotropic spatial resolution of the HCP data is 1.25 mm. For pre-training, we selected 100 subjects. In downstream tasks, we selected 43 subjects: 20 for fine-tuning, 3 for validation, and 20 for testing. Additionally, we utilized dMRI data from only 30 diffusion directions at b=1000 s/mm^2 to better align with clinical applications.

3.2 Data Preparation

To enhance the uniformity of gradient directions in dMRI data, we employed the MRtrix3 tool [16] to map the data onto a standard spherical space, followed by conversion back to the original dMRI space. To ensure consistency in the input data, we cropped the dMRI data into 64 × 64 × 64 patches, which were subsequently fed into the network for pre-training. Additionally, we normalize

Table 1. Quantitative NODDI results on the HCP dataset. 100% indicates fine-tuning with the full dataset, while 20% indicates fine-tuning with a portion of the dataset. The best results are in **bold**. We also performed a paired student's t-test, where * indicates a highly statistically significant improvement ($p <= 0.001$) over competing methods.

Method	Metric	100 %				20 %			
		ICVF	ISOVF	OD	Overall	ICVF	ISOVF	OD	Overall
HysparkMAE [14]	PSNR	18.83*	25.16*	20.32*	20.71*	18.57*	23.72*	19.03*	19.90*
	SSIM	0.738*	0.795*	0.657*	0.730*	0.682*	0.720*	0.549*	0.650*
	NRMSE	0.181*	0.181*	0.191*	0.184*	0.187*	0.213*	0.222*	0.202*
SelfMedMAE [23]	PSNR	19.58*	24.93*	20.94*	21.28*	18.44*	22.78*	19.60*	19.91*
	SSIM	0.737*	0.776*	0.698*	0.737*	0.680*	0.701*	0.612*	0.664*
	NRMSE	0.166*	0.186*	0.178*	0.173*	0.190*	0.238*	0.207*	0.202*
MAE [9]	PSNR	19.89*	25.25*	21.27*	21.60*	18.60*	23.76*	19.87*	20.24*
	SSIM	0.763*	0.798*	0.717*	0.760*	0.707*	0.730*	0.646*	0.694*
	NRMSE	0.161*	0.179*	0.171*	0.167*	0.186*	0.212*	0.201*	0.194*
LatentMAE [19]	PSNR	20.04*	26.02*	21.12*	21.73*	19.15*	23.96*	20.28*	20.69*
	SSIM	0.767*	0.805*	0.720*	0.764*	0.717*	0.787*	0.661*	0.688*
	NRMSE	0.158*	0.164*	0.174*	0.164*	0.175*	0.208*	0.192*	0.185*
DMAE	PSNR	**20.56**	**26.36**	**21.56**	**22.18**	**19.69**	**24.56**	**20.60**	**21.16**
	SSIM	**0.782**	**0.818**	**0.730**	**0.777**	**0.731**	**0.744**	**0.673**	**0.716**
	NRMSE	**0.149**	**0.158**	**0.166**	**0.156**	**0.164**	**0.194**	**0.185**	**0.175**

the b_{1000} signal using the b_0 signal. To evaluate our method, we selected two typical tasks in the dMRI field, microstructure estimation and super-resolution, as downstream tasks for the fine-tuning stage. For the microstructure estimation task, we used dMRI data from all gradient directions to generate gold standard Neurite Orientation Dispersion and Density Imaging (NODDI) via the AMICO framework [6]. In the super-resolution task, we generated low-resolution data by averaging adjacent voxels.

3.3 Model Training

Our framework is implemented in PyTorch and trained using the ADAM optimizer on a GTX 3090 GPU with 24GB memory. The encoder is based on the ViT-Base model [7]. During both pre-training and fine-tuning, we apply a warm-up strategy to adjust the learning rate. Additionally, we set the maximum learning rate for pre-training to 9×10^{-6}, the number of epochs to 1,000, and the batch size to 16. Building on this, we also set the patch size k to 16, the mask ratio of the patches ρ_{mask} to 75%, and γ to 0.1. For fine-tuning, we set the maximum learning rate to 5×10^{-4}, the number of epochs to 100, and the batch size to 4. Finally, mean squared error loss is used to train the model for microstructure estimation and super-resolution tasks.

3.4 Comparison Methods

We compare our method with several state-of-the-art approaches, including MAE [9], SelfMedMAE [23], LatentMAE [19], and HysparkMAE [14]. Since

MAE and LatentMAE are primarily designed for computer vision tasks, they adopt the same decoder architecture as ours, except for the q-space integration operation. In the decoder architectures of SelfMedMAE and HysparkMAE, we replaced the final linear layer with a three-stage convolutional module with activation functions to enhance the performance of regression tasks.

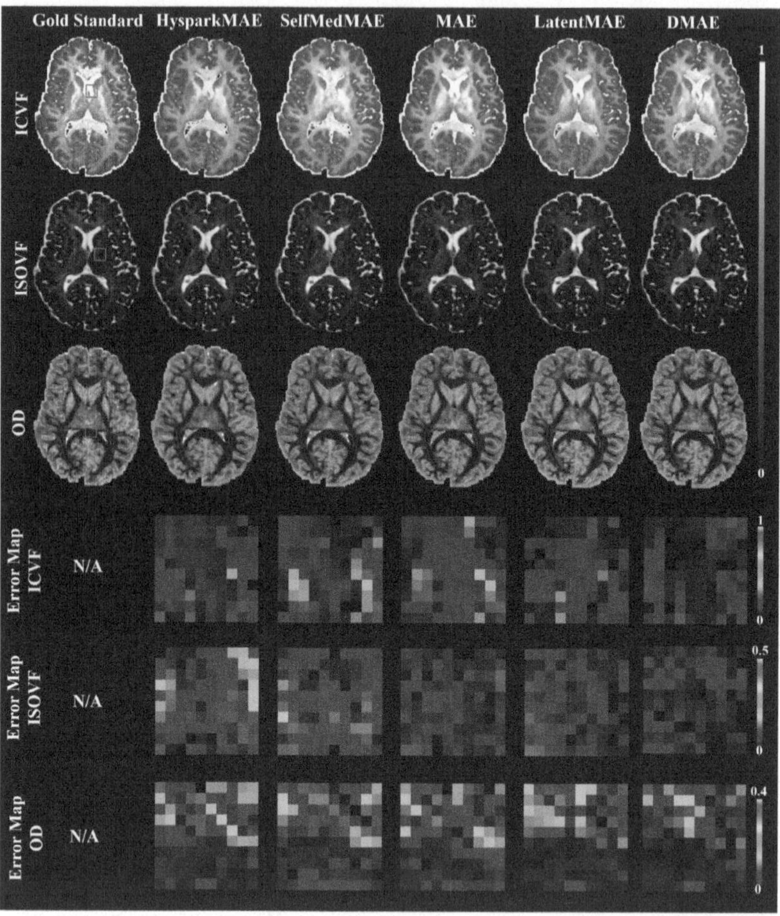

Fig. 4. Qualitative NODDI results on the HCP dataset. Predicted NODDI index maps (top) and close-up error maps (bottom). The results were generated for the single-shell data on HCP.

3.5 Microstructure Evaluation

We first performed downstream fine-tuning on the microstructure estimation task by freezing the pre-training encoder. In this process, only the decoder

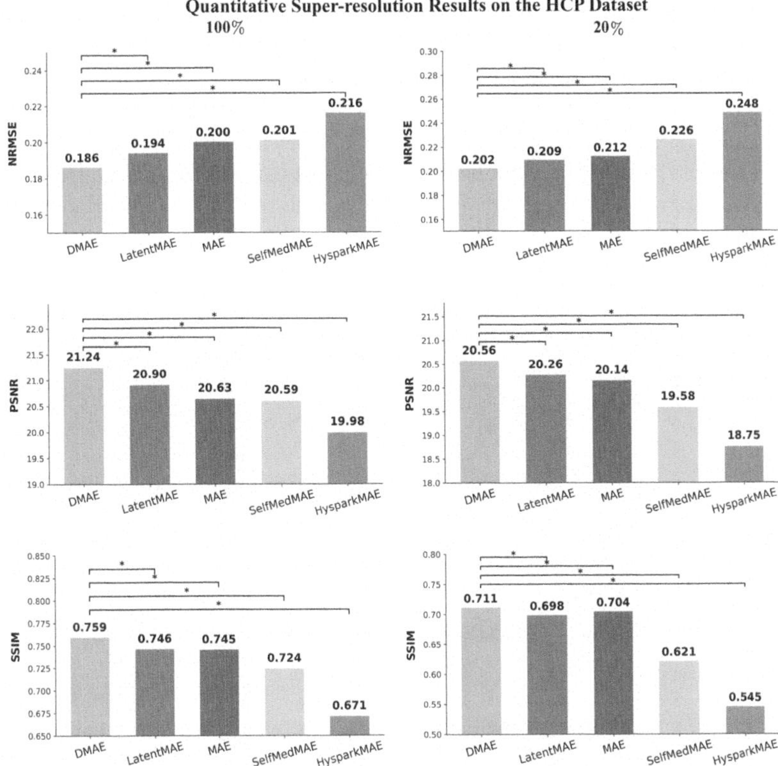

Fig. 5. Quantitative super-resolution results on the HCP dataset. 100% indicates fine-tuning with the full dataset, while 20% indicates fine-tuning with a portion of the dataset. We also performed a paired student's t-test, where * indicates a highly statistically significant improvement ($p <= 0.001$) over competing methods.

parameters were updated. To evaluate model performance, we selected NODDI as the target and predicted three key indices: Intra-Cellular Volume Fraction (ICVF), ISOtropic Volume Fraction (ISOVF), and Orientation Dispersion (OD), along with their combined performance (Overall). We used the Peak Signal-to-Noise Ratio (PSNR), Structural Similarity Index Measure (SSIM), and Normalized Root Mean Square Error (NRMSE) as evaluation metrics. Quantitative results, shown in Table 1, demonstrate that DMAE outperforms all competing methods across all metrics, whether trained on the full fine-tuning dataset or just 20% of it. Despite SelfMedMAE and HysparkMAE excelling in segmentation tasks, they perform relatively poorly in downstream dMRI tasks. A comparison between LatentMAE and MAE demonstrates the significant potential of latent-space MAE. Furthermore, by considering q-space information, our method further improves performance. Additionally, a comparison of MAE, SelfMedMAE, and HysparkMAE underscores the significant influence of encoder and decoder selection on the performance of downstream dMRI tasks.

Furthermore, paired student's t-test revealed that our method achieved a statistically significant improvement over competing methods. Additionally, Fig. 4 presents the visualization of predicted results and corresponding close-up error maps for all methods after fine-tuning the full dataset. It is evident that the NODDI maps generated by DMAE are most similar to the gold standard in terms of visual quality. The error visualizations in Fig. 4 reveal that our method consistently outperforms other competing methods, with reduced localized errors. This further indicates that our visualization results are closer to the gold standard.

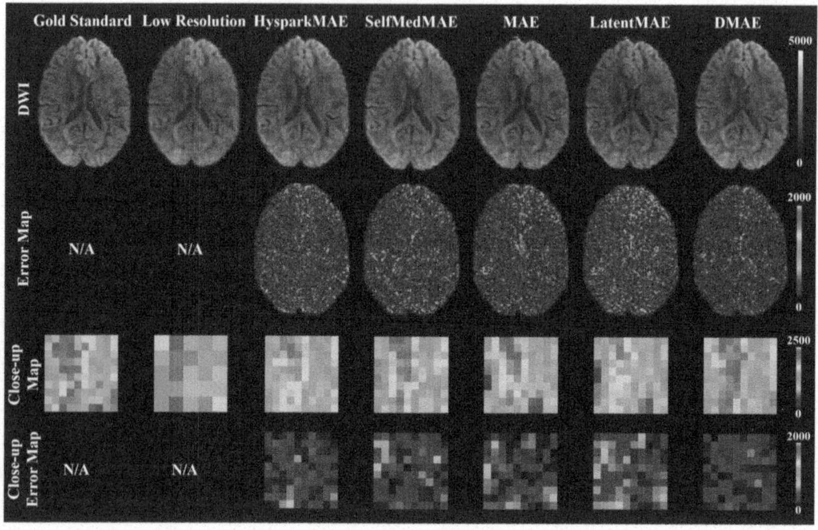

Fig. 6. Qualitative super-resolution results on the HCP dataset. Predicted high-resolution data index maps, error maps, and close-up maps. The results were generated for the single-shell data on HCP.

3.6 Super-Resolution Evaluation

We also conducted super-resolution experiments using the same setting as the microstructure estimation task. As shown in Fig. 5, DMAE consistently outperforms all competing methods, whether trained on the full fine-tuning dataset or 20% of it. Our method outperforms competing methods with a p-value < 0.001, as determined by paired student's t-test, underscoring the significant superiority of our approach in super-resolution. Additionally, Fig. 6 presents the visual results, error maps, and close-up views for DMAE and the competing methods using the full fine-tuning dataset. DMAE recovers more details and outperforms the others. The error maps in Fig. 6 clearly demonstrate that our method leads to a more accurate estimation, compared with other methods.

3.7 Ablation Study

To validate the effectiveness of the proposed masking strategy, pre-training strategy, and task-specific decoding, we conducted ablation experiments on the microstructure estimation task using the full fine-tuning dataset. Performance was evaluated using PSNR, SSIM, and NRMSE.

Table 2. Quantitative NODDI results of ablation study on the HCP dataset. The best results are in **bold**. We also performed a paired student's t-test, where * indicates a highly statistically significant improvement ($p <= 0.001$) over the ablation version.

NODDI	Metric	Mask strategy			Decoder design in fine-tuning			
		q-space	random	tube	A	B	C	D
ICVF	PSNR	20.26*	20.34*	**20.56**	19.61*	19.75*	19.96*	**20.56**
	SSIM	0.772*	0.760*	**0.782**	0.768*	0.771*	0.777*	**0.782**
	NRMSE	0.154*	0.153*	**0.149**	0.166*	0.163*	0.159*	**0.149**
ISOVF	PSNR	26.00*	26.06*	**26.36**	25.90*	25.83*	25.82*	**26.36**
	SSIM	0.805*	0.812*	**0.818**	0.800*	0.810*	0.813*	**0.818**
	NRMSE	0.164*	0.163*	**0.158**	0.166*	0.167*	0.168*	**0.158**
OD	PSNR	21.27*	21.35*	**21.56**	21.22*	21.25*	21.35*	**21.56**
	SSIM	0.720*	0.720*	**0.730**	0.726*	0.726*	0.724*	**0.730**
	NRMSE	0.171*	0.170*	**0.166**	0.175*	0.172*	0.169*	**0.166**
Overall	PSNR	21.88*	21.96*	**22.18**	22.54*	21.60*	21.74*	**22.18**
	SSIM	0.766*	0.768*	**0.777**	0.765*	0.769*	0.771*	**0.777**
	NRMSE	0.161*	0.159*	**0.156**	0.168*	0.166*	0.164*	**0.156**

Masking Strategy. To validate the effectiveness of the proposed masking strategy, we pre-trained using the q-space masking, random masking, and tube q-space masking. As shown in Table 2, the results demonstrate that the tube q-space masking outperforms the other two strategies across all metrics, with paired student's t-test analysis revealing significant differences. This suggests that the tube q-space masking is more effective for masking the x-q space.

Pre-training Strategy. To validate the effectiveness of the pre-training strategy, the results in Table 1 show that LatentMAE outperforms MAE, suggesting that latent-space MAE is more suitable for pre-training on dMRI data. Incorporating q-space information, DMAE consistently outperforms LatentMAE across all metrics, further highlighting the crucial role of q-space information.

Task-specific Decoding. For the downstream dMRI tasks, the decoder design is crucial. We explored various decoder options, with "A", "B", "C", and "D" representing UNETR [8], UNETR++ [13], decoder of DMAE except for QI, and decoder of DMAE with QI, respectively (shown in Table 2). The performance of "C" exceeds that of "A" and "B", indicating that the decoder of DMAE is better suited for downstream dMRI tasks. Additionally, the results in "D" outperform those in "C" across all metrics, demonstrating the effectiveness of QI in utilizing q-space information.

3.8 Conclusion

In this work, we propose the first MAE-based self-supervised learning framework tailored for downstream tasks in dMRI data. In DMAE, we create dMRI patches using a deliberately designed tube q-space masking strategy, which enables the model to simultaneously reconstruct information from both the x-space and q-space masks. Our DMAE further encodes these patches in a latent space to finish pre-training and learn high-level semantic representations of dMRI data. Finally, we design a task-specific decoder that allows our framework to effectively perform downstream dMRI tasks in the fine-tuning stage. The results from both quantitative and qualitative experiments confirm that our method significantly outperforms competing methods, providing more accurate predictions in two typical downstream dMRI tasks. The ablation study clearly illustrates the importance of our proposed pre-training strategy, tube q-space masking, and task-specific decoding in achieving the superior performance of our method. In future work, our objective is to explore additional downstream tasks in the dMRI field and validate our framework on larger-scale datasets.

References

1. Chatterjee, S., et al.: ShuffleUNet: super resolution of diffusion-weighted MRIs using deep learning. In: 2021 29th European Signal Processing Conference, pp. 940–944. IEEE (2021)
2. Chen, G., Dong, B., Zhang, Y., Lin, W., Shen, D., Yap, P.T.: XQ-SR: joint xq space super-resolution with application to infant diffusion MRI. Med. Image Anal. **57**, 44–55 (2019)
3. Chen, G., et al.: Estimating tissue microstructure with undersampled diffusion data via graph convolutional neural networks. In: International Conference on Medical Image Computing and Computer-Assisted Intervention, pp. 280–290. Springer (2020)
4. Chen, G., et al.: Hybrid graph transformer for tissue microstructure estimation with undersampled diffusion MRI data. In: International Conference on Medical Image Computing and Computer-Assisted Intervention, pp. 113–122. Springer (2022)
5. Chen, T., Kornblith, S., Norouzi, M., Hinton, G.: A simple framework for contrastive learning of visual representations. In: International Conference on Machine Learning, pp. 1597–1607. PMLR (2020)
6. Daducci, A., Canales-Rodríguez, E.J., Zhang, H., Dyrby, T.B., Alexander, D.C., Thiran, J.P.: Accelerated microstructure imaging via convex optimization (AMICO) from diffusion MRI data. Neuroimage **105**, 32–44 (2015)
7. Dosovitskiy, A.: An image is worth 16x16 words: Transformers for image recognition at scale. arXiv preprint arXiv:2010.11929 (2020)
8. Hatamizadeh, A., et al.: UNETR: transformers for 3D medical image segmentation. In: Proceedings of the IEEE/CVF Winter Conference on Applications of Computer Vision, pp. 574–584, January 2022

9. He, K., Chen, X., Xie, S., Li, Y., Dollár, P., Girshick, R.: Masked autoencoders are scalable vision learners. In: Proceedings of the IEEE/CVF Conference on Computer Vision and Pattern Recognition, pp. 16000–16009 (2022)
10. Oord, A.v.d., Li, Y., Vinyals, O.: Representation learning with contrastive predictive coding. arXiv preprint arXiv:1807.03748 (2018)
11. Ouyang, M., Dubois, J., Yu, Q., Mukherjee, P., Huang, H.: Delineation of early brain development from fetuses to infants with diffusion MRI and beyond. Neuroimage **185**, 836–850 (2019)
12. Sakaie, K., et al.: Multi-shell diffusion MRI of the fornix as a biomarker for cognition in Alzheimer's disease. Magn. Reson. Imaging **109**, 221–226 (2024)
13. Shaker, A.M., Maaz, M., Rasheed, H., Khan, S., Yang, M.H., Khan, F.S.: UNETR++: delving into efficient and accurate 3D medical image segmentation. IEEE Trans. Med. Imaging (2024)
14. Tang, F., et al.: Hyspark: hybrid sparse masking for large scale medical image pre-training. In: International Conference on Medical Image Computing and Computer-Assisted Intervention, pp. 330–340. Springer (2024)
15. Tarvainen, A., Valpola, H.: Mean teachers are better role models: weight-averaged consistency targets improve semi-supervised deep learning results. In: Advances in Neural Information Processing Systems vol. 30 (2017)
16. Tournier, J.D., et al.: Mrtrix3: a fast, flexible and open software framework for medical image processing and visualisation. Neuroimage **202**, 116137 (2019)
17. Van Essen, D.C., et al.: The WU-Minn human connectome project: an overview. NeuroImage **80**, 62–79 (2013)
18. Wasserthal, J., Neher, P.F., Hirjak, D., Maier-Hein, K.H.: Combined tract segmentation and orientation mapping for bundle-specific tractography. Med. Image Anal. **58**, 101559 (2019)
19. Wei, Y., Gupta, A., Morgado, P.: Towards latent masked image modeling for self-supervised visual representation learning. In: European Conference on Computer Vision, pp. 1–17. Springer (2025)
20. Xu, H., et al.: A registration-and uncertainty-based framework for white matter tract segmentation with only one annotated subject. In: 2023 IEEE 20th International Symposium on Biomedical Imaging, pp. 1–5. IEEE (2023)
21. Ye, C.: Tissue microstructure estimation using a deep network inspired by a dictionary-based framework. Med. Image Anal. **42**, 288–299 (2017)
22. Zhou, H.Y., et al.: nnFormer: volumetric medical image segmentation via a 3D Transformer. IEEE Trans. Image Process. (2023)
23. Zhou, L., Liu, H., Bae, J., He, J., Samaras, D., Prasanna, P.: Self pre-training with masked autoencoders for medical image classification and segmentation. In: 2023 IEEE 20th International Symposium on Biomedical Imaging, pp. 1–6. IEEE (2023)

Resolving Quantitative MRI Model Degeneracy in Self-supervised Machine Learning

Giulio V. Minore[1,2(✉)], Louis Dwyer-Hemmings[1,3,4], Timothy J.P. Bray[1,3,4], and Hui Zhang[1,5]

[1] Hawkes Institute, University College London, London, UK
giulio.minore.17@ucl.ac.uk
[2] Department of Medical Physics and Biomedical Engineering, University College London, London, UK
[3] Centre for Medical Imaging, University College London, London, UK
[4] Department of Imaging, University College London Hospital, London, UK
[5] Department of Computer Science, University College London, London, UK

Abstract. Quantitative MRI (qMRI) estimates tissue properties of interest from measured MRI signals. This process is conventionally achieved by model fitting, whose computational expense limits qMRI's clinical use, motivating recent development of machine learning-based methods. Self-supervised approaches are particularly popular as they avoid the pitfall of distributional shift that affects supervised methods. However, it is unknown how such methods behave if similar signals can result from multiple tissue properties, a common challenge known as model degeneracy. Understanding this is crucial for ascertaining the scope within which self-supervised approaches may be applied. To this end, this work makes two contributions. First, we demonstrate that model degeneracy compromises self-supervised approaches, motivating the development of mitigation strategies. Second, we propose a mitigation strategy based on applying appropriate constraining transforms on the output of the bottleneck layer of the autoencoder network typically employed in self-supervised approaches. We illustrate both contributions using the estimation of proton density fat fraction and R_2^* from chemical shift-encoded MRI, an ideal exemplar due to its exhibition of degeneracy across the full parameter space. The results from both simulation and *in vivo* experiments demonstrate that the proposed strategy helps resolve model degeneracy.

Keywords: quantitative MRI · model degeneracy · self-supervised machine learning

1 Introduction

Quantitative magnetic resonance imaging (qMRI) aims to improve on conventional MRI, which primarily relies on image contrast due to its signal being

I. Oguz et al. (Eds.): IPMI 2025, LNCS 15830, pp. 186–199, 2026.
https://doi.org/10.1007/978-3-031-96625-5_13

an aggregation of tissue properties. By disentangling these contributions to the signal, qMRI provides quantitative, more disease-specific measures of the underlying tissue properties, which increases the interpretability of scans, facilitating diagnosis and assessment of disease.

Conventionally, qMRI parameter estimation is performed through voxel-wise model fitting of the measured MRI signal. However, this is computationally expensive as it relies on iteratively and independently maximising some likelihood function for each voxel, reducing its clinical utility.

To improve on the computational requirements of conventional parameter estimation, machine learning, and especially deep learning-based methods have been proposed. Deep learning approaches generally employ supervised strategies [2,4,10,11,13], relying on ground truth parameter labels. However, their performance may be affected by distributional shift between the training and target measured signal. This could be caused by discrepancies between the distributions of training and target parameter combinations [9,13] or signal-to-noise ratio (SNR). To avoid any potential distributional shift, self-supervised methods have been proposed [1,11,16,23]. These do not require ground truth labels and can be used to train a network directly on the target data.

Despite these theoretical advantages, to the best of our knowledge, there have been no studies that have investigated how self-supervised methods would behave if similar signals can result from multiple tissue properties, a common challenge known as model degeneracy. This limits our understanding of the scope within which such strategies may be used. In this work, we first describe the impact of model degeneracy on self-supervised algorithms, then propose a method to mitigate its effect. These contributions are demonstrated both *in silico* and *in vivo* chemical shift-encoded MRI (CSE-MRI) data, which will serve as an exemplar of model degeneracy due to its presence throughout the parameter space.

2 Theory

In this section, following an introduction to the qMRI parameter estimation problem, we describe the effect of model degeneracy on self-supervised algorithms and the proposed method to mitigate this problem.

2.1 qMRI Parameter Estimation Problem

The aim of qMRI parameter estimation is to infer the unknown tissue properties $\boldsymbol{y} \in \mathbb{R}^{N_y}$ at a given voxel from the measured signal $\boldsymbol{S} \in \mathbb{R}^{N_S}$ at that voxel acquired with the settings $\boldsymbol{Z}$, where N_y is the number of distinct tissue properties and N_S is the number of signal measurements.

This problem has been typically solved by conventional fitting. This relies on fitting a biophysical forward model $\mathcal{M}$ that can predict the noise-free signal corresponding to some tissue properties $\tilde{\boldsymbol{y}}$, such that $\tilde{\boldsymbol{S}} = \mathcal{M}(\boldsymbol{Z}; \tilde{\boldsymbol{y}})$. The parameter estimates $\hat{\boldsymbol{y}}$ can then be found by maximising some appropriate likelihood function such that

$$\hat{\boldsymbol{y}} = \underset{\tilde{\boldsymbol{y}}}{\arg\max}\ \mathcal{L}(\mathcal{M}(\boldsymbol{Z}; \tilde{\boldsymbol{y}}) | \boldsymbol{S}). \tag{1}$$

In the presence of model degeneracy, the above likelihood will possess multiple local maxima, which must be resolved by fitting from multiple initial guesses to the parameter estimates.

2.2 Self-Supervised Method and the Impact of Model Degeneracy

Self-supervised algorithms have been recently proposed to accelerate qMRI parameter estimation [1,11,16,23]. Similar to the more common supervised alternatives, they aim to find a function $f_{\boldsymbol{\theta}} : \mathbb{R}^{N_S} \rightarrow \mathbb{R}^{N_y}$ parameterised by $\boldsymbol{\theta}$ (e.g., neural network weights) that can directly map a signal measured at a voxel to some tissue properties that are as close as possible to the underlying tissue properties at the voxel. However, unlike supervised alternatives, self-supervised algorithms use the predicted tissue properties $\tilde{\boldsymbol{y}} = f_{\boldsymbol{\theta}}(\boldsymbol{S})$ to reconstruct the noise-free signal using the forward model, such that $\tilde{\boldsymbol{S}} = \mathcal{M}(\boldsymbol{Z}; \tilde{\boldsymbol{y}})$.

This approach has been commonly considered as being based on a physics-based autoencoder (see Fig. 1). Similar to standard autoencoders, a neural network implements the encoder to map some input signal to the output of the bottleneck layer representing the parameter estimates $\tilde{\boldsymbol{y}}$. Unlike standard autoencoders, the decoder is not implemented with another neural network but with the forward model which as a result does not possess any learnable parameters. The loss is defined as a negative-likelihood between N measured signals $\boldsymbol{S}_i$ and predicted signals $\tilde{\boldsymbol{S}}_i = \mathcal{M}(\boldsymbol{Z}; f_{\boldsymbol{\theta}}(\boldsymbol{S}_i))$, such that

$$\ell = -\frac{1}{N}\sum_{i=1}^{N} \mathcal{L}(\tilde{\boldsymbol{S}}_i|\boldsymbol{S}_i) = -\frac{1}{N}\sum_{i=1}^{N} \mathcal{L}(\mathcal{M}(\boldsymbol{Z}; f_{\boldsymbol{\theta}}(\boldsymbol{S}_i))|\boldsymbol{S}_i). \tag{2}$$

While the effect of model degeneracy has been described for supervised methods [3,12], this has not yet been done for self-supervised algorithms. To help understand this effect, similar to [12], we begin with a simplifying theoretical analysis where we assume the strong form of model degeneracy, namely that there exists at least one subset $\boldsymbol{Y_d}$ in which different tissue properties $\boldsymbol{y}$ produce the same signal $\boldsymbol{S}_d$, such that

$$\forall\, \boldsymbol{y}_j, \boldsymbol{y}_k \in \boldsymbol{Y}_d,\ \boldsymbol{y}_j \neq \boldsymbol{y}_k \text{ and } \boldsymbol{S}_j = \boldsymbol{S}_k = \boldsymbol{S}_d. \tag{3}$$

Without loss of generality, assuming the first n samples belong to the degenerate subset, the loss function may be re-written as

$$\ell = -\frac{1}{N}\sum_{i=1}^{n} \mathcal{L}(\mathcal{M}(\boldsymbol{Z}; f_{\boldsymbol{\theta}}(\boldsymbol{S}_i))|\boldsymbol{S}_i) - \frac{1}{N}\sum_{i=n+1}^{N} \mathcal{L}(\mathcal{M}(\boldsymbol{Z}; f_{\boldsymbol{\theta}}(\boldsymbol{S}_i))|\boldsymbol{S}_i). \tag{4}$$

By construction, for $1 \leq i \leq n$, $\boldsymbol{S}_i = \boldsymbol{S_d}$. Since both the encoder (a neural network) and the decoder (a forward model) are functions and therefore cannot represent one-to-many mappings, the likelihood terms for the first n samples are identical.

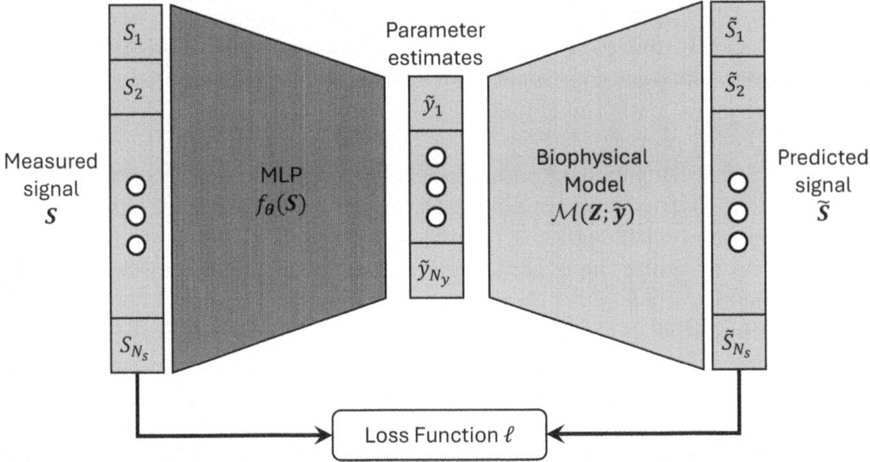

Fig. 1. Diagram of a physics-based autoencoder. A measured signal $\boldsymbol{S}$ is fed to a neural network which predicts tissue properties $\tilde{\boldsymbol{y}}$. These are then used to reconstruct the signal $\tilde{\boldsymbol{S}}$ using the forward model, which is used to calculate the self-supervised training loss.

Assuming the two terms in the loss can be independently minimised, any unique combination of tissue properties $\boldsymbol{y}$ in $\boldsymbol{Y}_d$ produces the signal that minimises the first term. Therefore, the network can learn to predict any individual combination in $\boldsymbol{Y}_d$ at random, neglecting all the other possible solutions. Given this result for the strong form of degeneracy, we hypothesise that a similar behaviour will manifest in the presence of the general form of degeneracy.

2.3 Proposed Method to Resolve Degeneracy

The proposed method is inspired by previous approaches developed for supervised algorithms [3,4] but with a novel adaptation to support the unique demand of self-supervised algorithms.

The existing approaches for supervised algorithms work by assuming that the parameter space can be subdivided such that within each sub-space degeneracy is no longer present. In [3], the parameter space is split into two; after rejecting one of the sub-spaces as containing only non-feasible parameter values, a neural network is trained with training labels drawn solely from the other sub-space. Later in [4], the parameter space is also split into two; but because both sub-spaces contain feasible parameter values, two neural networks are trained such that each sub-space has its own designated network trained with training labels drawn solely from the sub-space itself. At inference time, the outputs from the two networks are compared in terms of their likelihoods to choose the best parameter estimate. The present work builds on this general multi-network approach.

However, this approach cannot be directly applied to self-supervised methods. This is because training data for self-supervised and supervised methods are fundamentally different. For supervised methods, the training data consist of feature-label pairs, with the (input) feature being the signal and the (output) label being tissue properties. This allows the training data to be separated into distinct subsets with each containing labels belonging to some chosen subset of parameters. In contrast, the training data for self-supervised methods consist of feature-label pairs with both the (input) feature and the (output) label being the signal. As a result, the existing approaches for supervised methods are no longer applicable.

As an alternative, we propose to add a constraining transform to the bottleneck layer. While such transforms are commonly used in conventional fitting to ensure the physical plausibility of parameter estimates, here we take advantage of them to guarantee network predictions belong to a sub-space that does not exhibit degeneracy. To understand how this works, recall that for the existing self-supervised methods, the output of the bottleneck layer represents the predicted tissue properties such that $\tilde{\boldsymbol{y}} = f_{\boldsymbol{\theta}}(\boldsymbol{S})$. With the introduction of the constraining transform $\mathcal{C}$, the output following this transform now represents the predicted tissue properties such that $\tilde{\boldsymbol{y}} = \mathcal{C}(f_{\boldsymbol{\theta}}(\boldsymbol{S}))$.

3 Methods

This section describes an implementation of the proposed multi-network method for CSE-MRI data to demonstrate that model degeneracy may be resolved in self-supervised learning. This is performed both in simulation and *in vivo*, and a description of both datasets and experiments will be given.

3.1 An Example of qMRI Model Degeneracy: CSE-MRI

We now illustrate model degeneracy in CSE-MRI, which will serve as an exemplar to evaluate our proposed method to help resolve degeneracy. CSE-MRI is used to separate the contributions of water and fat to the signal in a single voxel, denoted ρ_W and ρ_F, typically from multi-echo gradient echo acquisition data. It aims to estimate proton density fat fraction (PDFF $= \frac{\rho_F}{\rho_W + \rho_F}$) and the R_2^* relaxation rate, which are useful biomarkers across a wide range of diseases and tissues [6,18,19,22,25,27]. Although their estimation has traditionally required complex data [14,21,28], magnitude-fitting techniques [5,26] have gained interest due to the phase data sometimes being unreliable or unavailable.

The forward model for the magnitude signal from CSE-MRI is given by

$$\mathcal{M}(t|S_0, \text{PDFF}, R_2^*) = S_0 \left| (1 - \text{PDFF}) + \text{PDFF} \sum_{k=1}^{K} r_k e^{(i2\pi f_{F,k} t)} \right| e^{(-tR_2^*)} \quad (5)$$

where S_0 is the signal at time $t = 0$, K is the number of spectral fat components, r_k and $f_{F,k}$ are respectively the relative amplitude and frequency shift of each spectral fat component and are defined *a priori*.

Assuming the MRI magnitude signal is Rician-distributed, the likelihood of a predicted signal $\tilde{\boldsymbol{S}}_i = \mathcal{M}(t|\tilde{\boldsymbol{y}}_i)$ in voxel i, reconstructed from predictions $\tilde{\boldsymbol{y}}_i$, given a measured signal $\boldsymbol{S}_i$ is [24]

$$\mathcal{L}_{\text{Rice}}(\tilde{\boldsymbol{S}}_i, \sigma^2|\boldsymbol{S}_i) = \prod_{k=1}^{N_{TE}} \frac{S_{ik}}{\sigma^2} \exp\Big(\frac{-(S_{ik}^2 + \tilde{S}_{ik}^2)}{2\sigma^2}\Big) I_0\Big(\frac{S_{ik}\tilde{S}_{ik}}{\sigma^2}\Big) \tag{6}$$

where N_{TE} is the number of echo times at which the signal is acquired, σ is the standard deviation of the underlying complex Gaussian noise and I_0 is the modified Bessel function of the first kind with order zero.

This forward model is known to exhibit systematic degeneracy with the presence of two local maxima in the likelihood function. These two maxima occur for either a low (closer to 0) or a high (closer to 1) PDFF, approximately opposite on the PDFF spectrum (see Fig. 2). These will henceforth be referred to as the water-dominant or fat-dominant solutions. Importantly, it has been shown that the PDFF space can be divided into two: a water-dominant space (PDFF < 0.58) and a fat-dominant one (PDFF ≥ 0.58), such that each degenerate pair is split into either the water-dominant space or the fat-dominant one [4,5,26]. As a result, within each of these two spaces, degeneracy is no longer present.

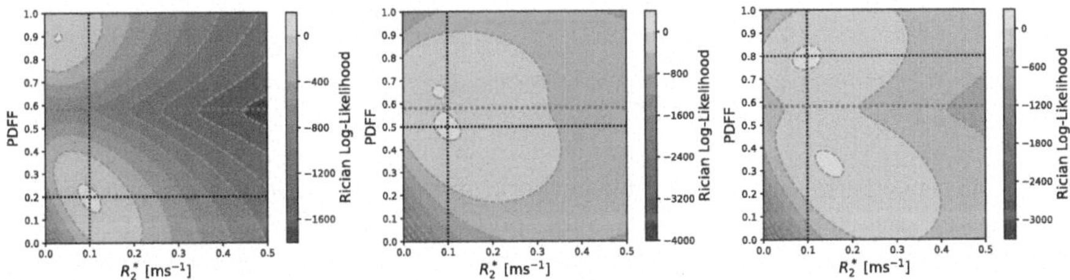

Fig. 2. Rician log-likelihood as a function of PDFF and R_2^* for different tissue parameter combinations. Ground truths are indicated as dotted black lines, representing $R_2^* = 0.1$ ms^{-1} and PDFF equal to 0.2 (left), 0.5 (centre) 0.8 (right). There are two maxima in likelihood, occurring above or below the red dotted switching line at PDFF $= 0.58$, giving rise to degeneracy.

While model degeneracy has been resolved in conventional fitting and supervised learning methods for magnitude CSE-MRI [4,5,26], to the best of our knowledge, no self-supervised technique has been developed for this problem. Here we apply the proposed method described in Sect. 2.3 to implement a dual network approach. For each network, we introduce an appropriate constraining transform for PDFF, such that each network respectively makes predictions in the water- or fat-dominant space. The constraining transforms are

$$\mathcal{C}(x) = \begin{cases} \frac{b}{2}(\sin(x) + 1) & \text{for the water-dominant network} \\ 1 - \frac{(1-b)}{2}(\sin(x) + 1) & \text{for the fat-dominant network} \end{cases} \tag{7}$$

where $b = 0.58$ corresponds to the boundary between the degenerate sub-spaces. For the water-dominant network (the 'water' network in short), its PDFF estimates are constrained to lie in the [0,b] interval, and for the fat-dominant network (the 'fat' network in short), its PDFF estimates are constrained to lie in the [b,1] interval. The S_0 and R_2^* parameters are also constrained to be non-negative by the squaring transform, to ensure that the predictions are physically plausible.

As in [4], the final parameter estimates are selected based on the candidate solution maximising the likelihood (see Fig. 3).

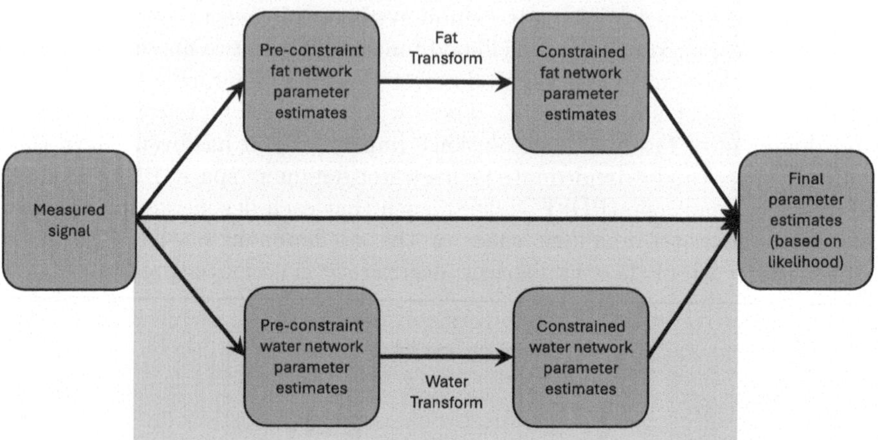

Fig. 3. Inference procedure using dual network approach. The measured signal is fed to the 'water' and 'fat' networks that make predictions in their respective parameter sub-space via the use of constraining transforms. The final parameter estimates are selected by comparing their likelihood.

3.2 Implementation

Architecture. The self-supervised encoder network architecture was chosen to resemble previous self-supervised work [1,23]. A multilayer perceptron takes a voxel-wise signal as input, which is then fed through five fully-connected layers, using exponential linear unit activation functions [7]. All but the last layer possess N_{TE} nodes, corresponding to the number of measurements, as was previously done in other self-supervised approaches. The final layer has three output nodes, corresponding to the three parameters of interest (namely S_0, PDFF and R_2^*).

Training. Self-supervised networks are directly trained using the target data on which predictions are made, both in *silico* and *in vivo*. This ensures that training and target data are identical, such that distributional shift cannot occur.

Networks are optimised by minimising the negative Rician log-likelihood loss [20] due to Rician distribution of magnitude signals. Training is performed using the Adam optimizer [17] with learning rate 0.001, minibatch size of 128, patience of 10 as in [1]. The code is available on GitHub[1].

3.3 Experiments

In Silico. We first investigate the effects of model degeneracy in synthetic data, where ground truth parameter values are known. To this end, we simulate signals with the forward model in Eq. (5), using the acquisition settings corresponding to the *in vivo* dataset described below. This synthetic data is generated by producing different combinations of PDFF and R_2^*, evenly sampling 100 PDFF values in the [0,1] interval and 50 R_2^* values in the [0,0.5] ms^{-1} interval. For each of these 100×50 combinations, 100 noisy signals are simulated by adding Gaussian noise to the underlying complex signal before taking its magnitude, such that the ground truth SNR is 60 (typical for CSE-MRI acquisitions [4,5, 26]).

We train single networks without constraining the PDFF predictions to lie within a specific sub-space (i.e. PDFF $\in [0,1]$) on synthetic data, where ground-truth tissue properties are known. We hypothesise that this naive implementation will unpredictably learn to output either the water-dominant or fat-dominant solutions. To evaluate the proposed method designed to ensure both the 'water' and 'fat' networks are learned, we also train two networks using the constraining transforms in Eq. (7).

Finally, the parameter estimation performance of the proposed method is evaluated in terms of accuracy (bias) and precision (standard deviation) on the synthetic dataset.

In Vivo. The *in vivo* dataset consists of the imaged pelvis of a healthy volunteer who gave informed consent. The data was acquired on a 3T Philips Ingenia scanner using a multi-echo 3D spoiled gradient echo sequence with 6 echoes (TE_1 = 1.15 ms, Δ_{TE} = 1.15 ms, TR = 8.5 ms, flip angle = 3°). 50 slices of thickness 1.5 mm were acquired, with matrix size 240 × 240 and voxel size 1.8 mm × 1.8 mm. To create pseudo-ground truths for the tissue properties *in vivo*, the data was supersampled by repeating the acquisition 4 times within a single scanning session, with each acquisition using two k-space averages to increase SNR. Conventional qMRI parameter estimation using the MAGORINO algorithm [5] was then performed on this supersampled dataset to provide pseudo-ground truths. In the same scanning session, the subject was also imaged using a single k-space average. This is representative of a clinical acquisition and serves as evaluation data to compare the parameter maps predicted by different qMRI methods to the pseudo-ground truths.

Parameter estimation is performed on the single k-space average *in vivo* image and compared to the pseudo-ground truths. A manually-selected threshold is used to remove background voxels, which are not of interest and may

[1] https://github.com/CIG-UCL/RAIDER-SSL.

affect the self-supervised procedure. We first use a naive self-supervised implementation using a single network to show that the network learns to predict tissue properties belonging to only one sub-space. The dual network approach is then implemented to help resolve model degeneracy. Supervised-based [4] and conventional fitting [5] are also performed, and the resulting PDFF, R_2^* and S_0 maps from these three methods are compared to the pseudo-ground truths.

4 Results

Figure 4 shows the *in silico* results of PDFF estimation for different self-supervised learning strategies. Using a single network approach, without constraining PDFF to lie within one of the two degenerate sub-spaces, results in the trained network to unpredictably make either water- or fat-dominant predictions. This phenomenon occurred regardless of whether a sub-space was rejected from the training data. That is, the network can learn to predict either solution when trained on data from the full, water-dominant or fat-dominant parameter spaces. However, training networks using the proposed constraining transforms ensures the networks make predictions in the selected parameter sub-space.

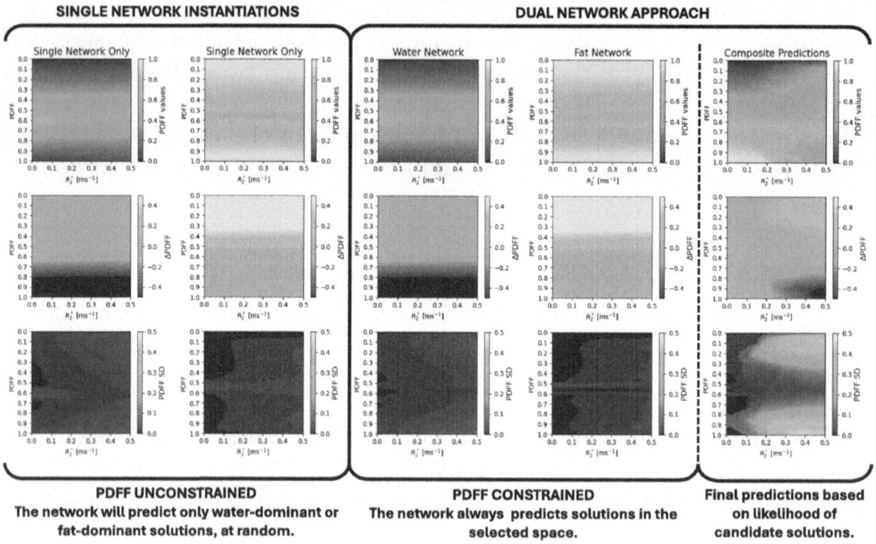

Fig. 4. Feasibility of self-supervised learning in the presence of model degeneracy *in silico*. Without the proposed constraining transforms, self-supervised networks can unpredictably learn to output either water- (1st column) or fat-dominant (2nd column) solutions. With the constraining transforms, the target networks are predictably learned (3th and 4th columns). The prediction that maximises likelihood is chosen as the estimate (5th column). Rows sequentially represent predicted PDFF values, bias and standard deviation.

Figure 5 shows that using a single self-supervised network does not resolve degeneracy *in vivo*, since only water-dominant solutions are predicted. By training constrained 'water' and 'fat' networks, the candidate solutions can be used to create composite predictions based on likelihood.

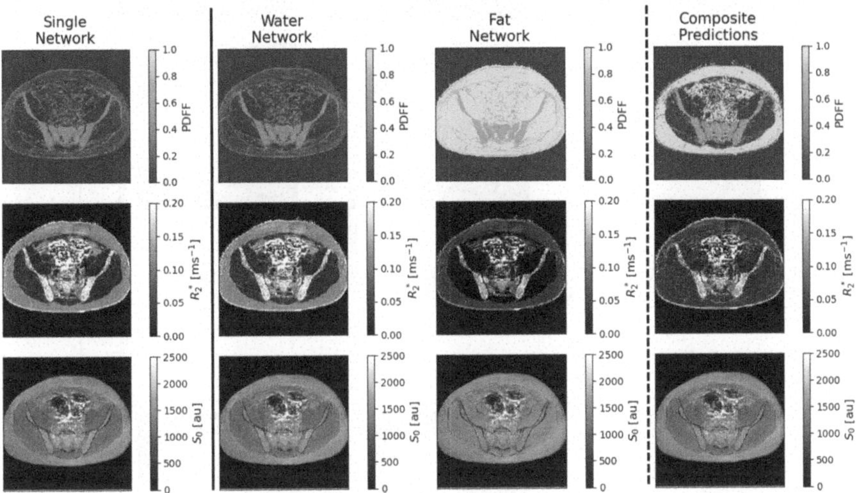

Fig. 5. Degeneracy mitigation *in vivo* using a dual network approach. A single network (leftmost) unpredictably output tissue properties in one degenerate sub-space. The proposed use of constraining transforms ensures 'water' and 'fat' networks are learned with certainty (middle two). The prediction that maximises likelihood is chosen as the estimate (rightmost).

Figure 6 shows the pseudo ground-truths acquired from the supersampled dataset, as well as the parameter maps estimated using self-supervised, supervised and conventional methods. Overall, the self-supervised method has lower bias than its supervised alternative for the clinically relevant PDFF and R_2^*, suggesting self-supervised strategies may be preferred over supervised ones, as they avoid distributional shift that can negatively affect parameter estimation. Indeed, self-supervised achieves lower PDFF bias in comparison to supervised, particularly visible in subcutaneous fat. In the same region, the supervised method exhibits larger R_2^* bias, while S_0 predictions are also larger, likely due to compensation for the higher predicted R_2^* decay. Furthermore, self-supervised and conventional techniques achieve similar performance across different tissue types. Although all three methods show large bias in the bowel, this is expected due to peristalsis (movement of the intestine due to involuntary muscle contractions).

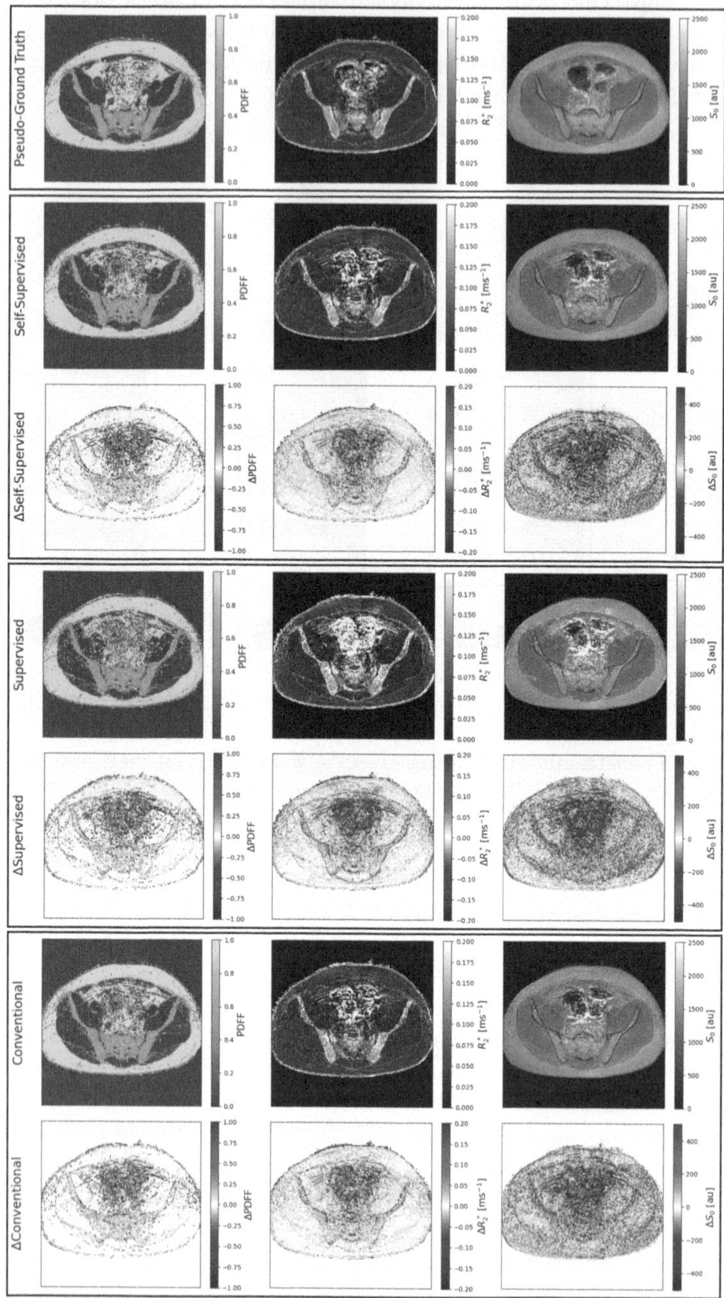

Fig. 6. Estimated parameter maps using self-supervised (2nd row), supervised (4th row) and conventional (6th row) methods. Their difference with the pseudo-ground truths obtained from the supersampled dataset (1st row) are also shown.

5 Discussion and Conclusions

In this work, we make two contributions to the understanding of the scope within which self-supervised methods may be applied for qMRI parameter estimation. First, we describe the behaviour of self-supervised algorithms in the presence of model degeneracy and confirm it in CSE-MRI data, both in simulation and *in vivo*. Neural networks trained on degenerate signals unpredictably learn to output a unique combination of tissue properties that produces that signal, compromising parameter estimation.

Second, we propose a method to help resolve degeneracy for self-supervised approaches by taking inspiration from previously developed strategies for supervised algorithms [3,4]. Both these strategies rely on the assumption that the parameter space can be split into sub-spaces, such that no degeneracy exists within each sub-space. Networks designated to a sub-space can then be trained by sampling labels from the sub-space itself to avoid the pitfalls of model degeneracy. However, these strategies cannot be directly applied to self-supervised methods due to the fundamental difference between the training data of these different machine learning strategies. On one hand, separating labels from degenerate sub-spaces may be effective for supervised learning since it relies on feature-label pairs of (input) features being the signal and (output) labels being tissue properties. On the other hand, both the features and labels of self-supervised algorithms are the signal, making this strategy inapplicable. Instead, we use a constraining transform to the bottleneck of the autoencoder, such that the estimated tissue properties lie in one of these sub-spaces. Using these constraining transforms and a dual network approach inspired by [4], we demonstrate that degeneracy can be mitigated using self-supervised methods in CSE-MRI. Two networks, respectively predicting water-dominant and fat-dominant tissue properties, produce two candidate solutions that can be compared based on likelihood to select the best tissue parameter estimate.

One may argue that the constraining transforms are not required because different networks may be learned by training multiple times until all possible variants are discovered. However, an obvious downside of this approach is an unnecessary increase in training time, which is not viable with this self-supervised implementation that requires to be retrained for different datasets.

Due to space limit, the present study has demonstrated our proposed method in only one qMRI example. However, model degeneracy is a common challenge in qMRI, from relaxometry [8] to diffusion [15]. Evaluating the broad applicability of our approach will be a key focus of future work.

Acknowledgments. Giulio Minore is supported by the EPSRC funded UCL Centre for Doctoral Training in Intelligent, Integrated Imaging in Healthcare (i4health) (EP/S021930/1) and the National Institute for Health Research (NIHR) Biomedical Research Centre (BRC) at University College London Hospitals. Timothy Bray receives support from the NIHR BRC (personal support, and grants BRC1121/HEI/TB/110410 and BRC1185/III/TB/101350).

Disclosure of Interests. The authors have no competing interests to declare that are relevant to the content of this article.

References

1. Barbieri, S., Gurney-Champion, O.J., Klaassen, R., Thoeny, H.C.: Deep learning how to fit an intravoxel incoherent motion model to diffusion-weighted MRI. Magn. Reson. Med. **83**(1), 312–321 (2020)
2. Bertleff, M., et al.: Diffusion parameter mapping with the combined intravoxel incoherent motion and kurtosis model using artificial neural networks at 3T. NMR Biomed. **30**(12) (2017)
3. Bishop, C.M., Roach, C.M.: Fast curve fitting using neural networks. Rev. Sci. Instrum. **63**(10), 4450–4456 (1992)
4. Bray, T.J.P., et al.: RAIDER: rapid, anatomy-independent, deep learning-based PDFF and R_2^* estimation using magnitude-only signals (2024), https://arxiv.org/abs/2403.01178
5. Bray, T., Bainbridge, A., Lim, E., Hall-Craggs, M.A., Zhang, H.: MAGORINO: magnitude-only fat fraction and R_2^* estimation with rician noise modeling. Magn. Reson. Med. **89**(3), 1173–1192 (2023)
6. Bray, T., Bainbridge, A., Punwani, S., Ioannou, Y., Hall-Craggs, M.A.: Simultaneous quantification of bone edema/adiposity and structure in inflamed bone using chemical shift-encoded MRI in spondyloarthritis. Magn. Reson. Med. **79**(2), 1031–1042 (2018)
7. Clevert, D., Unterthiner, T., Hochreiter, S.: Fast and accurate deep network learning by exponential linear units (elus). In: Bengio, Y., LeCun, Y. (eds.) 4th International Conference on Learning Representations, ICLR 2016, San Juan, Puerto Rico, 2–4 May 2016, Conference Track Proceedings (2016), http://arxiv.org/abs/1511.07289
8. Deoni, S., Rutt, B.K., Arun, T., Pierpaoli, C., Jones, D.K.: Gleaning multicomponent T_1 and T_2 information from steady-state imaging data. Magn. Reson. Med. **60**(6), 1372–1387 (2008)
9. Epstein, S.C., Bray, T., Hall-Craggs, M., Zhang, H.: Choice of training label matters: how to best use deep learning for quantitative MRI parameter estimation. Mach. Learn. Biomed. Imaging **2**, 586–610 (2024)
10. Golkov, V., et al.: q-space deep learning: Twelve-fold shorter and model-free diffusion MRI scans. IEEE Trans. Med. Imaging **35**(5), 1344–1351 (2016)
11. Grussu, F., Battiston, M., Palombo, M., Schneider, T., Wheeler-Kingshott, C., Alexander, D.C.: Deep learning model fitting for diffusion-relaxometry: a comparative study. In: Gyori, N., Hutter, J., Nath, V., Palombo, M., Pizzolato, M., Zhang, F. (eds.) Computational Diffusion MRI, pp. 159–172. Springer International Publishing, Cham (2021)
12. Guerreri, M., Epstein, S., Azadbakht, H., Zhang, H.: Resolving quantitative MRI model degeneracy with machine learning via training data distribution design. In: Frangi, A., de Bruijne, M., Wassermann, D., Navab, N. (eds.) Information Processing in Medical Imaging, pp. 3–14. Springer Nature Switzerland, Cham (2023)
13. Gyori, N.G., Palombo, M., Clark, C.A., Zhang, H., Alexander, D.C.: Training data distribution significantly impacts the estimation of tissue microstructure with machine learning. Magn. Reson. Med. **87**(2), 932–947 (2022)

14. Hernando, D., Kellman, P., Haldar, J.P., Liang, Z.P.: Robust water/fat separation in the presence of large field inhomogeneities using a graph cut algorithm. Magn. Reson. Med. **63**(1), 79–90 (2010)
15. Jelescu, I.O., Veraart, J., Fieremans, E., Novikov, D.S.: Degeneracy in model parameter estimation for multi-compartmental diffusion in neuronal tissue. NMR Biomed. **29**(1), 33–47 (2016)
16. Kaandorp, M., et al.: Improved unsupervised physics-informed deep learning for intravoxel incoherent motion modeling and evaluation in pancreatic cancer patients. Magn. Reson. Med. **86**(4), 2250–2265 (2021)
17. Kingma, D.P., Ba, J.: Adam: a method for stochastic optimization. In: Bengio, Y., LeCun, Y. (eds.) 3rd International Conference on Learning Representations, ICLR 2015, San Diego, CA, USA, 7-9 May 2015, Conference Track Proceedings (2015), http://arxiv.org/abs/1412.6980
18. Latifoltojar, A., et al.: Whole-body MRI quantitative biomarkers are associated significantly with treatment response in patients with newly diagnosed symptomatic multiple myeloma following bortezomib induction. Eur. Radiol. **27**(12), 5325–5336 (2017)
19. Morrow, J.M.F., et al.: MRI biomarker assessment of neuromuscular disease progression: a prospective observational cohort study. Lancet Neurol. **15**(1), 65–77 (2016)
20. Parker, C.S., Schroder, A., Epstein, S.C., Cole, J., Alexander, D.C., Zhang, H.: Rician likelihood loss for quantitative MRI using self-supervised deep learning (2023), https://arxiv.org/abs/2307.07072
21. Reeder, S.B., et al.: Water-fat separation with IDEAL gradient-echo imaging. J. Magn. Reson. Imaging **25**(3), 644–652 (2007)
22. Reeder, S.B., et al.: Quantification of liver iron overload with MRI: review and guidelines from the ESGAR and SAR. Radiology **307**(1), e221856–e221856 (2023)
23. Sen, S., et al.: ssVERDICT: self-supervised VERDICT-MRI for enhanced prostate tumor characterization. Magn. Reson. Med. **92**(5), 2181–2192 (2024)
24. Sijbers, J., den Dekker, A., Scheunders, P., Van Dyck, D.: Maximum-likelihood estimation of Rician distribution parameters. IEEE Trans. Med. Imaging **17**(3), 357–361 (1998)
25. Starekova, J., Hernando, D., Pickhardt, P.J., Reeder, S.B.: Quantification of liver fat content with CT and MRI: state of the art. Radiology **301**(2), 250–262 (2021)
26. Triay Bagur, A., Hutton, C., Irving, B., Gyngell, M.L., Robson, M.D., Brady, M.: Magnitude-intrinsic water-fat ambiguity can be resolved with multipeak fat modeling and a multipoint search method. Magn. Reson. Med. **82**(1), 460–475 (2019)
27. Yoon, J.H., et al.: Pancreatic steatosis and fibrosis: quantitative assessment with preoperative multiparametric MR imaging. Radiology **279**(1), 140–150 (2016)
28. Yu, H., et al.: Multiecho reconstruction for simultaneous water-fat decomposition and T_2^* estimation. J. Magn. Reson. Imaging **26**(4), 1153–1161 (2007)

Vision-Language Models

Knowledge-Enhanced Hyperbolic Language-Image Pretraining for Zero-Shot Learning

Linbin Han[1,2,3], Zhi Qiao[1], Xiantong Zhen[1], Jiahong Gao[2,3,4], and Zhen Qian[1(✉)]

[1] Institute of Intelligent Diagnostics, Beijing United-Imaging Research Institute, Beijing, China
zhen.qian@cri-united-imaging.com

[2] Center for MRI Research, AAIS, Peking University, Beijing, China

[3] Beijing City Key Lab for Medical Physics and Engineering, Institution of Heavy Ion Physics, School of Physics, Peking University, Beijing, China

[4] McGovern Institute for Brain Research, Peking University, Beijing, China

Abstract. In light of the inherent entailment relations between images and text, hyperbolic point vector embeddings, which leverage the hierarchical structure of hyperbolic space, have been used for visual-semantic representation, showing significant advantages in zero-shot learning tasks. However, unlike general image-text alignment tasks, the availability of high-quality paired image-text data in the medical domain is limited. This scarcity presents challenges for visual language models, hindering their ability to effectively comprehend free-text medical reports and the associated images. Moreover, many medical terms are complex, specialized, and abstract, and embeddings derived solely from raw imaging report texts often fail to generalize effectively. To address these challenges, we propose MkCLIPH, a hyperbolic space image-text alignment pre-training method that incorporates medical domain knowledge. MkCLIPH models the visual-semantic hierarchical partial order relationship through hierarchical entailment angle modeling and integrates medical domain knowledge as a prior to enhance the representation of medical image-text data. This improves generalization and interpretability. Experimental results demonstrate that our method outperforms baseline approaches in terms of interpretability and performance across a range of zero-shot tasks and datasets.

1 Introduction

Visual-language representation learning, which employs contrastive learning to align image and text features in a shared target space using large-scale image-text datasets, has become a cornerstone in various downstream applications, including zero-shot classification and zero-shot retrieval [20]. The adage "a picture is

L. Han and Z. Qiao: Equally Contributions.

I. Oguz et al. (Eds.): IPMI 2025, LNCS 15830, pp. 203–217, 2026.
https://doi.org/10.1007/978-3-031-96625-5_14

worth a thousand words" suggests that images often carry far more information than the sentences used to describe them, with sentences typically serving as summaries of the image content. This phenomenon is characterized as the visual-semantic hierarchy [27]. Incorporating this partial-order relationship, known as the visual-semantic hierarchy, into cross-modal modeling has been demonstrated to bolster the model's generalization capabilities [20] and to augment its interpretability [24].

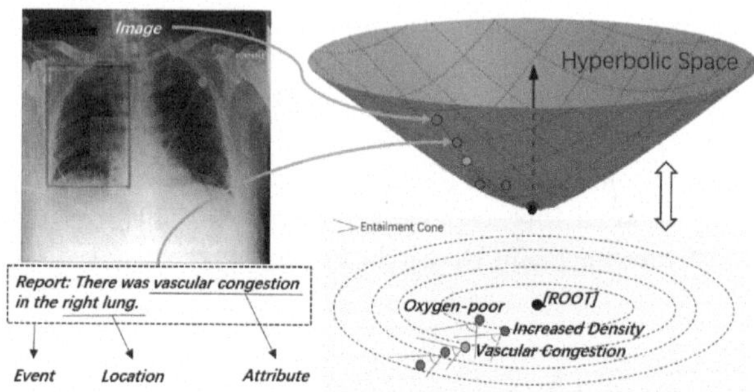

Fig. 1. The radiological assessment for this patient indicates right lung vascular congestion (the blue box in the figure identifies the location of the right lung, while the red heatmap displays the attention weights of individual image tokens in the Semantic-sensitive Image Feature Extraction). By projecting the image features into the hyperbolic space, we traverse from the projected node to the [ROOT] of hyperbolic space along the geodesic direction, outputting the nearest neighbor text embedding points. As shown in the lower right corner of the figure, the sequence is as follows: Image, Event, Vascular Congestion, Oxygen-poor, Increased Density, [ROOT]. (Color figure online)

In this context, Desai [7] introduced the MERU alignment framework, which models visual-semantic hierarchical features in hyperbolic space using entailment cone constraints to achieve effective image-text alignment. Furthermore, [22] proposed a similarity metric to mitigate the impact of spatial proximity for hierarchy distruption. Underspecified images and texts representation learning is modeled via hyperbolic embedding in [11]. However, the availability of high-quality paired image-text data in the medical domain remains limited, posing significant challenges for visual language models. This scarcity impedes their ability to effectively comprehend free-text medical reports and associate them with relevant images [3]. Moreover, many specialized medical terms are highly abstract (for instance, the term "infiltration" refers to white patches in the lung). As a result, features learned from raw medical reports alone often fail to generalize effectively to downstream tasks [32].

Several studies have attempted to incorporate domain-specific medical knowledge to improve the learning capacity of image-text representations [4,30,32]. Incorporating domain knowledge not only enriches the textual descriptions of imaging reports but also serves as prior knowledge that contextualizes highly abstract terms, thereby enhancing generalization. To achieve better generalization and interpretability of image-text representations, we model visual-semantic hierarchy in hyperbolic space while leveraging medical domain knowledge as prior information. Figure 1 demonstrates the embedding of our model, which expands the hierarchical partial order of visual semantics by further incorporating domain knowledge and unifying it within a hyperbolic space.

Firstly, in previous works [5,16,19,28], medical imaging reports were typically embedded as a whole into the alignment space. However, these reports often describe multiple conditions, each linked to specific regions or aspects of the images, and directly embedding them can lead to semantic confusion. Inspired by report generation [13], medical question answering [8] and representation learning [34], we use the structured medical reports composed of triplets: $< \mathrm{Attribute}, \mathrm{Location}, \mathrm{Existence} >$. In these triplets, attributes and locations are typically proper nouns in the medical field, and their corresponding concepts can be derived from a medical knowledge base. This structured data mitigates the semantic ambiguity of direct report embeddings, particularly by preventing the negative impact of rough negative sampling in contrastive learning. To further enhance the model's understanding, we introduce a medical knowledge-enhanced prompt module. This module leverages UMLS [2] and prompt engineering techniques to convert the triplets into natural text sets, including attribute concepts, location concepts, and event descriptions.

Secondly, we introduce the Medical Concept-Event Collaborative Embedding Block. This module uses a text encoder to obtain semantic embeddings for the text triplets described above. It then performs relational learning between the event text and knowledge text embeddings, thereby enhancing the event text embeddings with domain knowledge. Additionally, we incorporate a semantic-sensitive image feature extraction module that extracts event-relevant image features in conjunction with the knowledge-enhanced event text features. This reduces the semantic confusion caused by holistic embeddings. Furthermore, we extend the entailment relationships in hyperbolic alignment by introducing a knowledge-level entailment relationship and constructing a hierarchical hyperbolic entailment loss.

Finally, we use the extracted triplets from the medical reports as explicit semantic labels, which help distinguish between positive and negative samples, applying entailment and contrastive penalties to the negative samples. In summary, the key contributions of this work include:

- We introduce domain knowledge into hyperbolic space alignment, enhancing the ability to learn image-text representations and the understanding of fundamental medical concepts, thereby improving the generalization of representation learning.
- We propose a hierarchical hyperbolic entailment loss associates concepts, events, and image features in hyperbolic space, improving interpretability.

2 Preliminaries

2.1 Hyperbolic Geometry

Hyperbolic geometry is a non-Euclidean geometry with a constant negative curvature. In this study, we will use the Lorentz model on the upper half of a two-sheeted hyperboloid, as claimed in [17], comes with a simpler closed form of the geodesics and does not suffer from the numerical instabilities in approximating the distance. Lorentz model $\mathbb{H}^n$ processing a constant curvature $-c$ can be represented as a set of points $z \in \mathbb{R}^{n+1}$. Lets $z = [z_0, z_1, ..., z_n], z' = [z'_0, z'_1, ..., z'_n] \in \mathbb{H}^n$, the Lorentzian product $\langle z, z' \rangle_{\mathcal{L}} = -z_0 z'_0 + \sum_{i=1}^{n} z_i z'_i$. And, $\mathbb{H}^n = \{z \in \mathbb{R}^{n+1} : \langle z, z \rangle_{\mathcal{L}} = -1/c, \ c > 0\}$. The distance between z and z' is given by

$$d_{\ell}(z, z') = arccosh(-\langle z, z' \rangle_{\mathcal{L}}) \tag{1}$$

which is also the length of the geodesic that connects z and z'. We will refer to the one-hot vector $\mu_0 = [1/\sqrt{c}, 0, 0, 0...0] \in \mathbb{H}^n \subset \mathbb{R}^{n+1}$ as the origin of the hyperbolic space, which is denoted as [ROOT].

2.2 Tangent Space of Hyperbolic Space and Exponential Map

The tangent space at a point $\mu \in \mathbb{H}^n$ is a Euclidean space composed of vectors. Denoted by $T_\mu \mathbb{H}^n$, this tangent space represents the set of vectors in the same ambient space $\mathbb{R}^{n+1}$ where $\mathbb{H}^n$ is embedded. The vectors in $T_\mu \mathbb{H}^n$ satisfy an orthogonality condition relative to the Lorentzian product, defined as $T_\mu \mathbb{H}^n := \{u : \langle \mu, u \rangle_{\mathcal{L}} = 0\}$. This set can be visualized as the tangent space at the point μ on the forward hyperboloid sheet. Specifically, at the origin μ_0 of $\mathbb{H}^n$, the tangent space $T_{\mu_0} \mathbb{H}^n$ consists of vectors $v \in \mathbb{R}^{n+1}$. The norm $\|v\|_{\mathcal{L}}$, given by the Lorentzian inner product, simplifies to the Euclidean norm $\|v\|_2$, defined as $\|v\|_{\mathcal{L}} := \sqrt{\langle v, v \rangle_{\mathcal{L}}} = \|v\|_2$.

The exponential map provides a method for mapping a vector from a tangent space to its corresponding point on the surface of the hyperbolic space. For every $u \in T_\mu \mathbb{H}^n$, the exponential map $exp_\mu(u) : T_\mu \mathbb{H}^n \to \mathbb{H}^n$ allows us to project a vector u in $T_\mu \mathbb{H}^n$ onto $\mathbb{H}^n$ such that the distance from μ to the destination point of the map coincides with the Lorentzian norm $\|u\|_{\mathcal{L}}$ of u. In the context of hyperbolic space, the exponential map is given by the equation:

$$z = \exp_\mu(u) = \cosh(\|u\|_{\mathcal{L}})\mu + \sinh(\|u\|_{\mathcal{L}})\frac{u}{\|u\|_{\mathcal{L}}} \tag{2}$$

In this paper, we specifically consider exponential maps where μ represents the origin of the hyperboloid ($\mathrm{O} = [\sqrt{1/c}, \mathbf{0}]$).

3 Method

In this section, we first introduce the medical triplet data extracted from original medical imaging reports. Then, we present the overall modeling framework, including knowledge based triplet prompting, collaborative embedding,

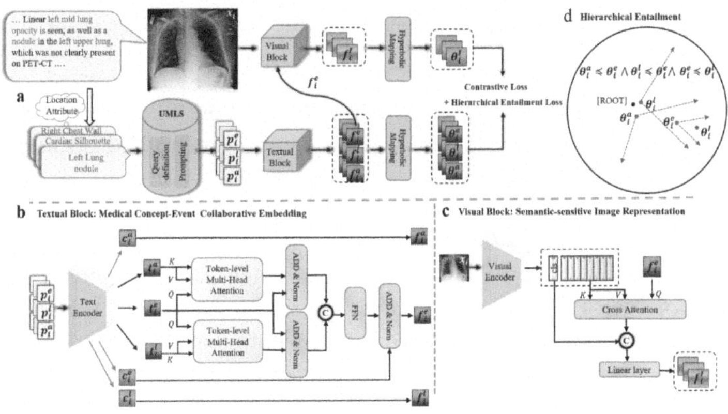

Fig. 2. Framework of Proposed Method

semantic-aware feature extraction, and hierarchical hyperbolic entailment loss function. Finally, we discuss model training and inference.

Triplet Data. Assume we have a training set with N samples, denoted as $D = \{(x_1, t_1), ..., (x_N, t_N)\}$, where x_i and t_i represent the i-th X-ray image and its corresponding medical report, respectively. Compared to natural text, the information in medical reports is more concise and comprehensive, as doctors typically indicate various abnormalities and their locations within the image. Some previous works [10,30] have been exploring to parse the medical report into a set of triplets and created public datasets, like ImaGenome [31], leveraging the UMLS [2] lexicon to cover most medical terms. Here, we directly use the structured dataset. Each triplet consists of ("Location", "Attribute", "Existence"), where "Attribute" refers to kinds of chest x-ray attribute (like findings, diseases, technical assessment, etc.). "Location" refers to anatomical structure mentioned in the radiology report (e.g., "right lower lobe"). Given a medical report t_i, the transformed set of triplets is S_i, i.e., $S_i = \{(a_k, l_k, r_k), k = 1, ..., ||S_i||\}$, where a_k, l_k, and r_k denote the attribute, location, and relation in the k-th triplet. During training, for each sample pair $< x_i, t_i >$, we randomly sample a triplet from S_i, which we simplify as $< x_i, (a_i, l_i, r_i) >$.

Knowledge-Based Triplet Prompting. The objective of this module is to transform the triplets (a_i, l_i, r_i) into textual sentences (P_i^a, P_i^l, P_i^e) by integrating domain-specific medical knowledge. Here, a_i and l_i represent general medical concepts, which are translated into sentences using the Unified Medical Language System (UMLS) [2] and prompt templates. Assume a_i='lung nodule', P_i^a = Prompt('lung nodule'): 'There exists {lung nodule} in this chest x-ray images. {lung nodule} refers to a focal, dense, round lesion with a clear or blurred margin, measuring less than or equal to 3 cm in diameter, seen on lung imaging.'. For location l_i, we similarly use prompt templates. The entire triplet denotes an event present in the image, therefor we design specific prompting template to convert it into P_i^e, like Prompt([‘lung nodule’, ‘left lung’, ‘present’]) = ‘This image shows there exists {lung nodule} in the {left lung}.’.

Concept-Event Collaborative Embedding. For the prompted triplets (P_i^a, P_i^l, P_i^e), we use BioClinicalBERT [1], a model pre-trained on medical text corpora from the MIMIC-CXR dataset [25], as the text encoder, generating token-level embeddings (t_i^a, t_i^l, t_i^e). As with the medical text embedding approach in PRIOR [6], we use the $[CLS]$ token as the global text features (c_i^a, c_i^l, c_i^e). To enhance the event feature with concept knowledge, we introduce a Concept-Event Collaborative Embedding module. The event feature t_i^e is related to the token-level enriched knowledge features of concepts feature t_i^a and t_i^l via token-level multi-head attention. The detailed procedure is illustrated in Fig. 2(b). Then, we can obtain the embedded features of concepts and event, (f_i^a, f_i^l, f_i^e).

Semantic-Sensitive Image Feature Extraction. To effectively extract features from image x_i, we adopt the widely used Vision Transformer (ViT) architecture for image feature extraction. Let $Encoder_{VIT}(x_i) = (\hat{x_{i_0}}, \hat{x_{i_1}}, ...\hat{x_{i_n}})$ represents the output of the image passed through VIT, where $x_{\hat{i_{global}}}$ is the output of [CLS] token, serving as the global feature of the image. Unlike natural images, pathological features often occupy only a portion of the entire image, making it difficult for global representations alone to capture relevant local semantic features. Therefore, previous works [6,9,16] have proposed using local feature extraction in conjunction with global features to capture semantically relevant local representations. In our approach, we introduce a semantic-sensitive image feature extraction module. Specifically, we leverage the commonly used self-attention mechanism [26] in cross-modal feature extraction. Using the outputs of ViT $(\hat{x_{i_0}}, ..., \hat{x_{i_n}})$ as keys (K) and values (V), and the embedding of triplet event f_i^e as the query (Q), we obtain the semantic-sensitive local image representation $x_{\hat{i_{local}}}$. By combining the global and local image representations, we derive the final image feature as $f_i^I = \text{ReLU}(Linear(x_{\hat{i_{global}}} + x_{\hat{i_{local}}}))$. The detailed procedure is illustrated in Fig. 2(c).

Hierarchical Hyperbolic Entailment. We begin by introducing the entailment loss [12,17] to reinforce the learning of partial order relationships. Due to the computational challenges and numerical instability associated with the Poincaré disk model, we adopt the Lorentz manifold-based hyperbolic entailment loss [7]. Suppose $x = [x_0, ..., x_n] \in \mathbb{H}^n, y = [y_0, ..., y_n] \in \mathbb{H}^n$, and x entails y $(x \preceq y)$. To measure the partial order relationship between x and y, the following loss function is employed:

$$L_{\text{en}}^P(x, y) = \max(0, \text{ext}(x, y) - \text{aper}(x)) \tag{3}$$

Here, referring to hyperbolic geometry in preliminaries, $\text{aper}(x)$ represents the entailment angle of x, with the calculation provided in Eq. 4, and $\text{ext}(x, y)$ denotes the angle between x and y, with the calculation given in Eq. 5. The main goal of the entailment loss is to identify and penalize cases where the entailment assumption is violated (i.e., when y lies outside the entailment cone of x).

$$\text{aper}(x) = \sin^{-1}\left(\frac{2K}{\sqrt{c(||x||^2 - x_0^2)}}\right) \tag{4}$$

In this equation, the constant $K = 0.1$ is used to set boundary conditions near the origin.

$$\text{ext}(x, y) = \cos^{-1}\left(\frac{y_0 + x_0 c\langle x, y\rangle_L}{\sqrt{||x|| - x_0^2}\sqrt{(c\langle x, y\rangle_L)^2 - 1}}\right) \tag{5}$$

Following the previous steps, we have obtained the triplet embeddings and the corresponding image embedding (f_i^a, f_i^l, f_i^e, f_i^I). We first project these embeddings from Euclidean space to hyperbolic space via using Eq. 2, resulting in ($\Theta_i^a \in \mathbb{H}^n$, $\Theta_i^l \in \mathbb{H}^n$, $\Theta_i^e \in \mathbb{H}^n$, $\Theta_i^I \in \mathbb{H}^n$). Based on the earlier entailment assumption, there are three levels of semantic entailment relationships: $\Theta_i^a \preceq \Theta_i^e$, $\Theta_i^l \preceq \Theta_i^e$, and $\Theta_i^e \preceq \Theta_i^I$. Among them, Therefore, for any positive sample pair (x_i, t_i), we define the hierarchical entailment loss as follows:

$$L_{Hen}^P(x_i, t_i) = L_{\text{en}}^P(\Theta_i^a, \Theta_i^e) + L_{\text{en}}^P(\Theta_i^l, \Theta_i^e) + L_{\text{en}}^P(\Theta_i^e, \Theta_i^I) \tag{6}$$

Modelling Negative Samples. In previous methods [5,7,16,20], image-text pairs from the same record were assumed to be positive samples, while pairs from different records were treated as negative samples. However, unlike natural datasets, many medical images or reports, especially those from normal cases, can be highly similar across different patients. Treating all unpaired samples as negative can disrupt the learned semantic structure and negatively affect the representations [14].

In contrast to earlier works, the triplets extracted from medical reports in our method are standardized and unified across the entire dataset. Additionally, the Semantic-Sensitive Image Feature Extraction module ensures that the image representations are strongly associated with the triplets. Therefore, in each epoch, the image-text pairs with the same triplet are considered positive samples, and those with different triplets are considered negative samples. Assuming that x does not entail y ($x \npreceq y$), we introduce a margin-based penalty inspired by the max-margin loss [27]. This penalty encourages the distance between negative samples to exceed a predefined margin, as shown below:

$$L_{\text{en}}^N(x, y) = \max(0, \epsilon_m - (\text{ext}(x, y) - \text{aper}(x))) \tag{7}$$

where ϵ_m is the margin for penalizing negative samples. Thus, for any negative sample pair ($\dot{x}_i$, $\dot{t}_i$), the corresponding hierarchical entailment loss is defined as:

$$L_{Hen}^N(\dot{x}_i, \dot{t}_i) = L_{\text{en}}^N(\dot{\Theta}_i^a, \dot{\Theta}_i^e) + L_{\text{en}}^P(\dot{\Theta}_i^l, \dot{\Theta}_i^e) + L_{\text{en}}^P(\dot{\Theta}_i^e, \dot{\Theta}_i^I) \tag{8}$$

Therefore, during training, the overall entailment loss is defined as:

$$\frac{1}{|\mathbb{P}|+|\mathbb{N}|}\left(\sum_{(x_i,t_i)\in\mathbb{P}} L^P_{Hen}(x_i,t_i)+\sum_{(\dot{x_i},\dot{t_i})\in\mathbb{N}} L^N_{Hen}(\dot{x_i},\dot{t_i})\right) \tag{9}$$

where $\mathbb{P}$ and $\mathbb{N}$ represent the sets of positive and negative samples, respectively.

Furthermore, similar to the solution in MERU [7], we adopt a contrastive loss approach using the Lorentzian distance instead of cosine similarity function to measure the similarity, as shown in Eq. 1. The resulting values are normalized by a temperature parameter τ and applied with the $Softmax$ operator. We compute the contrastive loss for all entailment cone constraint cases. The total loss is the average of these two losses.

4 Experiments

4.1 Training Details

Datasets: The MIMIC-CXR v2 dataset [25] includes over 227,000 paired image-report studies sourced from 65,379 patients. The Chest ImaGenome Dataset [31] provides structured data for 217,013 image-report pairs, extracting 1,256 combinations of relation annotations between 29 anatomical locations on chest X-rays (CXR) and 90 attributes, such as findings, diseases, technical assessments, and devices. We trained on these 217,013 pairs. For the 119 attributes and locations, we constructed a knowledge dictionary by gathering their medical descriptions from the Unified Medical Language System (UMLS) [2], which can be used for knowledge queries. During training, we primarily select frontal CXR images (AP or PA views) and apply random cropping, flipping, rotation, and other data augmentation techniques. The images are then resized to [224, 224].

Settings: We employ ViT-B [15] with a patch size of 16 as the image encoder, as it has demonstrated competitive performance in hyperbolic space [7]. Our initialization strategy for image/text encoders follows a similar style to MERU, with the exception of utilizing ClinicalBERT [1] as the text encoder. For MkCLIPH, we empirically set the curvature parameter c to 0.5. All experiments were conducted using two NVIDIA A40 GPUs and the PyTorch framework.

Table 1. Zero-shot image retrieval

Methods	Prec@3	Prec@5	Prec@10	NDCG@3	NDCG@5	NDCG@10
	Pretraining@Non-Medical Datasets					
CLIP	14.29	12.86	7.86	0.254	0.289	0.271
MERU	7.14	7.14	5.71	0.107	0.133	0.179
	Pretraining@Medical Datasets					
Gloria[9]	26.67	30.67	28	0.316	0.374	0.438
MedKLIP[30]	30.95	31.43	34.29	0.489	0.524	0.540
CLIP[20]	23.81	18.57	12.14	0.408	0.414	0.439
MERU[7]	33.33	30.0	27.86	0.417	0.452	0.465
MkCLIPH	**40.48**	**38.57**	**38.57**	0.534	**0.591**	0.584

Table 2. Zero-shot image classification

	Tuberculosis			ObjectCXR			CXR14 Multilabel		
Methods	AUC	F1	ACC	AUC	F1	ACC	Micro-AUC	Micro-F1	ACC
Pretraining@Non-Medical Datasets									
CLIP	0.512	0.286	0.166	0.481	0.5000	0.333	0.477	0.138	0.284
MERU	0.543	0.374	0.815	0.551	0.5001	0.333	0.484	0.139	0.279
Pretraining@Medical Datasets									
CLIP[20]	0.749	0.456	0.756	0.532	0.501	0.340	0.631	0.177	0.674
ConVIRT[33]	-	-	-	-	-	-	0.560	0.135	0.459
Gloria[9]	0.701	0.393	0.608	0.503	0.500	0.334	0.610	0.174	0.503
BioViL[3]	-	-	-	-	-	-	0.662	0.192	0.633
RGCT[23]	-	-	-	-	-	-	0.712	-	-
Xplainer[18]	-	-	-	-	-	-	0.717	-	-
MedKLIP[30]	0.755	0.464	0.782	0.544	0.501	0.370	0.726	0.244	0.796
MERU[7]	0.765	0.442	0.675	0.568	0.5003	0.373	0.636	0.196	0.673
MkCLIPH	**0.852**	**0.608**	**0.884**	**0.614**	**0.506**	**0.477**	**0.743**	**0.266**	**0.811**

Optimization: We adopt the AdamW optimizer with a weight decay of 0.2 and $(\beta_1, \beta_2) = (0.9, 0.98)$. Weight decay is disabled for all gains, biases, and learnable scalars. The models are trained for 40,000 iterations with a batch size of 128. The maximum learning rate is set to 1×10^{-5}, which is linearly increased for the first 500 iterations, followed by cosine decay to zero.

Baselines: We select four primary methods as baseline approaches, CLIP [20](a classic pre-training method for vision-language representation), Gloria [9](introducing global+local representation enhancement module to enhance the perception of local features), MedKLIP [30](introducing knowledge into the image-text alignment and treating the image-text alignment task as a weakly supervised classification task) and MERU [7](CLIP-like model, adopting hyperbolic space instead of Euclidean space to align image-text representations). CLIP, Gloria and MedKLIP implement the image-text representation pre-training in the Euclidean space, and MERU in hyperbolic space. Moreover, to further validate the performance of our model, we also adopt several methods that have emerged in recent years in medical image and text representation learning as baseline methods, including ConVIRT [33], Xplainer [18], BioViL [3], and RGCT [23].

4.2 Evaluation Task and Data

We evaluate MkCLIPH and all baselines on three categories of zero-shot tasks: binary classification, multi-label classification and text-image retrieval.

Zero-shot Image Classification The evaluation is conducted using three public datasets: (1) **Tuberculosis Chest X-ray dataset** [21], which contains 4200 frontal chest X-ray images. The task is to classify each image as either *normal* or *tuberculosis*. (2) **Object-CXR dataset**[1] is used to determine whether an image contains foreign objects, such as necklace. We used 6,600 images from the official training set for our evaluation. (3) **ChestXray14 dataset** [29], a multi-label

[1] https://2020.midl.io/challenges.html.

dataset containing 14 disease labels. We used all 25,203 images from the official test set for multi-label classification. For the binary classification tasks, we report the Area Under the Curve (AUC), F1 score and Accuracy (ACC). For the multi-label classification task, we provide Micro-AUC, Micro-F1 and Micro-ACC metrics.

Zero-shot Text-Image Retrieval: Following the practices of CLIP [20] and MERU [7], we also introduce the text-image retrieval task. For this purpose, we construct a text-image retrieval evaluation dataset. As described above, ChestXray14 [29] encompasses 14 different disease classes. Based on these class labels, we randomly extract 100 images for each class (exclusive), forming the ChestXray14 $\times$ 100 dataset, which consists of 1,400 images. For the text-image retrieval task, we evaluate performance using Top-k Precision (Prec@k) and Top-k Normalized Discounted Cumulative Gain (NDCG@k), where $k = 3, 5, 10$.

Table 3. Ablation Study of MkCLIPH

	CheXpert	Multilabel	Text2Image@10	
	Micro-AUC	Micro-F1	Prec	NDCG
MkCLIPH	0.743	0.266	38.57	0.584
1. w Freezing Text Encoder	0.675	0.247	17.86	0.495
2. w/o Concept-Event Collaborative Embedding	0.729	0.257	33.57	0.573
3. w/o Semantic-Sensitive Image Feature Extraction	0.517	0.156	18.57	0.421
4. w/o Hyperbolic Space Mapping	0.731	0.266	36.43	0.644

4.3 Quantitative Analysis

Zero-shot Image Classification. Table 2 demonstrates the following: (1) All models pre-trained on medical datasets outperform those pre-trained on large-scale general image-text datasets across various classification metrics, indicating the necessity of pre-training on domain-specific data. (2) MkCLIPH exhibits excellent transfer performance in the multi-label classification task. Compared to CLIP, MERU achieves higher accuracy. Compared to MedKLIP, MkCLIPH achieves higher accuracy. This indicates that utilizing hyperbolic space for image-text representation is more effective when modeling medical image-text data, which often involves visual-semantic hierarchy characteristics. MkCLIPH outperforms MERU across all metrics, demonstrating that incorporating domain knowledge enhances representation capabilities. Similar findings are observed when comparing MedKLIP and all other baselines in Euclidean space. (3) In the binary classification task based on ObjectCXR data, the overall performance of all models is significantly inferior to the other two classification tasks. This is primarily attributed to the scarcity of data involving foreign objects in the pre-training datasets. Our approach enhances the model's generalization ability through knowledge augmentation. While MedKLIP also leverages domain knowledge during representation learning, its adoption of a classification strategy limits the model's generalization capabilities.

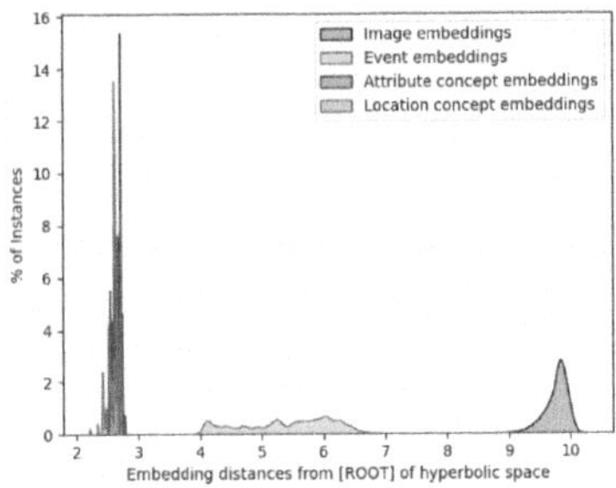

Fig. 3. Distribution of embedding distances from [ROOT]: We embed 160K training images and text from the MIMIC-CXR v2 dataset.

Zero-shot Retrieval. Table 1 presents the performance of all baseline models and MkCLIPH on the 'text-to-image retrieval' task. The results demonstrate that for retrieval tasks, representation learning in hyperbolic space consistently outperforms that in Euclidean space. Among all methods, MkCLIPH achieves the best retrieval performance. Moreover, MkCLIPH significantly improves the ranking quality of retrieval results compared to all baseline models. We speculate that this improvement is linked to the way knowledge is embedded, which enhances the model's ability to better understand both images and text. This enables more accurate matching between the two modalities. Additionally, models such as Gloria, CLIP, and MERU directly embed original text sentences, which may contain complex or ambiguous semantic information (e.g., multiple diagnoses or descriptions of various conditions). This confounding information can negatively affect the learning of accurate image-text representations. Similar trends were observed when comparing MedKLIP with Gloria.

Ablations Studies. In this section, we examine the impact of different design choices using MkCLIPH. Specifically, we trained five ablation models with default hyperparameters, and the results are presented in Table 3. From Table 3, we observe that: (1) Fine-tuning the pre-trained text encoder during training is essential for effective knowledge transfer. (2) The concept-Event collaborative embedding plays a crucial role in model training, especially for downstream tasks distinct from pre-training data. Its removal leads to significant performance drops on the Obj-CXR dataset, demonstrating its importance for model generalization.(3) Removing the semantic-sensitive image feature extraction module significantly degrades performance in diverse tasks due to the complexity of medical images, which may contain multiple diagnoses, leading to ambiguity with global representations alone. (4) Employing contrastive loss in Euclidean space without hyperbolic projection negatively impacts model performance, underscoring the necessity of hyperbolic space for enhanced results.

4.4 Qualitative Analysis

In this section, we conduct a qualitative exploration of the trained models to investigate how they capture the visual semantic hierarchy structure in image-text alignment tasks. Specifically, we utilize the concept of "Embedding distances from [ROOT]", introduced by [7], to examine the generality differences between text and image embeddings in hyperbolic space. This concept highlights that in a representation space that effectively captures the visual semantic hierarchy, text embeddings tend to be more general than image embeddings. As a result, text embeddings should be closer to the root node, [ROOT], in hyperbolic space, while image embeddings, being more specific, are typically situated farther away. Figure 3 visualizes the distance distributions among image, event, attribute concept, and location concept embeddings. Our model shows a hierarchical distribution of the domain concept embeddings, event embeddings, and image embeddings in hyperbolic space, which aligns with our expectations and suggests that it effectively captures the visual semantics hierarchy while integrating domain knowledge for partial order modeling. In addition, we perform text retrieval at intermediate steps while traversing from an image embedding to the [ROOT]. As shown in Fig. 4, for patient A with atelectasis in the left lower lung zone (highlighted by the blue box), the red heatmap indicates the attention weights from the Semantic-sensitive Image Feature Extraction module. This heatmap closely matches the blue box, providing more precise localization of the atelectasis. As we traverse from [ROOT] to the event, the retrieved texts become progressively more specific, aligning closely with the image.

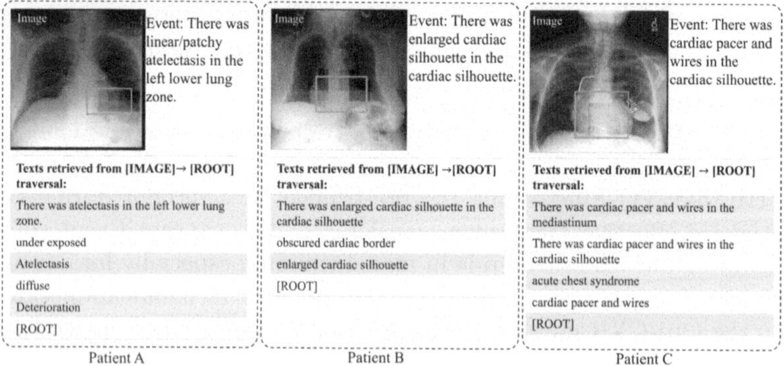

Fig. 4. Image traversals with our method: We perform text retrieval at intermediate steps while traversing from an image embedding to [ROOT] along the geodesic. The blue box in the figure highlights the location of the finding, while the red heatmap visualizes the attention weights of cross attention module in the Semantic-sensitive Image Feature Extraction module. (Color figure online)

5 Conclusion

This paper proposes MkCLIPH, a pre-training method for medical image-text alignment in hyperbolic space. By leveraging hierarchical entailment angle modeling and domain-specific knowledge, MkCLIPH enhances representation capabilities, improving generalization and interpretability. Experimental results demonstrate that MkCLIPH outperforms baseline methods in zero-shot tasks across various datasets.

References

1. Alsentzer, E., et al.: Publicly available clinical Bert embeddings (2019)
2. Bodenreider, O.: The unified medical language system (umls): integrating biomedical terminology. Nucleic Acids Res. **32**(Database issue), D267—70 (January 2004). https://doi.org/10.1093/nar/gkh061
3. Boecking, B., et al.: Making the most of text semantics to improve biomedical vision-language processing. In: The European Conference on Computer Vision (ECCV) (October 2022)
4. Chen, X., He, Y., Xue, C., Ge, R., Li, S., Yang, G.: Knowledge boosting: rethinking medical contrastive vision-language pre-training. In: Greenspan, H., Madabhushi, A., Mousavi, P., Salcudean, S., Duncan, J., Syeda-Mahmood, T., Taylor, R. (eds.) Medical Image Computing and Computer Assisted Intervention - MICCAI 2023, pp. 405–415. Springer Nature Switzerland, Cham (2023)
5. Cheng, P., Lin, L., Lyu, J., Huang, Y., Luo, W., Tang, X.: Prior: Prototype representation joint learning from medical images and reports. In: 2023 IEEE/CVF International Conference on Computer Vision (ICCV). pp. 21304–21314 (2023)
6. Cheng, P., Lin, L., Lyu, J., Huang, Y., Luo, W., Tang, X.: Prior: Prototype representation joint learning from medical images and reports. In: Proceedings of the IEEE/CVF International Conference on Computer Vision (ICCV), pp. 21361–21371 (October 2023)
7. Desai, K., Nickel, M., Rajpurohit, T., Johnson, J., Vedantam, S.R.: Hyperbolic image-text representations. In: Krause, A., Brunskill, E., Cho, K., Engelhardt, B., Sabato, S., Scarlett, J. (eds.) Proceedings of the 40th International Conference on Machine Learning. Proceedings of Machine Learning Research, vol. 202, pp. 7694–7731. PMLR (23–29 Jul 2023)
8. Hu, X., et al.: Expert knowledge-aware image difference graph representation learning for difference-aware medical visual question answering. In: Proceedings of the 29th ACM SIGKDD Conference on Knowledge Discovery and Data Mining, pp. 4156–4165. KDD '23, Association for Computing Machinery, New York, NY, USA (2023)
9. Huang, S.C., Shen, L., Lungren, M.P., Yeung, S.: Gloria: A multimodal global-local representation learning framework for label-efficient medical image recognition. In: 2021 IEEE/CVF International Conference on Computer Vision (ICCV), pp. 3922–3931 (2021). https://doi.org/10.1109/ICCV48922.2021.00391
10. Huang, X., Fang, Y., Lu, M., Yan, F., Yang, J., Xu, Y.: Dual-ray net: Automatic diagnosis of thoracic diseases using frontal and lateral chest x-rays. J. Med. Imaging Health Inform. **10**, 348–355 (2020). https://api.semanticscholar.org/CorpusID:209085265

11. Kim, W., Chun, S., Kim, T., Han, D., Yun, S.: Hype: hyperbolic entailment filtering for underspecified images and texts. In: Leonardis, A., Ricci, E., Roth, S., Russakovsky, O., Sattler, T., Varol, G. (eds.) Computer Vision - ECCV 2024, pp. 247–265. Springer Nature Switzerland, Cham (2025)
12. Le, M., Roller, S., Papaxanthos, L., Kiela, D., Nickel, M.: Inferring concept hierarchies from text corpora via hyperbolic embeddings. ACL **abs/1902.00913** (2019)
13. Li, M., Lin, B., Chen, Z., Lin, H., Liang, X., Chang, X.: Dynamic graph enhanced contrastive learning for chest x-ray report generation. In: 2023 IEEE/CVF Conference on Computer Vision and Pattern Recognition (CVPR), pp. 3334–3343 (2023). https://doi.org/10.1109/CVPR52729.2023.00325
14. Liu, B., et al.: Improving medical vision-language contrastive pretraining with semantics-aware triage. IEEE Trans. Med. Imaging **42**(12), 3579–3589 (2023). https://doi.org/10.1109/TMI.2023.3294980
15. Mu, N., Kirillov, A., Wagner, D., Xie, S.: Slip: self-supervision meets language-image pre-training. In: Computer Vision - ECCV 2022: 17th European Conference. Tel Aviv, Israel, October 23–27, 2022, Proceedings, Part XXVI, pp. 529–544. Springer-Verlag, Berlin, Heidelberg (2022)
16. Müller, P., Kaissis, G., Zou, C., Rueckert, D.: Joint learning of localized representations from medical images and reports. In: Computer Vision – ECCV 2022: 17th European Conference, Tel Aviv, Israel, October 23–27, 2022, Proceedings, Part XXVI, pp. 685–701. Springer-Verlag (2022)
17. Nickel, M., Kiela, D.: Learning continuous hierarchies in the Lorentz model of hyperbolic geometry. In: Dy, J., Krause, A. (eds.) Proceedings of the 35th International Conference on Machine Learning. Proceedings of Machine Learning Research, vol. 80, pp. 3779–3788. PMLR (10–15 Jul 2018)
18. Pellegrini, C., Keicher, M., Özsoy, E., Jiraskova, P., Braren, R., Navab, N.: Xplainer: From x-ray observations to explainable zero-shot diagnosis. In: International Conference on Medical Image Computing and Computer-Assisted Intervention, pp. 420–429. Springer (2023)
19. Qiao, Z., Han, L., Zhen, X., Gao, J., Qian, Z.: HYDEN: Hyperbolic density representations for medical images and reports. In: Proceedings of the 31st International Conference on Computational Linguistics, pp. 6285–6297. Association for Computational Linguistics, Abu Dhabi, UAE (Jan 2025)
20. Radford, A., et al.: Learning transferable visual models from natural language supervision. In: Meila, M., Zhang, T. (eds.) Proceedings of the 38th International Conference on Machine Learning. Proceedings of Machine Learning Research, vol. 139, pp. 8748–8763. PMLR (18–24 Jul 2021)
21. Rahman, T., et al.: Reliable tuberculosis detection using chest x-ray with deep learning, segmentation and visualization. IEEE Access **8**, 191586–191601 (2020). https://doi.org/10.1109/ACCESS.2020.3031384
22. Ramasinghe, S., Shevchenko, V., Avraham, G., Thalaiyasingam, A.: Accept the modality gap: An exploration in the hyperbolic space. In: 2024 IEEE/CVF Conference on Computer Vision and Pattern Recognition (CVPR), pp. 27253–27262 (2024)
23. Seibold, C., Reiß, S., Sarfraz, M.S., Stiefelhagen, R., Kleesiek, J.: Breaking with fixed set pathology recognition through report-guided contrastive training. In: International Conference on Medical Image Computing and Computer-Assisted Intervention, pp. 690–700. Springer (2022)
24. Selvaraju, R.R., Cogswell, M., Das, A., Vedantam, R., Parikh, D., Batra, D.: Grad-cam: Visual explanations from deep networks via gradient-based localization. In:

2017 IEEE International Conference on Computer Vision (ICCV), pp. 618–626 (2017). https://doi.org/10.1109/ICCV.2017.74

25. Shen, Lu; Johnson, A.E.W.P.T.J.L.L.W.F.M.G.M.M.M.B.E.S.P.C.L.A.G.M.R.G.: Mimic-iii, a freely accessible critical care database (2016)
26. Vaswani, A., et al.: Attention is all you need, pp. 6000–6010. NIPS'17, Curran Associates Inc., Red Hook, NY, USA (2017)
27. Vendrov, I., Kiros, R., Fidler, S., Urtasun, R.: Order-embeddings of images and language (2016)
28. Wang, F., Zhou, Y., Wang, S., Vardhanabhuti, V., Yu, L.: Multi-granularity cross-modal alignment for generalized medical visual representation learning. In: Proceedings of the 36th International Conference on Neural Information Processing Systems. NIPS '22, Curran Associates Inc. (2022)
29. Wang, X., Peng, Y., Lu, L., Lu, Z., Bagheri, M., Summers, R.M.: Chestx-ray8: Hospital-scale chest x-ray database and benchmarks on weakly-supervised classification and localization of common thorax diseases. In: 2017 IEEE Conference on Computer Vision and Pattern Recognition (CVPR), pp. 3462–3471 (2017)
30. Wu, C., Zhang, X., Zhang, Y., Wang, Y., Xie, W.: Medklip: medical knowledge enhanced language-image pre-training for x-ray diagnosis. In: 2023 IEEE/CVF International Conference on Computer Vision (ICCV), pp. 21315–21326 (2023). https://doi.org/10.1109/ICCV51070.2023.01954
31. Wu, J.T., Agu, N.N., Lourentzou, I., Sharma, A., Paguio, J.A., Yao, J.S., Dee, E.C., Mitchell, W., Kashyap, S., Giovannini, A., Celi, L.A., Moradi, M.: Chest imagenome dataset for clinical reasoning. CoRR **abs/2108.00316** (2021), https://arxiv.org/abs/2108.00316
32. Xiaoman Zhang, Chaoyi Wu, Y.Z.W.X., Wang, Y.: Knowledge-enhanced visual-language pre-training on chest radiology images. In: Nature Communation (October 2023)
33. Zhang, Y., Jiang, H., Miura, Y., Manning, C.D., Langlotz, C.P.: Contrastive learning of medical visual representations from paired images and text (2022)
34. Zhang, Z., Qiao, Z., Han, L., Yang, H., Qian, Z., Wu, J.: Hyperbolic vision language representation learning on chest radiology images. Health Inform. Sci. Syst. **13**(1), 27 (2025)

Structure Observation Driven Image-Text Contrastive Learning for Computed Tomography Report Generation

Hong Liu[2,3], Dong Wei[3], Qiong Peng[1], Yawen Huang[3], Xian Wu[3(✉)], Yefeng Zheng[3,4(✉)], and Liansheng Wang[1,2(✉)]

[1] School of Informatics, Xiamen University, Xiamen, China
qpeng@stu.xmu.edu.cn

[2] National Institute for Data Science in Health and Medicine, Xiamen University, Xiamen, China
liuhong@stu.xmu.edu.cn, lswang@xmu.edu.cn

[3] Jarvis Research Center, Tencent YouTu Lab, Shenzhen, China
{donwei,yawenhuang,kevinxwu,yefengzheng}@tencent.com

[4] Medical Artificial Intelligence Laboratory, Westlake University, Hangzhou, China
zhengyefeng@westlake.edu.cn

Abstract. Computed Tomography Report Generation (CTRG) aims to automate the clinical radiology reporting process, thereby reducing the workload of report writing and facilitating patient care. While deep learning approaches have achieved remarkable advances in X-ray report generation, their effectiveness may be limited in CTRG due to larger data volumes of CT images and more intricate details required to describe them. This work introduces a novel two-stage (structure- and report-learning) framework tailored for CTRG featuring effective structure-wise image-text contrasting. In the first stage, a set of learnable *structure-specific visual queries* "observe" corresponding structures in a CT image. The resulting observation tokens are contrasted with structure-specific textual features extracted from the accompanying radiology report with a structure-wise image-text contrastive loss. In addition, text-text similarity-based soft pseudo targets are proposed to mitigate the impact of false negatives, i.e., semantically identical image structures and texts from non-paired images and reports. Thus, the model learns structure-level semantic correspondences between CT images and reports. Further, a dynamic, diversity-enhanced negative queue is proposed to guide the network in learning to discriminate various abnormalities. In the second stage, the visual structure queries are frozen and used to select the critical image patch embeddings depicting each anatomical structure, minimizing distractions from irrelevant areas while reducing memory consumption. Also, a text decoder is added and trained for report generation. Our extensive experiments on two public datasets demonstrate that our framework establishes new state-of-the-art performance for CTRG in clinical efficiency, and its components are effective.

H. Liu and D. Wei—Contributed equally; H. Liu contributed to this work during an internship at Tencent.

I. Oguz et al. (Eds.): IPMI 2025, LNCS 15830, pp. 218–233, 2026.
https://doi.org/10.1007/978-3-031-96625-5_15

Keywords: Computed tomography report generation · Structure-wise image-text contrastive learning · Text-text similarity soft pseudo targets

1 Introduction

Diagnostic reports of widely used radiographs, such as X-ray and computed tomography (CT), play a crucial role in clinical practice, serving as essential documentation for healthcare and facilitating accurate diagnosis and timely treatment. Meanwhile, medical image interpretation requires domain expertise and careful observation, facing the potential risk of misdiagnosis in the presence of staff shortage and excessive workload [2].

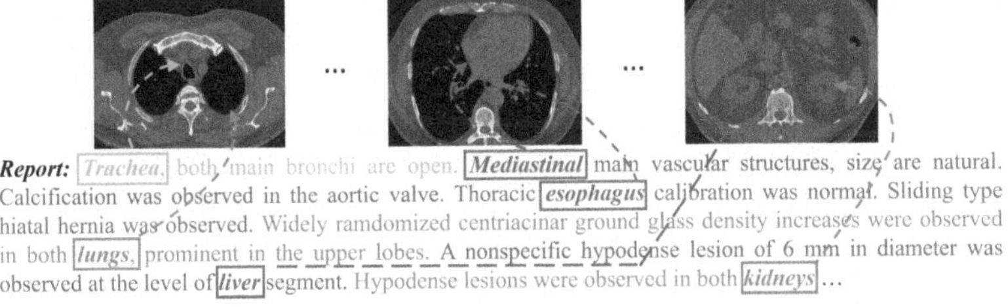

...
...
Report: Trachea, both main bronchi are open. Mediastinal main vascular structures, size are natural.
Calcification was observed in the aortic valve. Thoracic esophagus calibration was normal. Sliding type
hiatal hernia was observed. Widely ramdomized centriacinar ground glass density increases were observed
in both lungs, prominent in the upper lobes. A nonspecific hypodense lesion of 6 mm in diameter was
observed at the level of liver segment. Hypodense lesions were observed in both kidneys...

Fig. 1. An example CT image and corresponding report demonstrating highly structured text descriptions and visual correspondences.

Many approaches have attempted automatic medical report generation for 2D X-ray images with deep learning (DL) techniques [8,23,24,28]. These approaches have achieved promising advancements, but their transferability to 3D radiographs, such as CT, has yet to be validated. The task of CT report generation (CTRG) is more complex than X-ray report generation, mainly because of the combination of two factors: (1) substantially more data to process: e.g., a typical chest X-ray image is 512×512 to 1024×1024 pixels, whereas a typical thin-slice chest CT volume may contain a few hundred slices of 512×512 pixels; and (2) substantially more information to interpret: e.g., chest X-rays typically reveal tens of main findings [20], whereas chest CT images may concern more than 80 abnormalities [11]. As of now, CTRG is drawing increasing research interest. Tang et al. [41] proposed a self-attention-based scan localizer framework (SL-DG) for CTRG. To tackle the above-described challenges, SL-DG trained a scan localizer network based on disease classification using manually labeled medical terms as supervision, to select the critical slices in a CT sequence for report generation. However, manually labeling medical terms for CT images was labor-intensive. In addition, the limited supervision restricted the model's generality and usability, as only a few major abnormalities were annotated. Further, the 2D slice-based approach overlooked the inherent 3D context of CT volumes.

This work proposes a two-stage framework for CTRG featuring a novel structure-learning stage, where the model is trained with our proposed structure-level abnormality-enhanced contrastive learning objective. This objective is driven by anatomical structure observations derived from radiology reports, such that the model learns good representations for main structures in CT images. Unlike existing methods relying on well-prepared prior knowledge such as knowledge graphs [8,31] or manual abnormality annotations [41], our framework only requires high-level generic knowledge about *what* anatomical structures that a CT image depicts, e.g., a chest CT mainly depicts thorax, rib, lung, heart, pleura, liver, kidney, thyroid, etc. Figure 1 shows a CT image and corresponding report, demonstrating highly structured text descriptions and visual correspondences. Based on this knowledge, we propose to extract the most informative representations for each structure with a set of learnable structure-specific *visual queries* via cross-attention. Thus, the training becomes more focused and less computationally demanding. To leverage a broad source of supervision from the reports, we use a pretrained text encoder to extract structure-specific features from corresponding sentences. Then, learning of the structure-specific queries is supervised by a structure-wise image-text contrastive loss. In addition, we propose text-text similarity-based soft pseudo targets to mitigate the impact of false negatives, i.e., semantically identical image structures and texts from non-paired images and reports. A diversity-enhanced negative queue update strategy is proposed to guide the network in learning to discriminate various abnormalities. In the second stage, the learned visual encoder and structural queries are frozen. A text decoder is added and trained for report generation. The frozen visual structural queries extract structural representations and select the most representative image patch embeddings as the decoder's input. Extensive experiments on two public CTRG datasets demonstrate our framework's superior performance to existing state-of-the-art methods and the efficacy of its novel designs.

2 Related Work

Radiology Report Generation. Early studies directly applied various encoder-decoder models such as CNN-RNN [44], LRCN [10], and AdaAtt [33]. However, the lack of domain knowledge in these models limited their performance. To address this issue, Chen et al. [8] and Liu et al. [31] proposed incorporating medical prior knowledge into network training with a memory-driven transformer and a medical knowledge graph, respectively. Similarly, Li et al. [27] extracted normal and abnormal terminologies from the reports and used them as nodes to build a knowledge graph driving the report generation. In contrast, Yang et al. [51] introduced general and input-specific knowledge into training with a novel knowledge-enhanced multi-head attention mechanism. Although these methods made notable progress, the requirement of well-prepared/processed prior knowledge (e.g., in the form of a knowledge graph) limited their extensibility.

Another research trend is to improve the extraction of visual representations. Region-wise image features [38] have shown their effectiveness for fine-grained tasks, including report generation [42]. Yet training region-detection networks required a considerable annotation burden and complicated the process. Wang et al. [48] introduced an image-text matching branch to extract text-correlated visual features. Meanwhile, self-supervised pretraining methods have demonstrated strong representation learning capability [5,13] and efficacy in medical report generation [35]. The methods above have made remarkable progress in generating medical reports for 2D radiographs, typically chest X-rays. However, the transferability to 3D radiographs, such as CT, has yet to be validated.

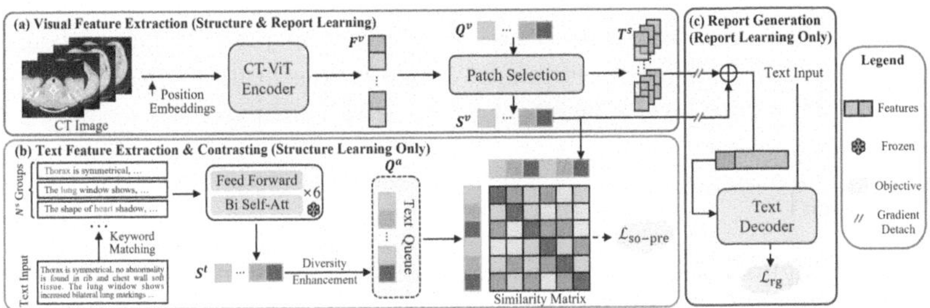

Fig. 2. Overview of our proposed framework.

Recently, researchers explored integrating semantic information to improve report generation through multi-task learning. Wang et al. [47] trained a multi-label classification network to guide report generation by extracting 768 high-frequency medical terms from RadGraph [21]. Jin et al. [22] proposed explicitly guiding the generation process with token prompts derived from diagnostic results obtained through classification. For 3D CTRG, Hamamci et al. [15] proposed enhancing the model with a cross-attention multimodal fusion module and hierarchical memory to incorporate longitudinal multimodal data. Chen et al. [9] introduced a disease prototype memory bank and a disease-aware attention module to capture diagnostic information. This information was used as guidance prompts to the LLaMA2-7B [43] model. Tang et al. [41] proposed a self-attention-based scan localizer framework (SL-DG) for CTRG. SL-DG involved training a scan localizer network with disease classification. The network was supervised by manually labeled medical terms, and was used to select critical slices in a CT sequence for report generation. While incorporating diagnostic results significantly improved the accuracy of generated reports, the limited supervision restricted the model's generality and usability, as only a few major abnormalities were annotated. In contrast, this work proposes directly extracting structure-level diagnostic information from raw reports via structure observation-driven image-text contrastive learning. Leveraging detailed structural information for contrastive learning, it aims to enhance the ability to generate accurate and comprehensive reports and improve the generality and usability of the model.

Medical Image-Language Contrastive Learning. Contrastive learning (CL) [6,17] aims to train models to distinguish between similar and dissimilar sample pairs by minimizing the distances between positive pairs and maximizing the distances between negative ones in the representation space. To improve the efficiency of CL, some works proposed to exploit the cross-modal local level [19] and disease-prototype level [45] semantic consistency. To mitigate the impact of false negatives, Byun et al. [3] proposed MAFA (MAnaging FAlse negatives) with connection mining to convert false negatives into positives and label smoothing for image-text CL loss. In the medical image domain, CL has been well explored and achieved encouraging results [4,12,30,40]. Liu et al. [30] proposed triaging image-report pairs into positive, negative, and additional neutral groups based on *global* inter-report similarity, to reduce false-negative pairs (i.e., semantically similar but from different cases) in contrastive pretraining. In contrast, we propose to learn *structure-wise local* image representations from reports via cross-modal structure-level image-text contrastive learning.

3 Method

Overview. Figure 2 shows an overview of our framework, including a structure- and a report-learning stage. As the premise, we use the CT-ViT [16] to extract image patch embeddings. In the structure-learning stage, a set of N^s learnable structure-specific visual queries Q^v extracts a set of subject-specific structure observation tokens S^v with an image patch selection layer. Meanwhile, N^s groups of structure-specific sentences are fed into a pretrained BERT-based model [14] to extract corresponding textual structure observation tokens S^t. To enforce consistency between the visual and textual structures, a structure observation driven image-text contrasting loss $\mathcal{L}_{\text{so-pre}}$ is enforced between S^v and S^t, where the negatives come from a dynamic diversity-enhanced negative text queue. In the report-learning stage, the visual encoder, structural queries, and patch selection layer are frozen. The patch selection layer is used to select the most informative image patch embeddings T^s for each primary structure, in addition to extracting S^v. Then, a text decoder, which receives T^s and S^v as input, is added and trained using a report-generation loss $\mathcal{L}_{\text{rg}}$ for the target task of CTRG.

Structure-wise Visual Feature Extraction. Considering the large dimensions of 3D CT images and intricate details in CT reports, simply adopting 2D image-based semantic alignment strategies (e.g., subject-level contrastive learning [30]) to our scenario could be suboptimal. This is because local, subtle coherence between image and text could be ignored in global alignment, which is critical in fine-grained tasks like CTRG. Huang et al. [19] proposed an attention framework to learn local representations by contrasting image sub-regions and report words. Medical CT reports are highly structured: they usually describe the same body parts from the same aspects—typically based on anatomical structures. Therefore, we propose to parse the reports at the *structure* level instead of word-wise, and align them with CT image patch embeddings accordingly. Below, we describe the image processing part first and will deal with the reports shortly.

As shown in Fig. 2(a), we extract the critical image patch embeddings corresponding to each structure with a set of structure-specific visual queries

$Q^v = [\boldsymbol{q}_i^v]_1^{N^s}$, where N^s is the number of considered structures. Denoting the embeddings of all image patches by $F^v = [\boldsymbol{v}_j]_1^{N^v}$, where N^v is the number of image patches, their cross-attention with the visual queries are computed:

$$A^v = \text{softmax}[Q^v W_0^v (F^v W_1^v)^T], \tag{1}$$

where W_0^v and W_1^v are linear projection matrices, and $A^v \in \mathbb{R}^{N^s \times N^v}$ is a pair-wise similarity matrix between the structure queries and image patches. Then, we obtain a set of visual, subject-specific structure observation embeddings S^v:

$$S^v = [\boldsymbol{s}_i^v]_1^{N^s} = A^v(F^v W_2^v), \tag{2}$$

where W_2^v is a linear projection matrix. S^v will be used for contrast shortly.

Structure-Wise Textual Token Extraction. We employ a pretrained text encoder [14] to extract embeddings for the textual structure observations. This encoder processes sentences that describe each specific structure, converting them into embeddings. The resulting embeddings, specifically the outputs of the class tokens for each structure, are collected as textual subject-specific structure observation tokens, denoted by $S^t = [\boldsymbol{s}_i^t]_1^{N^s}$ (Fig. 2(b)). Since we aim to align the visual encoder with a well-trained text encoder, we freeze the text encoder's parameters during training. To identify the sentences describing each structure, we implement a keyword-matching method: if a sentence contains any predefined keyword for a structure, it is considered to describe the structure. We also define an "others" category for sentences with no predefined keyword. Thus, not only does our report parsing rely on easily accessible high-level prior knowledge, but it is also straightforward to implement.

Cross-Modal Alignment. After obtaining the visual and textual subject-specific structure observation tokens, we align them by contrastive learning. Following [26], we learn a similarity function $\text{sim}(\boldsymbol{s}_i^v, \boldsymbol{s}_i^t) = g_v(\boldsymbol{s}_i^v)^T g_t(\boldsymbol{s}_i^t)$, where g_v and g_t are linear transformations mapping the embeddings to normalized lower-dimensional (512-d) representations. Then, we calculate the softmax-normalized image-to-text similarity:

$$p_m^{v2t}(\boldsymbol{s}_i^v) = \frac{\exp[\text{sim}(\boldsymbol{s}_i^v, \boldsymbol{s}_i^t)/\tau]}{\sum_{m=1}^{N^s N^q} \exp[\text{sim}(\boldsymbol{s}_i^v, \boldsymbol{s}_m^t)/\tau]}, \tag{3}$$

where τ is a learnable temperature parameter, $\boldsymbol{s}_m^t$ is a sample in a queue Q_a of textual tokens (Fig. 2(b)), and N^q is the queue length. Then, our structure observation driven image-text contrastive loss is defined as:

$$\mathcal{L}_{\text{so-itc}} = \mathbb{E}_{(\boldsymbol{s}_i^v, \boldsymbol{s}_i^t)}\left[\text{H}\left(\boldsymbol{y}^{v2t}(\boldsymbol{s}_i^v), \boldsymbol{p}^{v2t}(\boldsymbol{s}_i^v)\right)\right], \tag{4}$$

where H is the cross entropy, $\boldsymbol{p} = [p_m]_{m=1}^{N^s N^q}$, and $\boldsymbol{y}$ is the one-hot ground truth (1 for the positive pair and 0s for negative pairs). We define the positive pair as the visual and textual observation embeddings of the same subject and structure.

Soft Pseudo Targets. The one-hot label and training objective in Eq. (4) above penalize all negative pairs regardless of their correctness. Negative text for an image may also match the image's content, and vice versa. For example, the lung description of a patient with COVID-19 may also describe another patient's lung with COVID-19, although the text and image are from different subjects. To mitigate this issue, we devise a soft pseudo target based on the similarity between textual observation tokens. Concretely, the text-text similarity between two textual structure observation tokens is computed by $\text{sim}'(\boldsymbol{s}_i^t, \boldsymbol{s}_m^t) = g_t(\boldsymbol{s}_i^t)^T g_t(\boldsymbol{s}_m^t)$. Then, the soft pseudo target $\boldsymbol{q}^{t2t}$ can be obtained by replacing the similarity function sim in Eqn. (3) with sim$'$. Finally, we impose a Kullback-Leibler (KL) divergence loss between the normalized image-to-text similarities $\boldsymbol{p}^{v2t}$ and the soft target $\boldsymbol{q}^{t2t}$:

$$\mathcal{L}_{\text{so-kl}} = \mathbb{E}_{(\boldsymbol{s}_i^v, \boldsymbol{s}_i^t)} \left[\text{KL} \left(\boldsymbol{q}^{t2t}(\boldsymbol{s}_i^t) || \boldsymbol{p}^{v2t}(\boldsymbol{s}_i^v) \right) \right]. \tag{5}$$

The intuition is that, for a pair of visual and textual structure tokens of the same subject and structure, their similarities to other textual tokens are expected to be close, given the two modalities are well aligned.

The complete structure observation driven pretraining objective is:

$$\mathcal{L}_{\text{so-pre}} = (1 - \alpha)\mathcal{L}_{\text{so-itc}} + \alpha\mathcal{L}_{\text{so-kl}}, \tag{6}$$

where α is set to 0.5 empirically. Compared with global alignment, our $\mathcal{L}_{\text{so-pre}}$ aligns the structure-wise semantics between the images and reports to learn fine-grained representations.

Diversity-Enhanced Queue. During structure learning, we maintain a queue of textual structure observation tokens. To improve the efficiency of contrastive learning, we propose storing the most informative samples in the queue. Specifically, for each token $\boldsymbol{s}_i^t$ in a batch, we calculate the sum of its text-text similarities to all tokens in the queue: $S_i = \text{sum}(g_t(\boldsymbol{s}_i^t)^T g_t(\boldsymbol{s}_m^t))$. If its S_i value is smaller than the largest S of all tokens in the queue, it is enqueued, and the queue token with the largest S is dequeued. For practical implementation, we store the S values for tokens already in the queue instead of recomputing them. This selective process ensures that the queue contains diverse and informative samples for discriminative and contrastive learning while keeping the queue size relatively small to reduce computational overhead.

Report Learning and Inference. The preceding stage serves as a high-level medical prior informed, image-text pretraining in which a fine-grained, local structural image representation is learned. In the second stage (Fig. 2(c)), we freeze the image encoder, visual queries, and patch-selection layers and add and train a text decoder for report generation (note the text encoder in the preceding stage is no longer used). Our framework is agnostic to the exact text decoder used and applies to various text generation models. In this work, we experiment with a BERT decoder [25] and a large language model (LLM) LLaMA2-7B [43].[1] Formally, the model is trained with the chained next-token prediction objective:

[1] For LLaMA2-7B, we use LoRA [18] for parameter-efficient fine-tuning and a two-layer multilayer perception for visual feature projection.

$\mathcal{L}_{\text{rg}} = -\sum_{k=1}^{N^t} \log P(\boldsymbol{t}_k|\boldsymbol{t}_{1:k-1})$, where P is the probability of predicting the next token $\boldsymbol{t}_k$ conditioned on preceding ones of the target report.

To provide detailed information for comprehensive report statement generation, we consider two types of tokens as joint input to the text decoder: first, the image structural representations S^v to convey general information about each structure; second, the selected most informative image patch embeddings T^s to provide detailed information for each structure. Specifically, based on A^v, the pair-wise similarity matrix between the structure queries and image patches computed in Eq. (1), we identify and preserve K image patch embeddings most similar to each structural query, resulting in a total of $K \times N^s$ patch embeddings selected, denoted by T^s. K is set to 10 empirically (cf. Tables 3 and 4).

4 Experiments

Datasets. We conduct experiments on two public CT report generation datasets. **1) CT-RATE** [14] includes 25,692 non-contrast chest CT volumes with corresponding reports of 21,304 unique patients. Following [14], we standardize all CT volumes to a uniform voxel spacing of $0.75 \times 0.75 \times 1.5$ mm^3. The volumes are either center-cropped or padded to a consistent resolution of $480 \times 480 \times 240$ voxels. We use the official training set (24,128 volumes/20,000 patients) for training, and further split the official test set into a validation (360 volumes/300 patients) and a testing set (1,204 volumes/1,004 patients). The validation set is only used for model optimization, e.g., in ablation studies, whereas the test set is used for final performance reporting and comparison with other methods. **2) CTRG-Chest-548K** [41] includes 1,804 chest CT volumes and corresponding reports. We use the same data split as in [9]: 60% for training, 20% for validation, and 20% for testing. Following [9], each volume is resized to 256×256 voxels. For each case, we use the CT image and Findings section of the English report.

Evaluation Metrics. Following common practice in the literature (e.g., [9, 22]), we employ both natural language generation (NLG) and clinical efficacy (CE) metrics for evaluation. NLG metrics include BLEU [36], METEOR [1], and ROUGE-L [29]. CE metrics include precision, recall, and F1 score following [22] and [34]. For CT-RATE, we use the officially provided text classifier to extract 18 distinct types of abnormalities to calculate CE metrics. For CTRG-Chest-548K, following [9], we utilize a pretrained report labeler, CheXbert [39], to extract labels, which remains effective in labeling CT reports as shown by [9]. It should be noted that while some of the compared methods achieved high NLG metrics in our experiments, the sentences generated are highly repetitive and lack diversity and diagnostic value. This may result from the highly structured format of the reports in the two datasets. Thus, the generated reports closely follow the ground truth regarding wording, description sequence, etc., but fail to capture the diagnostic information expected for a medical report. Therefore, we mainly discuss the models' performance based on CE metrics below.

Implementation. The proposed framework is implemented using PyTorch (2.0.0) [37] and trained with four NVIDIA Tesla V100 GPUs with 32GB memory each. The optimizer is AdamW [32], with learning rates of 10^{-5} and 10^{-4} for structure and report learning. We set the warm-up ratio to 10% and use the linear learning rate scheduler after warm-up. The model is trained for 100K and 30K steps for structure and report learning, with a batch size of 8. The number of visual structural queries N^s is set to 10, corresponding to lung, trachea and bronchie, mediastinum and heart, esophagus, pleura, bone, thyroid, breast, abdomen, and others (e.g., thoracic cavity, prostate and so on, which are only mentioned in few reports), following the hierarchy of anatomical regions in [52]. The length N^q of the diversity-enhanced queue for the structure-wise image-text contrasting is empirically set to 1000 for each structure. Note that we only need to parse the original reports into structure-corresponding sentences for the structure-learning stage. The original reports are used for report learning and performance evaluation. The same preprocessing procedures are implemented for all compared methods for a fair comparison. including the keyword taxonomy for report parsing. Our implementation, including the keyword taxonomy for report parsing, is available at: https://github.com/ccarliu/CTRG

Comparison with State-of-the-Art (SOTA) Methods. We compare our framework to various SOTA methods, including those proposed for chest X-ray: R2Gen [8], R2GenCMN [7], M2KT [50], R2GenGPT [46], PromptMRG [22]; radiology pretraining methods: RadFM [49], CT-CLIP [14], GLoRIA [19]; and specialized CT report generation methods: CT2Rep [15], SL-DG [41], Dia-LLaMA [9]. Unless otherwise specified, we use the codes released by the authors. For the pretraining methods, we freeze the pretrained image encoders (i.e., checkpoints released by the authors [14,49] or trained with author-released codes [19]), and add and train the same BERT decoder [25] as our framework for report generation. For methods originally designed for 2D images, we replace their image encoders with the same one as ours to extract CT image representations. We optimize the performance of all compared methods on the validation data.

Table 1 shows the performance of all compared methods. As we can see, on both datasets, R2Gen, R2GenCMN, and M2KT yield generally low CE metrics, despite their reasonable NLG metrics. R2GenGPT shows some improvements in CE metrics by leveraging the capabilities of a large language model (LMM), but it has notably low NLG metrics. We conjecture this is because their global alignment strategies neglect vital sub-regions of CT images and thus fail to capture fine-grained features. PromptMRG achieves significant improvement in CE metrics after incorporating diagnostic classification results. Among the three pretrained models (RadFM, CT-CLIP, and GLoRIA), GLoRIA performs the best. It likely benefits from its localized correspondence modeling between images and reports, which can capture more detailed information for report generation. Among the models with specific designs for CT report generation (CT2Rep, SL-DG, and Dia-LLaMA), Dia-LLaMA yields decent performance. Its incorporation of diagnostic information and the capability of the LLM may benefit it. Finally, both implementations of our framework, using BERT (Ours-BERT) or LLaMA2-

Table 1. The performance of our model compared with SOTA methods on the test sets of CT-RATE [14] (left) and CTRG-Chest-548K [41] (right). *: cited from the original paper. †: our model with CT representation learned on CT-RATE.

CT-RATE [14]

Method	CE Metrics			NLG Metrics			
	Pre.	Rec.	F1	BL-1	BL-4	MTR	RG-L
R2Gen [8]	0.140	0.026	0.043	0.426	0.241	0.248	0.333
R2GenCMN [7]	0.187	0.094	0.091	0.431	0.243	0.266	0.350
M2KT [50]	0.323	0.120	0.156	0.456	**0.261**	0.255	0.348
R2GenGPT [46]	0.326	0.205	0.228	0.166	0.091	0.116	0.161
PromptMRG [22]	<u>0.353</u>	0.286	0.288	0.434	0.186	0.207	0.361
RadFM [49]	0.143	0.154	0.166	0.456	<u>0.259</u>	**0.278**	<u>0.382</u>
CT-CLIP [14]	0.217	0.129	0.145	0.455	<u>0.259</u>	<u>0.277</u>	**0.383**
GLoRIA [19]	0.298	0.251	0.240	0.448	0.220	0.270	0.346
CT2Rep [15]	0.206	0.073	0.099	0.447	0.251	0.275	0.377
Ours-BERT	0.321	**0.354**	**0.310**	**0.486**	0.254	0.275	0.351
Ours-LLaMA	**0.354**	<u>0.328</u>	<u>0.308</u>	<u>0.481</u>	0.225	0.266	0.318

CTRG-Chest-548K [41]

Method	CE Metrics			NLG Metrics			
	Pre.	Rec.	F1	BL-1	BL-4	MTR	RG-L
R2Gen [8]	0.211	0.124	0.146	0.332	0.229	0.217	**0.473**
R2GenCMN [7]	0.152	0.108	0.115	0.368	0.244	0.224	<u>0.451</u>
M2KT [50]	0.227	0.114	0.146	0.453	0.229	0.262	0.384
R2GenGPT [46]	0.273	0.130	0.175	0.259	0.147	0.263	0.275
PromptMRG [22]	0.311	0.347	0.301	0.467	0.236	0.231	0.383
RadFM [49]	0.395	0.353	0.341	0.471	0.253	0.244	0.396
CT-CLIP [14]	0.318	0.228	0.246	0.421	0.267	0.263	0.443
GLoRIA [19]	0.411	0.376	0.358	0.453	0.282	**0.280**	0.439
CT2Rep [15]	0.223	0.135	0.125	0.432	0.251	0.259	0.408
SL-DG* [41]	-	-	-	-	0.237	0.219	0.438
Dia-LLaMA* [9]	0.421	0.387	0.372	**0.512**	**0.296**	0.263	0.422
Ours-BERT	<u>0.434</u>	**0.413**	**0.393**	<u>0.501</u>	<u>0.294</u>	<u>0.265</u>	0.437
Ours-LLaMA	**0.435**	<u>0.410</u>	<u>0.387</u>	0.340	0.183	**0.280**	0.317
Ours-BERT†	0.503	0.496	0.468	0.511	0.297	0.266	0.419

7B (Ours-LLaMA) as the text decoder, achieve superior recall and F1 scores to other methods across the two datasets. Meanwhile, the precision scores of Ours-LLaMA are the highest, too. Interestingly, while Ours-BERT is also competent in the NLG metrics on both datasets, Ours-LLaMA is generally less effective in NLG metrics, especially on CTRG-Chest-548K. We conjecture that this discrepancy is due to the limited data available for effectively fine-tuning the LLM (CTRG-Chest-548K has only 1262 cases for training). It is worth emphasizing that our framework is built upon the easily accessible high-level prior knowledge of what structures a chest CT depicts. In contrast, PromptMRG, SL-DG, and Dia-LLaMA require term-level manual annotations for training. These results demonstrate the strong capability of our method in CTRG.

Figure 3 shows examples of generated CT reports for a case in the CT-RATE dataset. Compared with PromptMRG, our method generates more comprehensive diagnostic reports for various thoracoabdominal organs. In contrast, PromptMRG is more oriented to the abnormality annotations it uses for training and primarily includes abnormalities related to the heart and lungs (most of the abnormality annotations in this dataset are focused on the heart and lungs).

Transfer CT Representation. To assess the quality and generalizability of the CT representation learned by our structure-learning framework, we transfer the representation learned on CT-RATE to CTRG-Chest-548K for subsequent report learning. The bottom row of Table 1 demonstrates substantial performance improvements upon the representation learned on the small CTRG-Chest-548K dataset in all CE metrics and comparable performance for all NLG ones. We attribute the improvements to the better quality of the representation learned on the larger CT-RATE dataset. These results also validate the efficacy of our methodology in learning CT representation that is generalizable across domains.

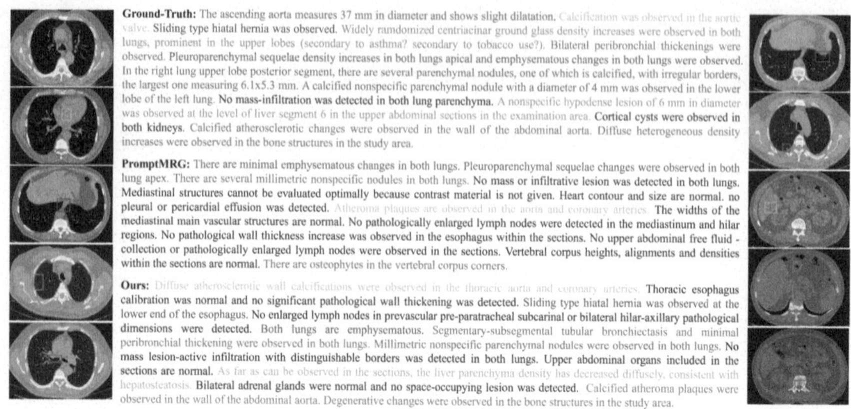

Fig. 3. Ground truth and example reports generated by PromptMRG [22] and our method for a case in CT-RATE [14]. Colored texts and color- and order-matching boxes in CT slices indicate clinic-relevant findings in the ground truth. Our method not only produces sentences with fine details but also identifies more observations than PromptMRG. CT slices are shown with varying windows to highlight lesions.

Table 2. Ablation study on the *validation* set of CT-RATE [14].

Ablation config.	pretrain			Fine-tune		CE Metrics ↑			NLG Metrics ↑			
	$\mathcal{L}_{\text{so-itc}}$	$\mathcal{L}_{\text{so-kl}}$	Diversity Queue	S^v	T^s	Pre.	Rec.	F1	BL-1	BL-4	MTR	RG-L
(a)	×	×	×	✓	×	0.221	0.115	0.124	**0.511**	**0.286**	0.272	**0.374**
(b)	✓	×	×	✓	×	0.307	0.317	0.300	0.497	0.266	0.272	0.357
(c)	✓	✓	×	✓	×	<u>0.325</u>	0.318	0.306	0.502	0.273	<u>0.273</u>	0.364
(d)	✓	✓	✓	✓	×	**0.334**	<u>0.323</u>	<u>0.309</u>	0.488	0.257	0.268	0.348
Full	✓	✓	✓	✓	✓	**0.334**	**0.356**	**0.320**	<u>0.509</u>	<u>0.282</u>	**0.277**	<u>0.371</u>

Ablation Study. We conduct ablative experiments to validate the efficacy of our framework's novel designs on the *validation* set of CT-RATE [14] with the BERT text decoder [25] (Table 2). The baseline (a) is our backbone network trained with $\mathcal{L}_{\text{rg}}$ alone, yielding the worst CE metrics. Row (b) adds the proposed structure learning with the structure observation-driven image-text contrastive loss $\mathcal{L}_{\text{so-itc}}$. It brings noticeable improvements upon the baseline, with at least 8.6% improvements for all CE metrics. Row (c) further enhances performance by introducing text-text similarity-based soft pseudo targets ($\mathcal{L}_{\text{so-kl}}$), improving CE metrics by 1.8%, 0.1%, and 0.6%. The diversity-enhanced negative queue also boosts performance in row (d), with improvements of 0.9%, 0.5%, and 0.3% in CE metrics. These two ablations validate the effectiveness of our proposed text-text similarity-based soft pseudo targets for contrastive learning, and the diversity-enhanced negative queue in learning diagnostic-related structure observation embeddings for report generation. Eventually, our full model achieves the

Table 3. Experimental results on the *validation* set of CT-RATE [14] in CE metrics, with the BERT text decoder [25]. Left: varying α (weight of $\mathcal{L}_{\text{so-kl}}$) with $K = 10$. Right: varying the number K of selected image patch embeddings for each structure with $\alpha = 0.2$.

α	Pre.	Rec.	F1	K	Pre.	Rec.	F1
0.0	0.318	0.335	0.307	0	**0.338**	0.323	0.309
0.1	**0.342**	0.331	0.313	5	0.331	0.332	0.310
0.2	0.334	**0.356**	**0.320**	10	0.334	**0.356**	**0.320**
0.3	0.321	0.334	0.306	15	0.328	0.342	0.307
0.4	0.312	0.328	0.301	20	0.322	0.318	0.303

Table 4. CE metrics (on the *validation* set), training memory consumption, and GMACs comparison on the CT-RATE [14] dataset by varying the number K of selected image patch embeddings for each structure. LLaMA2-7B [43] is used as the text decoder.

K	Pre.	Rec.	F1	Mem. (G)	GMACs
0	0.341	0.298	0.291	23.8	8730.9
5	0.353	0.316	0.306	24.5	9733.8
10	**0.355**	**0.322**	**0.309**	25.1	10760.8
15	0.339	0.321	0.302	25.8	11781.6
20	0.318	0.311	0.289	26.5	12806.5

best performance for all CE metrics with $\mathcal{L}_{\text{so-itc}}$, $\mathcal{L}_{\text{so-kl}}$ and the diversity queue for structure learning, and S^v and T^s for report learning.

Validate Designs of Our Method. Based on the validation set of CT-RATE, we determine 1) the optimal weight α for $\mathcal{L}_{\text{so-pre}}$ (Eqn. (6)) and 2) the optimal number K of selected image patch embeddings T^s for each structure, with the BERT text decoder [25]. We fix either while varying the other to reduce the search space. The results are shown in Table 3. Although the results look fairly stable, we select $\alpha = 0.2$ and $K = 10$ for final performance evaluation and comparison with other methods on the test data, due to their overall high performance in CE metrics compared with alternative values.

Effects of Patch Selection. Table 4 presents the CE metrics, training memory consumption, and GMACs (giga multiply-accumulate operations per second) for our model using LLaMA2-7B [43] as text decoder. Our model demonstrates commendable performance with $K = 10$, i.e., 110 total visual tokens (comprising 100 selected tokens T^s and 10 structure observation embeddings S^v) are fed to the decoder. This is significantly fewer than the original 4096 visual tokens produced by the visual encoder, which is beyond the limit of our hardware (maximum ~600 visual tokens when fine-tuning LLaMA2-7B with LoRA). Therefore, our structure-observation-driven visual feature extraction not only makes contrastive learning more focused on the main structures in chest CTs, but also effectively reduces computational overhead.

Report to Volume Retrieval. Following [14], we evaluate report-to-volume retrieval performance by computing the cosine similarity between the embeddings of target volumes and query texts. We report Recall@10, 50, and 100 and compare with CT-CLIP [14] in Table 5. Both variants of our method demonstrate superior retrieval performance compared with CT-CLIP, indicating that our

Table 5. Report-to-volume retrieval performance on the test set of CT-RATE [14]. *: cited from [14].

Method	Rec.@10	Rec.@50	Rec.@100
CT-CLIP [14]*	0.040	0.144	0.235
Ours w/o $\mathcal{L}_{\text{so-kl}}$	0.047	0.175	0.279
Ours	**0.049**	**0.189**	**0.296**

structure observation-driven image-text contrastive learning successfully captures the fine-grained subtle coherence between CT volumes and reports. Additionally, the declined performance of our model without $\mathcal{L}_{\text{so-kl}}$ (Ours w/o $\mathcal{L}_{\text{so-kl}}$) confirms the effectiveness of $\mathcal{L}_{\text{so-kl}}$ in enhancing cross-modal representation by mitigating the false-negative issue.

5 Conclusion

This paper presented a novel framework for learning to generate medical reports for 3D CT images. The framework featured effective structure-wise image-text learning built on high-level medical prior knowledge. Extensive experiments on chest CT demonstrated its superior performance to existing SOTA approaches and the efficacy of its novel designs. We plan to extend our framework for other volumetric imaging data. A limitation of this work was that the NLG metrics of our framework with the LLaMA2-7B decoder were not as good as those with the BERT decoder. This could result from the insufficiency of conventional NLG metrics in evaluating the more capable LLMs nowadays. We expect to assess LLMs' performance with better tools in the future.

Acknowledgments. This work was supported by the National Natural Science Foundation of China (Grant No. 62371409) and Fujian Provincial Natural Science Foundation of China (Grant No. 2023J01005).

References

1. Banerjee, S., Lavie, A.: METEOR: an automatic metric for MT evaluation with improved correlation with human judgments. In: Proceedings of the ACL Workshop on Intrinsic and Extrinsic Evaluation Measures for Machine Translation and/or Summarization, pp. 65–72 (2005)
2. Brady, A., Laoide, R.Ó., McCarthy, P., McDermott, R.: Discrepancy and error in radiology: concepts, causes and consequences. Ulst. Med. J. **81**(1), 3 (2012)
3. Byun, J., Kim, D., Moon, T.: MAFA: managing false negatives for vision-language pre-training. In: Proceedings of the IEEE/CVF Conference on Computer Vision and Pattern Recognition, pp. 27314–27324 (2024)
4. Chaitanya, K., Erdil, E., Karani, N., Konukoglu, E.: Contrastive learning of global and local features for medical image segmentation with limited annotations. Adv. Neural. Inf. Process. Syst. **33**, 12546–12558 (2020)
5. Chan, A., Ong, Y.S., Pung, B., Zhang, A., Fu, J.: CoCon: a self-supervised approach for controlled text generation. In: International Conference on Learning Representations (2021)
6. Chen, T., Kornblith, S., Norouzi, M., Hinton, G.: A simple framework for contrastive learning of visual representations. In: International Conference on Machine Learning, pp. 1597–1607. PMLR (2020)
7. Chen, Z., Shen, Y., Song, Y., Wan, X.: Cross-modal memory networks for radiology report generation. In: Zong, C., Xia, F., Li, W., Navigli, R. (eds.) Proceedings of the Annual Meeting of the Association for Computational Linguistics, pp. 5904–5914. Association for Computational Linguistics (2021)

8. Chen, Z., Song, Y., Chang, T.H., Wan, X.: Generating radiology reports via memory-driven transformer. In: Proceedings of the Conference on Empirical Methods in Natural Language Processing, pp. 1439–1449. Association for Computational Linguistics (2020)
9. Chen, Z., Luo, L., Bie, Y., Chen, H.: Dia-LLaMA: towards large language model-driven CT report generation. arXiv preprint arXiv:2403.16386 (2024)
10. Donahue, J., et al.: Long-term recurrent convolutional networks for visual recognition and description. In: Proceedings of the IEEE/CVF Conference on Computer Vision and Pattern Recognition, pp. 2625–2634 (2015)
11. Draelos, R.L., et al.: Machine-learning-based multiple abnormality prediction with large-scale chest computed tomography volumes. Med. Image Anal. **67**, 101857 (2021)
12. Feng, R., Zhou, Z., Gotway, M.B., Liang, J.: Parts2whole: self-supervised contrastive learning via reconstruction. In: Domain Adaptation and Representation Transfer, and Distributed and Collaborative Learning: Second MICCAI Workshop, DART 2020, and First MICCAI Workshop, DCL 2020, Held in Conjunction with MICCAI 2020, Lima, Peru, October 4–8, 2020, Proceedings 2, pp. 85–95. Springer (2020)
13. Gong, R., Han, X., Wang, J., Ying, S., Shi, J.: Self-supervised bi-channel transformer networks for computer-aided diagnosis. IEEE J. Biomed. Health Inform. **26**(7), 3435–3446 (2022)
14. Hamamci, I.E., et al.: A foundation model utilizing chest CT volumes and radiology reports for supervised-level zero-shot detection of abnormalities. CoRR (2024)
15. Hamamci, I.E., Er, S., Menze, B.: CT2Rep: automated radiology report generation for 3D medical imaging. In: International Conference on Medical Image Computing and Computer-Assisted Intervention, pp. 476–486. Springer (2024)
16. Hamamci, I.E., et al.: GenerateCT: text-guided 3D chest CT generation. CoRR (2023)
17. He, K., Fan, H., Wu, Y., Xie, S., Girshick, R.: Momentum contrast for unsupervised visual representation learning. In: Proceedings of the IEEE/CVF Conference on Computer Vision and Pattern Recognition, pp. 9729–9738 (2020)
18. Hu, E.J., et al.: LoRA: low-rank adaptation of large language models. In: International Conference on Learning Representations (2022)
19. Huang, S.C., Shen, L., Lungren, M.P., Yeung, S.: GLoRIA: a multimodal global-local representation learning framework for label-efficient medical image recognition. In: IEEE International Conference on Computer Vision, pp. 3942–3951 (2021)
20. Irvin, J., et al.: CheXpert: a large chest radiograph dataset with uncertainty labels and expert comparison. In: Proceedings of the AAAI Conference on Artificial Intelligence, vol. 33, pp. 590–597 (2019)
21. Jain, S., et al.: Radgraph: extracting clinical entities and relations from radiology reports. In: Neural Information Processing Systems Datasets and Benchmarks Track (Round 1) (2021)
22. Jin, H., Che, H., Lin, Y., Chen, H.: PromptMRG: diagnosis-driven prompts for medical report generation. In: Proceedings of the AAAI Conference on Artificial Intelligence, vol. 38, pp. 2607–2615 (2024)
23. Jing, B., Wang, Z., Xing, E.: Show, describe and conclude: on exploiting the structure information of chest X-ray reports. In: Korhonen, A., Traum, D., Màrquez, L. (eds.) Annual Meetings of the Association for Computational Linguistics, pp. 6570–6580. Association for Computational Linguistics (2019)

24. Li, C.Y., Liang, X., Hu, Z., Xing, E.P.: Knowledge-driven encode, retrieve, paraphrase for medical image report generation. In: Proceedings of the AAAI Conference on Artificial Intelligence, vol. 33, pp. 6666–6673 (2019)
25. Li, J., Li, D., Xiong, C., Hoi, S.: BLIP: bootstrapping language-image pre-training for unified vision-language understanding and generation. In: International Conference on Machine Learning, pp. 12888–12900. PMLR (2022)
26. Li, J., Selvaraju, R., Gotmare, A., Joty, S., Xiong, C., Hoi, S.: Align before fuse: vision and language representation learning with momentum distillation. Adv. Neural. Inf. Process. Syst. **34**, 9694–9705 (2021)
27. Li, M., Liu, R., Wang, F., Chang, X., Liang, X.: Auxiliary signal-guided knowledge encoder-decoder for medical report generation. World Wide Web **26**(1), 253–270 (2023)
28. Li, Y., Liang, X., Hu, Z., Xing, E.P.: Hybrid retrieval-generation reinforced agent for medical image report generation. In: Advances in Neural Information Processing Systems, vol. 31 (2018)
29. Lin, C.Y.: ROUGE: a package for automatic evaluation of summaries. In: Text Summarization Branches Out, pp. 74–81 (2004)
30. Liu, B., et al.: Improving medical vision-language contrastive pretraining with semantics-aware triage. IEEE Trans. Med. Imaging **42**(12), 3579–3589 (2023)
31. Liu, F., Wu, X., Ge, S., Fan, W., Zou, Y.: Exploring and distilling posterior and prior knowledge for radiology report generation. In: Proceedings of the IEEE/CVF Conference on Computer Vision and Pattern Recognition, pp. 13753–13762 (2021)
32. Loshchilov, I., Hutter, F.: Decoupled weight decay regularization. In: International Conference on Learning Representations (2019)
33. Lu, J., Xiong, C., Parikh, D., Socher, R.: Knowing when to look: adaptive attention via a visual sentinel for image captioning. In: Proceedings of the IEEE/CVF Conference on Computer Vision and Pattern Recognition, pp. 375–383 (2017)
34. Nicolson, A., Dowling, J., Koopman, B.: Improving chest X-ray report generation by leveraging warm starting. Artif. Intell. Med. **144**, 102633 (2023)
35. Pan, R., Ran, R., Hu, W., Zhang, W., Qin, Q., Cui, S.: S3-Net: a self-supervised dual-stream network for radiology report generation. IEEE J. Biomed. Health Inform. **28**(3), 1448–1459 (2024)
36. Papineni, K., Roukos, S., Ward, T., Zhu, W.J.: BLEU: a method for automatic evaluation of machine translation. In: Annual Meetings of the Association for Computational Linguistics, pp. 311–318 (2002)
37. Paszke, A., et al.: PyTorch: an imperative style, high-performance deep learning library. In: Advances in Neural Information Processing Systems, vol. 32 (2019)
38. Ren, S., He, K., Girshick, R., Sun, J.: Faster R-CNN: towards real-time object detection with region proposal networks. IEEE Trans. Pattern Anal. Mach. Intell. **39**(6), 1137–1149 (2016)
39. Smit, A., Jain, S., Rajpurkar, P., Pareek, A., Ng, A., Lungren, M.: CheXbert: combining automatic labelers and expert annotations for accurate radiology report labeling using BERT. In: Webber, B., Cohn, T., He, Y., Liu, Y. (eds.) Proceedings of the 2020 Conference on Empirical Methods in Natural Language Processing, pp. 1500–1519. Association for Computational Linguistics (2020)
40. Taleb, A., Kirchler, M., Monti, R., Lippert, C.: ContIG: self-supervised multimodal contrastive learning for medical imaging with genetics. In: Proceedings of the IEEE/CVF Conference on Computer Vision and Pattern Recognition, pp. 20908–20921 (2022)

41. Tang, Y., Yang, H., Zhang, L., Yuan, Y.: Work like a doctor: unifying scan localizer and dynamic generator for automated computed tomography report generation. Expert Syst. Appl. **237**, 121442 (2024)
42. Tanida, T., Müller, P., Kaissis, G., Rueckert, D.: Interactive and explainable region-guided radiology report generation. In: Proceedings of the IEEE/CVF Conference on Computer Vision and Pattern Recognition, pp. 7433–7442 (2023)
43. Touvron, H., et al.: Llama 2: open foundation and fine-tuned chat models. arXiv preprint arXiv:2307.09288 (2023)
44. Vinyals, O., Toshev, A., Bengio, S., Erhan, D.: Show and tell: a neural image caption generator. In: Proceedings of the IEEE/CVF Conference on Computer Vision and Pattern Recognition, pp. 3156–3164 (2015)
45. Wang, F., Zhou, Y., Wang, S., Vardhanabhuti, V., Yu, L.: Multi-granularity cross-modal alignment for generalized medical visual representation learning. Adv. Neural. Inf. Process. Syst. **35**, 33536–33549 (2022)
46. Wang, Z., Liu, L., Wang, L., Zhou, L.: R2GenGPT: radiology report generation with frozen LLMs. Meta-Radiology **1**(3), 100033 (2023)
47. Wang, Z., Tang, M., Wang, L., Li, X., Zhou, L.: A medical semantic-assisted transformer for radiographic report generation. In: International Conference on Medical Image Computing and Computer-Assisted Intervention, pp. 655–664. Springer (2022)
48. Wang, Z., Zhou, L., Wang, L., Li, X.: A self-boosting framework for automated radiographic report generation. In: Proceedings of the IEEE/CVF Conference on Computer Vision and Pattern Recognition, pp. 2433–2442 (2021)
49. Wu, C., Zhang, X., Zhang, Y., Wang, Y., Xie, W.: Towards generalist foundation model for radiology. arXiv preprint arXiv:2308.02463 (2023)
50. Yang, S., Wu, X., Ge, S., Zheng, Z., Zhou, S.K., Xiao, L.: Radiology report generation with a learned knowledge base and multi-modal alignment. Med. Image Anal. **86**, 102798 (2023)
51. Yang, S., Wu, X., Ge, S., Zhou, S.K., Xiao, L.: Knowledge matters: chest radiology report generation with general and specific knowledge. Med. Image Anal. **80**, 102510 (2022)
52. Zhang, X., Wu, C., Zhao, Z., Lei, J., Zhang, Y., Wang, Y., Xie, W.: RadGenome-Chest CT: a grounded vision-language dataset for chest CT analysis. arXiv preprint arXiv:2404.16754 (2024)

Hierarchical CLIPs for Fine-Grained Anatomical Lesion Localization from Whole-Body PET/CT Images

Mingyang Yu[1], Yaozong Gao[2], Yiran Shu[2], Yanbo Chen[2], Jingyu Liu[2], Caiwen Jiang[1], Kaicong Sun[1], Weifang Zhang[3], Yiqiang Zhan[2], Xiang Sean Zhou[2], Shaonan Zhong[4], Xinlu Wang[4], Meixin Zhao[3(✉)], and Dinggang Shen[1,2,5(✉)]

[1] School of Biomedical Engineering and State Key Laboratory of Advanced Medical Materials and Devices, ShanghaiTech University, Shanghai 201210, China
dgshen@shanghaitech.edu.cn
[2] Shanghai United Imaging Intelligence Co., Ltd., Shanghai, China
[3] Department of Nuclear Medicine, Peking University Third Hospital, Beijing 100191, China
zhaomeixin.student@sina.com
[4] Department of Nuclear Medicine, The First Affiliated Hospital of Guangzhou Medical University, Guangzhou 510000, China
[5] Shanghai Clinical Research and Trial Center, Shanghai 201210, China

Abstract. Accurate identification of lesions, including anatomical lesion localization, is critical for automated radiology report generation. However, this task is particularly challenging in whole-body PET/CT imaging due to large amount of diverse anatomical regions throughout the whole body. Existing studies mainly rely on anatomical detection or segmentation. These methods are generally limited to only a small subset of anatomical regions due to difficulty of manual segmentation and annotation for large set of anatomical regions in the training stage. To address this issue, we propose a hierarchical CLIP-based 3D model to precisely and efficiently identify 387 anatomical lesion locations within whole-body PET/CT scans. Our model is built on three strategies: (1) Hierarchical localization, based on which anatomical locations are identified from coarse to fine to improve localization accuracy, robustness, and scalability; (2) Semantic location augmentation, which incorporates anatomical knowledge of relative location to adjacent regions to encourage neighborhood preservation of text feature representations; and (3) Location ambiguity mitigation, which excludes penalties on the top K ambiguous localizations in a modified CLIP loss to alleviate the cases with lesions residing at the boundaries of multiple regions. Notably, this work is the first to achieve accurate, robust, and efficient whole-body anatomical lesion localization, with significant performance improvement compared to the SOTA methods on a large whole-body PET/CT dataset comprising 1748 subjects acquired from multiple scanner makers.

M. Yu, Y. Gao and Y. Shu—Contributed equally.

I. Oguz et al. (Eds.): IPMI 2025, LNCS 15830, pp. 234–246, 2026.
https://doi.org/10.1007/978-3-031-96625-5_16

Keywords: CLIP · Anatomical Lesion Localization · Whole-body PET/CT imaging

1 Introduction

Deep learning has been increasingly applied in medical applications [4,5,18]. Radiology reports are essential in clinical healthcare. Generating report from whole-body PET/CT scan is time-consuming, since radiologists must carefully review hundreds of slices. This leads to a clinical need for automated report generation. However, most existing report generation methods [1,9–12,16,21] use image captioning technique [16] that provides only general information and often lacks detailed description about precise anatomical lesion locations, reducing clinical utility and explainability. To achieve high-quality radiology report generation, it is essential to accurately identify anatomical location of each lesion following automatic lesion detection, as shown in Fig. 1(a). Different from lesion localization that matches predicted boxes or masks to annotations at voxel level, the anatomical lesion localization technique can map lesion to specific anatomical regions.

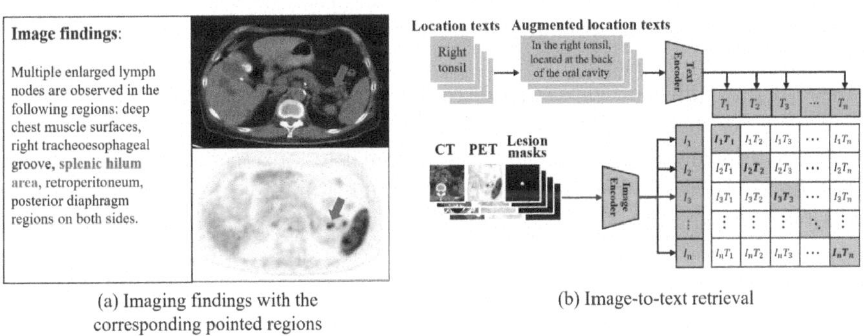

(a) Imaging findings with the corresponding pointed regions

(b) Image-to-text retrieval

Fig. 1. Anatomical lesion localization.

Deep learning-based segmentation or detection methods [13,20,24] can provide lesion positions at pixel level. Overlapping areas between lesion and organs can be determined as anatomical locations. These methods perform effectively in the cases containing limited number of potential anatomical locations, such as in chest CT scans by segmenting the lung into pulmonary lobes to locate nodules, and in mammography scans by dividing the breast into quadrants for lesion localization. However, for the case of whole-body PET/CT, there are a wide variety of anatomical sites to localize, posing huge challenges for previous methods. This is because training segmentation models for whole-body anatomical localization often requires a large amount of annotated labels, which are difficult and costly to have. Also, segmentation-based approaches may struggle to handle nearly 400 anatomical locations, as often mentioned in radiology reports, and they also rely on anatomical masks for localization, making it challenging to identify lesions situated in inter-organ spaces where no mask overlap exists. In addition, when localizing even a single lesion, segmentation must be performed for all anatomical structures, making segmentation-based methods extremely time-consuming.

To overcome the limitations of segmentation-based methods, we resort to Contrastive Language-Image Pretraining (CLIP) [17]. Particularly, we reformulate our task as an image-to-text retrieval problem, and then employ a CLIP-based model to precisely identify anatomical lesion locations in whole-body PET/CT images as shown in Fig. 1(b). Specifically, the CLIP aligns each image patch with its corresponding textual anatomical location in the feature space. By retrieving the most similar text embedding for a given image patch, the model can identify anatomical location, and achieve anatomical lesion localization. However, simple use of CLIP for anatomical lesion localization in whole-body PET/CT images can be challenging due to at least three reasons. First, localizing lesions among nearly 400 anatomical regions is difficult due to large variability in size and shape (while similar shape for left and right organs). Second, direct use of anatomical location terms (such as liver) is insufficient for image-to-text alignment, since it lacks anatomical knowledge such as spatial relationship for the model to better understand anatomical locations. Third, since radiology reports usually provide subjective descriptions especially for the cases when lesions span across multiple anatomical regions, i.e., the same lesion can be described in different ways, this introduces label ambiguity in model training.

To address these challenges, in this work we propose three strategies: (1) Hierarchical localization, to perform anatomical lesion localization in a coarse-to-fine hierarchy (e.g., from head to brain, tongue, and eyes) to leverage varying fields of view and spatial resolutions, which can simplify text retrieval among numerous location candidates and reduce task complexity; (2) Semantic location augmentation, to enrich textual anatomical locations with additional descriptions about its relative positions and spatial relation with neighboring organs, which enhances the CLIP by integrating anatomical knowledge and encourage the anatomically proximal locations to remain closely in the feature space even when target anatomical locations are significantly altered; (3) Location ambiguity mitigation, to address cases with lesions locating at boundaries of multiple anatomical regions by excluding penalties for the K most ambiguous lesion descriptions in the CLIP loss, which enhances the utility of our CLIP-based model for clinical usage.

The main contributions of this work are summarized as follows. (1) This is the first work to achieve anatomical lesion localization from 387 anatomical candidates for radiology reports. (2) Novel strategies for accurate and robust anatomical lesion localization are proposed, including hierarchical CLIP, anatomical knowledge-integrated text augmentation, and modified CLIP loss. (3) Our method has been evaluated on a large dataset including 1,748 whole-body PET/CT scans with 30,742 lesions from multiple scanner makers, with significant performance improvement over the SOTA models.

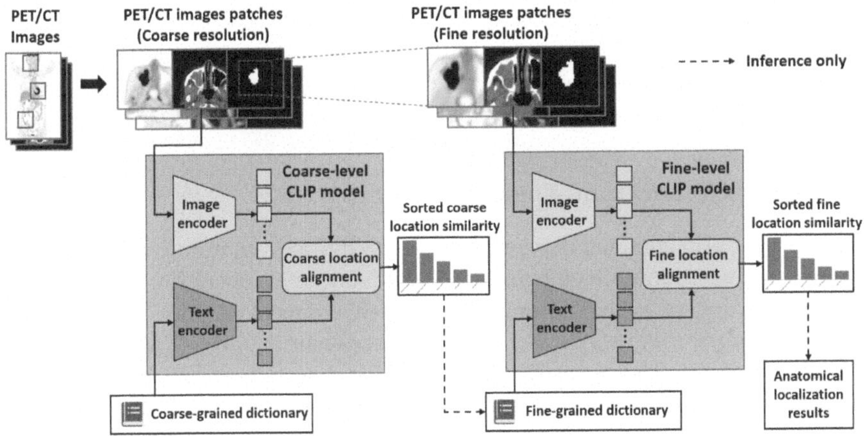

Fig. 2. Schematic illustration of our two-stage CLIP-based framework for training and inference. Specifically, the coarse-resolution patches have a size of (96, 96, 128) with a spacing of (2 mm, 2 mm, 2 mm), while the fine-resolution patches have a size of (64, 64, 64) with a spacing of (1.5 mm, 1.5 mm, 1.5 mm).

2 Method

Our model is built on a two-stage CLIP-based hierarchical framework for fine-grained anatomical lesion localization as illustrated in Fig. 2. The first stage uses coarse-resolution patches and a coarse-grained anatomical dictionary to coarsely locate the lesion regions, thus narrowing down candidates for more precise localization. The second stage uses fine-resolution patches and a fine-grained dictionary to identify the lesion locations more accurately. The detailed descriptions are given blow.

Hierarchical Localization. Inspired by the tree retrieval strategy [3], we sort anatomical locations in a hierarchical dictionary via a tree-structured layout as illustrated in Fig. 3(a). Nodes are arranged hierarchically in a coarse-to-fine scheme. Shallower nodes represent the broader regions such as "chest", while deeper nodes specify the finer sub-regions such as "left chest". Starting from the whole-body level, each anatomical region is progressively divided into finer sub-regions, enabling an ordered, efficient, and robust lesion localization.

Specifically, our hierarchical framework employs image patches with varying spatial resolutions and fields of view (FOV). In the first stage (using a coarse-level CLIP model), down-sampled image patches of paired PET, CT, and detected lesion labels are leveraged to find the top three most matched anatomical localizations from a coarse-grained dictionary, which contains totally 124 coarse regions. In the second stage (using a fine-level CLIP model), higher-resolution image patches with a narrower FOV are leveraged to identify more accurate subregion(s) of those top three coarse regions from a fine-grained dictionary, which contains totally 387 sub-regions. Inspired by [14], a label balancing strategy based on resampling is also applied to both coarse- and fine-grained stages to address the lesion imbalance issue. Our proposed hierarchical scheme significantly improves the accuracy and robustness of anatomical localization, along with high computational efficiency.

2.1 Semantic Location Augmentation

To enhance the alignment between image patches and anatomical descriptions using CLIP, we propose to incorporate additional anatomical knowledge about relative position of each anatomical site with adjacent organs. For example, instead of simple using terms like "liver", we augment it into a description of "in the liver, located in the upper right abdominal cavity, below the diaphragm, and above the stomach". This augmentation strategy is inspired by the fact that clinicians often consider *not only* the location of the lesion, *but also* its spatial relationship to the surrounding anatomical structures. Incorporating spatial description enables the model to better understand the circumstances, thus improving localization accuracy, particularly for the case of severe diseases with altered or obscured lesion appearances. For example, large lung tumors can appear strikingly similar with the shape of the anterior lateral portion of the liver when it is partially visible and surrounded by the lung cavity. By recognizing the unaffected areas described in the context, the morphological similarity between liver and lungs can be better distinguished, improving the retrieval accuracy even in the case of anatomical distortion. Our enriched spatial context achieves more context-aware representation and ensures anatomically-proximate locations to be distributed closely in the text feature space.

Figure 3(b) provides visual comparison of t-SNE [7] plots of the text embeddings in the first stage with and without using anatomical location augmentation. All anatomical locations are grouped into eight categories as marked in eight different colors, with each corresponding to one body region such as head, neck, and chest. The points in the plots represent the nodes within each category in the coarse-grained dictionary. For instance, the head region includes nodes such as eye, brain, and ear. As can be seen in Fig. 3(b-i), node embeddings without anatomical augmentation (using only anatomical terms) lead to separation of same body region in feature space. This indicates that the model fails to understand and capture the inherent spatial correlation of adjacent regions within each body region. This effect can lead to embedding shifts, especially when dealing with distorted anatomical structures caused by severe lesions. In contrast, as shown in Fig. 3(b-ii), anatomical augmentation improves the node embedding obviously, i.e., ensuring anatomical parts within the same body region to be distributed more tightly.

2.2 Location Ambiguity Mitigation

To address inherent ambiguity in anatomical localization for the cases when lesions locate at boundaries of multiple regions, we adopt a modified CLIP loss,

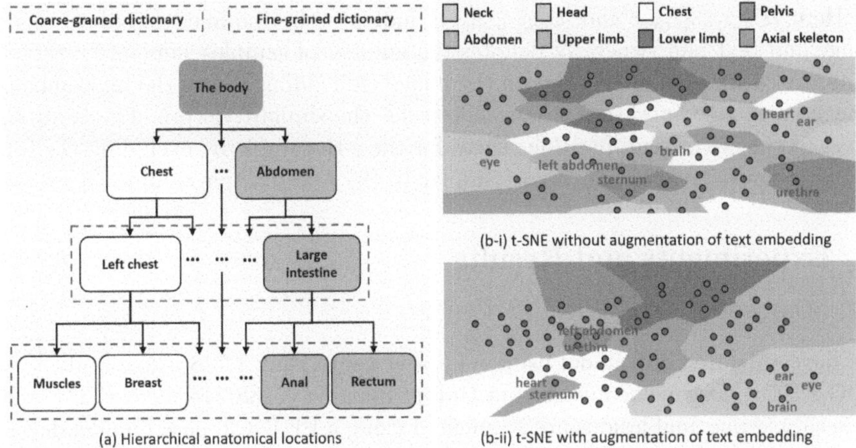

Fig. 3. (a) Hierarchical tree of the anatomical locations. (b) t-SNE visualization of the text embeddings in the first stage. Eight body parts are marked in different colors. The points represent the regions within each body part: (b-i) without using anatomical augmentation and (b-ii) with using anatomical augmentation.

based on semi-hard negative mining [19]. To be specific, different from the original InfoNCE loss, we exclude the top K difficult cases (i.e., $K = 2$ in our study) from the denominator, to avoid effect of negative samples with top K highest similarity scores. This also allows to ignore influence of ambiguous descriptions for some lesions with diagnosis difficulty. For example, as shown in Fig. 1(a), a lesion in the splenic hilum as marked with a red arrow could also be described as in the peripancreatic region or near the pancreatic tail. Such variability indicates that the ground truth may not strictly follow a one-to-one correspondence, but could include multiple valid annotations due to subjective choices in radiology reports. Penalizing these ambiguous "hardest negatives" can risk the training instability and also slow down the convergence. By excluding them and focusing on semi-hard negatives, our approach improves the training stability and accelerates the convergence. This operation can be achieved through a masked softmax function, ensuring that only semi-hard negatives contribute to the loss calculation as described below:

$$\mathcal{L}_{\text{CLIP}} = \frac{1}{2N} \left(\mathcal{L}_{\text{image-to-text}} + \mathcal{L}_{\text{text-to-image}}\right), \tag{1}$$

$$\mathcal{L}_{\text{image-to-text}} = -\frac{1}{N}\sum_{i=1}^{N} \log \frac{\exp\left(\text{sim}(f(x_i), g(y_i))/\tau\right)}{\sum\limits_{j \in N_i} \exp\left(\text{sim}(f(x_i), g(y_j))/\tau\right)}, \tag{2}$$

$$\mathcal{L}_{\text{text-to-image}} = -\frac{1}{N}\sum_{j=1}^{N} \log \frac{\exp\left(\text{sim}(f(x_j), g(y_j))/\tau\right)}{\sum\limits_{i \in N_j} \exp\left(\text{sim}(f(x_j), g(y_i))/\tau\right)}. \tag{3}$$

Here, $\mathcal{L}_{\text{image-to-text}}$ and $\mathcal{L}_{\text{text-to-image}}$ indicate the modified losses to align image and text pair, where N denotes the number of samples. $\text{sim}(x_i, y_j)$ represents the cosine similarity between the image embedding x_i and the text embedding y_j. τ is a temperature scalar that scales the similarity score. The term N_i refers to the set of semi-hard negatives for the i-th sample by excluding the top K hardest negative samples, most similar to the i-th positive sample.

3 Experiments and Results

3.1 Dataset

We have validated our model using a dataset comprising 1,748 whole-body ^{18}F-FDG PET/CT scans acquired from United Imaging (1288), Siemens (323), and GE (137). Abnormal lesions are segmented using a PET/CT-based model developed by United Imaging Intelligence (UII), and these segmented lesions are further reviewed by two experienced radiologists to remove false positives and irrelevant lesions. The remaining lesions were used for training (1,598 scans, 33,186 lesions), validation (50 scans, 1,056 lesions), and testing (100 scans, 2,502 lesions). In our experiments, we set 124 coarse-grained and 387 fine-grained anatomical locations.

To enhance data diversity, data augmentation techniques including random rotation and scaling are applied (with 50% probability) during the training. 3D lesion patches are extracted from PET, CT, and their corresponding segmentation masks. The PET and CT images are normalized to [0, 1] by min-max normalization. Text descriptions are augmented and truncated to a maximum length of 77 words.

3.2 Training Details

Training is conducted on an Nvidia L40 GPU with 48 GB GPU memory. Our model is trained for 300 epochs with each epoch containing 100 mini-batches. The mini-batch size is set as 48 for the coarse-level localization and 64 for the fine-level localization. The initial learning rate is set as 1×10^{-4} and is reduced by a factor of 10 every 100 epochs. Optimization is performed using the Adam optimizer with $\beta_1 = 0.9$, $\beta_2 = 0.999$, and a weight decay of 5×10^{-4}. The pre-trained text encoder in [17] is kept frozen throughout training.

3.3 Comparison Studies

To evaluate the effectiveness of our model, we conduct comparative experiments including both segmentation-based and classification-based methods. Specifically, we evaluate different image encoders, including three 3D convolutional networks, i.e., BasicNet-18 (ResNet-18 [15] without residual connection), ResNet-50 [6], and DenseNet-121 [2]) and a 3D Vision Transformer (3D ViT [23]). We primarily use accuracy to assess the performance of the investigated models. To achieve more detailed evaluation, we demonstrate the localization accuracy for the 8 regions of Hd (Head), Nk (Neck), Ch (Chest), Ab (Abdomen), Ax (Axial skeleton), UL (Upper Limb), LL (Lower Limb), and Pl (Pelvis) in Table 1. It should be noted that the dataset contains different numbers of lesions in the 8 regions (Head (34), Neck (584), Chest (885), Abdomen (344), Lower Limb (63), Axial Skeleton (304), Upper Limb (61), and Pelvis (227)), and the reported results are the average over the lesions.

Specifically, to compare with the segmentation-based method, we select TotalSegmentator [22], an open-source tool supporting 117 organ labeling using the state-of-the-art nnU-Net [8]. In TotalSegmentator, lesion localization is achieved by calculating the overlap between the segmented lesion mask and the organ mask. The organ with largest overlap is chosen as the anatomical localization. In fact, due to limitation of pre-defined organ masks, TotalSegmentator can only identify about 117 anatomical sites, prohibiting its applicability for fine-grained localization tasks. To ensure a fair comparison, we align the regions with those used in TotalSegmentator and exclud lesions in the regions not supported by TotalSegmentator, resulting in 870 lesions in 117 regions for testing. As shown in Table 1 (under "Sites = 117"), our method significantly outperforms TotalSegmentator in overall accuracy, achieving 85.4% compared to 81.0% by TotalSegmentator ($p < 0.05$, chi-square test). Moreover, unlike the segmentation-based TotalSegmentator, which requires at least 77 s for whole-body organ segmentation regardless of the number of existing lesions, our model performs direct lesion localization without segmentation in only 34 milliseconds per lesion until all the lesions are localized. In average, our model takes about 0.71 s for conducting localization for all the lesions across whole body. Additionally, our method can scale up to 387 anatomical sites or even more, enabled by both hierarchical localization setup and semantic location augmentation.

Table 1. Quantitative comparison for anatomical localization in terms of average accuracy (%), where accuracy is defined as requiring all keywords to match.

Sites	Method	Total	Hd	Nk	Ch	Ab	Ax	UL	LL	Pl
117	TotalSegmentator	81.0	80.0	78.3	**88.9**	**84.6**	71.9	73.9	85.7	100.0
	Our Method	**85.4**	**100.0**	**95.7**	85.1	81.5	**83.9**	**100.0**	**100.0**	**100.0**
387	BasicNet-18	64.4	61.8	65.1	74.4	48.0	69.4	73.8	65.1	76.7
	BasicNet-18*	73.6	76.5	87.8	72.7	57.3	67.1	83.6	57.1	75.3
	ResNet-50*	72.5	64.7	87.2	71.1	56.4	68.1	77.0	49.2	76.7
	DenseNet-121*	67.6	67.6	79.3	67.9	49.4	62.2	83.6	46.0	72.2
	ViT-Base-16*	67.4	64.7	77.2	67.8	48.5	63.2	78.7	60.3	74.5
	Our Method	**84.2**	**79.4**	**96.1**	**83.4**	**68.6**	**83.6**	**90.2**	**76.2**	**78.9**

Methods marked by * use a hierarchical structure for a fair comparison.

With regard to the classification-based models, we compare with several image encoders with multi-modal inputs to identify lesion locations, including

BasicNet-18, ResNet-50, DenseNet-121, and a 3D adaptation of Vision Transformer (3D ViT [23]). Initially, we train classifiers to directly predict all 387 sites. However, due to severe class imbalance and large number of anatomical sites, these approaches yield low localization accuracy. For example, BasicNet-18 achieves only 64.4% accuracy without hierarchical localization as shown in Table 1 (under "Sites = 387"). To fairly compare with these methods, we incorporate hierarchical localization into the image classifiers. This strategy significantly improves the performance, with BasicNet-18* being enhanced from 64.4% to 73.6% in accuracy. Among all these methods, our model obtains a statistically significant accuracy improvement of 84.2% ($p < 0.05$, chi-square test), particularly in regions such as the neck (96.1%) and chest (83.4%), which validates the superiority of our framework.

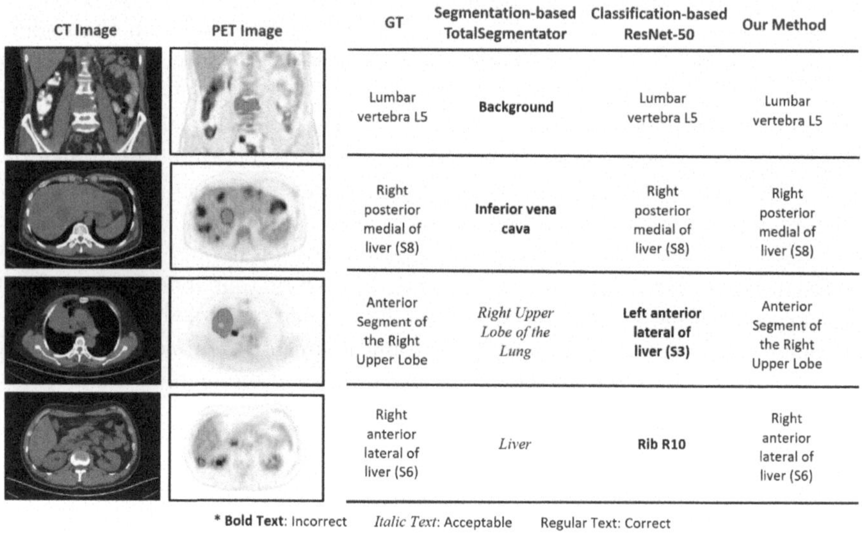

CT Image	PET Image	GT	Segmentation-based TotalSegmentator	Classification-based ResNet-50	Our Method
		Lumbar vertebra L5	**Background**	Lumbar vertebra L5	Lumbar vertebra L5
		Right posterior medial of liver (S8)	**Inferior vena cava**	Right posterior medial of liver (S8)	Right posterior medial of liver (S8)
		Anterior Segment of the Right Upper Lobe	*Right Upper Lobe of the Lung*	**Left anterior lateral of liver (S3)**	Anterior Segment of the Right Upper Lobe
		Right anterior lateral of liver (S6)	*Liver*	**Rib R10**	Right anterior lateral of liver (S6)

* **Bold Text**: Incorrect *Italic Text*: Acceptable Regular Text: Correct

Fig. 4. Qualitative comparison among segmentation-based TotalSegmentator, classification-based ResNet-50, and the proposed model in four representative cases.

Besides quantitative evaluation, qualitative comparison is also conducted, with the results shown in Fig. 4. Four representative cases are provided to show the superiority of our model, with the first two cases over the segmentation-based TotalSegmentator, and the last two cases over the classification-based ResNet-50. Specifically, in the first case, TotalSegmentator misclassifies a bone lesion in the lumbar vertebra L5 as background due to cancer erosion, whereas our method correctly identifies the lesion location. In the second case, a lesion in the porta hepatis is mis-localized as the inferior vena cava by TotalSegmentator due to spatial proximity. In contrast, both our method and the classification-based ResNet-50 can accurately localize the lesion as the porta hepatis. In the

Table 2. Ablation study of the proposed method in terms of accuracy (%) across different body regions, where accuracy requires all keywords to match.

Method	Total	Hd	Nk	Ch	Ab	Ax	UL	LL	Pl
B	78.7	70.6	89.2	78.1	61.9	78.6	86.9	68.3	**81.1**
B+H	80.2	61.8	92.1	78.4	64.8	81.3	90.2	68.3	77.5
B+H+S	82.1	70.6	95.2	81.5	64.8	79.9	**93.4**	**79.4**	79.7
B+H+S+L	**84.2**	**79.4**	**96.1**	**83.4**	**68.6**	**83.6**	90.2	76.2	78.9

third case, classification-based ResNet-50 mis-localizes a lesion in the right lung to the liver due to its similarity of appearance to the hepatic dome. However, our method accurately identifies the lesion within the pulmonary segment by leveraging anatomical knowledge. In the fourth case, a hepatic boundary lesion near the rib is incorrectly localized to the rib by classification-based ResNet-50, whereas our approach correctly recognizes the specific liver segment. Visual evaluation further verifies effectiveness of our method compared to both the segmentation-based and classification-based methods.

3.4 Ablation Study

To accurately identify anatomical lesion locations, our framework integrates three key strategies: 1) Hierarchical Localization (H), 2) Semantic Location Augmentation (S), and 3) Location Ambiguity Mitigation (L). Also, we employ the conventional CLIP model as the baseline method (B) that aligns coarse-resolution image patches with all the anatomical locations in a single step, without leveraging hierarchical CLIP and other enhancements. We primarily use accuracy to assess the impact of each strategy. The results are summarized in Table 2. Similarly, the whole body is divided into eight regions for detailed evaluation: Hd (Head), Nk (Neck), Ch (Chest), Ab (Abdomen), Ax (Axial skeleton), UL (Upper Limb), LL (Lower Limb), and Pl (Pelvis).

The baseline method (B) achieves a total accuracy of 78.7%, with relatively good performance in regions such as neck (89.2%) and pelvis (81.1%), but performing poorly in areas like abdomen (61.9%). Incorporating Hierarchical Localization (B+H) improves the total accuracy from 78.7% to 80.2% ($p < 0.05$, chi-square test), benefiting particularly the regions such as neck and axial skeleton, where coarse localization helps narrow down the lesion positions. Incorporating Semantic Location Augmentation (B+H+S) further enhances the total accuracy from 80.2% to 82.1% ($p< 0.05$, chi-square test), leveraging spatial relationships among neighboring anatomical structures to reduce retrieval errors. This improvement is especially evident in regions like upper limb (93.4%) and lower limb (79.4%), where augmented spatial descriptions mitigate anatomical variability. Ultimately, incorporating Location Ambiguity Mitigation (B+H+S+L) results in the highest accuracy of 84.2% ($p< 0.05$, chi-square test), achieving significant gain in regions such as head (79.4%) and neck (96.1%). Location ambiguity mitigation effectively addresses location ambiguity in anatomically complex areas.

3.5 Zero-Shot Localization

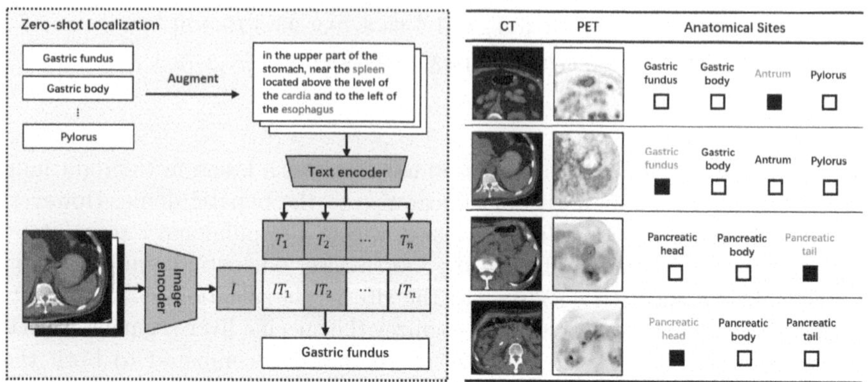

Fig. 5. Zero-shot anatomical localization for further fine-grained localization of lesions in the stomach and pancreas. The green font indicates the gold standard for precise localization. (Color figure online)

Besides, we conduct additional experiments to evaluate our model on unseen anatomical locations, specifically sub-regions of the stomach (gastric fundus, gastric body, antrum, pylorus) and pancreas (pancreatic head, pancreatic tail, pancreatic body). To localize these unseen sub-regions, we augment their textual descriptions with spatial relationship to known anatomical landmarks.

As shown in Fig. 5, our model, trained on general regions such as the stomach and pancreas with captured anatomical knowledge, can successfully refine localization of unseen sub-regions. For example, the gastric fundus is augmented with the description "in the upper part of the stomach, near the spleen, above the cardia, and to the left of the esophagus", where these landmarks, such as the spleen, cardia, and esophagus, are known from the training set. By leveraging these enriched descriptions and the anatomical knowledge learned during the training, the model accurately localizes sub-regions not present in the training data, such as the gastric fundus, antrum, and pancreatic tail. This zero-shot capability, enabled by the model's understanding of anatomical relationship, is crucial for clinical applications where anatomical descriptions can vary across institutions, and offers a scalable and robust solution for whole-body anatomical lesion localization.

4 Conclusion

In this work, we have presented a hierarchical CLIP-based model to address the challenging fine-grained whole-body anatomical lesion localization in PET/CT images. Our model has two major merits: (1) accurate and robust identification of 387 anatomical lesion localizations; and (2) promising zero-shot capability. These merits are mainly attributed to the three proposed strategies, i.e., (1) coarse-to-fine anatomical localization, (2) semantic location augmentation by integrating additional anatomical knowledge of spatial relation, and (3) location ambiguity mitigation by excluding penalties on negative pairs with top K highest similarity scores. Our model has been validated on a large whole-body PET/CT dataset consisting of 1748 subjects from multiple scanner makers. Experimental results show that our model outperforms both the segmentation-based and classification-based models significantly. It is worth noting that our model can complete one anatomical lesion localization within 34 milliseconds, while the segmentation-based TotalSegmentator has to perform whole-body segmentation in advance and takes at least 77 s even for localizing only one lesion. This demonstrates efficiency of our model. In summary, we have presented a precise, scalable, and efficient solution to whole-body anatomical lesion localization for PET/CT images, providing a significant step to automated radiology report generation.

Acknowledgments.. This work was supported in part by National Natural Science Foundation of China (grant numbers 82441023, U23A20295, 62131015, 82394432), the China Ministry of Science and Technology (S20240085, STI2030-Major Projects-2022ZD0209000, STI2030-Major Projects-2022ZD0213100), Shanghai Municipal Central Guided Local Science and Technology Development Fund (No. YDZX20233100001001), The Key R&D Program of Guangdong Province, China (grant number 2023B0303040001), HPC Platform of ShanghaiTech University, the special fund of Beijing Clinical Key Specialty Construction Program, P. R. China (2022), China Medical Health Development Foundation, Peking University Third Hospital || United-Imaging Research Institution Intelligential Imaging Joint Research & Development Center Foundation, and Key Clinical Project of Peking University Third Hospital (BYSYZD2019038 and BYSYZD2023016). We also acknowledge the Department of Nuclear Medicine, The First Affiliated Hospital of Guangzhou Medical University, for providing validation data.

References

1. Allaouzi, I., Ben Ahmed, M., Benamrou, B., Ouardouz, M.: Automatic caption generation for medical images. In: Proceedings of the 3rd International Conference on Smart City Applications, pp. 1–6 (2018)
2. Anwar, S.M., Majid, M., Qayyum, A., Awais, M., Alnowami, M., Khan, M.K.: Medical image analysis using convolutional neural networks: a review. J. Med. Syst. **42**, 1–13 (2018)
3. Bard, J.B.: The AEO, an ontology of anatomical entities for classifying animal tissues and organs. Front. Genet. **3**, 18 (2012)
4. Cui, Z., et al.: TSegNet: an efficient and accurate tooth segmentation network on 3D dental model. Med. Image Anal. **69**, 101949 (2021)
5. Fan, J., Cao, X., Wang, Q., Yap, P.T., Shen, D.: Adversarial learning for mono-or multi-modal registration. Med. Image Anal. **58**, 101545 (2019)

6. Hara, K., Kataoka, H., Satoh, Y.: Can spatiotemporal 3D CNNs retrace the history of 2D CNNs and imagenet? In: Proceedings of the IEEE conference on Computer Vision and Pattern Recognition, pp. 6546–6555 (2018)
7. Hinton, G., Van Der Maaten, L.: Visualizing data using t-SNE journal of machine learning research. J. Mach. Learn. Res. **9**, 2579–2605 (2008)
8. Isensee, F., Jaeger, P.F., Kohl, S.A., Petersen, J., Maier-Hein, K.H.: nnU-Net: a self-configuring method for deep learning-based biomedical image segmentation. Nat. Methods **18**(2), 203–211 (2021)
9. Kaur, N., Mittal, A., Singh, G.: Methods for automatic generation of radiological reports of chest radiographs: a comprehensive survey. Multimedia Tools Appl. **81**(10), 13409–13439 (2022)
10. Liao, Y., Liu, H., Spasić, I.: Deep learning approaches to automatic radiology report generation: a systematic review. Inform. Med. Unlocked **39**, 101273 (2023)
11. Messina, P., et al.: A survey on deep learning and explainability for automatic report generation from medical images. ACM Comput. Surv. (CSUR) **54**(10s), 1–40 (2022)
12. Monshi, M., Poon, J., Chung, V.: Deep learning in generating radiology reports: a survey. Artif. Intell. Med. **106**, 101878 (2020)
13. Nasrullah, N., Sang, J., Alam, M.S., Mateen, M., Cai, B., Hu, H.: Automated lung nodule detection and classification using deep learning combined with multiple strategies. Sensors **19**(17), 3722 (2019)
14. Nitesh, V.C.: Smote: synthetic minority over-sampling technique. J. Artif. Intell. Res. **16**(1), 321 (2002)
15. Ou, X., et al.: Moving object detection method via resnet-18 with encoder-decoder structure in complex scenes. IEEE Access **7**, 108152–108160 (2019)
16. Pavlopoulos, J., Kougia, V., Androutsopoulos, I.: A survey on biomedical image captioning. In: Proceedings of the Second Workshop on Shortcomings in Vision and Language, pp. 26–36 (2019)
17. Radford, A., et al.: Learning transferable visual models from natural language supervision. In: International Conference on Machine Learning, pp. 8748–8763. PMLR (2021)
18. Ren, X., et al.: Interleaved 3D-CNNs for joint segmentation of small-volume structures in head and neck CT images. Med. Phys. **45**(5), 2063–2075 (2018)
19. Schroff, F., Kalenichenko, D., Philbin, J.: Facenet: a unified embedding for face recognition and clustering. In: Proceedings of the IEEE Conference on Computer Vision and Pattern Recognition, pp. 815–823 (2015)
20. Shen, L., Margolies, L.R., Rothstein, J.H., Fluder, E., McBride, R., Sieh, W.: Deep learning to improve breast cancer detection on screening mammography. Sci. Rep. **9**(1), 12495 (2019)
21. Sloan, P., Clatworthy, P., Simpson, E., Mirmehdi, M.: Automated radiology report generation: a review of recent advances. IEEE Rev. Biomed. Eng. (2024)
22. Wasserthal, J., et al.: Totalsegmentator: robust segmentation of 104 anatomic structures in CT images. Radiol. Artif. Intell. **5**(5) (2023)
23. Wu, C., Zhang, X., Zhang, Y., Wang, Y., Xie, W.: Towards generalist foundation model for radiology. arXiv preprint arXiv:2308.02463 (2023)
24. Yan, K., Wang, X., Lu, L., Summers, R.M.: DeepLesion: automated mining of large-scale lesion annotations and universal lesion detection with deep learning. J. Med. Imag. **5**(3), 036501–036501 (2018)

Multi-view and Multi-scale Alignment for Contrastive Language-Image Pre-training in Mammography

Yuexi Du[1], John A. Onofrey[1,2,3], and Nicha C. Dvornek[1,2(✉)]

[1] Departments of Biomedical Engineering, Yale University, New Haven, CT, USA
nicha.dvornek@yale.edu

[2] Radiology and Biomedical Imaging, Yale University, New Haven, CT, USA

[3] Urology, Yale University, New Haven, CT, USA

Abstract. Contrastive Language-Image Pre-training (CLIP) demonstrates strong potential in medical image analysis but requires substantial data and computational resources. Due to these restrictions, existing CLIP applications in medical imaging focus mainly on modalities like chest X-rays that have abundant image-report data available, leaving many other important modalities under-explored. Here, we propose one of the first adaptations of the full CLIP model to mammography, which presents significant challenges due to labeled data scarcity, high-resolution images with small regions of interest, and class-wise imbalance. We first develop a specialized supervision framework for mammography that leverages its multi-view nature. Furthermore, we design a symmetric local alignment module to better focus on detailed features in high-resolution images. Lastly, we incorporate a parameter-efficient fine-tuning approach for large language models pre-trained with medical knowledge to address data limitations. Our multi-view and multi-scale alignment (MaMA) method outperforms state-of-the-art baselines for three different tasks on two large real-world mammography datasets, EMBED and RSNA-Mammo, with only 52% model size compared with the largest baseline. (The code is available at https://github.com/XYPB/MaMA.)

1 Introduction

Contrastive learning [4,12,13] has become one of the most popular self-supervised representation learning paradigms due to its intuitive concept and robust performance. Recently, the introduction of natural language signals to contrastive learning [34] has given rise to modern visual-language models [21, 22,25]. Contrastive Language-Image Pre-training (CLIP) [34] has also been applied in the medical imaging domain [10,15,45,47–49,51], showing promising improvement in medical image understanding when large-scale datasets are available [10,16,19,49]. However, the CLIP model in the natural image domain usually demands billion-level image-text pairs to be properly trained [34,40,40,41], which is almost impossible in the medical domain due in part to privacy and security concerns. Existing medical CLIP methods either build general-purpose models with many anatomical sites and modalities from public databases [10,49] or

I. Oguz et al. (Eds.): IPMI 2025, LNCS 15830, pp. 247–262, 2026.
https://doi.org/10.1007/978-3-031-96625-5_17

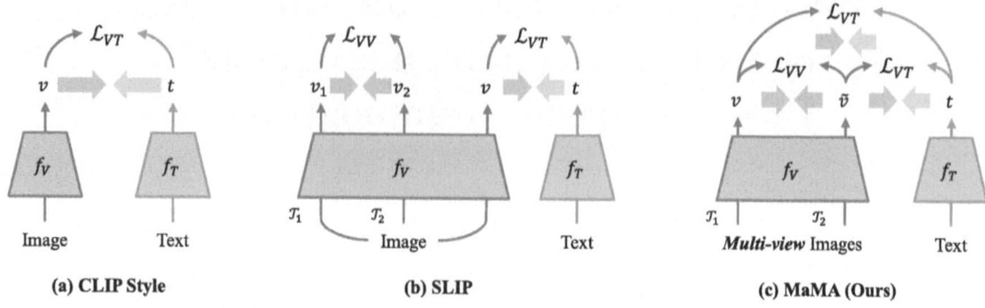

Fig. 1. Comparison of Visual-Language Contrastive Learning Frameworks. (a) CLIP [34] style; (b) SLIP [30] style; (c) Our proposed MaMA, which aligns image-image and image-text features, exploiting the multi-view nature of mammography.

focus on single modalities with large-scale (still less than 1M) datasets, *e.g.*, chest X-ray or pathology images [15,20,44–48,51,54]. Thus, other imaging modalities, like mammography, have yet to fully benefit from visual-language pre-training.

Mammography is a critical medical imaging modality for screening and diagnosis of breast cancer, one of the most commonly diagnosed cancers globally and a leading cause of cancer-related mortality in women [42]. While visual-language pre-training (VLP) has the potential to improve mammography interpretation, there are two major obstacles: 1) *Limited data and annotation*: While recent work has introduced a large-scale mammography image and tabular dataset of more than 110,000 patients [17], no corresponding clinical reports are available.2) *Nature of mammography*: Each mammography study usually contains four views of the same patient: left and right side, each with craniocaudal (CC) and mediolateral oblique (MLO) views, giving rise to the critical properties of *bilateral asymmetry* [7] and *ipsilateral correspondence* [28]. Bilateral asymmetry means images from different sides of the same patient can contain different information, *e.g.*, density, calcification, and mass findings. Ipsilateral correspondence means different views of the same side share similar information from different viewpoints. Clinicians consider both properties and all four images at once when reading a study. Meanwhile, lesions of interest are often relatively small compared with the high-resolution (~2,000-by-2,000 pixels) mammograms, which further challenges a model's ability to focus on local details. This pixel-level imbalance compounds the problem of image-level imbalance, in which the vast majority of mammograms will not contain cancer. While recent works [5,11] attempt to address these issues using VLP, they either simply fine-tune a pre-trained CLIP model with a small amount of data [5] or apply CLIP with same side mammograms and hand-crafted prompts [11], rather than leveraging the full mammography domain information.

To address these challenges, we propose a novel

*M*ulti-view *a*nd *M*ulti-scale *A*lignment *i.e.*, *MaMA*, contrastive language-image pre-training framework that exploits the multi-view property of mammography and aligns multi-scale features simultaneously. Our work offers the following contributions: **1) Multi-view Design**: We extend the CLIP-style method to leverage the unique multi-view nature of mammography images, introducing (i) inter-study image-to-image contrastive loss, and (ii) symmetric image-text loss. **2) Symmetric Local Alignment (SLA)**: Designed for the relatively small ROIs in mammography, the SLA module improves model understanding of local features without needing dense annotation. **3) Efficient Large Language Model (LLM) as Text Encoder**: Using a parameter-efficient fine-tuned LLM improves understanding of the text while addressing data scarcity. **4) Other Contributions**: We propose two important strategies specifically for mammography VLP: (i) a template-based method to generate structured captions from tabular data that mimics realistic clinical reports and (ii) meta-information masking augmentation to mitigate zero-shot performance loss when training with complex captions.

We validate our method on two large-scale mammography datasets, EMBED [17] and RNSA-Mammo [3], with multiple settings and compare SOTA medical CLIP methods. The proposed method surpasses all the baselines with only 52% model size, showing promise on multiple mammography-related tasks.

2 Related Works

Medical Visual-Language Pre-training. Existing medical VLP approaches can be divided into two types. The first type trains a general-purpose medical CLIP model with large, multi-site, multi-modality datasets derived from PubMed [10,49], scaling dataset size but relying on a vanilla CLIP design [34]. While these models show promise in generalizing across various sites, they can be suboptimal compared to modality-specific models due to the lack of tailored designs for particular imaging modalities. The second type focuses on chest X-ray [15,44–48,51,54], using large datasets like MIMIC-CXR [19] or CheXpert [16], and often relies on single-view images. Some also require full clinical reports [44,45,54], complicating adoption. Recently, a CLIP-based method for mammography was introduced, fine-tuning a pre-trained CLIP model with a multi-view aggregation module [5]. However, it does not employ contrastive pre-training, does not address pixel-level imbalance, and cannot correlate reports with fine-grained ROIs. Moreover, it only uses a few thousand private cases. Another work proposed Mammo-CLIP [11] for pre-training on mammograms but ignored their multi-scale nature and was trained on less than 50k images, limiting generalization and increasing domain shift risk.

Multi-view Contrastive Learning. Methods like SLIP [30] and DeCLIP [23] combine image-image and image-text loss to achieve robust learning. Similar ideas have been applied to 3D shape recognition [6,38] and action recognition [36]. In mammography [9,24,39], multi-view consistency helps learn high-level shared information. While Mammo-CLIP [11] has attempted to learn

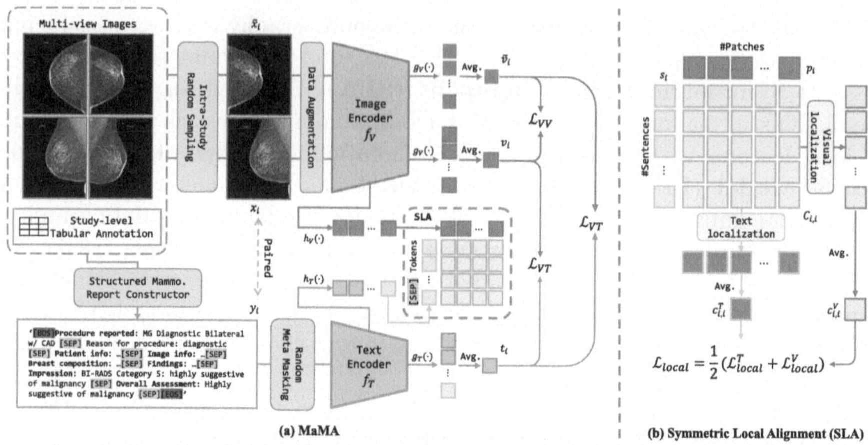

Fig. 2. Proposed Multi-view and Multi-scale (MaMA) VLP Framework. (a) We utilize the multi-view information of mammography to conduct symmetric image-image and image-text contrastive learning. (b) We localize the most relevant sentence for each image patch and the most relevant patch for each sentence and align these matched local features via symmetric local alignment.

intra-side multi-view in a contrastive way, it ignores the correspondence between different sides.

Unsupervised Local Contrastive Learning. Correlating dense visual representations with fine-grained semantics is crucial for tasks like segmentation. Recent unsupervised methods [15,26,27,36,45,46,50,53] have tried to address this challenge. Some rely on pre-trained detectors [50], others aggregate dense similarity scores into image-level contrastive learning [18,26,46,53], while some perform token-level matching with high computational costs [15,36,45]. However, none of these learn the explicit local correspondence score for medical images.

3 Method

3.1 Structured Report Construction

Different from datasets that provide paired images with corresponding clinical reports, *e.g.*, MIMIC-CXR [19], large-scale mammography datasets with the full report available are rare. Rather, existing datasets in this domain [3,17,31] mainly provide a tabular annotation with the anonymized meta-information and the clinical findings, *e.g.*, breast density, calcification findings, and Breast Imaging Reporting and Data System (BI-RADS) assessment category [37]. Clinical findings serve as cross-validation evidence for the final diagnosis. Using a CLIP-style [52] caption with only the simple class label will result in a highly simplified caption and limit the model's understanding of the image due to missing details.

We propose a fully automatic template-based caption construction pipeline following the standard clinical report structure [32] (Fig. 2(a)). We first create a report template with segments describing **study procedure**, **patient meta-information**, **image meta-information**, **breast composition**, **findings**, **clinical impression** and the **overall assessment** in a clinical report style. Each segment contains keywords that are then replaced with the corresponding information in the tabular data to build a complete clinical report for each image. Our motivation for the structured template is that it can provide more contextual information to the LLM than data in a tabular format.

Meta-information Masking. The increased information from the meta-data may be memorized by the model during pre-training and result in learning shortcuts for the model decision. We propose a *data augmentation* method that randomly masks each meta-information keyword with probability m when constructing the caption to force the model to focus on disease-related information.

3.2 Multi-view VLP

Our multi-view contrastive VLP framework optimizes both image-to-image and symmetric image-to-text contrastive loss (Fig. 1(a)).

Multi-view Visual Contrastive Loss. We first optimize the contrastive loss between the multi-view images (Fig. 2(a)). Let $\mathcal{D} = \{(x_i, y_i),\ i = 0, 1, \ldots, N\}$ be a multimodal dataset, where there are N individual images x_i and corresponding text captions y_i. We define a study to include the data from the same imaging session for a patient, including one or more image-text pairs. For a random image-text pair $(x_i, y_i) \in \mathcal{D}$, we uniformly sample another image $\tilde{x}_i$ from the same study that x_i belongs to as the positive sample. Note that $\tilde{x}_i$ could be x_i as the augmented view of the same image is naturally a positive sample. We augment both images with random data augmentation and feed into the vision encoder f_V and d-dimensional global projection head g_V followed by average pooling to get the visual embedding $v_i, \tilde{v}_i \in \mathbb{R}^d$, *i.e.*, $v_i = \text{avg}(g_V(f_V(x_i)))$. We then compute the similarity for each pair of visual embeddings and optimize the InfoNCE [4] loss for v_i in a mini-batch of size B:

$$\mathcal{L}_{VV}(v_i, \tilde{v}_i) = \log \frac{\exp(sim(v_i, \tilde{v}_i)/\tau_1)}{\sum_{j=1}^{B} \exp(sim(v_i, v_j)/\tau_1)}, \text{ where } sim(v_i, v_j) = \frac{v_i^T v_j}{\|v_i\| \|v_j\|}, \tag{1}$$

where τ_1 is the visual temperature and v_j is the j-th visual embedding in the batch. Considering the ipsilateral correspondence, it is natural to treat images on the same side as positive samples of each other, as the features, *e.g.*, tumors, present in one view, are also present in the other view. On the other hand, even if considering bilateral asymmetry for images from different sides, they still share much high-level information such as patient-level features (*e.g.*, breast structure) and similar breast density. Introducing multi-view mammography contrastive learning forces the model to learn semantically similar features from images within the same study. This also provides a stronger self-supervised signal than using random augmented images. We follow SimCLR [4] to build the image-image contrastive learning framework for simplicity.

Symmetric Visual-Text Contrastive Loss. While existing methods like SLIP [30] also optimize both image-image and image-text contrastive loss, we note there is a potential contradiction between these two objectives when computed for different samples (Fig. 1(b)), *i.e.*, $\mathcal{L}_{VV}$ and $\mathcal{L}_{VT}$ are independent and the extra image will increase memory cost. To address this, we propose symmetrically optimizing $\mathcal{L}_{VT}$ and re-using v_i.

We feed caption y_i to the tokenizer and text encoder f_T and then the text global projection head g_T with average pooling to get the text embedding $t_i \in \mathbb{R}^d$. We optimize the following CLIP [34] loss (Fig. 2(a)):

$$\mathcal{L}_{VT}(v_i, t_i) = -\frac{1}{2}(\log \frac{\exp(sim(v_i, t_i)/\tau_2)}{\sum_{j=1}^{B} \exp(sim(v_i, t_j)/\tau_2)} + \log \frac{\exp(sim(t_i, v_i)/\tau_2)}{\sum_{j=1}^{B} \exp(sim(t_i, v_j)/\tau_2)}), \quad (2)$$

where τ_2 is the learnable CLIP temperature. We symmetrically compute $\mathcal{L}_{VT}$ for both v_i and $\tilde{v}_i$ using the same t_i. Namely, we minimize the semantic distance between two views of the same study and the corresponding report simultaneously. Even if the information in y_i is not completely matched with $\tilde{x}_i$, *e.g.*, different side and view information, they still share a large overlap in patient-level information. This encourages the model to learn the shared patient-level features via $\mathcal{L}_{VT}(\tilde{v}_i, t_i)$ while focusing on diagnosis-related information by $\mathcal{L}_{VT}(v_i, t_i)$.

3.3 Symmetric Local Alignment (SLA)

Mammography usually contains high-frequency details and the ROI is usually very small. These properties require a higher resolution for the deep learning method to work properly. It also challenges the model's ability to extract important local information and filter out less meaningful background and tissue unrelated to diagnosis. To address these challenges, we propose a symmetric local alignment (SLA) module. Specifically, the SLA module allows the model to determine the local correspondence between each sentence and image patch (Fig. 2(b)).

We start with extracting local features from input (x_i, y_i). We feed the image and caption to the vision encoder f_V and text encoder f_T respectively, followed by corresponding local projection head h_V and h_T without pooling to produce output feature sequence $v_i^{local} \in \mathbb{R}^{N_V \times d}$ and $t_i^{local} \in \mathbb{R}^{N_T \times d}$, where N_V and N_T are the length of visual tokens and text tokens, respectively. We then extract sentence-level features by selecting the embedding corresponding to the `[SEP]` token, which results in a sequence of sentence embeddings $s_i \in \mathbb{R}^{S \times d}$, where S is the number of sentences. We extract the image patch-level features by removing the `[CLS]` tokens, resulting in a sequence of patch embeddings $p_i \in \mathbb{R}^{P \times d}$, where P is the number of patches. We then compute the sentence-patch correspondence matrix $C_{i,i} \in \mathbb{R}^{S \times P}$ in the form of cosine similarity, which reveals the relevance between local patches and each sentence in the report. However, we cannot directly supervise the learning of this matrix since we have no access to the dense local correspondence. Thus, we aggregate the patch-sentence level correspondence matrix $C_{i,i}$ to an image-report level similarity score. We start by localizing the patch that has the highest correspondence for each sentence. Namely, we find the most relevant region in the image for each sentence. We call this process *Visual Localization*. We then average the similarity score for each sentence to obtain a correspondence score which describes the similarity of the most relevant patch for the whole report $c_{i,i}^V = \frac{1}{S}\sum_j \max_k C_{i,i}(j,k)$, where $C_{i,i}(j,k)$ is the similarity between the j-th sentence and the k-th patch. Similarly, we conduct *Text Localization* by finding the most similar sentence for each patch and averaging it to get a score for the similarity of the most relevant sentence for the whole image $c_{i,i}^T = \frac{1}{P}\sum_k \max_j C_{i,i}(j,k)$. We compute the aggregated visual and text local scores for all p and s in the mini-batch and optimize the InfoNCE [13] loss:

$$\mathcal{L}_{local}^V(i) = -\frac{1}{2}(\log\frac{\exp(c_{i,i}^V/\tau_{local})}{\sum_{j=1}^B \exp(c_{i,j}^V/\tau_{local})} + \log\frac{\exp(c_{i,i}^V/\tau_{local})}{\sum_{j=1}^B \exp(c_{j,i}^V/\tau_{local})}), \tag{3}$$

where τ_{local} is the local temperature. $\mathcal{L}_{local}^T$ is defined similarly. The final local loss will then be $\mathcal{L}_{local} = \frac{1}{2}(\mathcal{L}_{local}^V + \mathcal{L}_{local}^T)$. We note that optimizing $\mathcal{L}_{local}$ from the beginning of the training can lead to unstable behavior as the initial visual and language embeddings are noisy. Thus, we add this loss after δ optimization steps, after the global CLIP loss $\mathcal{L}_{VT}$ is converged.

The intuition behind this design is to mimic the process of radiologic interpretation of a medical image in the real world. On the one hand, in mammography, the clinician will look for the image regions and local features that appear most suspicious for cancer. On the other hand, the clinician will write the radiology report in a few sentences based on the findings across the whole image, while matching each description with a specific feature of the image. Our proposed SLA gives the model the ability to perceive fine-grain local image detail with sentence-level description. The derived local similarity map could also be used as a guide of the relevance between specific image details and each sentence in the provided report and therefore improve the interpretability of the model.

3.4 Overall Pre-training Target

The overall pre-training objective function of our method is given by Eq. (4):

$$\mathcal{L}(v_i, \tilde{v}_i, t_i) = \mathcal{L}_{VV}(v_i, \tilde{v}_i) + \mathcal{L}_{VT}(v_i, t_i) + \mathcal{L}_{VT}(\tilde{v}_i, t_i) + w\mathcal{L}_{local}. \tag{4}$$

We set $w = 0.0$ in the first $\delta = 8,000$ training steps and $w = 1.0$ afterward.

3.5 LLM with PEFT as Text Encoder

Lastly, we incorporate parameter-efficient fine-tuning (PEFT) of a pre-trained LLM as our text encoder rather than using the pre-trained BERT encoder [1].

Table 1. BI-RADS Prediction Results on EMBED. We report balanced accuracy (bACC) and AUC (in %) of BI-RADS prediction for each method under zero-shot, linear-probing, and full fine-tune settings. We evaluate the effect of training data size on linear probing. * denotes use of the official pre-trained weights. The best and second-best results are in bold and underlined, respectively. Our method is shaded in gray.

Methods	Zero-shot		Linear Probing						Full Fine-tune	
	100%		1%		10%		100%		100%	
	bACC	AUC	bACC	AUC	bACC	AUC	bACC	AUC	bACC	AUC
Vision only										
Random-ViT [8]	-	-	20.84	57.15	20.68	61.54	22.10	61.76	22.87	62.59
DiNOv2-ViT [33]	-	-	22.63	61.83	25.17	66.00	29.33	70.11	30.83	71.73
CLIP style										
CLIP [34]	23.05	59.81	26.66	70.35	31.65	74.98	34.35	74.11	34.25	71.61
SLIP [30]	24.14	67.47	22.94	64.43	27.86	69.48	30.93	71.95	21.75	61.96
ConVIRT [51]	25.27	65.13	24.62	65.09	30.38	73.33	31.27	74.03	34.54	74.05
MGCA [45]	26.55	63.76	23.66	64.19	30.11	72.24	30.27	72.54	34.15	73.89
MM-MIL [46]	21.78	62.41	25.85	67.16	30.94	71.99	35.11	76.12	33.05	71.26
Mammo-CLIP-B2* [11]	16.68	56.27	23.89	62.92	23.96	66.53	24.26	66.41	25.04	69.01
Mammo-CLIP-B5* [11]	16.19	55.28	23.91	64.40	24.81	69.44	24.25	69.61	25.03	70.26
MaMA	**31.04**	**74.83**	**28.46**	**70.63**	**35.12**	**75.98**	**39.75**	**77.50**	**40.31**	**77.36**

Using a pre-trained LLM with strong domain knowledge can help improve the model's understanding of the text caption and provide a more robust supervised signal for the visual-language pre-training. Moreover, PEFT (*e.g.*, LoRA [14]) can greatly reduce the cost of adapting an LLM when computing resources are limited and maintain strong performance after fine-tuning. Adapting an LLM with PEFT thus has the potential to greatly improve performance while reducing trainable parameters and GPU memory costs compared to learning the commonly adopted BERT-style encoder. To adapt the decoder-only LLM, we use the last non-padding tokens as the caption representation.

4 Experiments

4.1 Pre-training Setup

Dataset. We pre-trained our model on the Emory **EMBED** [17] dataset, which is one of the largest open mammography datasets. It contains 364,564 2D mammography images for more than 20k patients collected from 4 hospitals. The dataset provides tabular annotation about the patient, imaging meta-information, and corresponding study-level findings, *e.g.*, density and BI-RADS. We split it into train/validation/test partitions, each with 70%/10%/20% patients. All the images are resized and padded to 518 × 518 without changing the aspect ratio to balance the memory cost and batch size requirement.

Implementation Details. We initialize our image and text encoder with DiNOv2-ViT-B [33] and BioMedLM [2] respectively, and adapt LoRA [14] to the text encoder for efficient fine-tuning. The meta-information masking ratio

Table 2. Density Prediction Results on EMBED. We report balanced accuracy (bACC) and AUC (in %) of density prediction for each method under zero-shot, linear-probing, and full fine-tune settings. We evaluate the effect of training data size on linear probing. * denotes use of official pre-trained weights. The best and second-best results are in bold and underlined, respectively. Our method is shaded in gray.

Methods	Zero-shot		Linear Probing						Full Fine-tune	
	100%		1%		10%		100%		100%	
	bACC	AUC	bACC	AUC	bACC	AUC	bACC	AUC	bACC	AUC
Vision only										
Random-ViT [8]	-	-	45.81	72.83	45.11	72.62	47.01	72.92	68.47	88.61
DiNOv2-ViT [33]	-	-	66.71	89.18	70.80	90.46	71.20	90.47	59.54	88.61
CLIP style										
CLIP [34]	73.56	92.37	74.64	91.50	75.00	90.62	75.97	92.39	77.47	**93.69**
SLIP [30]	**75.45**	92.17	73.24	91.56	74.79	92.37	75.23	92.46	64.72	86.37
ConVIRT [51]	64.85	87.66	74.34	92.21	74.95	92.56	74.74	92.58	77.93	93.60
MGCA [45]	69.00	88.36	71.43	90.83	72.25	91.21	72.20	91.24	77.74	93.64
MM-MIL [46]	69.73	89.07	74.23	91.96	76.69	93.34	75.77	91.65	75.92	92.59
Mammo-CLIP-B2* [11]	51.38	79.69	66.62	87.84	66.18	87.98	66.39	87.98	73.76	91.24
Mammo-CLIP-B5* [11]	47.04	71.47	68.90	89.28	69.46	89.47	69.34	89.51	73.62	91.28
MaMA	75.40	**93.46**	**76.26**	**93.11**	**78.11**	**93.62**	**78.09**	**93.65**	**78.02**	93.65

m is 0.8. We train each model with the AdamW optimizer [29] using a learning rate of 4E-5, weight-decay of 0.1, and cosine annealing scheduler for 40k steps. A warm-up for 4k steps is applied. The SLA loss is added after $\delta = 8$k steps. We use a batch size of 144 on 4 A5000 GPUs with BFloat-16 precision. Experiments using twice larger image input sizes or batch sizes did not change performance.

4.2 Experimental Setup

Tasks and Datasets. We primarily evaluate on the **EMBED** [17] dataset for both BI-RADS assessment (7 classes) and breast density (4 classes) prediction where the distribution of labels is highly imbalanced. We further sub-sample 7,666 images for BI-RADS prediction and 7,301 images for density prediction from the test split following the dataset distribution for more realistic evaluation. We use all the images with BI-RADS 5 and 6 in the BI-RADS test set to avoid bias due to insufficient data. We use the open **RSNA-Mammo** [3] dataset with 54k images as out-of-domain evaluation for binary cancer detection. We split it into train/test sets with 85%/15% data. Given the extremely imbalanced distribution, we report balanced accuracy (bACC) and AUC as our primary metrics. We also report sensitivity and specificity of the RSNA-Mammo cancer detection.

Evaluation Settings. We evaluate all methods under zero-shot, linear probing, and full fine-tuning settings. We provide the patient and imaging meta-information in EMBED zero-shot evaluation since it is readily available before the clinician's diagnosis. For linear probing, we attach a linear classifier and fine-tune it using 1%, 10%, or 100% of the training data following prior work

Table 3. Cancer Detection Results on RSNA-Mammo [3]. We evaluate linear probing and fully fine-tuned settings for the cancer prediction task. We report balanced accuracy (bACC), AUC, sensitivity (SEN), and specificity (SPE) (in %). * denotes the use of official pre-trained weights. The best and second-best results are in bold and underlined, respectively. Our method is shaded in gray.

Methods	Zero-shot		Linear Probing				Full Fine-tune			
	bACC	AUC	bACC	AUC	SEN	SPE	bACC	AUC	SEN	SPE
Vision only										
Random-ViT [8]	-	-	51.90	56.34	72.60	31.21	56.71	57.62	**77.88**	35.53
DiNOv2-ViT [33]	-	-	63.23	68.59	59.62	66.84	55.12	58.18	70.19	40.06
CLIP style										
CLIP [34]	55.74	57.85	63.89	70.28	58.17	69.61	56.86	61.20	69.23	44.49
SLIP [30]	54.96	57.19	62.48	67.51	**78.37**	46.60	56.74	60.05	63.94	49.53
ConVIRT [51]	53.04	55.04	65.89	70.70	66.83	64.96	54.53	69.85	11.06	**98.01**
MGCA [45]	55.11	55.89	60.79	67.45	71.15	50.43	55.99	68.67	14.90	97.07
MM-MIL [46]	53.42	53.55	64.02	70.67	58.17	**69.86**	59.97	65.04	57.21	62.73
Mammo-CLIP-B2* [11]	57.35	58.11	61.50	65.76	66.35	56.57	63.98	69.46	69.23	59.00
Mammo-CLIP-B5* [11]	54.97	59.90	62.47	67.19	62.50	61.67	64.57	70.94	66.35	63.07
MaMA	**60.84**	**63.55**	**67.50**	**73.99**	72.60	62.40	**65.20**	**73.01**	67.31	63.10

[15,44,45,47,48,51]. This data efficiency study with linear probing focuses on the quality of the pre-trained embedding and helps demonstrate the difference between each VLP method. For full fine-tuning, we again attach a linear classifier and fine-tune the whole vision model using 100% of the training data. Our learning rate is set to 5E-4 and weight decay to 1E-3 using the SGD optimizer with cosine annealing scheduler for 8k steps with batch size 36. A warm-up of 100 steps is applied.

Baselines. As one of the first contrastive language-image pre-training methods for mammography, we compare two different styles of baselines: 1) *Vision only*: ViT [8] models with random initialization and DiNOv2 [33] pre-training. 2) *CLIP style*: **CLIP** [35]; **SLIP** [30]; **MM-MIL** [46], which learns local image-language relationships via a multiple instance learning paradigm; **ConVIRT** [51], one of the first chest X-ray specific CLIP models; **MGCA** [45], which applies multi-granularity feature alignment; and **Mammo-CLIP** [11]. We pre-train and fine-tune all the CLIP style methods except Mammo-CLIP on our dataset with the same settings and same pre-trained DiNOv2 ViT. For Mammo-CLIP, we use the official pre-trained weights with EfficientNet [43] backbone and the official fine-tuning settings. We choose not to compare medical VLP methods that adapt domain-specific design and require annotations not present in our dataset [44, 47,48] or performed worse than our baselines in other studies [15,54]. We also do not compare related work that has no released implementation [5,26].

4.3 Results

BI-RADS Prediction. We report the performance on the EMBED BI-RADS prediction in Table 1. We note that MaMA achieves the best overall performance

Table 4. Model Ablation. Ablation of different designs on the EMBED BI-RADS prediction with balanced accuracy (bACC) and AUC (in %). The best and second-best results are highlighted in bold and underlined. Our full method is shaded in gray.

Methods				Zero-shot		Linear Probing		Full Fine-tune	
SLA	Symm. $\mathcal{L}_{VT}$	$\mathcal{L}_{VV}$	PEFT-LLM	bACC	AUC	bACC	AUC	bACC	AUC
	✓	✓	✓	29.28	71.16	38.71	<u>77.50</u>	30.55	70.69
✓		✓	✓	<u>31.03</u>	<u>72.79</u>	<u>39.57</u>	77.39	<u>39.47</u>	<u>76.23</u>
✓	✓		✓	27.32	70.18	37.21	**77.95**	23.78	63.97
✓	✓	✓		23.88	62.84	38.96	77.43	22.29	63.77
✓	✓	✓	✓	**31.04**	**74.83**	**39.75**	<u>77.50</u>	**40.31**	**77.36**

Table 5. Multi-view ablation. Different multi-view contrastive strategies.

Methods	EMBED BI-RADS					
	Zero-shot		Linear Probing		Full Fine-tune	
	bACC	AUC	bACC	AUC	bACC	AUC
Same Image	30.48	73.95	39.70	**77.73**	39.35	76.44
Intra-side	<u>30.71</u>	<u>74.21</u>	**39.93**	77.41	35.17	76.09
Intra-study w/o self	29.33	73.21	38.20	77.49	<u>39.80</u>	<u>76.65</u>
Intra-study	**31.04**	**74.83**	<u>39.75</u>	<u>77.50</u>	**40.31**	**77.36**

Table 6. Caption ablation. Different text caption construction strategies.

Methods	EMBED BI-RADS					
	Zero-shot		Linear Probing		Full Fine-tune	
	bACC	AUC	bACC	AUC	bACC	AUC
CLIP-style	**35.99**	**77.66**	37.74	77.25	24.00	65.35
Tabular-style	22.53	58.74	<u>38.85</u>	<u>77.40</u>	<u>38.43</u>	<u>76.47</u>
No Meta Mask	27.19	68.20	36.94	76.33	24.06	64.85
Struct. Cap.	<u>31.04</u>	<u>74.83</u>	**39.75**	**77.50**	**40.31**	**77.36**

in all three evaluation settings. Our method outperforms the SoTA baselines by more than 7% AUC in the zero-shot classification. Meanwhile, MaMA also shows the best data efficiency in the linear probing setting. It shows a non-trivial improvement of more than 4% of balanced accuracy when using full training data. Even when using only 1% of training data, *e.g.*, less than 10 images for BI-RADS categories 5 and 6, it beats the best baselines by 2% bACC. A similar trend can be found in the full fine-tuning setting, where the gap is more than 6% bACC. Mammo-CLIP [11] performs poorly in all 3 settings here, especially in the zero-shot setting. This is likely because the official pre-trained models were pre-trained with much less in-house screening-only mammography data (only have BI-RADS category 0–2), even if using a higher image resolution.

Density Prediction. We present the results on EMBED breast density prediction in Table 2. Our MaMA model surpassed all the baselines again on most of the metrics. Because the breast density distribution is relatively more balanced, it is reasonable to see a reduced gap between the baselines and our method. Yet, our method still performs either the best or the second best under each evaluation. This demonstrates our model's capability of generalizing on different tasks.

Out-of-Domain Data Analysis. We report performance on the out-of-domain RSNA-Mammo dataset in Table 3. Since RSNA-Mammo [3] is an extremely imbalanced dataset (48:1 negative to positive), full fine-tuned models may suffer from overfitting and thus perform worse than linear probing. We note our model performs best in terms of balanced accuracy and AUC with a notable gap under

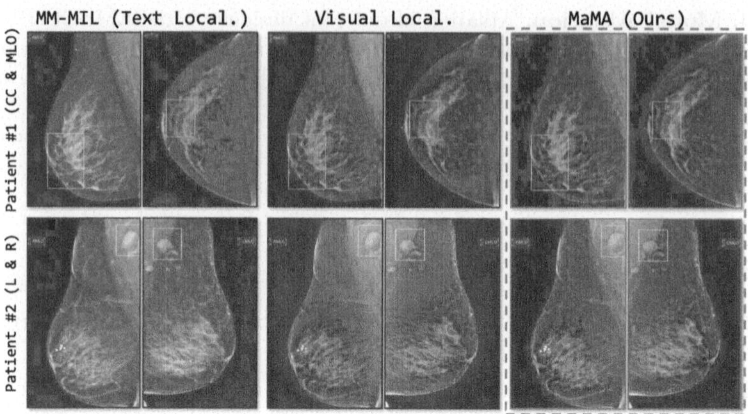

Fig. 3. Local Similarity Maps Overlaid on Mammograms. We visualize the learned local similarity map for the "Impressions" sentence on test mammograms from EMBED [17] for MM-MIL [46], our method with only visual localization, and our full method. Heat maps are normalized to [0,1]. The first row shows mammograms from the same side but different view; the second row shows mammograms from the same view but different side. The white boxes denote the dataset-provided annotated ROIs [17].

all 3 settings. While some baselines outperform our model in either sensitivity or specificity, we note these models are not informative, *i.e.*, they tend to collapse and predict the majority of images to one of the classes. This will lead to a high score in one of the sensitivity or specificity metrics and low performance in the other. In contrast, our approach shows reasonable results for both metrics with both sensitivity and specificity greater than 60%. Mammo-CLIP [11] shows a more reasonable performance in this task since it only contains screening mammography and is closer to the pre-training domain of this baseline.

4.4 Ablation Experiments

Model Design. We ablate the influence of each component in Table 4. We note each component has an important contribution to the overall model performance, as removing any one resulted in inferior performance. We note that the baseline without PEFT-LLM instead employs BioClinicalBERT [1] and shows a clear drop in zero-shot performance, which validates the importance of using a PEFT-LLM. However, this model still performs well on the linear probing and full fine-tuning tasks, which demonstrates the effectiveness of our other design choices.

Multi-view Ablation. We ablate the multi-view sampling strategy here by using: 1) the same image, 2) an intra-side image, 3) an intra-study image except for the self-augmented view, and 4) all intra-study images and augmented view (Table 5). We note that training with only one image loses the multi-view understanding. Using only intra-side images only considers ipsilateral correspondence and performs worse. Ignoring self-augmented views also degrades performance.

Caption Ablation. We evaluate different caption construction strategies in Table 6. We note that a CLIP style caption that only focuses on class labels shows a better zero-shot performance, but degenerates greatly under linear probing and full fine-tuning. Meanwhile, tabular captions with only key information show the worst zeros-shot performance. Captions without meta-information masking will fail with zero-shot settings since it learned to focus mainly on clinical-irrelevant meta-information during pre-training. Our full design, using a structural caption with meta-information masking augmentation, shows the best performance.

Local Attention Visualization. We visualize the learned local patch-sentence similarity map in Fig. 3. We visualize the similarity map for the "Impression" sentence, which includes the most important diagnosis information. We note that our methods generally have a better localization quality. The model can accurately locate the high-density and tumor-related regions. The MM-MIL [46] with only text localization failed to detect the ROI. Optimizing only visual localization resulted in a vague and inaccurate correspondence map. Our method also shows a multi-view correspondence here compared with the other baselines, especially in the same side image for patient 1.

5 Discussion and Conclusion

In this work, we presented a novel multi-view and multi-scale alignment contrastive language-image pre-training method for mammography. We proposed utilizing the multi-view nature of mammography and providing local image-sentence correspondence to help address the challenges of small ROIs and high image resolution and provide fine-grained visual clues for decisions. The proposed method greatly outperforms multiple existing baselines. Still, we have yet to evaluate other downstream tasks like object detection and segmentation. We will further aim to improve the performance for better clinical usability.

Acknowledgement. This work was supported by NIH grant R21EB032950.

References

1. Alsentzer, E., et al.: Publicly available clinical BERT embeddings. In: Proceedings of the 2nd Clinical Natural Language Processing Workshop, Minneapolis, Minnesota, USA, pp. 72–78. Association for Computational Linguistics (2019). https://doi.org/10.18653/v1/W19-1909, https://www.aclweb.org/anthology/W19-1909
2. Bolton, E., et al.: BiomedLM: a 2.7 b parameter language model trained on biomedical text. arXiv preprint arXiv:2403.18421 (2024)
3. Carr, C., et al.: RSNA screening mammography breast cancer detection (2022). https://kaggle.com/competitions/rsna-breast-cancer-detection
4. Chen, T., Kornblith, S., Norouzi, M., Hinton, G.: A simple framework for contrastive learning of visual representations. In: International Conference on Machine Learning, pp. 1597–1607. PMLR (2020)

5. Chen, X., et al.: MAMMO-clip: leveraging contrastive language-image pre-training (clip) for enhanced breast cancer diagnosis with multi-view mammography. arXiv preprint arXiv:2404.15946 (2024)
6. Delitzas, A., et al.: Multi-clip: contrastive vision-language pre-training for question answering tasks in 3d scenes. arXiv preprint arXiv:2306.02329 (2023)
7. Donnelly, J., et al.: Asymmirai: interpretable mammography-based deep learning model for 1-5-year breast cancer risk prediction. Radiology **310**(3), e232780 (2024)
8. Dosovitskiy, A., et al.: An image is worth 16x16 words: transformers for image recognition at scale. arXiv preprint arXiv:2010.11929 (2020)
9. Du, Y., Hooley, R.J., Lewin, J., Dvornek, N.C.: Sift-DBT: self-supervised initialization and fine-tuning for imbalanced digital breast tomosynthesis image classification. arXiv preprint arXiv:2403.13148 (2024)
10. Eslami, S., Meinel, C., De Melo, G.: Pubmedclip: how much does clip benefit visual question answering in the medical domain? In: Findings of the Association for Computational Linguistics: EACL 2023, pp. 1181–1193 (2023)
11. Ghosh, S., Poynton, C.B., Visweswaran, S., Batmanghelich, K.: Mammo-clip: a vision language foundation model to enhance data efficiency and robustness in mammography. arXiv preprint arXiv:2405.12255 (2024)
12. Grill, J.B., Strub, F., Altché, F., Tallec, C., Richemond, P., Buchatskaya, E., Doersch, C., Avila Pires, B., Guo, Z., Gheshlaghi Azar, M., et al.: Bootstrap your own latent-a new approach to self-supervised learning. Adv. Neural. Inf. Process. Syst. **33**, 21271–21284 (2020)
13. He, K., Fan, H., Wu, Y., Xie, S., Girshick, R.: Momentum contrast for unsupervised visual representation learning. arXiv preprint arXiv:1911.05722 (2019)
14. Hu, E.J., et al.: Lora: low-rank adaptation of large language models. arXiv preprint arXiv:2106.09685 (2021)
15. Huang, S.C., Shen, L., Lungren, M.P., Yeung, S.: Gloria: a multimodal global-local representation learning framework for label-efficient medical image recognition. In: Proceedings of the IEEE/CVF International Conference on Computer Vision, pp. 3942–3951 (2021)
16. Irvin, J., et al.: Chexpert: a large chest radiograph dataset with uncertainty labels and expert comparison. In: Proceedings of the AAAI Conference on Artificial Intelligence, vol. 33, pp. 590–597 (2019)
17. Jeong, J.J., et al.: The emory breast imaging dataset (embed): a racially diverse, granular dataset of 3.4 million screening and diagnostic mammographic images. Radiol. Artif. Intell. **5**(1), e220047 (2023)
18. Ji, Z., Shaikh, M.A., Moukheiber, D., Srihari, S.N., Peng, Y., Gao, M.: Improving joint learning of chest x-ray and radiology report by word region alignment. In: Machine Learning in Medical Imaging: 12th International Workshop, MLMI 2021, Held in Conjunction with MICCAI 2021, Strasbourg, France, September 27, 2021, Proceedings 12. pp. 110–119. Springer (2021)
19. Johnson, A.E., et al.: Mimic-cxr-jpg, a large publicly available database of labeled chest radiographs. arXiv preprint arXiv:1901.07042 (2019)
20. Lai, Z., Li, Z., Oliveira, L.C., Chauhan, J., Dugger, B.N., Chuah, C.N.: Clipath: fFine-tune clip with visual feature fusion for pathology image analysis towards minimizing data collection efforts. In: Proceedings of the IEEE/CVF International Conference on Computer Vision, pp. 2374–2380 (2023)
21. Li, J., Li, D., Savarese, S., Hoi, S.: Blip-2: bootstrapping language-image pre-training with frozen image encoders and large language models. In: International Conference on Machine Learning, pp. 19730–19742. PMLR (2023)

22. Li, J., Li, D., Xiong, C., Hoi, S.: Blip: bootstrapping language-image pre-training for unified vision-language understanding and generation. In: International Conference on Machine Learning, pp. 12888–12900. PMLR (2022)
23. Li, Y., et al.: Supervision exists everywhere: a data efficient contrastive language-image pre-training paradigm. arXiv preprint arXiv:2110.05208 (2021)
24. Li, Z., et al.: Domain generalization for mammography detection via multi-style and multi-view contrastive learning. In: Medical Image Computing and Computer Assisted Intervention–MICCAI 2021: 24th International Conference, Strasbourg, France, September 27–October 1, 2021, Proceedings, Part VII 24, pp. 98–108. Springer (2021)
25. Liu, H., Li, C., Wu, Q., Lee, Y.J.: Visual instruction tuning. In: Advances in Neural Information Processing Systems, vol. 36 (2024)
26. Liu, J., et al.: MLIP: medical language-image pre-training with masked local representation learning. arXiv preprint arXiv:2401.01591 (2024)
27. Liu, S., et al.: Grounding DINO: marrying DINO with grounded pre-training for open-set object detection. arXiv preprint arXiv:2303.05499 (2023)
28. Liu, Y., Zhang, F., Chen, C., Wang, S., Wang, Y., Yu, Y.: Act like a radiologist: towards reliable multi-view correspondence reasoning for mammogram mass detection. IEEE Trans. Pattern Anal. Mach. Intell. **44**(10), 5947–5961 (2021)
29. Loshchilov, I., Hutter, F.: Decoupled weight decay regularization. arXiv preprint arXiv:1711.05101 (2017)
30. Mu, N., Kirillov, A., Wagner, D., Xie, S.: Slip: self-supervision meets language-image pre-training. In: European Conference on Computer Vision, pp. 529–544. Springer (2022)
31. Nguyen, H., Pham, H., Le, L., Dao, M., Lam, K.V.C.: An open dataset of chest x-rays with radiologist annotations. PhysioNet https://doi.org/10.13026/3akn-b287 (2021)
32. Onken, M., Eichelberg, M., Riesmeier, J., Jensch, P.: Digital imaging and communications in medicine. In: Biomedical Image Processing, pp. 427–454. Springer (2010)
33. Oquab, M., et al.: Dinov2: learning robust visual features without supervision (2023)
34. Radford, A., et al.: Learning transferable visual models from natural language supervision. In: International Conference on Machine Learning, pp. 8748–8763. PMLR (2021)
35. Radford, A., et al.: Language models are unsupervised multitask learners. OpenAI blog **1**(8), 9 (2019)
36. Shah, K., Shah, A., Lau, C.P., de Melo, C.M., Chellappa, R.: Multi-view action recognition using contrastive learning. In: Proceedings of the IEEE/CVF Winter Conference on Applications of Computer Vision, pp. 3381–3391 (2023)
37. Sickles, E.A., D'Orsi, C.J., Bassett, L.W., et al.: ACR BI-RADS mammography. In: ACR BI-RADS Atlas, Breast Imaging Reporting and Data System. American College of Radiology, Reston, VA, 5th edn. (2013)
38. Song, D., Fu, X., Nie, W., Li, W., Liu, A.: MV-clip: multi-view clip for zero-shot 3d shape recognition. arXiv preprint arXiv:2311.18402 (2023)
39. Sun, L., et al.: Breast mass classification based on supervised contrastive learning and multi-view consistency penalty on mammography. IET Biometrics **11**(6), 588–600 (2022)
40. Sun, Q., Fang, Y., Wu, L., Wang, X., Cao, Y.: Eva-clip: Improved training techniques for clip at scale. arXiv preprint arXiv:2303.15389 (2023)

41. Sun, Q., Wang, J., Yu, Q., Cui, Y., Zhang, F., Zhang, X., Wang, X.: Eva-clip-18b: scaling clip to 18 billion parameters. arXiv preprint arXiv:2402.04252 (2024)
42. Sung, H., et al.: Global cancer statistics 2020: globocan estimates of incidence and mortality worldwide for 36 cancers in 185 countries. CA: Can. J. Clin. **71**(3), 209–249 (2021)
43. Tan, M., Le, Q.: Efficientnet: Rethinking model scaling for convolutional neural networks. In: International Conference on Machine Learning, pp. 6105–6114. PMLR (2019)
44. Wan, Z., et al.: Med-unic: unifying cross-lingual medical vision-language pre-training by diminishing bias. In: Advances in Neural Information Processing Systems, vol. 36 (2024)
45. Wang, F., Zhou, Y., Wang, S., Vardhanabhuti, V., Yu, L.: Multi-granularity cross-modal alignment for generalized medical visual representation learning. Adv. Neural. Inf. Process. Syst. **35**, 33536–33549 (2022)
46. Wang, P., Wells, W.M., Berkowitz, S., Horng, S., Golland, P.: Using multiple instance learning to build multimodal representations. In: International Conference on Information Processing in Medical Imaging, pp. 457–470. Springer (2023)
47. Wang, Z., Wu, Z., Agarwal, D., Sun, J.: Medclip: contrastive learning from unpaired medical images and text. arXiv preprint arXiv:2210.10163 (2022)
48. Wu, C., Zhang, X., Zhang, Y., Wang, Y., Xie, W.: Medklip: medical knowledge enhanced language-image pre-training. medRxiv, pp. 2023–01 (2023)
49. Zhang, S., et al.: Biomedclip: a multimodal biomedical foundation model pretrained from fifteen million scientific image-text pairs. arXiv preprint arXiv:2303.00915 (2023)
50. Zhang, Y., Ma, Z., Gao, X., Shakiah, S., Gao, Q., Chai, J.: Groundhog: grounding large language models to holistic segmentation. arXiv preprint arXiv:2402.16846 (2024)
51. Zhang, Y., Jiang, H., Miura, Y., Manning, C.D., Langlotz, C.P.: Contrastive learning of medical visual representations from paired images and text. In: Machine Learning for Healthcare Conference, pp. 2–25. PMLR (2022)
52. Zhao, Z., et al.: Clip in medical imaging: a comprehensive survey. arXiv preprint arXiv:2312.07353 (2023)
53. Zheng, K., et al.: Dreamlip: language-image pre-training with long captions. arXiv preprint arXiv:2403.17007 (2024)
54. Zhou, H.Y., Lian, C., Wang, L., Yu, Y.: Advancing radiograph representation learning with masked record modeling. arXiv preprint arXiv:2301.13155 (2023)

Interpretable Few-Shot Retinal Disease Diagnosis with Concept-Guided Prompting of Vision-Language Models

Deval Mehta[1,2(✉)], Yiwen Jiang[1,2], Catherine Jan[3,4], Mingguang He[4,5,6,7], Kshitij Jadhav[8], and Zongyuan Ge[1,2,9]

[1] AIM for Health Lab, Faculty of IT, Monash University, Melbourne, Australia
deval.mehta@monash.edu

[2] Faculty of Engineering, Monash University, Melbourne, Australia

[3] Centre for Eye Research Australia, Royal Victorian Eye and Ear Hospital, East Melbourne, VIC, Australia

[4] Ophthalmology, Department of Surgery, The University of Melbourne, Melbourne, VIC, Australia

[5] Research Centre for SHARP Vision (RCSV), The Hong Kong Polytechnic University, Kowloon, Hong Kong

[6] Centre for Eye and Vision Research (CEVR), 17W Hong Kong Science Park, New Territories, Hong Kong

[7] Department of Optometry and Vision Sciences, The University of Melbourne, Melbourne, Australia

[8] Koita Centre of Digital Health, Indian Institute of Technology Bombay, Mumbai, India

[9] Airdoc-Monash Research Lab, Monash University, Melbourne, Australia

Abstract. Recent advancements in deep learning have shown significant potential for classifying retinal diseases using color fundus images. However, existing works predominantly rely exclusively on image data, lack interpretability in their diagnostic decisions, and treat medical professionals primarily as annotators for ground truth labeling. To fill this gap, we implement two key strategies: extracting interpretable concepts of retinal diseases using the knowledge base of GPT models and incorporating these concepts as a language component in prompt-learning to train vision-language (VL) models with both fundus images and their associated concepts. Our method not only improves retinal disease classification but also enriches few-shot and zero-shot detection (novel disease detection), while offering the added benefit of concept-based model interpretability. Our extensive evaluation across two diverse retinal fundus image datasets illustrates substantial performance gains in VL-model based few-shot methodologies through our concept integration approach, demonstrating an average improvement of approximately **5.8%** and **2.7%** mean average precision (mAP) for 16-shot learning and zero-shot (novel class) detection respectively. Our method marks a pivotal step towards interpretable and efficient retinal disease recognition for real-world clinical applications.

D. Mehta and Y. Jiang—Authors contributed equally.

I. Oguz et al. (Eds.): IPMI 2025, LNCS 15830, pp. 263–277, 2026.
https://doi.org/10.1007/978-3-031-96625-5_18

Keywords: fundus image · prompt learning · concepts · few-shot

1 Introduction

Globally, retinal disorders are significant contributors to ocular morbidity and visual impairment [35]. Retinal diseases like age-related macular degeneration (AMD) and diabetic retinopathy (DR) are the leading cause of irreversible blindness in older populations and working-age individuals in developed countries [7,16]. Notably, the initial stages of AMD and DR often present with only subtle symptoms but can progress to severe, irreversible optic nerve damage if not addressed promptly [18]. Thus, the importance of early detection and accurate diagnosis of retinal diseases is crucial in averting permanent alterations to vision.

Recent advancements in diagnostic technologies, including Fundus photography (for retina, optic disc and macula), Optical Coherence Tomography (OCT), and Fundus Fluorescein Angiography (FFA), have significantly increased the accessibility of routine eye disease screenings within healthcare and ophthalmology practices, thereby enhancing the potential for early intervention [26]. Of these techniques, color fundus imaging (illumination of retina by white light) remains the most practical and straightforward method for ocular examination due to its efficiency and simplicity. Especially, the advent of compact, portable cameras has bolstered the utility of color fundus imaging in teleophthalmology applications, particularly in remote areas [2]. Nonetheless, the manual analysis of fundus images demands considerable expertise and time from ophthalmologists, a challenge that has been exacerbated during the recent public health crisis [24].

Several studies have demonstrated the efficacy of utilizing deep learning (DL) techniques for automated analysis of color fundus images, successfully identifying prevalent retinal diseases such as DR, glaucoma, and AMD [1,8,20]. However, detecting rare eye diseases is a challenge due to data scarcity. Recent DL techniques have extended to the diagnosis of less common disorders, including Coat's disease and retinoschisis [9,13]. There have also been few works which concentrate on developing few-shot capability (i.e., using only a few examples of a disease category during training) of DL models to detect both common and rare retinal diseases [5,23,28]. Developing few-shot capability is vital for diagnosing rare diseases where sample scarcity hinders effective training of DL models.

Authors in [28] developed an unsupervised probabilistic model to detect rare eye diseases based on the latent feature embeddings of a trained Convolutional Neural Network (CNN) on common eye diseases, whereas [23] adapted prototype networks for episodic learning to learn an enhanced latent space representation for few-shot detection. Beyond enhancing feature space representation, certain studies have focused on leveraging knowledge distillation across various eye diseases and implementing distinct sampling approaches to detect rare diseases and cultivate few-shot learning capabilities. For example, the relational subset knowledge distillation [13] divided the long-tailed data into multiple subsets based on the region and phenotype features of eye diseases and employed knowledge distillation across these subsets. Similarly, the hierarchical knowledge guided learning [14] demonstrated that prior relational knowledge between different ophthalmic diseases represented in a hierarchical manner along with instance-wise class balanced sampling could enhance rare disease detection.

However, despite advancements in few-shot learning using color fundus images, existing approaches face two limitations - 1) They rely solely on color fundus imagery, employing strategies like knowledge transfer, subsampling, or enhancing latent space disease feature representation. 2) They lack interpretability, relying primarily on GradCAM-based attention map explanations [30] and employ domain knowledge only limited to ground truth disease labels. These approaches overlook the rich domain knowledge that could inform model development, especially the explicit consideration of disease-specific concepts and descriptive attributes, commonly used by experts to reach a diagnosis.

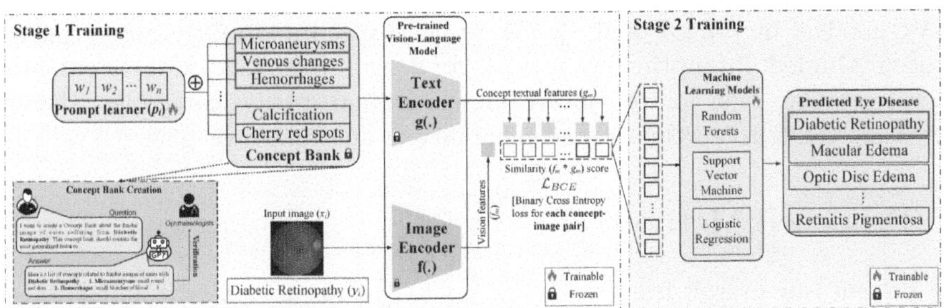

Fig. 1. Our framework consists of two stages: **Stage 1** focuses on developing a concept bank through GPT and validation by ophthalmologists, followed by concept-guided prompt learning of VL models for concept prediction. **Stage 2** trains machine learning models to classify disease categories based on Stage 1's concepts.

Our Contributions: In this work, we tackle the above two challenges in two steps: first, by constructing a concept bank that encapsulates domain knowledge specific to retinal eye diseases from color fundus images; and second, by incorporating this knowledge into the training of vision-language (VL) models, such as CLIP [29]. Constructing concept banks manually is a labor-intensive and resource-demanding process that requires highly specialized expertise. To address this critical limitation, our approach leverages the extensive knowledge base of large language models, such as GPT-3 [4], to generate a concept bank for retinal eye diseases. This automated process significantly reduces the dependency on resource-intensive manual efforts while capturing diverse detailed medical knowledge. To ensure reliability and eliminate inaccuracies, our generated concept bank is validated by expert ophthalmologists, transforming their role from exhaustive creation to efficient verification. This streamlined process not only enhances accuracy but also improves scalability, making it an efficient approach for integrating accurate domain knowledge into medical AI systems.

Subsequently, we incorporate this concept bank into the model training process by leveraging a prompt-tuning paradigm to integrate domain knowledge with VL models. Prompt learning, a recent trend in Natural Language Processing (NLP), has emerged as an efficient approach for adapting VL models to downstream tasks. Recently, prompt learning techniques such as Context Optimization (CoOp) [41], Conditional Context Optimization (CoCoOp) [42], and Ad-CLIP [34] have demonstrated their effectiveness in classification tasks, particularly in few-shot learning scenarios. Despite their efficiency these methods lack interpretability. Building on these advancements, we integrate prompt learning with our concept bank to develop an interpretable concept-guided prompt-tuning paradigm, enabling effective prediction of disease-specific concepts and categories.

We evaluate our framework on two color fundus imaging datasets - an in-house dataset comprising of 29 retinal diseases and the public RFMiD [25] dataset. Our evaluation results demonstrate significant mean average precision (mAP) improvements in both few-shot and zero-shot detection capability of VL models by **5.8%** and **2.7%** respectively. Importantly, our proposed Concept-Guided Prompting framework for Vision-Language Models enhances interpretability, even for previously unseen retinal disease categories, thereby increasing the trustworthiness and reliability of automated diagnostic systems.

2 Proposed Method

We present our overall framework in Fig. 1, consisting of two training stages. The first stage involves predicting concepts within the input image, while the second stage focuses on determining the disease category from these predicted concepts.

2.1 Stage 1: Concept-Guided Prompting of Vision-Language Models

We first describe the process of building the concept bank, which is necessary for concept-guided training.

Accurate Construction of Retinal Diseases Concept Bank: Large Language Models (LLMs) like GPT-4, trained on substantial medical knowledge, have achieved scores comparable to passing the United States Medical Licensing Examination [10] and have been shown to produce concept banks with greater factuality and groundability than human-written content in specific domains [39]. We leverage the embedded knowledge of LLMs to simplify the creation of our retinal disease concept bank, reducing dependence on specialists, as shown in Fig. 1.

Figure 2 illustrates our two adaptable prompting templates for extracting key concepts related to retinal diseases. Template 1 explicitly requests concepts for a specified condition, while Template 2 generates diagnostic concepts by comparing the condition to a normal retinal fundus image. To ensure consistency, each prompt is used twice, and only overlapping or common concepts across all

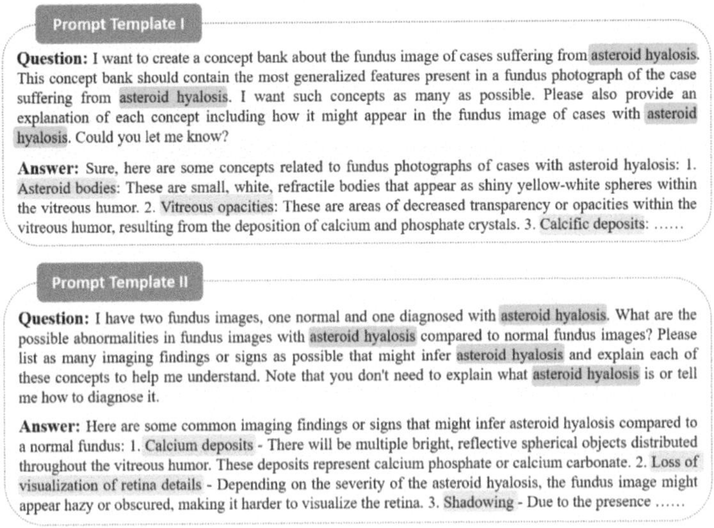

Prompt Template I

Question: I want to create a concept bank about the fundus image of cases suffering from asteroid hyalosis. This concept bank should contain the most generalized features present in a fundus photograph of the case suffering from asteroid hyalosis. I want such concepts as many as possible. Please also provide an explanation of each concept including how it might appear in the fundus image of cases with asteroid hyalosis. Could you let me know?

Answer: Sure, here are some concepts related to fundus photographs of cases with asteroid hyalosis: 1. Asteroid bodies: These are small, white, refractile bodies that appear as shiny yellow-white spheres within the vitreous humor. 2. Vitreous opacities: These are areas of decreased transparency or opacities within the vitreous humor, resulting from the deposition of calcium and phosphate crystals. 3. Calcific deposits: ......

Prompt Template II

Question: I have two fundus images, one normal and one diagnosed with asteroid hyalosis. What are the possible abnormalities in fundus images with asteroid hyalosis compared to normal fundus images? Please list as many imaging findings or signs as possible that might infer asteroid hyalosis and explain each of these concepts to help me understand. Note that you don't need to explain what asteroid hyalosis is or tell me how to diagnose it.

Answer: Here are some common imaging findings or signs that might infer asteroid hyalosis compared to a normal fundus: 1. Calcium deposits - There will be multiple bright, reflective spherical objects distributed throughout the vitreous humor. These deposits represent calcium phosphate or calcium carbonate. 2. Loss of visualization of retina details - Depending on the severity of the asteroid hyalosis, the fundus image might appear hazy or obscured, making it harder to visualize the retina. 3. Shadowing - Due to the presence ......

Fig. 2. Design of two prompt templates for constructing the retinal disease concept bank. The figure illustrates the prompts for the retinal disease of Asteroid Hyalosis with adaptable disease-specific parts. These two templates prompt LLMs twice, generating concepts that are later validated by ophthalmologists.

prompts are retained. This process is applied across all retinal conditions in our dataset to construct a comprehensive concept bank.

To mitigate risks of misinformation [15] and ensure accuracy, the generated concepts are validated by ophthalmologists. For example, as shown in Fig. 2, concepts for Asteroid Hyalosis generated by Template 1 include asteroid bodies, vitreous opacities, and calcific deposits, while Template 2 identifies calcium deposits, loss of retinal detail visualization, and shadowing. Overlapping concepts such as calcium deposits, vitreous opacities, and asteroid bodies were selected as representative and confirmed by ophthalmologists as attributes arising from calcium and phosphate crystal deposits commonly associated with Asteroid Hyalosis [3,36].

Our in-house dataset includes 77 concepts spanning 29 retinal diseases, while the RFMiD dataset contains 119 concepts across 46 retinal diseases, with several categories sharing common concepts. This streamlined and validated approach ensures high accuracy, efficiency, and reliability in generating concept banks for medical AI applications.

Vision Language (VL) Models: We commence with an overview of VL models, specifically highlighting CLIP [29]. CLIP features two distinct encoders: an image encoder and a text encoder, as illustrated in Fig. 1. The image encoder, typically based on Resnet50 [12] or ViT [6], learns a low-dimensional feature representation of the input image. Conversely, the text encoder, built on the

Transformer architecture [37], processes natural language input (usually "a photo of [CATEGORY NAME]") to generate text representations. CLIP is then trained to align these two embedding spaces by enhancing the cosine similarity for matching image-text pairs and reducing it for non-matching ones. Importantly, CLIP has been trained on a vast dataset of 400 million image-text pairs, enabling it to capture a wide range of visual concepts transferable to downstream tasks. Thus, we utilize the pre-trained CLIP encoders for their broad feature representation capabilities, keeping them frozen during our model training as depicted in Fig. 1.

Integrating Concepts with VL Models: Let $X = \{x_i\}_{i=1}^{N}$ represent the set of N input images. Each image x_i belongs to a ground truth disease category $y_i \in \{1, 2, \ldots, K\}$ of the total K categories and is also associated with a subset of concepts C_i from the concept bank $C = \{c_1, c_2, \ldots, c_E\}$, containing E concepts. In Fig. 1, the input image x_i's concepts and disease category are marked in brown. Then, the few-shot learning task involves classifying x_i correctly with minimal training examples of y_i.

Drawing on recent advancements in prompt learning techniques such as CoOp [41] and CoCoOp [42], which enhance few-shot detection in vision-language models, we devise a set of learnable prompt vectors (p_t). In our approach, this prompt is uniquely terminated (see ablation study for different termination token positions) with the concept name, given by below.

$$p_t = [w_1][w_2]...[w_M][CONCEPT] \tag{1}$$

where $[CONCEPT]$ represents a single concept from the concept bank set C, and each $[w_i] \in \{\ 1, 2, ...M\}$ is a vector with the same dimension as word embeddings (i.e., 512 for CLIP) and M is a hyperparameter specifying the number of prompt tokens. Each of the 512 vector prompt tokens $[w_1, .., w_M]$ are randomly initialized and represent words in CLIP's 49,152-size vocabulary. These tokens are shared across all concepts. **The prompt tokens and ML models are the only trainable components as marked in Fig.** 1.

The prompt p_t is then given to CLIP's frozen text encoder g(·), generating textual concept features $g_m = g(p_t)$. Concurrently, the input image x_i is processed by CLIP's frozen image encoder to yield the visual features $f_m = f(x_i)$.

After extracting textual concept features and visual features, similarity scores ($f_m \odot g_m$) are calculated, as depicted in Fig. 1. Given an image contains multiple concepts, its visual features must align with the textual features of these concepts, framing this as a multi-label classification problem. Therefore, the concepts are passed one-by-one as a CONCEPT TOKEN in Eq. 1 and then disease-specific concepts are matched via BCE loss as highlighted in Fig. 1 and given by below.

$$Concept_{loss} = -\frac{1}{E}\sum_{j=1}^{E}[c_j \log(\hat{c}_j) + (1 - c_j)\log(1 - \hat{c}_j)] \tag{2}$$

where E represent the total number of concepts, c_j indicates the ground truth presence (1 for present, 0 for absent) of the jth concept in image x_i, $\hat{c}_j$ denotes the predicted probability for the jth concept, and $Concept_{loss}$ is the mean BCE loss across all concepts.

In Fig. 1, the input image classified as Diabetic Retinopathy (DR), features key concepts like Microaneurysms, Hard exudates, and Hemorrhages, highlighted in brown. Our optimization aims to predict these concepts accurately, using BCE loss to identify these three concepts (as 1) and exclude others (as 0). For images with multiple diseases, all pertinent concepts are targeted for prediction. The optimization goal is to minimize $Concept_{loss}$, with gradients backpropagated through the text encoder $g(\cdot)$, thereby training the prompt learner parameters via matching concept features between the image and text.

2.2 Stage 2: Predicting Disease Category from Concepts

Following stage 1's concept prediction training, we utilize concept logits for disease categorization using the concept-bottleneck model (CBM) [19]. The key idea behind CBM is the introduction of an intermediate concept bottleneck layer before the final classification stage, allowing the model to predict interpretable concepts alongside the final categories. This design significantly enhances the interpretability of deep learning models by providing insights into how specific concepts contribute to the predictions. For mapping concept logits to disease categories, we employ machine learning techniques such as Random Forest (RF), Logistic Regression (LR), and Support Vector Machine (SVM). Although end-to-end mapping with a Multi-layer Perceptron (MLP), as suggested by CBM [19], is an alternative, our experiments demonstrate that traditional ML models outperform MLPs for our dataset, as discussed in the ablation study later. We adhere to standard training procedures for RF, LR, and SVM, detailed below.

$$\hat{y}_i = \frac{1}{K}\sum_{k=1}^{K} f_{i,k}(c_x) \tag{3}$$

$$\hat{y}_i = \frac{1}{1+e^{-(w_i \cdot c_x + b_i)}} \tag{4}$$

$$\hat{y}_i = sign(w_i \cdot c_x + b_i) \tag{5}$$

where $\hat{y}_i$ denotes the predicted probability for the ith disease category, with c_x being the input logits for image x_i's predicted concepts from stage 1. For LR and SVM, w_i and b_i represent the weight and bias, respectively, whereas, $f_{i,k}$ is the kth decision tree's prediction in the RF model.

2.3 Training and Implementation Details

To construct the initial concept bank, we utilized the GPT-3.5-Turbo Model API. Our method was trained on the ViT-B16 CLIP pre-trained model, incorporating standard data augmentations—random flip, center crop, and horizontal translation—and resizing input images to 224×224. We started with a learning rate of 1×10^{-3}, applying a cosine learning rate scheduler, a warm-up phase of 5 epochs, and a constant schedule type. Training utilized a batch size of 32, SGD optimizer, and spanned 100 epochs across both in-house and public datasets. Following existing long-tailed works [13,14], the evaluation metric was mean average precision (mAP), with few-shot and zero-shot detection experiments averaged over five runs with different random seeds for training shots. Experiments were conducted on PyTorch with an NVIDIA A100 GPU.

3 Datasets

In-House Dataset: Our in-house dataset comprises 4,535 color fundus images (captured with Topcon TRC-NW8 camera) across 29 disease categories, annotated for ground truth retinal disease by two senior ophthalmologists. This long-tailed dataset predominantly features diseases like Macular edema, Central Retinal vein occlusion, and Diabetic Retinopathy, with fewer instances of rarer conditions like Anterior ischemic optic neuropathy, Choroiditis, Choroidal hemangioma, and Macular dystrophy. The dataset is divided into training, validation, and test sets, containing 2928, 727, and 880 images respectively.

RFMiD - The RFMiD dataset [25] contains 3,200 fundus images across 46 disease categories, annotated by two senior retinal experts. This long-tailed dataset primarily features common diseases like Diabetic Retinopathy, Media Haze, Optic Disc Cupping, but includes scarce instances of rare diseases such as Coloboma and Hemorrhagic Retinopathy. The dataset is pre-divided into training, validation, and testing sets with 1920, 640, and 640 images respectively. It's a multi-label dataset, with 749 images bearing multiple labels and 2,451 images with a single label. Our approach treats this dataset with a multi-label classification loss function.

4 Results

We evaluate our proposed framework in both few-shot and zero-shot settings, benchmarking its performance against state-of-the-art prompt learning approaches - CoOp [42], CoCoOp [41], ProDA [21], MaPLe [17] for vision-language (VL) models. Additionally, we conduct interpretability analyses to assess the framework's effectiveness and perform comprehensive ablation studies to identify the optimal hyperparameters for our approach.

4.1 Few-Shot Classification

In few-shot classification, we randomly select n images per category from the training fold and test it on all the images from the test fold. Table 1 shows that image-only baseline performs the poorest, with improvements seen in fine-tuned

CLIP, further enhanced by prompt learning, and maximized with our concept-guided approach. RFMiD dataset results are lower than our in-house dataset due to its multi-label classification complexity. Notably, our MaPLe + Concept method outperforms others in both datasets, benefiting from multi-modal prompts that bolster vision-language generalization. The average increment of all our concept-guided prompt learning methods increases is **6.45%** and **5.15%** for 16-shot detection for our In-house and RFMiD dataset respectively.

Table 1. Few-shot classification enhancement for different n shots (in terms of $mean_{std}$ mAP of 5 random seeds) with concept-guided training across prompt learning methods on in-house and public datasets (best viewed in zoom).

Method/Dataset/Shots	In-house dataset				RFMiD			
	n = 2	n = 4	n = 8	n = 16	n = 2	n = 4	n = 8	n = 16
Image-only (ViTB-16)	$29.36_{1.68}$	$32.41_{1.97}$	$35.67_{1.62}$	$38.71_{1.43}$	$27.58_{1.85}$	$30.46_{1.37}$	$33.59_{1.5}$	$35.64_{1.26}$
CLIP [29] (linear probing)	$36.75_{2.07}$	$41.84_{2.23}$	$44.73_{1.93}$	$48.38_{0.98}$	$34.27_{2.19}$	$38.42_{1.95}$	$41.75_{2.1}$	$46.81_{1.22}$
CoOp [42]	$39.43_{1.32}$	$47.71_{1.5}$	$51.52_{2.05}$	$55.64_{1.45}$	$37.89_{2.06}$	$40.63_{1.72}$	$48.93_{1.49}$	$52.41_{1.45}$
CoCoOp [41]	$39.36_{1.38}$	$47.49_{1.57}$	$51.06_{1.56}$	$53.97_{1.20}$	$37.75_{2.12}$	$40.44_{1.89}$	$47.61_{1.39}$	$51.42_{1.40}$
ProDA [21]	$39.72_{1.53}$	$47.24_{1.87}$	$51.31_{1.47}$	$55.02_{1.33}$	$37.61_{1.56}$	$41.43_{1.51}$	$48.79_{1.55}$	$53.18_{1.21}$
MaPLe [17]	$57.64_{2.28}$	$58.36_{1.02}$	$62.58_{1.45}$	$66.32_{1.27}$	$49.44_{1.76}$	$51.84_{1.89}$	$58.62_{1.51}$	$61.53_{1.65}$
CoOp + Concepts (Ours)	$40.9_{2.83}$	$51.43_{1.32}$	$55.8_{1.46}$	$61.67_{0.33}$	$39.27_{1.73}$	$44.83_{2.24}$	$52.25_{1.64}$	$57.47_{1.41}$
CoCoOp + Concepts (Ours)	$40.61_{2.28}$	$50.94_{1.5}$	$55.39_{1.68}$	$61.26_{0.59}$	$38.96_{1.37}$	$45.58_{1.30}$	$51.46_{1.21}$	$57.26_{1.32}$
ProDA + Concepts (Ours)	$42.16_{1.61}$	$52.08_{1.51}$	$56.26_{1.98}$	$63.14_{1.25}$	$41.91_{1.66}$	$45.59_{1.38}$	$54.62_{1.53}$	$59.35_{1.10}$
MaPLe + Concepts (Ours)	$\mathbf{59.47}_{1.55}$	$\mathbf{61.9}_{0.88}$	$\mathbf{65.80}_{1.8}$	$\mathbf{70.67}_{1.32}$	$\mathbf{50.57}_{1.49}$	$\mathbf{55.4}_{2.37}$	$\mathbf{60.57}_{1.96}$	$\mathbf{65.07}_{1.94}$
Average Increment	1.75	3.89	4.19	6.45	2.01	4.26	3.74	5.15

4.2 Base to Novel Generalization (Zero-Shot Detection)

Our framework's zero-shot detection capability is evaluated through base-to-novel generalization, training on common base categories and testing on unseen novel categories, which are underrepresented in the dataset. Models were trained with 16 shots from base categories, defined as the most common first half in each dataset, while novel categories were excluded from training. This resulted in 15 base and 14 novel categories for our in-house dataset, and 23 each for RFMiD.

The entire concept bank, representing knowledge of all diseases, was included during training, but no novel category samples were used. Testing focused solely on novel categories, making the task highly challenging due to the long-tailed nature of the datasets and the use of limited base samples, simulating real-world clinical conditions without fine-tuning on novel samples.

Table 2 shows that our concept-based approach enhances prompt learning, boosting mAP by **2.69%** and **2.71%** on the in-house and RFMiD datasets.

Table 2. Zero-shot detection enhancement via concept-guided training in prompt learning methods on in-house and public datasets. Performance presented as $mean_{std}$.

Method/Dataset	In-house dataset		RFMiD	
	weighted f1	mAP	weighted f1	mAP
Image-only (ViTB-16)	$15.63_{1.23}$	$18.47_{0.66}$	$14.82_{0.46}$	$16.79_{0.51}$
CLIP [29] (linear probing)	$19.46_{0.65}$	$21.52_{1.07}$	$16.48_{1.29}$	$18.33_{1.38}$
CoOp [42]	$21.19_{0.65}$	$23.85_{0.98}$	$18.96_{1.18}$	$19.42_{1.04}$
CoCoOp [41]	$25.82_{0.54}$	$26.95_{1.19}$	$19.71_{0.67}$	$21.36_{0.54}$
ProDA [21]	$25.16_{1.27}$	$25.73_{0.55}$	$19.64_{0.77}$	$20.92_{0.70}$
MaPLe [17]	$26.71_{1.19}$	$28.38_{0.62}$	$20.49_{1.59}$	$21.61_{1.66}$
CoOp + Concepts (Ours)	$24.31_{0.96}$	$25.29_{0.98}$	$20.74_{1.5}$	$22.63_{1.18}$
CoCoOp + Concepts (Ours)	$27.93_{0.52}$	$29.41_{1.09}$	$21.87_{0.84}$	$23.5_{0.99}$
ProDA + Concepts (Ours)	$27.29_{1.19}$	$28.17_{1.26}$	$21.35_{0.8}$	$23.86_{0.53}$
MaPLe + Concepts (Ours)	$\mathbf{30.61}_{1.06}$	$\mathbf{32.8}_{0.8}$	$\mathbf{22.39}_{0.77}$	$\mathbf{24.17}_{0.66}$
Average Increment	2.81	2.69	1.89	2.71

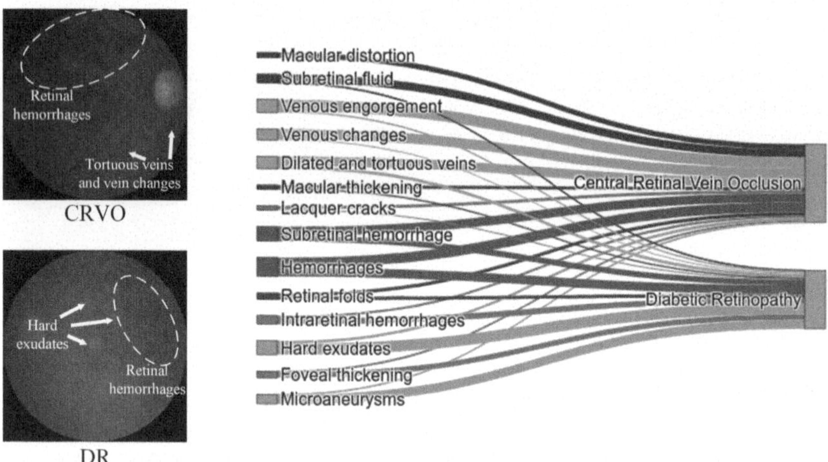

Fig. 3. Fundus images illustrating key representative attributes of Diabetic Retinopathy (DR) and Central Retinal Vein Occlusion (CRVO) and a sankey diagram depicting the flow of the concepts relevant to DR and CRVO learnt by our framework. We present the top 5 and the bottom 5 concepts associated with each DR and CRVO based on the contribution scores of the concepts to DR and CRVO from our LR model. The scores are averaged and normalized over all the test samples of DR and CRVO.

4.3 Interpretability: From Concepts to Diseases

Our stage 2 training using Logistic Regression provides clarity on concept's contribution to the predicted disease, thus enhancing the interpretability of diagnosis. Concept representation enhances interpretability of predicted categories demonstrated in two scenarios: distinguishing similar categories in a few-shot setting (Fig. 3) and detecting novel categories in a zero-shot setting (Fig. 4).

Few-Shot Distinction: We differentiate Diabetic Retinopathy (DR) and Central Retinal Vein Occlusion (CRVO), two retinal vascular diseases [27] which exhibit overlapping features like hemorrhages and subretinal hemorrhage [31]. DR-specific features include hard exudates and microaneurysms [40], while CRVO is characterized by venous changes and engorgements [11]. Figure 3 highlights these distinctions, where concept weights, averaged and normalized across test samples of DR and CRVO, are represented by the width of the flow and boxes. Shared concepts, such as hemorrhages, exhibit strong associations with both diseases, while distinguishing concepts (e.g., venous changes for CRVO, hard exudates for DR) have high associations with their respective conditions and minimal flow to the other. Unrelated concepts show negligible association with either condition.

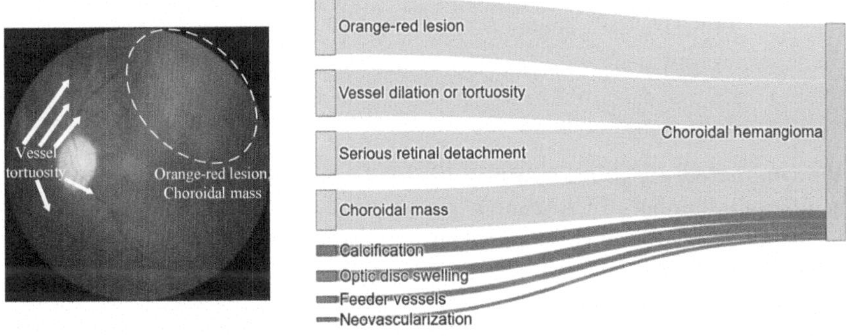

Fig. 4. Fundus image illustrating key representative attributes of Choroidal hemangioma and a sankey diagram depicting the flow of the concepts relevant to the novel category of Choroidal hemangioma in zero-shot detection setting of our framework. We present the top 4 and the bottom 4 concepts associated with choroidal hemangioma learnt by our model. The scores are averaged and normalized over all the test samples choroidal hemangioma.

Zero-Shot Detection: In Fig. 4, we present the scenario for a rare condition choroidal hemangioma [33]. We selected the top 4 most relevant and the bottom 4 least relevant concepts according to their contribution averaged over all the test samples of choroidal hemangioma. Our framework accurately identifies key concepts, including orange-red lesion [38], vessel dilation [22], serous retinal detachment, and choroidal mass, with strong associations depicted by wide flows. Less relevant concepts, such as calcification, feeder vessels, and optic disc swelling [32], are correctly identified with minimal association, demonstrating the framework's ability to generalize and handle unseen categories effectively.

4.4 Ablation Study for Concept Token Position and Number of Learnable Tokens

Building on prior work [42] highlighting the impact of class token position in prompt learning, we examine how the [CONCEPT] token's placement affects our retinal disease diagnosis framework. We test three positions: START $\{0\}$, MIDDLE $\{ceil(M/2)\}$, and END $\{M-1\}$, where M is the number of learnable tokens. As shown in Table 3, the END position yields the best performance, likely due to the model summarizing preceding tokens for a more comprehensive concept representation.

Table 3. Ablation study on [CONCEPT] position in our prompt learner, showing average mAP from 5 random seeds for 16-shot classification on our In-house dataset.

Method/Position of Token	START	MIDDLE	END
CoOp + Concepts (Ours)	59.91	58.35	**61.67**
CoCoOp + Concepts (Ours)	58.35	61.08	**61.26**
ProDA + Concepts (Ours)	62.93	63.07	**63.14**
MaPLe + Concepts (Ours)	69.46	68.31	**70.67**

We conducted an ablation study (Table 4) to assess performance variation with different numbers of learnable prompt tokens (2 to 64, in powers of 2). Performance improved with more tokens but saturated at 32.

Table 4. Ablation study of number of learnable tokens in our prompt learner, showing average mAP from 5 random seeds for 16-shot classification on In-house dataset.

Method/Number of learnable tokens	2	4	8	16	32	64
CoOp + Concepts (Ours)	58.36	59.47	60.58	60.91	**61.67**	61.05
CoCoOp + Concepts (Ours)	57.24	58.39	60.27	60.92	**61.26**	60.53
ProDA + Concepts (Ours)	59.97	60.65	62.41	63.06	**63.14**	62.93
MaPLe + Concepts (Ours)	67.94	68.33	70.04	70.18	**70.67**	70.24

4.5 Ablation Study for Stage 2 Training Strategy

Table 5 presents our exploration of various machine learning methods for the stage 2 training strategy. We compared Logistic Regression (LR), Random Forest (RF), Support Vector Machine (SVM), and an end-to-end training strategy for n-shot classification. The results demonstrate that Logistic Regression consistently outperformed the other methods across all n-shot settings.

Table 5. Evaluation of Stage 2 models using Stage 1 MaPLe + Concepts, showing average mAP from 5 random seeds for n-shot classification on In-house dataset.

Method/Number of Shots	n = 2	n = 4	n = 8	n = 16
Logistic Regression	**59.47**	**61.9**	**65.8**	**70.67**
Support Vector Machine	57.42	58.64	61.94	65.28
Random Forests	57.31	58.47	60.62	66.41
MLP (end-to-end training with Stage 1)	56.68	58.21	59.93	64.52

5 Conclusion

In this study, we present an efficient approach to achieve interpretable few-shot classification and zero-shot detection of retinal diseases by incorporating concept-guided prompt learning into vision-language models. Our framework provides an efficient way of creating concept bank of retinal diseases and integrating these domain knowledge into vision-language models, significantly enhancing their performance and providing added benefits of interpretability. Our research marks a significant step forward in developing highly effective and interpretable models for retinal disease diagnosis, with potential applicability to other medical image analysis domains, paving the way for real-world clinical implementation.

References

1. Araujo, T., et al.: Dr| graduate: uncertainty-aware deep learning-based diabetic retinopathy grading in eye fundus images. Med. Image Anal. **63**, 101715 (2020)
2. Bahl, A., Rao, S.: Diabetic retinopathy screening in rural India with portable fundus camera and artificial intelligence using eye mitra opticians from essilor india. Eye **36**(1), 230–231 (2022)
3. Bergren, R.L., Brown, G.C., Duker, J.S.: Prevalence and association of asteroid hyalosis with systemic diseases. Am. J. Ophthalmol. **111**(3), 289–293 (1991)
4. Brown, T., et al.: Language models are few-shot learners. Adv. Neural. Inf. Process. Syst. **33**, 1877–1901 (2020)
5. Chen, Y., Guo, X., Pan, Y., Xia, Y., Yuan, Y.: Dynamic feature splicing for few-shot rare disease diagnosis. Med. Image Anal. **90**, 102959 (2023)
6. Dosovitskiy, A., et al.: An image is worth 16x16 words: Transformers for image recognition at scale. arXiv preprint arXiv:2010.11929 (2020)
7. Friedman, D.S., O'Colmain, B.J., Munoz, B., Tomany, S.C., McCarty, C., De Jong, P., Nemesure, B., Mitchell, P., Kempen, J., et al.: Prevalence of age-related macular degeneration in the united states. Arch. Ophthalmol. **122**(4), 564–572 (2004)
8. Fu, H., Cheng, J., Xu, Y., Zhang, C., Wong, D., Liu, J., Cao, X.: Disc-aware ensemble network for glaucoma screening from fundus image. IEEE Trans. Med. Imaging **37**(11), 2493–2501 (2018)
9. Gao, M., et al.: Discriminative ensemble meta-learning with co-regularization for rare fundus diseases diagnosis. Med. Image Anal. **89**, 102884 (2023)
10. Gilson, A., et al.: How does chatgpt perform on the united states medical licensing examination? the implications of large language models for medical education and knowledge assessment. JMIR Med. Educ. **9**(1), e45312 (2023)

11. Hayreh, S.S., Zimmerman, M.B.: Fundus changes in central retinal vein occlusion. Retina **35**(1), 29–42 (2015)
12. He, K., Zhang, X., Ren, S., Sun, J.: Deep residual learning for image recognition. In: Proceedings of the IEEE Conference on Computer Vision and Pattern Recognition, pp. 770–778 (2016)
13. Ju, L., Wang, X., Wang, L., Liu, T., Zhao, X., Drummond, T., Mahapatra, D., Ge, Z.: Relational subsets knowledge distillation for long-tailed retinal diseases recognition. In: Medical Image Computing and Computer Assisted Intervention–MICCAI 2021: 24th International Conference, Strasbourg, France, September 27–October 1, 2021, Proceedings, Part VIII 24, pp. 3–12. Springer (2021)
14. Ju, L., Yu, Z., Wang, L., Zhao, X., Wang, X., Bonnington, P., Ge, Z.: Hierarchical knowledge guided learning for real-world retinal disease recognition. IEEE Trans. Med. Imaging (2023)
15. Kaddour, J., Harris, J., Mozes, M., Bradley, H., Raileanu, R., McHardy, R.: Challenges and applications of large language models. arXiv preprint arXiv:2307.10169 (2023)
16. Kempen, J.H., et al.: The prevalence of diabetic retinopathy among adults in the united states. Archives of ophthalmology (Chicago, Ill.: 1960) **122**(4), 552–563 (2004)
17. Khattak, M.U., Rasheed, H., Maaz, M., Khan, S., Khan, F.S.: Maple: multi-modal prompt learning. In: Proceedings of the IEEE/CVF Conference on Computer Vision and Pattern Recognition, pp. 19113–19122 (2023)
18. Kocur, I., Resnikoff, S.: Visual impairment and blindness in Europe and their prevention. Br. J. Ophthalmol. **86**(7), 716–722 (2002)
19. Koh, P.W., et al.: Concept bottleneck models. In: International Conference on Machine Learning, pp. 5338–5348. PMLR (2020)
20. Li, T., Bo, W., Hu, C., Kang, H., Liu, H., Wang, K., Fu, H.: Applications of deep learning in fundus images: a review. Med. Image Anal. **69**, 101971 (2021)
21. Lu, Y., Liu, J., Zhang, Y., Liu, Y., Tian, X.: Prompt distribution learning. In: Proceedings of the IEEE/CVF Conference on Computer Vision and Pattern Recognition, pp. 5206–5215 (2022)
22. Lupidi, M., et al.: New insights on circumscribed choroidal hemangioma: "bench to bedside". Graefes Arch. Clin. Exp. Ophthalmol. **262**(4), 1093–1110 (2024)
23. Murugappan, M., Prakash, N., Jeya, R., Mohanarathinam, A., Hemalakshmi, G., Mahmud, M.: A novel few-shot classification framework for diabetic retinopathy detection and grading. Measurement **200**, 111485 (2022)
24. Nair, A.G., Gandhi, R.A., Natarajan, S.: Effect of covid-19 related lockdown on ophthalmic practice and patient care in India: results of a survey. Indian J. Ophthalmol. **68**(5), 725 (2020)
25. Pachade, S., et al.: Retinal fundus multi-disease image dataset (rfmid): a dataset for multi-disease detection research. Data **6**(2), 14 (2021)
26. Panwar, N., et al.: Fundus photography in the 21st century–a review of recent technological advances and their implications for worldwide healthcare. Telemed. e-Health **22**(3), 198–208 (2016)
27. Park, S.S.: Cell therapy applications for retinal vascular diseases: diabetic retinopathy and retinal vein occlusion. Investigative ophthalmology & visual science **57**(5), ORSFj1–ORSFj10 (2016)
28. Quellec, G., Lamard, M., Conze, P.H., Massin, P., Cochener, B.: Automatic detection of rare pathologies in fundus photographs using few-shot learning. Med. Image Anal. **61**, 101660 (2020)

29. Radford, A., et al.: Learning transferable visual models from natural language supervision. In: International Conference on Machine Learning, pp. 8748–8763. PMLR (2021)
30. Selvaraju, R.R., Cogswell, M., Das, A., Vedantam, R., Parikh, D., Batra, D.: Grad-cam: visual explanations from deep networks via gradient-based localization. In: Proceedings of the IEEE International Conference on Computer Vision, pp. 618–626 (2017)
31. Sen, P., Nunez do Rio, J.M., Bagchi, A., Sivaprasad, S.: Differences in macular microvascular changes between eyes with central retinal vein occlusion and proliferative diabetic retinopathy. Eye **35**(11), 3170–3172 (2021)
32. Shanmugam, P.M., Ramanjulu, R.: Vascular tumors of the choroid and retina. Indian J. Ophthalmol. **63**(2), 133–140 (2015)
33. Shields, C.L., Shields, J.A., Augsburger, J.J.: Choroidal osteoma. Surv. Ophthalmol. **33**(1), 17–27 (1988)
34. Singha, M., Pal, H., Jha, A., Banerjee, B.: Ad-clip: adapting domains in prompt space using clip. In: Proceedings of the IEEE/CVF International Conference on Computer Vision, pp. 4355–4364 (2023)
35. Thapa, R., Khanal, S., Tan, H.S., Thapa, S.S., van Rens, G.H.M.B.: Prevalence, pattern and risk factors of retinal diseases among an elderly population in Nepal: the Bhaktapur retina study. Clinical Ophthalmology, pp. 2109–2118 (2020)
36. Tripathy, K.: Asteroid hyalosis. N. Engl. J. Med. **379**(8), e12 (2018)
37. Vaswani, A., et al.: Attention is all you need. Advances in neural information processing systems **30** (2017)
38. Witschel, H., Font, R.L.: Hemangioma of the choroid. a clinicopathologic study of 71 cases and a review of the literature. Survey Ophthalmol. **20**(6), 415–431 (1976)
39. Yang, Y., Panagopoulou, A., Zhou, S., Jin, D., Callison-Burch, C., Yatskar, M.: Language in a bottle: Language model guided concept bottlenecks for interpretable image classification. In: Proceedings of the IEEE/CVF Conference on Computer Vision and Pattern Recognition (CVPR), pp. 19187–19197 (2023)
40. Yau, J.W., et al.: Global prevalence and major risk factors of diabetic retinopathy. Diabetes Care **35**(3), 556–564 (2012)
41. Zhou, K., Yang, J., Loy, C.C., Liu, Z.: Conditional prompt learning for vision-language models. In: Proceedings of the IEEE/CVF Conference on Computer Vision and Pattern Recognition, pp. 16816–16825 (2022)
42. Zhou, K., Yang, J., Loy, C.C., Liu, Z.: Learning to prompt for vision-language models. Int. J. Comput. Vision **130**(9), 2337–2348 (2022)

Full Conformal Adaptation of Medical Vision-Language Models

Julio Silva-Rodríguez[1](✉), Leo Fillioux[2], Paul-Henry Cournède[2], Maria Vakalopoulou[2], Stergios Christodoulidis[2], Ismail Ben Ayed[1], and Jose Dolz[1]

[1] ÉTS Montréal, Montréal, Canada
julio-jose.silva-rodriguez@etsmtl.ca
[2] CentraleSupélec, Université Paris-Saclay, Gif-sur-Yvette, France

Abstract. Vision-language models (VLMs) pre-trained at large scale have shown unprecedented transferability capabilities and are being progressively integrated into medical image analysis. Although its discriminative potential has been widely explored, its reliability aspect remains overlooked. This work investigates their behavior under the increasingly popular split conformal prediction (SCP) framework, which theoretically guarantees a given error level on output sets by leveraging a labeled calibration set. However, the zero-shot performance of VLMs is inherently limited, and common practice involves few-shot transfer learning pipelines, which cannot absorb the rigid exchangeability assumptions of SCP. To alleviate this issue, we propose *full conformal adaptation*, a novel setting for jointly adapting and conformalizing pre-trained foundation models, which operates transductively over each test data point using a few-shot adaptation set. Moreover, we complement this framework with SS-Text, a novel training-free linear probe solver for VLMs that alleviates the computational cost of such a transductive approach. We provide comprehensive experiments using 3 different modality-specialized medical VLMs and 9 adaptation tasks. Our framework requires *exactly* the same data as SCP, and provides consistent relative improvements of up to 27% on set efficiency while maintaining the same coverage guarantees. Code is available: https://github.com/jusiro/FCA

Keywords: VLMs · Transfer learning · Conformal prediction

1 Introduction

Large pre-trained vision-language models (VLMs), such as CLIP [32] or ALIGN [16], are becoming increasingly popular in a plethora of computer vision problems [14,25,38], exhibiting unprecedented transfer capabilities to downstream tasks. Nevertheless, as these models gain traction in safety-critical scenarios, such as healthcare [24,36,39,47], ensuring their reliability and safety is paramount to prevent potential failures and thus minimize risks to human life. Therefore, the safe deployment of VLMs in high-stakes tasks requires not only precise discriminative performance but also robust uncertainty estimates.

I. Oguz et al. (Eds.): IPMI 2025, LNCS 15830, pp. 278–293, 2026.
https://doi.org/10.1007/978-3-031-96625-5_19

Conformal Prediction (CP) [1,42–44] is a machine learning framework that provides model agnostic, and *distribution-free*, finite-sample validity guarantees for handling reliability, and which has recently been adopted in modern neural networks [3,20,27,34,48]. Specifically, CP defines a non-conformity score function to produce a finite prediction set for a given unseen data point, which includes the true label with a specified confidence level. For example, in the medical context, CP can ensure that the correct diagnosis (e.g., presence of a given disease) is included in the predictive sets 95% of the time.

Split conformal prediction (SCP) [22,29,44] is arguably the most popular CP approach for uncertainty quantification, enjoying widespread adoption. SCP relies on a *labeled calibration subset* to find a threshold on the non-conformity score that ensures a desired marginal coverage level on test data, whose distribution must be, at least, *exchangeable.* w.r.t. calibration. Despite its increasing popularity, the rigid assumptions of SCP limit its successful deployment for foundation models. Even though VLMs have shown promising zero-shot capabilities, they usually require an adaptation stage to perform properly, e.g., in low-prevalence concepts or under domain shifts [32,41], where the few-shot paradigm is becoming popular [38]. However, the problem becomes evident when, if the non-conformity scores are re-adjusted based on such limited calibration data, the covariate shift breaks the exchangeability principle w.r.t. unseen examples. This raises the following question: *can we adapt modern VLMs to downstream tasks yet ensure coverage guarantees on new data without relying on multiple, data-expensive, labeled subsets?* To address this question, we revisit full conformal prediction (FCP) [1,35], which trains the model in a transductive fashion to ensure exchangeable reliability measures. However, its usability in modern deep neural networks is unclear since FCP requires multiple model fits for each query sample—concretely, one for each label—which explains the broader adoption of SCP. In this work, we accommodate this transductive setting to overcome the limitations observed in prior literature. The main contributions are:

- We introduce *full conformal adaptation* (FCA), a novel framework that operates transductive to adapt pre-trained VLMs and create conformalized outputs with validity guarantees, based on a few-shot adaptation subset.
- Furthermore, we complement FCA with a *training-free* linear probing solver, SS-Text, which reduces 150× the computational burden of adapting VLMs compared to SoTA approaches while providing robust performance.
- Comprehensive experiments on 9 public datasets covering 3 different medical image modalities demonstrate the proposed approach's superiority compared to SCP, and the potential of conformal prediction in medical image analysis.

2 Related Work

Transfer Learning in VLMs. VLMs exhibit outstanding general zero-shot capabilities [32]. However, this good performance vanishes when specific domains and concepts are represented with low frequency during pre-training [41]. This limitation motivates its data-efficient, few-shot adaptation to novel classes [11, 12, 38, 50, 52]. For example, Prompt Learning aims to optimize a set of learnable input tokens for each task [12, 52]. Black-box Adapters, which directly operate on feature representations, have emerged as a more efficient alternative, e.g., see [14, 36] for a direct comparison. Indeed, advanced linear probing methods that combine visual and text knowledge [14, 23, 38, 49] currently yield the best results. For instance, CLAP [38] follows a constrained optimization objective, whereas LP++ [14] combines visual and text logits through trainable blending parameters. In contrast to prior literature, which mostly studies VLM adaptation from a discriminative standpoint, our work explores its overlooked reliability.

Conformal Prediction in Vision. Current trends on CP for image classification leverage black-box predictors using split conformal prediction [3, 6, 9, 34, 40]. These works are usually evaluated on general vision tasks; thus, their performance is yet to be explored in medical imaging. From the technical side, recent efforts are devoted to better non-conformity measures, which account for better adaptability and set efficiency. For example, LAC [27] creates predictive sets by directly using the raw class probabilities. Adaptive Prediction Sets (APS) [34] computes the non-conformity score by accumulating the sorted softmax values in descending order, and its regularized extension RAPS [3] integrates explicit penalties to enforce small sets. However, these works employ models trained on an independent and identically distributed (i.i.d.) dataset w.r.t. testing, an unrealistic scenario in the era of foundation models.

3 Background

3.1 Zero-Shot Models

Contrastive Vision-Language Models. VLMs usually follow the CLIP [32] setting and are trained on large heterogeneous datasets to encode similar representations between paired image and text information. CLIP-alike models comprise a vision encoder, $f_\theta(\cdot)$, and a text encoder, $f_\phi(\cdot)$, which project data points into an ℓ_2-normalized D-dimensional shared embedding space, yielding the corresponding visual, $\mathbf{v} \in \mathbb{R}^{D\times 1}$, and text, $\mathbf{t} \in \mathbb{R}^{D\times 1}$, embeddings. These models provide strong representations, which can be transferred in a black-box manner [11, 14, 23, 28, 38, 49, 50]. Given a pre-computed image feature and class-wise prototypes, $\mathbf{W} = (\mathbf{w}_c)_{1\leq c\leq C}$, with $\mathbf{w}_c \in \mathbb{R}^{D\times 1}$, and C the number of target categories, probabilities can be computed as:

$$p_c(\mathbf{W}) = \frac{\exp(\mathbf{v}^\top \mathbf{w}_c/\tau)}{\sum_{j=1}^{C}\exp(\mathbf{v}^\top \mathbf{w}_j/\tau)}, \tag{1}$$

where τ is a temperature parameter learned during the pre-training, $\mathbf{v}^\top\mathbf{w}$ is the dot product, equivalent to cosine similarity, as vectors are ℓ_2-normalized, and $\mathbf{p}(\mathbf{W}) = (p_c(\mathbf{W}))_{1\leq c\leq C}$ corresponds to the predicted probabilities vector.

Zero-Shot Inference. Contrastive VLMs enable zero-shot predictions, i.e., no need to explicitly learn $\mathbf{W}$, by embedding a textual description for each label, so-called prompt. Thus, given a set of J text prompts, $\{\{\mathbf{t}_{cj}\}_{j=1}^{J}\}_{c=1}^{C}$, a common practice is to obtain a zero-shot prototype for each target category by computing the average of the ℓ_2-normalized text embeddings for each class, $\mathbf{t}_c = \frac{1}{J}\sum_{j=1}^{J}\mathbf{t}_{cj}$. These prototypes can be readily integrated in Eq. 1, such that $\mathbf{W}^* = (\mathbf{t}_c)_{1\leq c\leq C}$.

3.2 Conformal Prediction

Preliminaries. Let us define a multi-class image classification task composed of image and label data pairs $(\mathbf{x}, y)$, randomly sampled from a joint distribution $\mathcal{P}_{\mathcal{XY}}$. Also, we denote a black-box deep network, $\pi(\cdot)$, which outputs probability assignments for each category, $\mathbf{p} = \pi(\mathbf{x})$, in a label space, $\mathcal{Y} = \{1, 2, ..., C\}$, and which is trained on a base subset of R samples from $\mathcal{XY}$, $\mathcal{D}_{\text{train}}= \{(\mathbf{x}_r, y_r)\}_{r=1}^{R}$.

Conformal Prediction. CP [44] aims to produce prediction sets containing the ground truth label with a user-specified probability. Formally, the goal is to construct a set-valued mapping $\mathcal{C} : \mathcal{X} \rightarrow 2^C$, such that:

$$\mathcal{P}(Y \in \mathcal{C}(\mathbf{x})) \geq 1 - \alpha, \tag{2}$$

where $\alpha \in (0, 1)$ denotes the desired error rate, and $\mathcal{C}(\mathbf{x}) \subset \mathcal{Y}$ is the prediction set. Equation 2 is known as the *coverage guarantee* [44], and is *marginal* over $\mathcal{XY}$.

Split Conformal Prediction. SCP [29] enables deploying coverage guarantees for any pre-trained predictor [22]. This framework solely requires access to N labeled calibration data points, $\mathcal{D}_{\text{cal}}= \{(\mathbf{x}_i, y_i)\}_{i=1}^{N}$, and M test data samples, $\mathcal{D}_{\text{test}}= \{(\mathbf{x}_m)\}_{m=N+1}^{N+M}$, both drawn from i.i.d or exchangeable distributions [44]. The SCP procedure is as follows: *i*) *First* a non-conformity measure $s_i = \mathcal{S}(\pi(\mathbf{x}_i), y_i)$ is evaluated, where s_i is a measure of deviation; *ii*) *Second*, the 1-α quantile of the non-conformity score is determined from calibration data, which will serve as a confidence threshold to satisfy a given coverage:

$$\hat{s} = \inf\left[s : \frac{|i \in \{1, ..., N\} : s_i \leq s|}{N} \geq \frac{\lceil (N+1)(1-\alpha)\rceil}{N}\right]; \tag{3}$$

iii) *Third*, for each testing sample, the non-conformity scores are calculated, and the prediction set comprises labels whose non-conformity score falls within $\hat{s}$:

$$\mathcal{C}(\mathbf{x}) = \{y \in \mathcal{Y} : \mathcal{S}(\pi(\mathbf{x}), y) \leq \hat{s}\}. \tag{4}$$

4 Methods

4.1 Problem Statement

Pitfalls of SCP in VLMs. A relevant corpus of recent works is popularizing SCP in deep vision networks [3,8,40]. Nevertheless, prior art focuses on narrow scenarios where the black-box model has been intensively trained on a dataset that is in-distribution to calibration/testing. However, this scenario is unrealistic in the era of foundation models. Their zero-shot performance is limited to the pre-training concept frequency and therefore requires adaptation [41]. This adaptation is usually carried out in the few-shot data regime [32,38] leveraging a small balanced support set of $N = K \cdot |\mathcal{Y}|$, typically with $K \leq 16$. With such limited supervision, SCP is naturally constrained. If the model is adapted in the support set, then the new scores would not be exchangeable w.r.t. test data and, as shown in Table 1, the marginal coverage in Eq. 2 will be violated.

Transductive Conformal Prediction. To address the above limitations, we leverage well-established knowledge in transductive prediction, particularly *full conformal prediction* [1,30,35]. FCP provides a more flexible approach in which the model is trained transductively for each test data point $\mathbf{x}_m$ jointly with the training set without requiring calibration data. The intuition is that the true category for $\mathbf{x}_m$ lies within $\mathcal{Y}$. Thus, if the model $\pi(\cdot)$ is trained for each possible label assignment, i.e., $\pi(\cdot)^y : y_m = y \in \mathcal{Y}$, then the score distribution obtained from the correct label will be exchangeable to training score distribution. Therefore, the label-specific non-conformity measure given the trained model, $s_i^y = \mathcal{S}(\pi_i^y(\mathbf{x}), y_i)$, can be used to compute the $1 - \alpha$ quantile from the training score distribution, $\hat{s}^y$, similarly to Eq. 3, and output predictive sets:

$$\mathcal{C}(\mathbf{x}) = \{y \in \mathcal{Y} : s^y \leq \hat{s}^y\}, \tag{5}$$

whose output sets satisfy the coverage guarantee in Eq. 2 if fitting $\pi(\cdot)$ is invariant to permutations, and assuming exchangeable data distributions [44].

Nevertheless, FCP has been typically discarded due to its expensive computations, which require C model training for each test image, being of impractical deployment when using modern deep networks. Indeed, the FCP procedure involves fitting the whole network on each label $y \in \mathcal{Y}$ and each test point, i.e., $\mathcal{D}_{train} = \{(\mathbf{x}_1, y_1), ..., (\mathbf{x}_r, y_r), ...(\mathbf{x}_R, y_R), (\mathbf{x}_m, y)\}$.

4.2 Full Conformal Adaptation

There exist two limitations that prohibit FCP deployment in modern vision-language models: *i*) accessing the pre-training dataset, and *ii*) the computational burden of training largely parametrized networks for each test sample. In the following, we describe *full conformal adaptation*. This novel framework tackles such issues and allows for deploying reliable yet precise solutions in modern vision-language models. An overview is presented in Fig. 1, and we describe its key methodological components above:

- *Using an adaptation set.* Instead of training the model from scratch, we propose to *adapt* it using a small (few-shot) subset, similar to the one employed for calibration in SCP, i.e., $\mathcal{D}_{\text{adapt}} \equiv \mathcal{D}_{\text{cal}}$, thus omitting the pre-training data.
- *Transfer learning.* Instead of updating the whole network parameters, we propose operating over the black-box features extracted from the adaptation set, $\{(\mathbf{v}_i, y_i)\}_{i=1}^N$, and the new test data point, $\mathbf{v}_m$. Thus, model training in the transductive conformal setting can be reduced to finding the optimum linear probe for each explored label, $\mathbf{W}^y$, such that $s_i^y = \mathcal{S}(\mathbf{p}_i(\mathbf{W}^y), y_i)$.

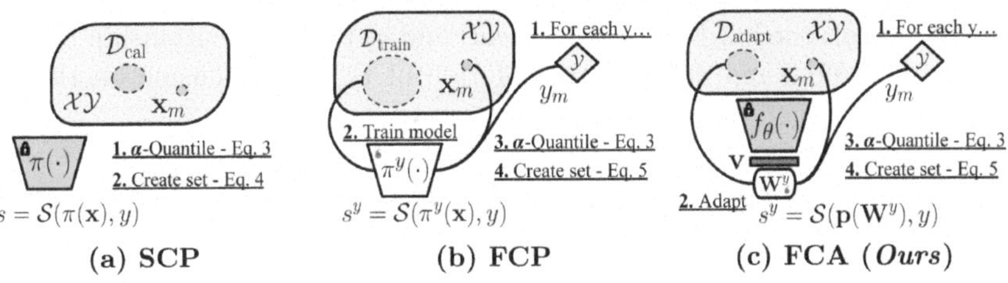

Fig. 1. Overview of conformal frameworks: (a) split conformal prediction [29], (b) Full conformal prediction [1], and (c) full conformal adaptation (*Ours*).

The proposed setting satisfies the coverage guarantees as FCP (see [1,30, 35]) but translates the exchangeability assumption to the feature space. Also, note FCA requires *exactly* the same data sources as SCP, yet is more data efficient, as it allows both adaptation and conformalization. However, popular linear probing strategies might still be time-expensive—see Fig. 3(b). Motivated by this observation, we propose a novel solver for finding $\mathbf{W}^y$ more efficiently next.

4.3 SS-Text: VLMs Adaptation at Speed-Light

Constrained Linear Probing. Currently, VLMs are adapted through logistic regression techniques, which combine visual supervisory signals with text knowledge [11,14,23,38,49]. Generally, such methods can be interpreted from a constrained optimization standpoint, in which the new classifier, $\mathbf{W}$, is optimized to minimize a cross-entropy loss and remain close to the class-text prototypes:

$$\min_{\mathbf{W}} \; \mathcal{L}(\mathbf{W}) = -\frac{1}{N}\sum_{i=1}^{N}\sum_{c=1}^{C} y_{ic} \, \ln(p_{ic}(\mathbf{W})) + \frac{\lambda}{2}\sum_{c=1}^{C} ||\mathbf{w}_c - \mathbf{t}_c||_2^2. \tag{6}$$

Popular choices in the literature involve solving Eq. 6 via gradient descent, which usually provides satisfactory results since the overall problem is convex. However, the computational effort of computing gradient steps limits its feasibility as a solution in the proposed transductive setting. To alleviate this issue, we propose a novel *training-free* solver to find an approximate solution.

Training-Free Approximation. We can disentangle Eq. 6 as the sum of two terms, $\mathcal{L} = g_1 + g_2$, for some $\lambda > 0$, such that:

$$g_1 = -\frac{1}{N}\sum_{i=1}^{N}\sum_{c=1}^{C} y_{ic} \; (\mathbf{v}^\top \mathbf{w}_c/\tau) + \frac{\lambda}{2}\sum_{c=1}^{C} ||\mathbf{w}_c - \mathbf{t}_c||_2^2, \tag{7}$$

$$g_2 = \frac{1}{N}\sum_{i=1}^{N} \ln\left(\sum_{j=1}^{C} \exp(\mathbf{v}^\top \mathbf{w}_j/\tau)\right), \tag{8}$$

where g_1 controls the hard label assignment, and g_2 the soft relationship between categories. We now approximate the solution of Eq. 6 by focusing on the first term and minimizing its gradients w.r.t. each $\mathbf{w}_c$:

$$\frac{\partial g_1}{\partial \mathbf{w}_c} = -\frac{1}{N}\sum_{i=1}^{N}(y_{ic}\ \mathbf{v}/\tau) + \lambda(\mathbf{w}_c - \mathbf{t}_c). \tag{9}$$

Given any $\lambda > 0$, the expression in g_1 is convex w.r.t each $\mathbf{w}_c$, as it is the sum of linear and convex functions. Hence, we can calculate its minimum as:

$$\mathbf{w}_c^* = \arg\min_{\mathbf{w}_c} \frac{\partial g_1}{\partial \mathbf{w}_c} = \frac{1}{\lambda N \tau}\sum_{i=1}^{N} y_{ic}\mathbf{v} + \mathbf{t}_c. \tag{10}$$

Note that this solution, denoted as SimpleShot-Textual (SS-Text), is a linear combination of textual (*right*) and visual (*left*) prototypes. Hence, we fix $\lambda = 1/(N\tau)$ motivated by: *i)* omitting the effect of τ, and *ii)* diminishing the effect of textual priors as more support data is available.

5 Experiments

5.1 Setup

Medical Vision-Language Models. Three major medical image modalities are employed in this work: histology, fundus, and chest X-ray (CXR) images. **Histology**: CONCH [24] is used as a histology-specialized model, using a customized larger-scale ViT-B/16 visual backbone. **Ophtalmology:** FLAIR [39], the first VLM for retina imaging, is selected. **CXR**: We used CONVIRT [51] pre-trained on MIMIC [18]. FLAIR and CONVIRT follow the same architectural design: the text encoder is BioClinicalBERT [2], and the vision encoder is ResNet-50.

Downstream Tasks. The selected tasks contain 4 to 19 categories and tackle fine-grained tissue/disease classification or grading scenarios. **Histology**: the datasets involve three different organs and cancer types: colon in NCT-CRC [19], prostate Gleason grading in SICAPv2 [37], and SkinCancer [21]. **Ophtalmology**: the popular diabetic retinopathy (DR) grading is included, using MESSIDOR [7]. Analogously, myopic maculopathy staging in MMAC [31], and fine-grained disease detection in FIVES [17], are assessed. **CXR**: detection of different diseases from frontal CXR is targeted using popular datasets: a subset of five CheXpert [15] categories, employed in [47], 19 fine-grained categories in NIH-LT [13,45], and COVID detection in pneumonia cases [5,33].

Conformal Prediction Algorithms. Three popular non-conformity scores are employed: LAC [27], and two adaptive approaches, APS [34], and RAPS [3], to generate prediction sets at error rates of $\alpha \in \{0.1, 0.05\}$. The hyper-parameters in RAPS are set to $k_{\text{reg}} = 1$ and $\lambda = 0.001$. These values provided stable performance in [3].

Experimental Protocol and Metrics. The adaptation of medical VLMs assumes the presence of a few-shot adaptation subset, with $K \in \{1, 2, 4, 8, 16\}$. We define a realistic, validation-free setting in which this unique support set adapts the VLM and calibrates conformal prediction methods. It is worth mentioning that this implies that the model selection, i.e., how the adaptation methods are trained, does not rely on validation data. Thus, the training hyper-parameters are constant across tasks. Three conformal prediction settings are explored in the context of VLMs: *i*) **SCP**, the standard setting studied in vision classifiers [3,6,8,40] which operates over zero-shot predictions obtained following Sect. 3.1 to produce conformal sets, *ii*) **Adapt+SCP**, which first adjusts an adapter on top of pre-trained features and then follows the SCP over the new output predictions, using the same support set for adaptation and calibration; and *iii*) The proposed **full conformal adaptation**, which jointly adapts and calibrates in a transductive fashion, as detailed in Sect. 4.2. For all these strategies, CP is evaluated when using at least $K \geq 4$, which ensures an appropriate quantile search in Eq. 3. Test data is obtained from each dataset and remains a constant partition, and the support set is retrieved from the training subset. All experiments are repeated 20 times using different random seeds. Class-wise balanced accuracy (ACA) is included to evaluate the discriminative performance, as suggested in [26]. Also, figures of merit typically employed in conformal prediction settings are computed: coverage ("Cov."), average set size ("Size"), and class-conditioned coverage gap ("CCV") [8].

Adaptation Methods. SS-Text is compared to recent black-box strategies for adapting VLMs. First, gradient-based trained solutions are used. For vanilla linear probing (LP), we follow ZS-LP in [38]. CLAP [38] and LP++ [14,36] are also integrated, which are trained using full-batch gradient descent for 300 epochs as in [38], using SGD optimizer with momentum of 0.9. The initial learning rate in LP and CLAP is fixed to 0.1 as in [38], using a cosine-scheduler decay. For LP++, the learning rate is data-driven, as proposed by the authors. Furthermore, training-free solvers are included: SimpleShot [46], as a vision-only method; and vision-language strategies such as TIP-Adapter [50] (with α_{tip} and β_{tip} fixed to 1.0) and LP++ [14,36] initializations.

5.2 Conformal Prediction Results

This section evaluates the performance of the proposed FCA framework compared to the more traditional SCP or Adapt+SCP. Note that we leverage SS-Text as an adaptation strategy since, in this section, we are only interested in validating the conformalization strategy.

Adapting VLMs with Accuracy and Reliability. Results in Table 1 show a near-satisfied average coverage using SCP. However, the results are similar for all non-conformity measures, i.e., LAC, APS, and RAPS. This observation showcases the limited effect of recent efforts devoted to better adaptive scores in the context of zero-shot medical VLMs. When the VLMs are adapted, the accuracy is improved by an average of +16.9 points, demonstrating the necessity

Table 1. Conformal prediction results using 16 shots, with three popular non-conformity scores, and two error levels, $\alpha \in \{0.05, 0.1\}$. Average across modalities and tasks. "↓" indicates smaller is better. Red highlights unsatisfied error rates.

	Method		$\alpha = 0.10$			$\alpha = 0.05$		
		ACA↑	Cov.	Size↓	CCV↓	Cov.	Size↓	CCV↓
LAC	SCP	50.2	0.890	3.99	9.96	0.951	4.88	5.68
	Adapt+SCP	**67.1**$_{+16.9}$	0.842	**2.40**$_{-1.59}$	11.17$_{+1.21}$	0.921	**3.07**$_{-1.81}$	6.87$_{+1.19}$
	FCA (*Ours*)	**67.1**$_{+16.9}$	0.896	2.91$_{-1.08}$	**8.38**$_{-1.58}$	0.952	3.56$_{-1.32}$	**5.02**$_{-0.66}$
APS	SCP	50.2	0.900	4.05	9.59	0.952	4.88	5.54
	Adapt+SCP	**67.1**$_{+16.9}$	0.858	**2.56**$_{-1.49}$	8.57$_{-1.02}$	0.924	**3.19**$_{-1.69}$	6.08$_{+0.54}$
	FCA (*Ours*)	**67.1**$_{+16.9}$	0.898	3.06$_{-0.99}$	**6.12**$_{-3.47}$	0.949	3.67$_{-1.21}$	**4.24**$_{-1.30}$
RAPS	SCP	50.2	0.901	4.16	9.55	0.952	5.12	5.57
	Adapt+SCP	**67.1**$_{+16.9}$	0.856	**2.55**$_{-1.61}$	8.64$_{-0.91}$	0.923	**3.17**$_{-1.95}$	6.12$_{+0.55}$
	FCA (*Ours*)	**67.1**$_{+16.9}$	0.898	3.05$_{-1.11}$	**6.21**$_{-3.34}$	0.951	3.66$_{-1.46}$	**4.23**$_{-1.34}$

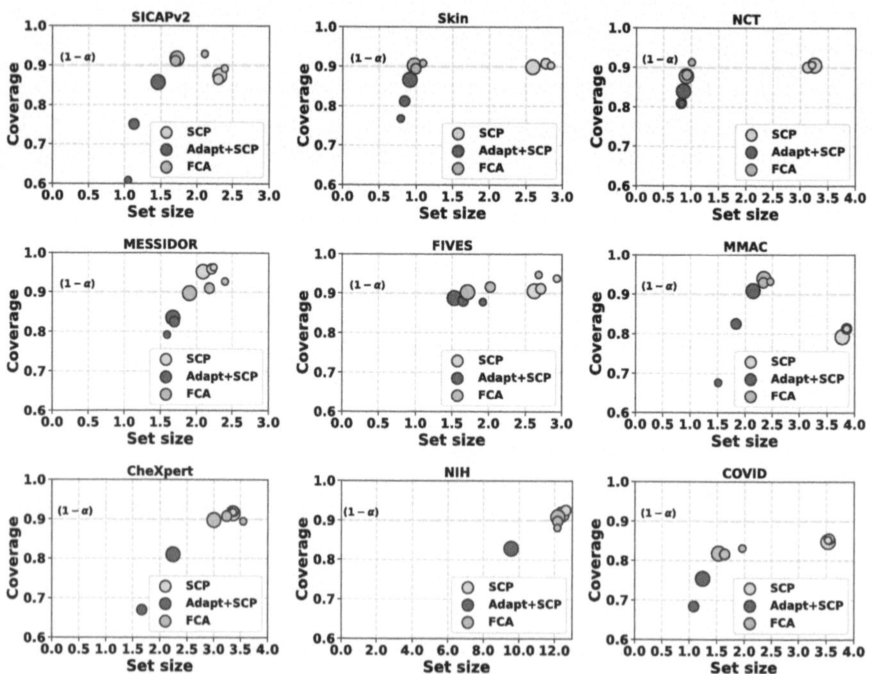

Fig. 2. Conformal prediction results per dataset. Results were obtained using $\alpha = 0.10$, and LAC [27]. Each dot represents the performance, with increasing size correlated with the number of shots for adaptation, i.e., $K \in \{4, 8, 16\}$.

of transfer learning. Even though Adapt+SCP reduces the produced set sizes, *leveraging the available shots for both adaptation and conformalization naively penalizes coverage.* Thus, its produced sets are unreliable. In contrast, the proposed FCA excels in both aspects. First, it provides discriminative performance on par with the standard adaptation. Second, it keeps the marginal coverage at the desired error level, reducing the produced sets by nearly 27% compared to SCP. FCA also improves the more fine-grained class-conditional coverage, decreasing its violation up to ~15% for LAC and ~35% for adaptive scores.

Detailed, Per-dataset Findings. When observing the results in Fig. 2, one can notice that exact marginal coverage is usually not achieved—not even for SCP, e.g., in SICAPv2 or MMAC. In contrast to the main corpus of conformal prediction evaluating CP in vision, medical image classification poses more challenging scenarios to ensure the exchangeability principle. First, testing data comes from patients who are naturally inaccessible during calibration. Second, the label-marginal distribution of the testing data is unknown, and typical few-shot adaptation pipelines assume a balanced support set. Combined with the finite sampling error, these facts explain such observation and showcase the necessity of developing CP techniques tailored to medical image analysis. However, it is worth mentioning that the proposed FCA framework consistently improves average coverage satisfaction (especially when compared with the naive Adapt+SCP strategy) while improving the prediction sets by large margins.

5.3 Few-Shot Adaptation Benchmark

This section explores the performance, in the discriminative aspect, of the proposed SS-Text solver compared to SoTA linear probing strategies. We also evaluate the feasibility of each method to serve as a solver for full conformal adaptation, which refers to its efficiency, measured in terms of how many fits (evaluating one category for a query image) they perform per second in such a setting.

Discriminative Performance. Figure 3(a) demonstrates that the proposed training-free SS-Text is a competitive solution in terms of accuracy. This is especially the case for the low-shot scenario, in which SS-Text outperforms popular recent solutions, such as CLAP [38] or LP++ [14]. Also, SS-Text is robust when the number of support samples increases, where basic linear probe baselines provide competitive results. We argue that these results are the product of the explored realistic setting, in which no validation data is accessible for defining a fine-grained scheduler and early-stopping intensive gradient-based techniques. When trained on a fixed scheduler, methods such as CLAP or LP++ might suffer in finding a homogeneous optimal configuration for all datasets.

Efficiency. Figure 3(b) showcases the extreme efficiency of SS-Text when performing sample-wise predictions within the proposed full conformal adaptation setting, being approximately 150× faster than gradient-based approaches. For example, if using a gradient-based linear probe strategy (LP), the inference of each image in the Skin dataset ($M = 30{,}000$; $C = 16$) would require at least

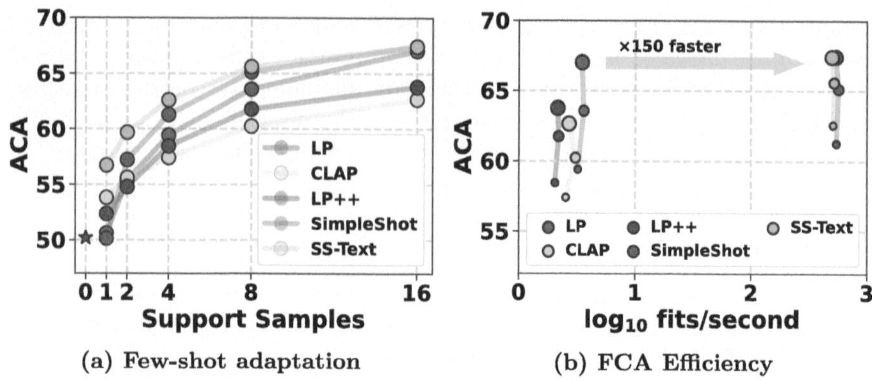

Fig. 3. Few-shot adaptation performance (a) and efficiency analysis (b). Efficiency, regarding full conformal adaptation predictions: images per class - per second. The dot size in (b) indicates the number of shots, i.e., $K \in \{4, 8, 16\}$.

2.8 s per image (in contrast to 30 ms with SS-Text), which accumulates approximately 23 h for the whole dataset, thus being infeasible its deployment in FCA. These figures underscore the necessity of efficient, training-free solvers to provide trustworthy sets in real-world scenarios.

5.4 In-Depth Studies

Role of λ. We fixed $\lambda = 1/(N_{\mathcal{T}})$ in Eq. 10, assuming that the more supervision signals we leverage for adaptation, the less effect of the constraint is desired. Table 2 assesses this adaptive model selection, compared to fixing it for different values, concretely $\lambda \in \{0.1, 1.0, 10\}$. The proposed adaptive strategy achieves the best performance balance between low-shot (e.g., $K \in \{1, 2, 4\}$), and a larger number of supervisory signals (e.g., $K \in \{8, 16\}$). This favorable comparison also extends to the strategy proposed in CLAP [38], where the authors proposed a class-wise multiplier based on zero-shot performance on the support set.

Comparison to Training-Free Baselines. SS-Text is also compared to popular training-free baselines in Table 2 (*top*). The proposed method consistently outperforms TIP-Adapter [50] and LP++ (training-free version) [14]. Also, compared to SimpleShot [46], SS-Text shows favorable performance. Note that SimpleShot does not rely on any text information, and thus, its performance on the low-shot regime is hampered. Such results showcase the importance of integrating text supervision for the few-shot adaptation of medical VLMs. Note that the performance gap between SimpleShot and SS-Text decreases with the amount of data since SS-Text naturally approximates SimpleShot with large N values.

Qualitative Assessment. We now depict some figures highlighting the utility of conformal prediction for medical image analysis. Concretely, we focus on two grading tasks for which well-known inter-observer variability exists. Figure 4 introduces the set size distribution and the frequency of class appearance within

Table 2. Study on how to fix λ in SS-Text solver (*bottom*) and comparison with training-free baselines (*top*). Average accuracy across nine datasets.

Method	Setting	$K=1$	$K=2$	$K=4$	$K=8$	$K=16$
Zero-shot [32]	(only text)	50.2	50.2	50.2	50.2	50.2
SimpleShot [46]	(only vision)	50.1	57.2	61.3	65.1	**67.4**
TIP-Adapter [50]	(training-free)	55.5	55.5	60.2	62.2	63.2
LP++ [14]	(training-free)	50.7	51.0	51.4	52.1	53.2
SS-Text	Fixed $\lambda = 0.1/\tau$	55.1	59.0	61.5	63.6	64.3
SS-Text	Fixed $\lambda = 1.0/\tau$	53.5	54.1	54.5	54.7	54.6
SS-Text	Fixed $\lambda = 10/\tau$	51.2	51.2	51.2	51.1	51.2
SS-Text	$\lambda_c \simeq$ zero-shot perf. [38]	51.4	58.4	62.6	**65.7**	**67.4**
SS-Text (*Ours*)	$\lambda = 1/(N\tau)$	**56.7**	**59.7**	**62.6**	65.6	**67.4**

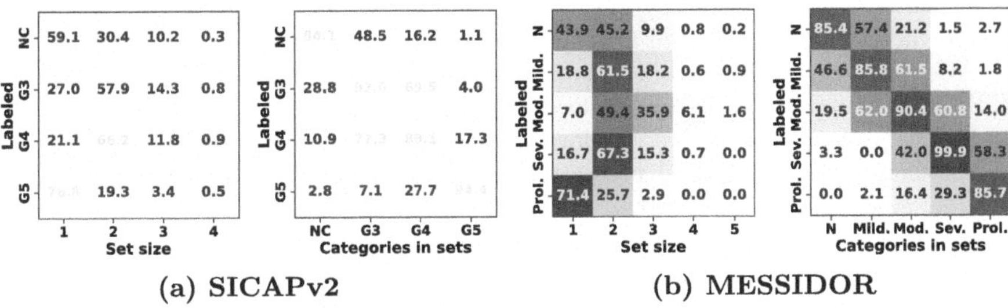

(a) SICAPv2 **(b) MESSIDOR**

Fig. 4. Qualitative evaluation of conformal prediction for grading tasks, i.e., Gleason for histology (a) and diabetic retinopathy for retina (b). Left: set size distribution per class. **Right**: label frequency in sets corresponding for each category. Results using FCA with $K = 16$, SS-Text solver, and $\alpha = 0.1$.

the predicted sets for each category. First, one can notice that smaller sets are usually retrieved for the more extreme and "easy" categories, while they tend to increase in categories that present similar patterns. Also, the occurrence matrices unveil an overlap between the latter classes, similar to typically observed inter-rater variability for these tasks [4,10]. It is worth remembering that such sets present empirical guarantees of giving the correct category for 90% of the cases for tasks with relatively smaller accuracy, i.e., 69% and 60%. Thus, the conformal prediction setting smoothly provides information regarding uncertainties natural from grading tasks.

6 Discussion

While the proposed FCA has proven remarkable properties compared to SCP, it also presents some limitations. First, it provides theoretical guarantees under exchangeability, which might be unrealistic to achieve in certain medical image analysis scenarios. Second, the framework requires one linear probe fit for each test image and category, which might be prohibitive in dense prediction tasks or when considering many classes. However, the latter scenario is least common in medical image tasks, where only tenths of tags are usually addressed at a time.

Acknowledgments. This work was funded by the Natural Sciences and Engineering Research Council of Canada (NSERC) and *Fonds de recherche du Québec* (FRQ). We thank Calcul Québec and Compute Canada. We gratefully acknowledge the DATAIA program for supporting Jose Dolz as a visiting professor at Université Paris-Saclay.

Disclosure of Interests. The authors have no competing interests to declare that are relevant to the content of this article.

References

1. Gammerman, A., Volodya Vovk, V.V.: Learning by transduction. In: Conference on Uncertainty in Artificial Intelligence, pp. 148–156 (1998)
2. Alsentzer, E., et al.: Publicly available clinical BERT embeddings. In: Clinical Natural Language Processing Workshop (2019)
3. Angelopoulos, A.N., Bates, S., Jordan, M., Malik, J.: Uncertainty sets for image classifiers using conformal prediction. In: International Conference on Learning Representations (ICLR) (2020)
4. Arvaniti, E., et al.: Automated gleason grading of prostate cancer tissue microarrays via deep learning. Sci. Rep. **8** (2018)
5. Chowdhury, M., et al.: Can ai help in screening viral and covid-19 pneumonia? IEEE Access **8**, 132665–132676 (2020)
6. Correia, A.H., Massoli, F.V., Louizos, C., Behboodi, A.: An information theoretic perspective on conformal prediction. In: Advances in Neural Information Processing Systems (NeurIPS) (2024)
7. Decencière, E., et al.: Feedback on a publicly distributed image database: the messidor database. Image Anal. Stereol. **33**, 231–234 (2014)
8. Ding, T., Angelopoulos, A., Bates, S., Jordan, M., Tibshirani, R.J.: Class-conditional conformal prediction with many classes. In: Advances in Neural Information Processing Systems (NeurIPS), vol. 36 (2023)
9. Einbinder, B.S., Romano, Y., Sesia, M., Zhou, Y.: Training uncertainty-aware classifiers with conformalized deep learning. In: Advances in Neural Information Processing Systems (NeurIPS) (2022)
10. Galdran, A., Dolz, J., Chakor, H., Lombaert, H., Ayed, I.B.: Cost-sensitive regularization for diabetic retinopathy grading from eye fundus images. In: Medical Image Computing and Computer Assisted Intervention (MICCAI), pp. 1–7, October 2020
11. Gao, P., et al.: Clip-adapter: better vision-language models with feature adapters. Int. J. Comput. Vis. (2023)
12. Yao, H., Rui Zhang, C.X.: Visual-language prompt tuning with knowledge-guided context optimization. In: Proceedings of the IEEE/CVF Conference on Computer Vision and Pattern Recognition (CVPR) (2023)
13. Holste, G., et al.: Long-tailed classification of thorax diseases on chest x-ray: A new benchmark study. In: MICCAI Workshop on Data Augmentation, Labelling, and Imperfections, pp. 22–32 (2022)

14. Huang, Y., Shakeri, F., Dolz, J., Boudiaf, M., Bahig, H., Ayed, I.B.: Lp++: a surprisingly strong linear probe for few-shot clip. In: Proceedings of the IEEE/CVF Conference on Computer Vision and Pattern Recognition (CVPR), pp. 23773–23782 (2024)
15. Irvin, J., et al.: Chexpert: a large chest radiograph dataset with uncertainty labels and expert comparison. In: Proceedings of the AAAI Conference on Artificial Intelligence, vol. 33, pp. 590–597 (2019)
16. Jia, C., et al.: Scaling up visual and vision-language representation learning with noisy text supervision. In: International Conference on Machine Learning (ICML), pp. 4904–4916 (2021)
17. Jin, K., et al.: Fives: a fundus image dataset for artificial intelligence based vessel segmentation. Sci. Data **9**, 475 (2022)
18. Johnson, A., et al.: MIMIC-CXR, a de-identified publicly available database of chest radiographs with free-text reports. Sci. Data **6**, 317 (2019)
19. Kather, J.N., Halama, N., Marx, A.: 100,000 histological images of human colorectal cancer and healthy tissue. Zenodo10 **5281** (2018)
20. Kim, B., Xu, C., Barber, R.: Predictive inference is free with the jackknife+-after-bootstrap. Adv. Neural. Inf. Process. Syst. **33**, 4138–4149 (2020)
21. Kriegsmann, K., et al.: Deep learning for the detection of anatomical tissue structures and neoplasms of the skin on scanned histopathological tissue sections. Front. Oncol. **12**, 1022967 (2022)
22. Lei, J., G'Sell, M., Rinaldo, A., Tibshirani, R.J., Wasserman, L.: Distribution-free predictive inference for regression. J. Am. Stat. Assoc. **113**(523), 1094–1111 (2018)
23. Lin, Z., Yu, S., Kuang, Z., Pathak, D., Ramanan, D.: Multimodality helps unimodality: Cross-modal few-shot learning with multimodal models. In: Proceedings of the IEEE/CVF Conference on Computer Vision and Pattern Recognition (CVPR) (2023)
24. Lu, M.Y., et al.: A visual-language foundation model for computational pathology. Nat. Med. **30**, 863–874 (2024)
25. Luo, H., Bao, J., Wu, Y., He, X., Li, T.: Segclip: patch aggregation with learnable centers for open-vocabulary semantic segmentation. In: International Conference on Machine Learning (ICML), pp. 23033–23044. PMLR (2023)
26. Maier-Hein, L., et al.: Metrics reloaded: recommendations for image analysis validation. Nature Methods **21** (2024)
27. Mauricio Sadinle, J.L., Wasserman, L.: Least ambiguous set-valued classifiers with bounded error levels. J. Am. Stat. Assoc. **114**(525), 223–234 (2019)
28. Ouali, Y., Bulat, A., Martinez, B., Tzimiropoulos, G.: Black box few-shot adaptation for vision-language models. In: Proceedings of the IEEE/CVF International Conference on Computer Vision (ICCV) (2023)
29. Papadopoulos, H., Proedrou, K., Vovk, V., Gammerman, A.: Inductive confidence machines for regression. In: European Conference on Machine Learning (ECML)., pp. 345–356 (2002)
30. Proedrou, K., Nouretdinov, I., Vovk, V., Gammerman, A.: Transductive confidence machines for pattern recognition. In: European Conference on Machine Learning (ECML), pp. 381–390 (2002)
31. Qian, B., et al.: A competition for the diagnosis of myopic maculopathy by artificial intelligence algorithms. JAMA Ophthalmology **142**(11), 1006–1015 (2024)
32. Radford, A., et al.: Learning transferable visual models from natural language supervision. In: International Conference on Machine Learning (ICML), pp. 8748–8763 (2021)

33. Rahman, T., et al.: Exploring the effect of image enhancement techniques on covid-19 detection using chest x-ray images. Comput. Biol. Med. **132**, 104319 (2021)
34. Romano, Y., Sesia, M., Candes, E.: Classification with valid and adaptive coverage. In: Advances in Neural Information Processing Systems (NeurIPS), vol. 33, pp. 3581–3591 (2020)
35. Saunders, C., Gammerman, A., Vovk, V.: Transduction with confidence and credibility. In: International Joint Conference on Artificial Intelligence (IJCAI), pp. 722–726 (1999)
36. Shakeri, F., et al.: Few-shot adaptation of medical vision-language models. In: Medical Image Computing and Computer-Assisted Intervention (MICCAI), pp. 553–563 (2024)
37. Silva-Rodríguez, J., Colomer, A., Sales, M.A., Molina, R., Naranjo, V.: Going deeper through the gleason scoring scale: an automatic end-to-end system for histology prostate grading and cribriform pattern detection. Comput. Methods Programs Biomed. **195**, 105637 (2020)
38. Silva-Rodríguez, J., Hajimiri, S., Ayed, I.B., Dolz, J.: A closer look at the few-shot adaptation of large vision-language models. In: Proceedings of the IEEE/CVF Conference on Computer Vision and Pattern Recognition (CVPR), pp. 23681–23690 (2024)
39. Silva-Rodríguez, J., Chakor, H., Kobbi, R., Dolz, J., Ayed, I.B.: A foundation language-image model of the retina (flair): encoding expert knowledge in text supervision. Med. Image Anal. **99**, 103357 (2025)
40. Stutz, D., Dvijotham, K.D., Cemgil, A.T., Doucet, A.: Learning optimal conformal classifiers. In: International Conference on Learning Representations (ICLR) (2022)
41. Udandarao, V., et al.: No "zero-shot" without exponential data: pretraining concept frequency determines multimodal model performance. In: Advances in Neural Information Processing Systems (NeurIPS) (2024)
42. Vovk, V.: Conditional validity of inductive conformal predictors. In: Proceedings of the Asian Conference on Machine Learning, vol. 25, pp. 475–490 (2012)
43. Vovk, V.: Transductive conformal predictors. In: Artificial Intelligence Applications and Innovations, pp. 348–360 (2013)
44. Vovk, V., Gammerman, A., Shafer, G.: Algorithmic Learning in a Random World. Springer, January 2005
45. Wang, X., Peng, Y., Lu, L., Lu, Z., Bagheri, M., Summers, R.: Chestx-ray8: hospital-scale chest x-ray database and benchmarks on weakly-supervised classification and localization of common thorax diseases. In: Proceedings of the IEEE/CVF Conference on Computer Vision and Pattern Recognition (CVPR), pp. 3462–3471 (2017)
46. Wang, Y., Chao, W.L., Weinberger, K.Q., van der Maaten, L.: Simpleshot: revisiting nearest-neighbor classification for few-shot learning. In: arXiv preprint arXiv:1911.04623 (2019)
47. Wang, Z., Wu, Z., Agarwal, D., Sun, J.: Medclip: contrastive learning from unpaired medical images and text. In: Empirical Methods in Natural Language Processing (EMNLP), pp. 1–12 (10 2022)
48. Xu, C., Xie, Y.: Conformal prediction interval for dynamic time-series. In: International Conference on Machine Learning, pp. 11559–11569. PMLR (2021)
49. Yu, T., Lu, Z., Jin, X., Chen, Z., Wang, X.: Task residual for tuning vision-language models. In: Proceedings of the IEEE/CVF Conference on Computer Vision and Pattern Recognition (CVPR), pp. 10899–10909 (2023)

50. Zhang, R., Fang, R., Zhang, W., Gao, P., Li, K., Dai, J., Qiao, Y., Li, H.: Tip-adapter: Training-free clip-adapter for better vision-language modeling. In: European Conference on Computer Vision (ECCV), pp. 1–19 (11 2022)
51. Zhang, Y., Jiang, H., Miura, Y., Manning, C.D., Langlotz, C.P.: Contrastive learning of medical visual representations from paired images and text. In: Machine Learning for Healthcare (MHLC), pp. 1–24 (2022)
52. Zhou, K., Yang, J., Loy, C.C., Liu, Z.: Learning to prompt for vision-language models. Int. J. Comput. Vis. (2022)

A Reality Check of Vision-Language Pre-training in Radiology: Have We Progressed Using Text?

Julio Silva-Rodríguez(✉), Jose Dolz, and Ismail Ben Ayed

ÉTS Montréal, Montreal, Canada
julio-jose.silva-rodriguez@etsmtl.ca

Abstract. Vision language pre-training has recently gained popularity as it allows learning rich feature representations using large-scale data sources. This paradigm has quickly made its way into the medical image analysis community. In particular, there is an impressive amount of recent literature developing vision-language models for radiology. However, the available medical datasets with image-text supervision are scarce, and medical concepts are fine-grained, involving expert knowledge that existing vision-language models struggle to encode. In this paper, we propose to take a prudent step back from the literature and revisit supervised, unimodal pre-training, using fine-grained labels instead. We conduct an extensive comparison demonstrating that unimodal pre-training is highly competitive and better suited to integrating heterogeneous data sources. Our results also question the potential of recent vision-language models for open-vocabulary generalization, which have been evaluated using optimistic experimental settings. Finally, we study novel alternatives to better integrate fine-grained labels and noisy text supervision. Code and weights are available: https://github.com/jusiro/DLILP.

Keywords: Vision-language pre-training · Transfer learning · Radiology

1 Introduction

The recent advancements in deep learning have yielded remarkable outcomes to enhance computer-aided medical image analysis [23]. Nevertheless, these have been classically hampered by the necessity of using large labeled datasets for training successful specific solutions, which may not generalize properly under domain drifts [9]. Currently, there is a paradigm shift led by multimodal foundation models. Such visual understanding models are pre-trained for specific domains using large dataset assemblies and heterogeneous learning objectives. In this way, foundation models learn rich generalizable features that can be efficiently adapted to downstream tasks [36]. These conditions are ideal for the widespread adoption of deep-learning solutions to clinical institutions, with limited data and computational resources [26,47,50]. In particular, vision-language

I. Oguz et al. (Eds.): IPMI 2025, LNCS 15830, pp. 294–309, 2026.
https://doi.org/10.1007/978-3-031-96625-5_20

contrastive pre-training methods such as CLIP [29] have revolutionized the computer vision field. These approaches train a joint multimodal space, in which text and visual data representations are aligned. Using web-mined data, CLIP gathers a collection of 400M image-text pairs for pre-training and has shown impressive generalization capabilities when transferred on various downstream computer-vision tasks. Driven by CLIP's popularity, vision-language models are also paving the way for building strong medical foundation models in different application domains, such as histology [13], retina [37], and radiology [53]. In particular, radiology, and more concretely, chest X-ray image understanding, has been an essential focus of this emergent literature since radiology text reports are the *de facto* raw supervisory information easily accessible from medical clinical records. A myriad of recent vision-language models, such as Convirt [52], REFERS [53], GlorIA [12], MedCLIP [45], medKLIP [46], and others [3,41,49,51], attest this trend. Many of these recent works are published in the top vision conferences or prestigious journals, advocating a paradigm shift in radiology imaging interpretation, driven by contrastive image-text pre-training. However, as we show, the potential to leverage large transferable vision models through more classical approaches, such as unimodal pre-training, has been severely underestimated.

Relying on text supervision for vision pre-training in medical domains faces several challenges. First, available datasets are orders of magnitude smaller compared to natural images. For example, these models are mainly built upon the textual information available in MIMIC [17], which assembles solely 257K image-text pairs. Second, as discussed in [3], medical linguistics are highly specialized and contain domain-specific structures. These include negations (e.g. "*there is no consolidation*"), expressions of uncertainty (e.g., "*possibly progressing to pneumonia*"), spatial relations (e.g., "*bilateral heterogeneous airplane opacities*"), hierarchical relationships (e.g., "*infection*" → "*pneumonia*"), or abbreviations. Although some efforts have been devoted to regularize the training to focus on this information [3,54], vision-language pre-training struggles to properly encode such expert knowledge. Indeed, this is not only the case of medical knowledge. As recent studies show, vision-language models might struggle to properly codify basic spatial information [18] or fine-grained vision-text correspondences [39]. Thus, in addition to text supervision, recent works [45,46,51] have proposed using label information for aligning better image and text representations. These labels are obtained through entity extraction NLP methods, such as CheXpert-labeler [14] or RadGraph [15], and follow radiologist-designed rule-based algorithms able to encode text reports to concrete labels through expert knowledge—*note that these labels do not require costly manual image annotation.* Indeed, before the wave of vision-language models, these labels represented the predominant supervision for training dataset-specific deep learning models for chest X-rays, and an important number of datasets (e.g., CheXpert [38], NIH [43], or PadChest [4]) included primarily image-label information. Nonetheless, even though these datasets contain fine-grained labels, supervised pre-training is being surprisingly overlooked in the current literature, even as a baseline to measure actual progress in the field. Based on these observations, we present the following contributions:

1. We challenge the status quo of current contrastive vision-language models (VLMs) for visual comprehension of chest X-rays (CXR), advocating for revisiting **supervised pre-training**. In particular, we focus on evaluating their zero- and few-shot transferability on a broad 7-task benchmark.
2. We demonstrate (see *Observation 1*) that such unimodal pre-training is a largely competitive solution, able to integrate larger heterogeneous sources.
3. In addition, we offer a critical view of the current trends in evaluating the zero-shot capabilities of CXR VLMs to novel diseases (see *Observations 2 and 3*). Concretely, we show that local unspecific findings drive textual disease prototypes, and VLMs fail to distinguish between overlapping conditions.
4. Finally, we investigate approaches for effectively integrating labels and noisy textual information. Concretely, we propose a novel **D**isentangled **L**anguage-**I**mage-**L**abel **P**re-training, **DLILP**. Unlike existing strategies, DLILP offers a robust trade-off for zero-shot generalization to both known and novel and is scalable to combining image-label and image-text datasets.

2 Related Work

Pre-training and Adapting Visual Recognition Models. Current computer vision applications are fueled by transferring rich pre-trained representations learned on large-scale datasets. Traditionally, pre-training has been driven by human-annotated data for a given set of heterogeneous categories, such as ImageNet [6], via standard cross-entropy or supervised contrastive [19] objectives. More recently, leveraging large-scale datasets with text supervision has gained increasing interest within the computer vision community. In particular, foundation models such as CLIP [29] or ALIGN [16] have shown great success in zero-shot generalization and efficient transfer learning following multi-modal contrastive learning. To also integrate discriminative, label-driven information, UniCL [48] proposed a unified framework by aligning image, text, and label spaces into the same optimization criteria. While UniCL showed superior performance to its supervised or only-text counterparts, concurrent studies [8,21,32,44] have pointed out that transfer learning from supervised pre-training should be done carefully, as specific optimization criteria and network architectures can substantially impact its performance. Concretely, using softmax cosine similarities, trainable temperature scaling, and an MLP projection during pre-training, as commonly used in contrastive pre-training objectives, are key factors for the proper transferability of such models [32].

Large-Scale Vision Models in CXRs. Transfer learning from natural to medical domains, and in particular to radiography images, has been a largely adopted and successful strategy [43] that speeds up convergence and discriminative performance when the training data is limited [30]. To bridge the gap between natural and radiology domains, leveraging large unsupervised datasets via self-supervised learning [2,22] has been exhaustively explored. More recently, the emergence of open-access datasets with radiology reports, i.e., MIMIC [17], has fueled the progress of multimodal models. For example, pre-trained models such as ConVIRT [52] and REFERS [53] demonstrated that incorporating semantic information via language leads to better transferrable features, whereas CheXzero [40] showed radiologist-level performance zero-shot disease recognition. Different strategies are currently being explored to improve pre-training, which include spatial alignment enhancement, i.e., GLoRIA [12] and MGCA [41], masking [54], or using soft similarity matrices [24]. On the other hand, BioViL [3] instead focuses on improving text understanding using domain-specific pre-training of the text encoder. Furthermore, a relevant body of recent literature [45,49,51] explores the integration of supervised labeled datasets to provide larger-scale models. For example, MedCLIP [45] proposed aligning unpaired images and texts through labels, via an asymmetrical soft similarity matrix. CXR-CLIP [49] transforms categorical supervision to text using prompt templates. MedKLIP [46] and KED [51] incorporate domain knowledge and explicitly align the learned representations in the label space. Despite the great efforts devoted to visual-language learning, supervised (i.e., unimodal) pre-training has been surprisingly overlooked, and its potential compared to vision-language models remains unexplored.

From Text to Labels in Radiology Reports. Supervision in chest radiographs naturally comes from text descriptions, which are carried out during clinical routine. These can be accessed in massive amounts from clinical records, and serve as a source to avoid time-consuming image labeling from experienced radiologists. Thanks to the joint effort between radiologists and NLP scientists, several named-entity recognition (NER) tools have been developed, such as Negbio [28] or Chexpert-labeler [14], which are able to extract labels, e.g., diseases and lesions, from text reports. NER algorithms have become the *de facto* solution for labeling large-scale CXR datasets, such as NIH [43], CheXpert [14], MIMIC [17], or PadChest [4]. Although these labels could be imperfect, NER algorithms are highly data-efficient [25]. Moreover, current entity extraction methods are validated on a wide number of conditions (e.g., 14 for CheXpert [14], 20 for NIH [11], or 96 for PadChest [4]). Hence, NER methodologies are continuously improving, and current solutions such as RadGraph [15], RadText [42], or X-Raydar-NLP [7] show promising capabilities.

3 Methodology

3.1 Preliminaries

Problem Setup. We define a quadruplet-wise data format, that generally describes the information available in an assembly of N chest X-ray samples, with text and label supervision, $\mathcal{D}_{ILT} = \{(\mathbf{X}_n, \mathbf{y}_n^{\text{img}}, \mathbf{T}_n, \mathbf{y}_n^{\text{text}})\}_{n=1}^{N}$. $\mathbf{X} \in \mathbb{R}^{\Omega}$ denotes a CXR 2D image, with Ω its spatial domain, and $\mathbf{T} \in \mathcal{T}$ its associated text description. Furthermore, $\mathbf{y} = (y_1, ..., y_c, ..., y_C)$ is a multi-label vector for a set of C *base* categories, such that $y_c \in \{0, 1\}$. Note that for one sample n, the label information associated with the image, $\mathbf{y}_{n,c}^{\text{img}}$, and text description, $\mathbf{y}_{n,c}^{\text{text}}$,

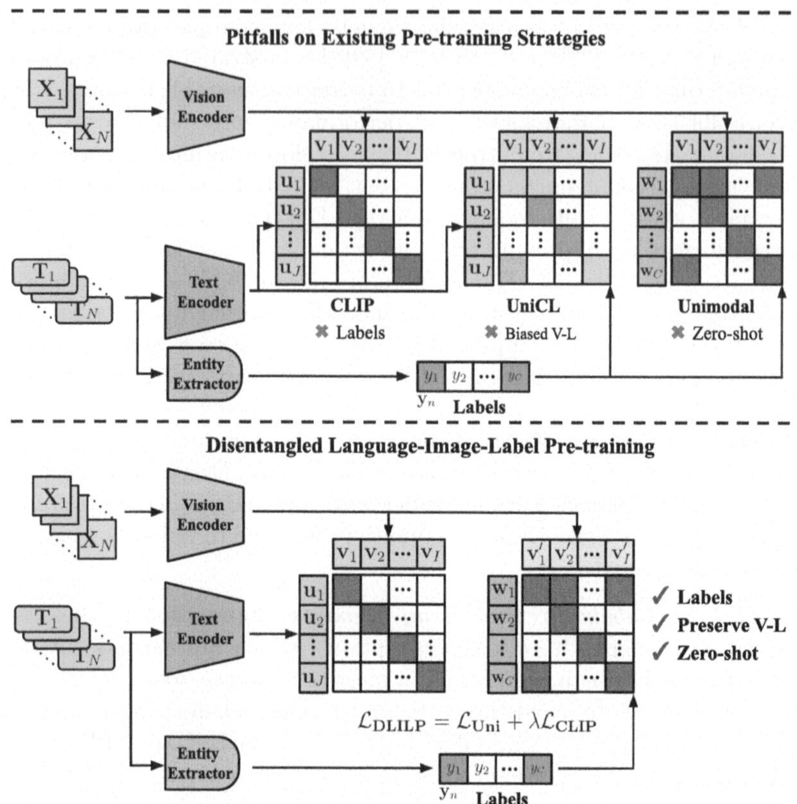

Fig. 1. Training transferable vision models. Radiology reports include text descriptions, from which labels are extracted through entity extractor methods. Previous methods struggle to align language-image-label information without compromising zero-shot generalization—see Sect. 3.2. We propose **DLILP**, a **D**isentangled **L**anguage-**I**mage-**L**abel **P**re-training that exploits text and label supervision in separate feature projections, described at Sect. 3.3.

might be different. $\mathbf{T}_n$ represents an individual sentence of the whole radiology report. Thus, an individual text description can represent semantic information related only to a subset of the categories that are found in the image. Given an assembly of datasets, $\mathcal{D}$, the objective is to **learn a strong visual representation model, specialized for CXR image understanding (see Fig.** 1).

Dual-Encoder Architectures. Let $\theta = \{\theta_f(\cdot), \theta_p(\cdot)\}$ denote the vision encoder, with $\theta_f(\cdot)$ a feature extractor and $\theta_p(\cdot)$ a projection head. The feature extractor $\theta_f(\cdot)$ yields a vision feature representation $\tilde{\mathbf{v}} \in \mathbb{R}^{D_{\mathbf{v}}} : \tilde{\mathbf{v}}_i = \theta_f(\mathbf{X}_i)$ of an input image $\mathbf{X}_i$, with $D_{\mathbf{v}}$ the dimension of the visual feature space. Similarly, let $\phi = \{\phi_f(\cdot), \phi_p(\cdot)\}$ denote the text encoder, $\phi_f(\cdot)$ being a feature extractor and $\phi_p(\cdot)$ a projection head. The feature extractor $\phi_f(\cdot)$ provides a text embedding $\tilde{\mathbf{u}} \in$

$\mathbb{R}^{D_\mathbf{u}} : \tilde{\mathbf{u}}_j = \phi_f(\mathbf{T}_j)$ of an input text $\mathbf{T}_j$, with $D_\mathbf{u}$ denoting the dimensionality of text features. Each of the projection heads, $\theta_p(\cdot)$ and $\phi_p(\cdot)$, maps the independent modality representations into a joint unit hyper-sphere space: $\mathbf{v} = \frac{\theta_p(\tilde{\mathbf{v}})}{||\theta_p(\tilde{\mathbf{v}})||}$ and $\mathbf{u} = \frac{\phi_p(\tilde{\mathbf{u}})}{||\phi_p(\tilde{\mathbf{u}})||}$. In this normalized space, the similarity between image $\mathbf{X}_i$ and text description $\mathbf{T}_j$ is evaluated by the cosine similarity, $\mathbf{v}_i^\top \mathbf{u}_j$, where $\top$ denotes the transpose operator. Optimizing dual-encoder architectures jointly relies on constraining the learned representations to match their textual counterparts and dis-match unpaired ones. The learning process is usually performed in mini-batched stochastic gradient descent. In each step, a batch of indices is randomly retrieved from the assembly dataset, such that $\mathcal{B} \subset \{1, \ldots, N\}$.

3.2 Pitfalls on Existing Pre-training Strategies

CLIP [29]. Designed for image-text datasets, the learning objective aims to guide paired data to produce similar representations and push-away embedding representations from any unpaired image-text or text-image pair. The one-to-one mapping considers a bidirectional contrastive learning objective, $\mathcal{L}_{\text{CLIP}} = \mathcal{L}_{\text{CLIP}}^{\text{i2t}} + \mathcal{L}_{\text{CLIP}}^{\text{t2i}}$, whose components are defined as:

$$\mathcal{L}_{\text{CLIP}}^{\text{i2t}}(\theta, \phi, \tau | \mathcal{B}) = -\sum_{i \in \mathcal{B}} \log \frac{\exp(\mathbf{v}_i^T \mathbf{u}_i / \tau)}{\sum_{j \in \mathcal{B}} \exp(\mathbf{v}_i^T \mathbf{u}_j / \tau))}, \tag{1}$$

$$\mathcal{L}_{\text{CLIP}}^{\text{t2i}}(\theta, \phi, \tau | \mathcal{B}) = -\sum_{j \in \mathcal{B}} \log \frac{\exp(\mathbf{v}_j^T \mathbf{u}_j / \tau)}{\sum_{i \in \mathcal{B}} \exp(\mathbf{v}_i^T \mathbf{u}_j / \tau))}. \tag{2}$$

Even though CLIP loss has proven to be a powerful tool for leveraging large-scale datasets with associated text supervision with minimum supervisory effort, it lacks the fine-grained information that can be found in the form of labels, which the text encoder is assumed to learn. While this does not pose any particular problem in general vision problems, in specialized domains such as medical imaging, with limited data and complex semantics, the text encoder struggles to encode this information efficiently.

UniCL [48] attempts to unify the learning objective across image, text, and label spaces. This is done by modifying the one-to-one similarity matrix in CLIP to a soft-labeled target, by positively pairing images and texts with their labeled categories. The overall training objective, $\mathcal{L}_{\text{UniCL}} = \mathcal{L}_{\text{UniCL}}^{\text{i2t}} + \mathcal{L}_{\text{UniCL}}^{\text{t2i}}$, is defined as:

$$\mathcal{L}_{\text{UniCL}}^{\text{i2t}}(\theta, \phi, \tau | \mathcal{B}) = -\sum_{i \in \mathcal{B}} \frac{1}{|P_{\text{i2t}}(i)|} \sum_{i' \in P_{\text{i2t}}(i)} \log \frac{\exp(\mathbf{u}_i^T \mathbf{v}_{i'} / \tau)}{\sum_{j \in \mathcal{T}_{\mathcal{B}}} \exp(\mathbf{u}_i^T \mathbf{v}_j / \tau)}, \tag{3}$$

$$\mathcal{L}_{\text{UniCL}}^{\text{t2i}}(\theta, \phi, \tau | \mathcal{B}) = -\sum_{j \in \mathcal{B}} \frac{1}{|P_{\text{t2i}}(j)|} \sum_{j' \in P_{\text{t2i}}(j)} \log \frac{\exp(\mathbf{u}_{j'}^T \mathbf{v}_j / \tau)}{\sum_{i \in \mathcal{X}_{\mathcal{B}}} \exp(\mathbf{u}_i^T \mathbf{v}_j / \tau)}, \tag{4}$$

where $|\cdot|$ denotes the cardinality of a given set, and $P_{\text{i2t}}(i)$ and $P_{\text{t2i}}(j)$ represent indices of positive-paired cross-modal representations for each image and text in the batch $\mathcal{B}$, respectively. For the multi-label scenario in CXRs, aligned pairs should contain at least one overlapping category, such that:

$$P_{\text{i2t}}(i) = \{i' | (i' \in \mathcal{B}, \exists c | y_{i',c}^{\text{text}} = y_{i,c}^{\text{img}} = 1)\},$$

$$P_{\text{t2i}}(j) = \{j' | (j' \in \mathcal{B}, \exists c | y_{j',c}^{\text{img}} = y_{j,c}^{\text{text}} = 1)\}.$$

Although UniCL loss encourages learning both a discriminative and semantic-rich feature space, our empirical evidence (see Sect. 4.2, *Observation 2*) suggests that **label information biases vision-language alignment**. In the case of using a reduced set of labeled categories, as is usually the case in medical domains, the learned representations might fail to capture other information contained in text descriptions, thus worsening their discriminative performance on unseen scenarios during label alignment, i.e., zero-shot predictions.

Unimodal Supervised Learning. A classical alternative to pre-train a large-scale vision model using labeled datasets is standard supervised pre-training. In this case, the text encoder is replaced by a linear embedding layer $\tilde{\mathbf{W}}^{C \times D_p}$, with D_p the dimensionality of the projection layer of the visual encoder. In addition, class prototypes are ℓ_2-normalized, such that $\mathbf{W} = \frac{\tilde{\mathbf{W}}}{||\tilde{\mathbf{W}}||}$. In the multi-label scenario, class-wise scores are computed using the sigmoid activation function, $\hat{y} = \sigma(\mathbf{W}^\top \mathbf{v}/\tau)$, and learning is driven by the binary cross entropy loss:

$$\mathcal{L}_{\text{Uni}}(\theta, \tau, \mathbf{W}|\mathcal{B}) = -\sum_{i \in \mathcal{B}} \frac{1}{C} \sum_{c} (y_{i,c}^{\text{img}} \cdot \log(\hat{y}_{i,c}) + (1 - y_{i,c}^{\text{img}}) \cdot \log(1 - \hat{y}_{i,c})). \quad (5)$$

This solution is largely more computationally efficient as it does not involve using a text encoder. In addition, it does not require prompt engineering for generalization on the base categories. A limitation, however, is that only-vision (i.e., unimodal) models lack the capability of zero-shot predictions in novel categories.

3.3 Disentangled Language-Image-Label Pre-training

To address the limitations of label alignment in vision-language pre-training, we propose a **D**isentangled **L**anguage-**I**mage-**L**abel **P**re-training (DLILP) strategy.

Training. Image-label and image-text supervision are incorporated into different subspaces of the learned vision representation. In particular, label supervision is driven by the cross-entropy loss, similar to the Unimodal pre-training, whereas we adopt CLIP loss for image-text alignment. To do so, we define two different projection layers, $\theta_p^{\text{I-L}}$ and $\theta_p^{\text{I-T}}$, which produce ℓ_2-normalized feature spaces. Formally, the DLILP optimization criteria can be defined as follows:

$$\mathcal{L}_{\text{DLILP}} = \mathcal{L}_{\text{Uni}}(\{\theta_f, \theta_p^{\text{I-L}}\}, \tau^{\text{I-L}}, \mathbf{W}|\mathcal{B}) + \lambda \cdot \mathcal{L}_{\text{CLIP}}(\{\theta_f, \theta_p^{\text{I-T}}\}, \phi, \tau^{\text{I-T}}|\mathcal{B}), \quad (6)$$

where λ is a blending hyper-parameter that balances the relative importance of vision-language and vision-label pre-training. Note that we train separate temperature scaling parameters, $\tau^{\text{I-L}}$ and $\tau^{\text{I-T}}$, for each term.

Table 1. Frontal Chest X-ray datasets assembly. We compiled open-access datasets for training and evaluation. Green-colored categories indicate novel classes not explicitly used during CheXpert and MIMIC pre-training.

Pre-train	**#Imgs**	**Text**	**#C**	**Categories**
CheXpert (C) [14]	191,026	-	14	[NF, ECard, Card, LLes, LOp, Edem, Cons,
MIMIC (M) [17]	154,595	✓	14	PnMo, Atel, PnTh, PlEff, PlOt, Fract, Dev]
PadChest (P) [4]	96,201	-	84	(see code)
Evaluation	**#Train**	**#Test**	**#C**	**Categories**
$\text{CheXpert}_{5\times200}$	1,000	1,000	5	[Atel, Card, Cons, Edem, PlEff]
$\text{MIMIC}_{5\times200}$	1,000	1,000	5	[Atel, Card, Cons, Edem, PlEff]
RSNA [34]	8,400	3,600	2	[NF, PnMo]
SSIM [35]	4800	1200	2	[NF, PnTh]
COVID [5,31]	1,200	4,000	4	[Normal, Covid, PnMo, LOp]
NIH-LT [11,43]	920	920	20	[Atel, Card, PlEff, Inf, Mass, Nod, PnMo, PnTh, Cons, Edem, Emph, Fib, PlThi, PnPer, PnMed, SubEm, TAor, CalAor, NF]
VinDr [27]	2,000	2,000	5	[NF, Bro, BrPn, BrLi, PnMo]

Inference. DLILP allows robust generalization over known categories using the learned class prototypes, $\mathbf{W}$, and the image-label projection. In the case of novel categories, zero-shot predictions using engineered text prompts can also be computed, using the unbiased image-text projection of the vision encoder, and the prototypes obtained using the text encoder.

4 Experiments

4.1 Setup

Datasets. Frontal chest X-ray open-access datasets are employed to train and evaluate the transferability of large-scale pre-trained models. Table 1 depicts a summary. To address the **pre-training** stage, we used large datasets such as MIMIC (M) [17] and CheXpert (C) [14]. The 14 *base categories* ($\mathcal{B}$) labeled in the CheXpert dataset are considered for label alignment during pre-training. PadChest (P) [4], containing 84 different labeled findings, is used when specified. Labels are extracted from text-only datasets using CheXpert-labeler [14]. For label-only datasets, the text is obtained using a template similar to [40,48]. To **evaluate** the capabilities of the resulting models, we used seven different datasets: MIMIC [17], CheXpert [14], SSIM [35], RSNA [34], NIH [43], VinDr [27], and COVID [14]. Some of these datasets include *novel diseases* ($\mathcal{N}$), which have not been explicitly used during image-label alignment in the pre-training.

Vision-Language Architecture. We designed both encoders following relevant prior literature in the topic [45,46,49,52]. In particular, we used ResNet-50 [10] pre-trained on ImageNet [6] as a vision encoder, θ, and BioClinicalBERT [1] as the text encoder. All feature projections, i.e., $\theta_p(\cdot)$ and $\phi_p(\cdot)$, $\theta_p^{\text{I-L}}$ and $\theta_p^{\text{I-T}}$, are linear layers of 512 output features, following prior works [29,45].

Large-Scale Training. The vision and text encoders are trained using a batch size of 128 images of 224×224 pixels. AdamW is used as the optimizer, with a weight decay of 10^{-5}, and a base learning rate of 10^{-4}. Cosine scheduler decay is applied for 30 epochs, with an initial first warm-up epoch. The 10% of the training subset is sampled for validation. The same data augmentation used in prior related literature [45] is applied: random horizontal flips, rotations up to 5 degrees, scaling between $[0.9, 1.1]$ factor ranges, and color jittering with brightness and contrast ratios from $[0.8, 1.2]$. Validation loss is monitored epoch-wise during training, and early-stopping is applied with a margin of 5 epochs, saving the best model weights. For DLILP, the λ hyper-parameter is set to 0.1.

Transferability. The transfer capabilities of each pre-training strategy are evaluated in the zero- and few-shot regimes. **a) Zero-shot**: for CLIP and UniCL frameworks, text-driven class-wise prototypes are obtained using an assembly of text prompts, as in [45]. For the Unimodal pre-training, only zero-shot classification on known categories is possible by retrieving the class weights of the target categories, $\mathbf{W}_c$. For DLILP, we follow a hybrid approach, using image-text or image-label projections, depending on whether the target category is known, as detailed in Sect. 3.3. Finally, class-wise scores are obtained in all cases by computing softmax cosine similarity between class prototypes and projected vision features. **b) Linear probing**: we use the vision features before the projection layer, $\tilde{\mathbf{v}}$, to train a linear classifier. Concretely, the same solver proposed in CLIP [29] is used. The adaptation is performed in the popular few-shot regime [33], in which only $K = \{1, 2, 4, 8, 16\}$ images per class are available for adaptation.

Evaluation Protocol. Experiments are repeated using 5 different random seeds. When evaluating using base-only ($\mathcal{B}$) or novel-only ($\mathcal{N}$) target diseases classification, the corresponding subset of categories is separated for adaptation and evaluation. Average class-wise accuracy (ACA) is used as a metric, as in [45].

Baselines. The transferability of the proposed strategies is compared to relevant SoTA models. We gathered the pre-trained weights (when available) and conducted transferability experiments. Concretely, GlorIA [12], MedCLIP [45], BioVIL [3], MedKLIP [46], and KED [51] are included. MedCLIP, MedKLIP, and KED, include label alignment during model pre-training. In particular, MedCLIP pre-training follows a training objective similar to UniCL's.

4.2 Main Results

Observation 1: Unimodal Leads to More Scalable Transferability Than Existing Vision-Language Models. We compare the few-shot transferability of the different pre-training strategies over the 7 downstream datasets. Figure 2(a) includes transferability results with increasing pre-training data, which show that CLIP loss struggles to scale properly when adding label-only datasets (see M+C to M+C+P). In contrast, supervised cross-entropy constantly improves w.r.t. the amount of data available (see M+C or M+C+P). Also, Fig. 2(b) shows few-shot transferability results when pre-trained with M+C

Table 2. Generalization/Transferability results. Performance of different pre-training strategies disentangling known ($\mathcal{B}$) and new findings ($\mathcal{N}$).

	CheXp	MIMIC	SSIM	RNSA	NIH_{LT}		VinDR		**Avg.**		
	$\mathcal{B}$	$\mathcal{B}$	$\mathcal{B}$	$\mathcal{B}$	$\mathcal{B}$	$\mathcal{N}$	$\mathcal{B}$	$\mathcal{N}$	$\mathcal{B}$	$\mathcal{N}$	Avg.
(a) Zero-shot generalization											
CLIP	51.50	49.70	77.80	63.04	40.98	29.10	68.66	32.20	58.61	**30.65**	44.63
UniCL	45.40	46.60	75.30	90.86	57.66	9.10	73.16	42.20	64.83	25.65	45.24
Unimodal	42.80	47.40	77.20	94.60	61.70	-	65.80	-	**64.92**	-	-
DLILP	49.50	48.60	77.90	93.50	60.80	29.10	54.20	31.10	64.08	30.10	**47.09**
(b) Linear probing transferability ($K = 16$)											
CLIP	54.50	49.60	69.10	93.20	46.52	32.50	71.68	38.20	64.10	35.35	49.73
UniCL	53.10	50.90	65.58	93.78	46.50	27.52	71.32	37.54	63.53	32.53	48.03
Unimodal	54.20	53.70	67.68	94.36	47.16	33.20	75.34	37.44	65.41	35.32	50.37
DLILP	55.60	54.50	72.74	93.82	50.66	32.24	71.36	40.76	**66.45**	**36.50**	**51.48**

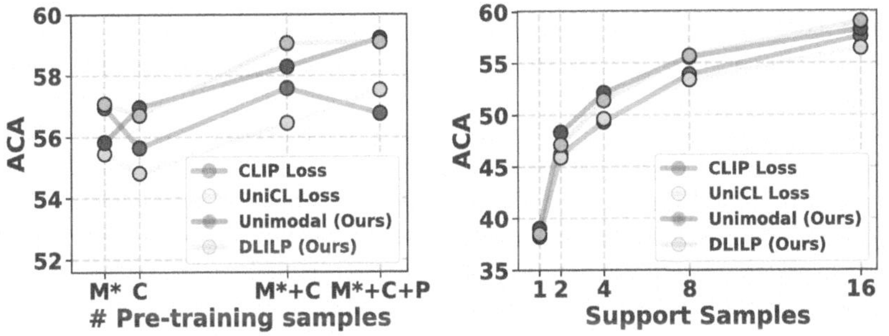

(a) Data scalability

(b) Few-shot adaptation

Fig. 2. Transferability. (a) Effect of increasing pre-training data (K = 16); (b) Few-shot adaptation. Average for 7 tasks. M: MIMIC; C: CheXpert; P: PadChest.

datasets for different shots. Again, Unimodal offers better adaptation than CLIP and UniCL ($K \geq 2$).

Observation 2: Label Alignment During Vision-Language Pre-Training Might Produce Biased Joint Representations. We now study the capability of each pre-training strategy to generalize to novel categories. Results in Table 2 disentangle the zero-shot and linear probing performance between base and new findings. **a) Zero-shot**: UniCL archives average improvements (+6.2%) compared to the original CLIP on known categories thanks to the label information incorporated. However, it largely degrades the performance when evaluated on novel categories (−5.0%). Interestingly, Unimodal pre-training offers the best results for base categories. Note that this strategy is

Table 3. Zero-shot on COVID dataset.

		MedCLIP	MedKLIP	CLIP	UniCL	Unimodal	DLILP
2-class	Disease name	74.1	51.8	69.6	80.5	-	77.0
	Description*	78.8	82.9	74.2	83.7	**85.1**	81.6
4-class	Disease name	40.5	20.2	32.7	45.5	-	36.6
	Description*	42.9	32.5	48.8	44.8	**51.6**	50.0

*"*patchy or confluent, band like ground-glass* ***opacity*** *or* ***consolidation***"

more computationally efficient since does not require training any text encoder. Also, this method does not require heuristic prompt engineering to define the zero-shot text prompts properly, thanks to using learned prototypes. **b) Linear probing**: Again, Unimodal pre-training is largely a competitive solution, with slightly better overall performance compared to CLIP loss (+0.6%). Note that UniCL loss does not show any benefits, reinforcing the inconvenience of this label-driven loss.

Observation 3: The Zero-Shot Capabilities of CXR Vision-Language Models Have Been Overestimated. Prior literature, i.e., MedCLIP [45] and MedKLIP [46], have defended the effectiveness of vision-language pre-training to generalize to unseen diseases thanks to text-driven predictions. These experiments have been typically carried out in the COVID dataset by differentiating between normal and COVID scans using text descriptions (see Table 3). However, this description contains lesions that appeared in the pre-training stage. These findings (i.e., opacities and consolidations) are unspecific [20] and may be correlated with other pneumonia conditions. We extend this benchmark to four categories (see Table 1) in Table 3. In this scenario, the overall performance degrades greatly. More interestingly, following the same zero-shot prediction strategy, we can obtain class-wise prototypes for the Unimodal pre-training by selecting the weights corresponding to the findings in the description. The visual prompt for COVID would be the average embedding between the pre-trained prototypes for opacity and consolidation. Surprisingly, this option outperforms the designed text prompts in VLMs. These observations, combined with the limited generalization observed for novel diseases in Table 2(a), question the advancements claimed in recent literature for open-vocabulary generalization.

DLILP Performance. Although existing vision-language pre-training alternatives offer limited contributions compared to Unimodal, the proposed DLILP objective shows interesting properties. First, DLILP shows better scalability concerning data integration over baseline VLMs (see Fig. 2(a)). Second, DLILP demonstrates robust zero-shot generalization across both base and new categories (see Table 2), with the best average performance across both sets for both zero-shot (+1.9% over UniCL) and few-shot (+3.5% over UniCL).

SoTA Comparison. Table 4 is introduced without base/new disentanglement since prior models might present different base categories (e.g., MedKLIP [46]

Table 4. Available vision-language models transferability. Linear probing results ($K = 16$) for SoTA pre-trained models.

Method	Data	CheXp	MIMIC	SSIM	RNSA	NIH	VinDR	COVID	**Avg.**
MedKLIP [46]	M	34.30	32.60	64.82	88.18	14.04	26.34	68.04	46.90
KED [51]	M	42.50	40.20	66.04	92.12	19.40	26.18	73.24	51.38
BioVIL [3]	M	46.70	43.80	73.68	94.08	21.22	26.20	62.46	52.59
Unimodal	M	51.80	51.30	68.04	93.42	21.20	27.68	77.40	55.83
DLILP	M	53.30	52.90	69.80	93.78	25.34	26.84	77.62	**57.08**
GlorIA [12]	C	46.00	41.60	66.30	91.16	18.78	23.02	72.92	51.40
Unimodal	C	52.30	48.20	71.52	93.88	24.20	29.14	79.48	**56.96**
MedCLIP [45]	M+C	54.40	50.50	69.48	94.20	20.98	27.80	72.30	55.67
CXR-CLIP [49]	M+C	52.20	46.10	69.34	92.00	25.90	26.26	76.82	55.52
Unimodal	M+C	54.20	53.70	67.68	94.36	26.20	30.26	81.62	58.29
DLILP	M+C	55.60	54.50	72.74	93.82	26.72	28.98	81.02	**59.05**

or KED [51]). Unimodal obtains the best results, whose average improvements w.r.t. top competitors range [2.6%, 5.6%]. This observation applies also to models including label information, such as MedCLIP [45], MedKLIP [46], or KED [51].

4.3 Ablation Studies

What Features to Transfer? We evaluate two possibilities: using the features extracted by the vision encoder, $\tilde{v}$, or the ones projected, v. Using the first ones improves base CLIP loss transferability (+2.3%), but especially label-driven learning losses, i.e. UniCL (+2.6%), Unimodal (+2.6%), and DLILP (+3.2%).

DLILP Configuration. Table 5(a) motivates disentangling image-label and image-text supervisory signals in different projections.

On the Effect of λ**.** Table 5(b) studies λ in Eq. 6. Small values of λ offer the best base/novel average performance. Comparing these results to Table 2(b), one could find that λ values between 0.1 and 10 offer average gains to all baselines.

Table 5. DLILP configuration. Linear probe ($K = 16$), across datasets.

(a) Projections	$\{\theta_p\}$	$\{\theta_p^{\text{I-L}}, \theta_p^{\text{I-T}}\}$
Base	65.2	**66.5**$_{(+1.3)}\uparrow$
Novel	35.4	**36.5**$_{(+1.1)}\uparrow$
Avg.	50.3	**51.5**$_{(+1.2)}\uparrow$

(b) Effect of λ	0	0.1	1	10
Base	65.4	66.5	65.9	64.8
Novel	35.3	36.5	36.8	36.1
Avg.	50.4	**51.5**	51.4	50.4

5 Discussion

This work addresses large-scale pre-training for CXR image classification. In this topic, fine-grained labels extracted with specialized entity extraction methods are usually the only available information. However, current literature mostly focuses on (noisy) vision-language pre-training, following CLIP's popularity. As we observe in this work, current experimental designs mask the actual transferability of such networks, especially w.r.t. novel diseases. Indeed, when properly compared with classical unimodal pre-training, such approaches showcase limited advantages. We would want to emphasize that this work does not aim to neglect the unarguable progress made in multimodal learning, e.g., in related topics such as medical report generation. On the contrary, this paper aims to point out better evaluation designs (e.g. differentiating $\mathcal{B}/\mathcal{N}$ conditions) and establish adequate baselines to measure the progress of pre-training strategies in the field, where Unimodal and DLILP are to be taken into consideration.

Acknowledgments. This work was funded by the Natural Sciences and Engineering Research Council of Canada (NSERC). We thank Calcul Québec and Compute Canada.

Disclosure of Interests. The authors have no competing interests to declare that are relevant to the content of this article.

References

1. Alsentzer, E., et al.: Publicly available clinical BERT embeddings. In: Clinical Natural Language Processing Workshop (2019)
2. Azizi, S., et al.: Big self-supervised models advance medical image classification. In: Proceedings of the IEEE International Conference on Computer Vision (ICCV), pp. 3458–3468 (2021)
3. Boecking, B., et al.: Making the most of text semantics to improve biomedical vision–language processing. In: European Conference on Computer Vision (ECCV), pp. 1–21 (2022)
4. Bustos, A., Pertusa, A., Salinas, J.M., de la Iglesia-Vayá, M.: Padchest: a large chest x-ray image dataset with multi-label annotated reports. Med. Image Anal. **66**, 101797 (2020)
5. Chowdhury, M., et al.: Can ai help in screening viral and covid-19 pneumonia? IEEE Access **8**, 132665–132676 (2020)

6. Deng, J., Dong, W., Socher, R., Li, L.J., Li, K., Fei-Fei, L.: Imagenet: a large-scale hierarchical image database. In: Proceedings of the IEEE/CVF Conference on Computer Vision and Pattern Recognition (CVPR), pp. 248–255 (2009)
7. Dicente Cid, Y., et al.: Development and validation of open-source deep neural networks for comprehensive chest x-ray reading: a retrospective, multicentre study. The Lancet Digital Health **6** (12 2023)
8. Feng, Y., Jiang, J., Tang, M., Jin, R., Gao, Y.: Rethinking supervised pre-training for better downstream transferring. In: International Conference on Learning Representations (ICLR) (2022)
9. Finlayson, S.G., Subbaswamy, A., Singh, K., Bowers, J., Kupke, A., Zittrain, J., Kohane, I.S., Saria, S.: The clinician and dataset shift in artificial intelligence. N. Engl. J. Med. **385**, 283–286 (2021)
10. He, K., Zhang, X., Ren, S., Sun, J.: Deep residual learning for image recognition. In: Proceedings of the Conference on Computer Vision and Pattern Recognition (CVPR), pp. 1–12, December 2016
11. Holste, G., et al.: Long-tailed classification of thorax diseases on chest x-ray: A new benchmark study. In: MICCAI Workshop on Data Augmentation, Labelling, and Imperfections, pp. 22–32 (2022)
12. Huang, S.C., Shen, L., Lungren, M.P., Yeung, S.: Gloria: a multimodal global-local representation learning framework for label-efficient medical image recognition. In: Proceedings of the IEEE International Conference on Computer Vision (ICCV), pp. 3942–3951 (2021)
13. Huang, Z., Bianchi, F., Yuksekgonul, M., Montine, T., Zou, J.: A visual-language foundation model for pathology image analysis using medical twitter. Nat. Med. **29**, 1–10 (2023)
14. Irvin, J., Rajpurkar, P., et al.: Chexpert: a large chest radiograph dataset with uncertainty labels and expert comparison. In: Proceedings of the AAAI Conference on Artificial Intelligence 33, pp. 590–597 (2019)
15. Jain, S., et al.: Radgraph: Extracting clinical entities and relations from radiology reports. In: NeurIPS Datasets and Benchmarks Track (2021)
16. Jia, C., et al.: Scaling up visual and vision-language representation learning with noisy text supervision. In: International Conference on Machine Learning (ICML), pp. 1–13 (2021)
17. Johnson, A., et al.: Mimic-cxr, a de-identified publicly available database of chest radiographs with free-text reports. Sci. Data **6**, 317 (2019)
18. Kamath, A., Hessel, J., Chang, K.W.: What's "up" with vision-language models? investigating their struggle with spatial reasoning. In: Empirical Methods in Natural Language Processing (EMNLP) (2023)
19. Khosla, P., et al.: Supervised contrastive learning. In: Larochelle, H., Ranzato, M., Hadsell, R., Balcan, M., Lin, H. (eds.) Advances in Neural Information Processing Systems (NeurIPS), vol. 33, pp. 18661–18673 (2020)
20. Kong, W., Agarwal, P.P.: Chest imaging appearance of covid-19 infection. Radiology: Cardiothoracic Imaging **2**(1) (2020)
21. Kornblith, S., Chen, T., Lee, H., Norouzi, M.: Why do better loss functions lead to less transferable features? In: Advances in Neural Information Processing Systems (NeurIPS) (2021)
22. Krishnan, R., Rajpurkar, P., Topol, E.: Self-supervised learning in medicine and healthcare. Nature Biomed. Eng. **6**, 1–7 (2022)
23. Litjens, G., et al.: A survey on deep learning in medical image analysis. Medical Image Analysis **42** (2017)

24. Liu, B., Lu, D., Wei, D., Wu, X., Wang, Y., Zhang, Y., Zheng, Y.: Improving medical vision-language contrastive pretraining with semantics-aware triage. IEEE Trans. Med. Imaging **42**, 3579–3589 (2023)
25. McDermott, M., Hsu, T., Weng, W.H., Ghassemi, M., Szolovits, P.: Chexpert++: Approximating the chexpert labeler for speed, differentiability, and probabilistic output. In: Machine Learning for Healthcare (MHLC) (2020)
26. Moor, M., Banerjee, O., Abad, Z.S.H., Krumholz, H.M., Leskovec, J., Topol, E.J., Rajpurkar, P.: Foundation models for generalist medical artificial intelligence. Nature **616**, 259–265 (4 2023)
27. Nguyen, H.Q., et al.: Vindr-cxr: an open dataset of chest x-rays with radiologist's annotations. Sci. Data **9** (2022)
28. Peng, Y., Wang, X., Lu, L., Bagheri, M., Summers, R., Lu, Z.: Negbio: a high-performance tool for negation and uncertainty detection in radiology reports. In: AMIA Informatics Research (2018)
29. Radford, A., et al.: Learning transferable visual models from natural language supervision. In: International Conference on Machine Learning (ICML), pp. 8748–8763 (2021)
30. Raghu, M., Zhang, C., Kleinberg, J., Bengio, S.: Transfusion: Understanding transfer learning for medical imaging. In: Advances in neural information processing systems (NeurIPS), pp. 1–11 (2019)
31. Rahman, T., et al.: Exploring the effect of image enhancement techniques on covid-19 detection using chest x-ray images. Comput. Biol. Med. **132**, 104319 (2021)
32. Sariyildiz, M.B., Kalantidis, Y., Alahari, K., Larlus, D.: No reason for no supervision: Improved generalization in supervised models. In: International Conference on Learning Representations (ICLR) (2023)
33. Shakeri, F., et al.: Few-shot adaptation of medical vision-language models. In: Medical Image Computing and Computer Assisted Intervention (MICCAI), pp. 553–563 (2024)
34. Shih, G., et al.: Augmenting the national institutes of health chest radiograph dataset with expert annotations of possible pneumonia. Radiology: Artificial Intelligence **1** (2019)
35. SIIM-ACR: SIIM-ACR Pneumothorax Segmentation Kaggle Challenge. https://siim.org/page/pneumothorax_challenge
36. Silva-Rodríguez, J., Hajimiri, S., Ayed, I.B., Dolz, J.: A closer look at the few-shot adaptation of large vision-language models. In: Proceedings of the IEEE/CVF Conference on Computer Vision and Pattern Recognition (CVPR), pp. 23681–23690 (2024)
37. Silva-Rodríguez, J., Chakor, H., Kobbi, R., Dolz, J., Ayed, I.B.: A foundation language-image model of the retina (flair): Encoding expert knowledge in text supervision. Med. Image Anal. **99**, 103357 (2025)
38. Smit, A., Jain, S., Rajpurkar, P., Pareek, A., Ng, A., Lungren, M.: Combining automatic labelers and expert annotations for accurate radiology report labeling using BERT. In: Proceedings of the 2020 Conference on Empirical Methods in Natural Language Processing (EMNLP), pp. 1500–1519 (2020)
39. Tang, Y., Yamada, Y., Zhang, Y.M., Yildirim, I.: When are lemons purple? the concept association bias of vision-language models. In: Empirical Methods in Natural Language Processing (EMNLP) (2023)
40. Tiu, E., Talius, E., Patel, P., Langlotz, C., Ng, A., Rajpurkar, P.: Expert-level detection of pathologies from unannotated chest x-ray images via self-supervised learning. Nature Biomed. Eng. **6**, 1–8 (2022)

41. Wang, F., Zhou, Y., Wang, S., Vardhanabhuti, V., Yu, L.: Multi-granularity cross-modal alignment for generalized medical visual representation learning. In: Advances in Neural Information Processing Systems (NeurIPS) (2022)
42. Wang, S., Lin, M., Ding, Y., Shih, G., lu, Z., Peng, Y.: Radiology text analysis system (radtext): Architecture and evaluation. In: IEEE International Conference on Healthcare Informatics (ICHI), vol. 2022, pp. 288–296 (2022)
43. Wang, X., Peng, Y., Lu, L., Lu, Z., Bagheri, M., Summers, R.: Chestx-ray8: Hospital-scale chest x-ray database and benchmarks on weakly-supervised classification and localization of common thorax diseases. In: Proceedings of the IEEE/CVF Conference on Computer Vision and Pattern Recognition (CVPR), pp. 3462–3471 (2017)
44. Wang, Y., Tang, S., Zhu, F., Bai, L., Zhao, R., Qi, D., Ouyang, W.: Revisiting the transferability of supervised pretraining: an mlp perspective. In: Proceedings of the IEEE/CVF Conference on Computer Vision and Pattern Recognition (CVPR) (2022)
45. Wang, Z., Wu, Z., Agarwal, D., Sun, J.: Medclip: Contrastive learning from unpaired medical images and text. In: Empirical Methods in Natural Language Processing (EMNLP), pp. 1–12 (2022)
46. Wu, C., Zhang, X., Zhang, Y., Wang, Y., Xie, W.: Medklip: Medical knowledge enhanced language-image pre-training for x-ray diagnosis. In: Proceedings of the IEEE/CVF International Conference on Computer Vision (ICCV), pp. 21372–21383 (2023)
47. Wójcik, M.A.: Foundation models in healthcare: opportunities, biases and regulatory prospects in Europe. EGOVIS **13429**, 32–46 (2022)
48. Yang, J., Li, C., Zhang, P., Xiao, B., Liu, C., Yuan, L., Gao, J.: Learning transferable visual models from natural language supervision. In: Proceedings of the IEEE/CVF Conference on Computer Vision and Pattern Recognition (CVPR), pp. 19141–19151 (2022)
49. You, K., et al.: Cxr-clip: toward large scale chest x-ray language-image pre-training. In: Medical Image Computing and Computer Assisted Intervention (MICCAI), pp. 101–111 (2023)
50. Zhang, S., Metaxas, D.: On the challenges and perspectives of foundation models for medical image analysis. Med. Image Anal. **91**, 102996 (2024)
51. Zhang, X., Wu, C., Zhang, Y., Xie, W., Wang, Y.: Knowledge-enhanced visual-language pre-training on chest radiology images. Nat. Commun. **14**(1), 4542 (2023)
52. Zhang, Y., Jiang, H., Miura, Y., Manning, C.D., Langlotz, C.P.: Contrastive learning of medical visual representations from paired images and text. In: Machine Learning for Healthcare (MHLC), pp. 1–24 (2022)
53. Zhou, H.Y., Chen, X., Yinghao, Z., Luo, R., Wang, L., Yu, Y.: Generalized radiograph representation learning via cross-supervision between images and free-text radiology reports. Nature Mach. Intell. **4**, 1–9 (2022)
54. Zhou, H.Y., Lian, C., Wang, L., Yu, Y.: Advancing radiograph representation learning with masked record modeling. In: International Conference on Learning Representations (ICLR) (2023)

Shape Analysis

ToothForge: Automatic Dental Shape Generation Using Synchronized Spectral Embeddings

Tibor Kubík[1,2(✉)], François Guibault[1], Michal Španěl[2], and Hervé Lombaert[1]

[1] Polytechnique Montréal, Montreal, Canada
ikubik@fit.vut.cz
[2] Brno University of Technology, Brno, Czech Republic

Abstract. We introduce *ToothForge*, a spectral approach for automatically generating novel 3D teeth, effectively addressing the sparsity of dental shape datasets. By operating in the spectral domain, our method enables compact machine learning modeling, allowing the generation of high-resolution tooth meshes in milliseconds. However, generating shape spectra comes with the instability of the decomposed harmonics. To address this, we propose modeling the latent manifold on *synchronized* frequential embeddings. Spectra of all data samples are aligned to a common basis prior to the training procedure, effectively eliminating biases introduced by the decomposition instability. Furthermore, synchronized modeling removes the limiting factor imposed by previous methods, which require all shapes to share a common fixed connectivity. Using a private dataset of real dental crowns, we observe a greater reconstruction quality of the synthetized shapes, exceeding those of models trained on unaligned embeddings. We also explore additional applications of spectral analysis in digital dentistry, such as shape compression and interpolation. ToothForge facilitates a range of approaches at the intersection of spectral analysis and machine learning, with fewer restrictions on mesh structure. This makes it applicable for shape analysis not only in dentistry, but also in broader medical applications, where guaranteeing consistent connectivity across shapes from various clinics is unrealistic. The code is available at https://github.com/tiborkubik/toothForge.

Keywords: 3D tooth shape generation · Digital dentistry · Spectral shape learning · Geometric deep learning

1 Introduction

3D representations provide important insight into the analysis of anatomical shapes. In digital dentistry, they enable advances in diagnosis, treatment planning, and patient-specific prosthetics design. For tasks within this domain, such as automatic crown design, data-driven approaches have been proposed [11,24,28], offering automated solutions to improve the efficiency of lab technicians.

I. Oguz et al. (Eds.): IPMI 2025, LNCS 15830, pp. 313–326, 2026.
https://doi.org/10.1007/978-3-031-96625-5_21

However, dental shape datasets are typically limited in size due to reasons such as privacy concerns and expensive annotations. In addition, they are imbalanced due to anatomical factors. For example, third molars are less commonly represented due to their frequent extraction in clinical practice. These persistent real-world challenges are unlikely to change, hindering the potential of data-driven tasks that rely on rich datasets to train robust models. Synthetic data generation has emerged as a solution to address data scarcity and class imbalance in 2D domain [5]. Our work explores synthetic shape generation to enhance the performance of learning-based analysis of 3D digital dentistry, motivated and resolved by what follows.

Tackling Shape Dataset Scarcity is Challenging. Although synthetic data generation has been demonstrated in the 2D medical image domain [7], generating synthetic data for 3D shapes introduces unique difficulties. Non-Euclidean 3D shapes require intricate geometric operations without the benefit of an underlying regular grid. Maintaining spatial relationships further increases computational complexity [5]. The requirement of generating new samples *on-the-fly* adds to the challenge, particularly for 3D shape data.

Choosing the Appropriate 3D Represention is Key. The choice of data representation plays a critical role in designing machine learning models for shape analysis. A diverse range of shape representations has been explored, including projection-based methods [13,23] and 3D grid approaches [18,25]. Although projection-based techniques allow for the adoption of image-domain architectures, they may introduce ambiguities in selecting optimal projections. Volumetric methods enable straightforward 3D convolution designs but face high memory demands. Other approaches operate directly on the mesh structure [4,10,12], which preserves connectivity information but requires subsampling that leads to the loss of fine anatomical details. This is a critical problem for accurate medical shape analysis. Point cloud representations are a compelling alternative. They eliminate the need for connectivity information, relying solely on point-wise features [19,20,27,29]. This representation has shown effectiveness in various applications, including 3D generative modeling [1] and the development of anatomical statistical shape models [2]. However, operating in the spectral domain enables shape analysis to operate in the frequential domain, unlocking more compact and efficient approaches when analyzing complex shapes [21,22].

The Key Lies in Shape Spectra. Spectral coefficients encode a shape geometry through its intrinsic properties, often requiring only a limited set of harmonics to capture key features effectively. Additionally, spectral representations allow for truncating higher-frequency components as shown in Fig. 1. This reduces data dimensionality while retaining essential characteristics, making learning tasks computationally more efficient. These benefits contrast with spatial representations, which often scale inefficiently with resolution and lose significant information during stratification. What is more, spectral representations are inherently ordered, reducing the need for expensive neighbor searches in high-dimensional feature spaces. The effectiveness of spectral approaches has been demonstrated in various learning tasks [3,8,9,16]. These harmonics are however inherently unstable, and training on such coefficients introduces unwanted distortions into the network. We address this issue by using synchronized coefficients during training to eliminate the bias. We also explore spectral analysis tools in dental imaging for tasks such as tooth compression.

Using Spectral Generative Models on Medical Shapes is Challenging. Although prior work on spectral shape analysis exists [21], the concept of modeling latent spaces based on spectral coefficients for generating novel shapes was first introduced in [14]. Here, the authors trained a spectral autoencoder (AE) on a dataset of human poses and showed promising results in shape interpolation. However, its application to medical shape data remains challenging for at least two reasons. Firstly, AEs lack the regularization required to ensure a smooth latent space, which is particularly important for small-scale datasets such as those in medical domains. Without this, sampling in latent space voids yields implausible reconstructions. Our model design choice is β-VAE with cyclical annealing schedule [6]. We chose a β-VAE over an AE or a standard VAE for its ability to balance reconstruction accuracy and feature disentanglement. This allows better control over the trade-off between geometric fidelity and a smooth latent space, which is essential for generating plausible shapes. More complex models such as GANs or diffusion models were avoided due to their data-intensive nature and training instability, making them less suitable for low-scale datasets.

More critically, the framework in [14] and its follow-up work [15] requires fixed connectivity of all training meshes. This makes it unapplicable in real-world medical scenarios, where shapes scanned at different clinics using various intra-oral scanners have different connectivity. To address this, we introduce *spectral synchronization* during training, aligning all shape spectra to a common base. This alignment ensures that the model learns from spectral coefficients in a shared spectral space, regardless of connectivity. It also minimizes the bias caused by the instability of harmonics, producing more reliable predictions.

Our Contributions in Spectral Shape Learning and Digital Dentistry. We introduce an efficient and accurate approach to real-time generation of synthetic tooth shapes utilising spectral analysis. Beyond exploring low-pass filtering for tooth crowns, spectral alignment, and potentionally correspondence, we are the first to utilize modal coefficients to train a spectral autoencoder in digital dentistry. A key contribution of our work is the use of synchronized embeddings during training, a novel approach that eliminates noise arising from the inherent instability of harmonics. This advancement also enables the use of datasets with varying mesh connectivity, advancing the state-of-the-art in this field. Another important outcome of this solution, which arises naturally from training on synchronized frequency coefficients, is the vertex-wise correspondence among all generated shapes. We show how critical design choices ensure both high fidelity in the generated shapes and smoothness in the latent space. Comparative benchmarks highlight the superiority of our method over spatial approaches.

2 Methodology

We first outline the principles of differential operators and spectral analysis on manifolds. Building on this, we discuss an approach that utilizes spectral coefficients as features for generative model training. We address limitations inherent in this method by introducing the concept of spectral synchronization and latent space regularization. This enhances the versatility of the framework in real-world medical scenarios such as tooth generation.

2.1 Spectral Decomposition on Teeth Shapes

Discrete Manifolds. In the discrete setting, a manifold $\mathcal{X}$ is sampled at n points in $\mathbb{R}^3$, and its approximation is given by a triangular mesh $(\mathcal{V}, \mathcal{E}, \mathcal{F})$, where $\mathcal{V} = \{v, \cdots, v_n\}$ is a set of the sampled points $v = (x, y, z) \in \mathbb{R}^3$, $\forall v \in \mathcal{V}$, edges $\mathcal{E} \subseteq \mathcal{V}^2$ and faces $\mathcal{F} \subseteq \mathcal{V}^3$. Based on linear FEM, the discretization of the Laplace-Beltrami operator $\Delta_{\mathcal{X}} f$ takes the form of an $n \times n$ sparse matrix $\mathbf{L} = \mathbf{A}^{-1}\mathbf{W}$. Here, the *mass matrix* $\mathbf{A}$ is a sparse diagonal matrix of elements $a_i = \frac{1}{3}\sum_{jk:(ijk)\in\mathcal{F}} A_{i,j,k}$ where $A_{i,j,k}$ denotes the triangle area formed by vertices i, j and k. $\mathbf{W}$ is a symmetric matrix of edge-wise weights (equivalent to classical *cotangent formula* [17]):

$$\mathbf{W}_{ij} = \begin{cases} -(cot\alpha_{ij} + cot\beta_{ij}) & i \neq j, v_j \in N_1(v_i), \\ \sum_{v_j \in N_1(v_i)} (cot\alpha_{ij} + cot\beta_{ij}) & i = j, \\ 0 & v_j \notin N_1(v_i), \end{cases} \tag{1}$$

where $N_1(v_i)$ is the 1-ring neighborhood of the vertex $v_i \in \mathcal{V}$, and α_{ij} and β_{ij} are angles opposite to the edge formed by vertices $v_i \in \mathcal{V}$ and $v_j \in \mathcal{V}$.

Spectral Analysis on Meshes. In Riemannian geometry, the orthogonal eigenbasis of the Laplacian obtained by the eigendecomposition process is used to define an analogy of the Fourier transform. In matrix notation, it is defined as a square matrix $\mathbf{L} = \mathbf{\Phi}\mathbf{\Lambda}\mathbf{\Phi}^T$. Here, $\mathbf{\Lambda}$ is a diagonal matrix of real, non-negative eigenvalues $\lambda_i \in \mathbb{R}$, $\mathbf{\Lambda} = \text{diag}(\lambda_1, \ldots, \lambda_n)$, where $\lambda_1 = 0 \leq \cdots \leq \lambda_n$. Note that the eigenvalues are increasingly ordered due to the inverse of $\mathbf{A}$ in the Laplacian calculation. $\mathbf{\Phi}$ is a matrix of the corresponding eigenvectors $\mathbf{\Phi} = (\phi_i, \ldots, \phi_n)$, such that $\mathbf{L}\phi_i = \lambda_i\phi_i$, $\forall i = 1, \ldots, n$, $\phi_i \in \mathbb{R}^n$. The low frequencies correspond to the smallest eigenvalues, encoding coarse shape information (such as crown length of incisors). Large eigenvalues correspond to high-frequency shape information such as molar cusp morphologies, exhibiting rapid oscillations. In addition, any function defined at the vertices of a mesh can be represented as:

$$g = \sum_{i=1}^{n} \langle f, \phi_i \rangle \phi_i = \mathbf{\Phi}\mathbf{\Phi}^T f. \tag{2}$$

Here, $\langle f, \phi_i \rangle$ represent *spectral coefficients*, frequency coordinates that contain information about the geometry of the original vertices in a compressed way.

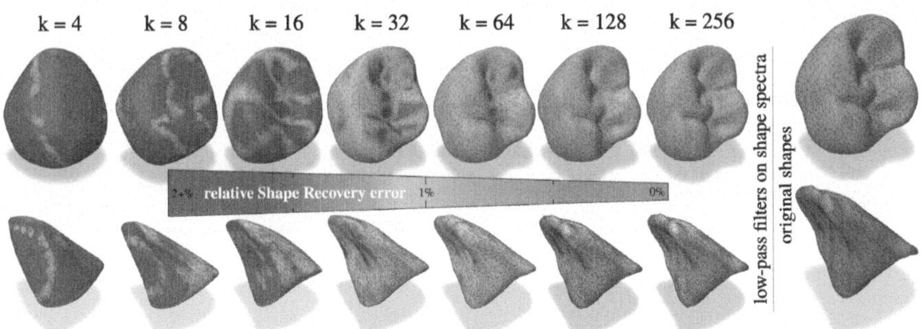

Fig. 1. Example of low-pass filtering on molar and incisor shapes. Higher spectral components preserve finer geometric details. With as low as 128 coefficients, the recovered shapes preserve most of the anatomical details.

A low-dimensional representation of a mesh can be obtained by keeping only the leading spectral coefficients. We consider all mesh vertices as a matrix $\mathbf{V} \in \mathbb{R}^{n \times 3}$ whose columns define their x, y and z positions in space. Given $\mathbf{V}$, we can reconstruct an approximate version of the mesh by using the k-truncated eigenbasis:

$$\mathbf{V}^k = \sum_{i=1}^{k} \langle \mathbf{V}, \phi_i \rangle \phi_i = \mathbf{\Phi}_k \mathbf{\Phi}_k^T \mathbf{V}. \tag{3}$$

To quantify the information loss, one can measure the *Shape Recovery (SR) error* defined as the mean squared error between vertex positions of the original and smoothed shape. Examples of tooth approximations are provided in Fig. 1.

2.2 Spectral Autoencoding: The Baseline Approach

Spectral Autoencoder (*SAE-LP-k*), a framework originally proposed in [14], utilizes the concepts from spectral analysis. SAE-LP-k processes 3D meshes by first transforming their spatial vertices into spectral coefficients. This is achieved by projecting each vertex onto its corresponding eigenspace using a truncated set of k eigenvectors, thus forming k-banded spectral coefficients. The computation of these coefficients serves as input to the autoencoder. Mathematically, the spectral coefficients $\mathbf{C}_k$ using k-truncated eigenbasis $\mathbf{\Phi} \in \mathbb{R}^{n \times k}$ for a mesh with n vertices are computed as:

$$\mathbf{C}_k = \mathbf{\Phi}_k^T \mathbf{V}, \tag{4}$$

where $\mathbf{V} \in \mathbb{R}^{n \times 3}$ represents the spatial coordinates of the mesh vertices. This approach addresses issues such as the high number and disordered arrangement of vertices, making the network invariant to rotation, scale, and translation.

Spectral autoencoder function consists of an encoder $e_\theta(z|\mathbf{C}_k)$ and a decoder $d_\gamma(\hat{\mathbf{C}}_k|z)$. $e_\theta : \mathbb{R}^{k \times 3} \rightarrow \mathbb{R}^d$ maps the spectral coefficients to a single deterministic

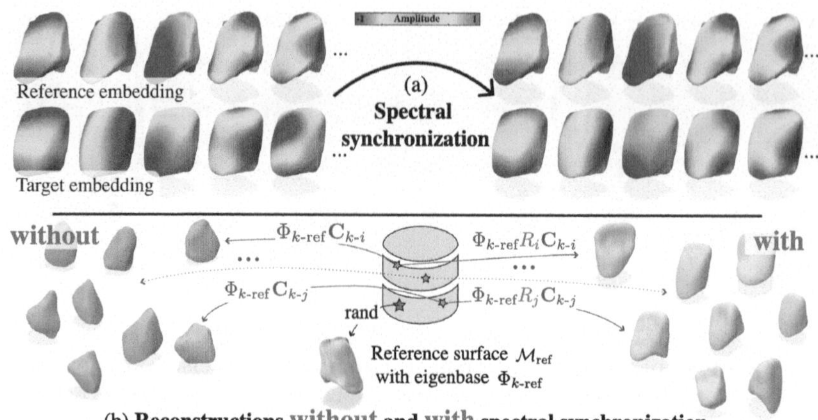

Fig. 2. Impact of spectral synchronization. We perform spectral synchronization across all shapes in the dataset as depicted in (a). This process standardizes the harmonics, ensuring they all *speak a common language*. As illustrated in (b), ommiting this step leads to incorrect reconstructions, introducing significant noise into the network when trained on unaligned data.

point in d-dimensional latent space, and $d_\gamma : \mathbb{R}^d \to \mathbb{R}^{k\times 3}$ reconstructs the coefficients back to the spectral domain. The reconstruction loss $\mathcal{L}_{\text{rec}} = \|\mathbf{C}_k - \hat{\mathbf{C}}_k\|^2$ ensures fidelity to the original representation within the spectral domain, without reverting to the spatial domain. The parameters of the encoder and decoder are both computed by convolutional neural networks: since the array of spectral coefficients is already ordered, the network is able to learn meaningful features by sliding a convolution kernel over coefficients. The learnable (un)pooling procedure is also simple and resembles classical (un)pooling operations.

2.3 Eliminating Constant Connectivity Constraints via Spectral Synchronization

While the spectral autoencoder effectively learns a latent representation from modal coefficients, it operates under the restrictive assumption that all shapes in the training dataset have identical connectivity. This assumption allows for a direct computation of spectral coefficients using a common k-truncated eigenbase $\mathbf{\Phi}_{k-\text{ref}}$ since the per-vertex correspondence is well defined among all shapes within the dataset. However, this requirement is unrealistic for medical datasets, where meshes exhibit variable connectivities. Moreover, since the exact spectrum is rarely known and is usually approximated numerically, this introduces several issues, leading to incompatible bases between meshes. *Sign flips* can occur, as all scalar multiples of an eigenfunction are contained in the same eigenspace. Arbitrary basis could be constructed of *higher dimensional eigenspaces*. *Switching of eigenfunctions* can occur, where two close eigenvalues can switch their order due to numerical instabilities or geometry deformations.

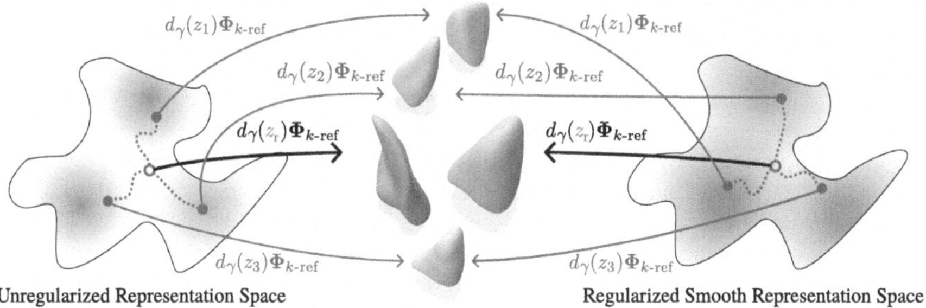

Fig. 3. Effect on regularization of latent space modeled on synchronized spectral coefficients. The process denoted by the solid arrows $d_\gamma(z)\Phi_{k-\text{ref}}$ involves decoding the latent vector and projecting the result through a common basis into the spatial domain. If z is sampled from a void region of the manifold, the resulting geometry becomes corrupted. Regularization reduces the likelihood of such cases, ensuring more reliable outputs.

We utilize a robust computation of $\Phi_{k-\text{ref}}$ to develop a method that remains invariant to connectivity and vertex number differences among shapes within the dataset, and that suppresses any biases caused by instable computation of harmonics. Given a dataset of meshes $\{\mathcal{M}_i\}_{i=1}^N$ each with its respective k-truncated basis Φ_{k-i} and coefficients $\mathbf{C}_{k-i}$, we randomly select a mesh $\mathcal{M}_{\text{ref}}$ and compute its truncated basis $\Phi_{k-\text{ref}} \in \mathbb{R}^{n \times k}$. We assume the existence of a spectral transformation R_i with dimensionality of $k \times k$ for each shape $\mathcal{M}_i$:

$$R_i = ((\Phi_{k-\text{ref}})^T \Phi_{k-\text{ref}})^{-1} (\Phi_{k-\text{ref}})^T \Phi_i c_i, \quad \forall i \in \{1, \ldots, N\}, \tag{5}$$

where c_i is an unknown correspondence map that matches the rows of $\Phi_{k-\text{ref}}$ with the equivalent rows of Φ_i. The transformation relies on identifying the correspondence map c_i. This map is optimized by minimizing the L_2-norm difference between the aligned spectral coefficients:

$$\|\Phi_{k-\text{i}} c_i \mathbf{C}_i - \Phi_{k-\text{ref}} R_i \mathbf{C}_i\|^2, \tag{6}$$

where $\mathbf{C}_i$ is the spectral coefficient array of $\mathcal{M}_i$. Symmetry is enforced by adding the inverse mapping. After alignment, the spectral coefficients of all shapes are represented in the common reference basis, enabling consistent downstream processing. Additionally, c_i provides vertex-wise correspondence between each shape and the reference. The bias introduced during training without synchronization is illustrated in Fig. 2.

2.4 Smooth Latent Manifold Design on Synchronized Spectral Coefficients with β-VAEs

While (synchronized) spectral autoencoders offer a powerful means of compressing and reconstructing spectral coefficients of 3D shapes, their use for generating novel samples given a sparse initial dataset is limited. When working with

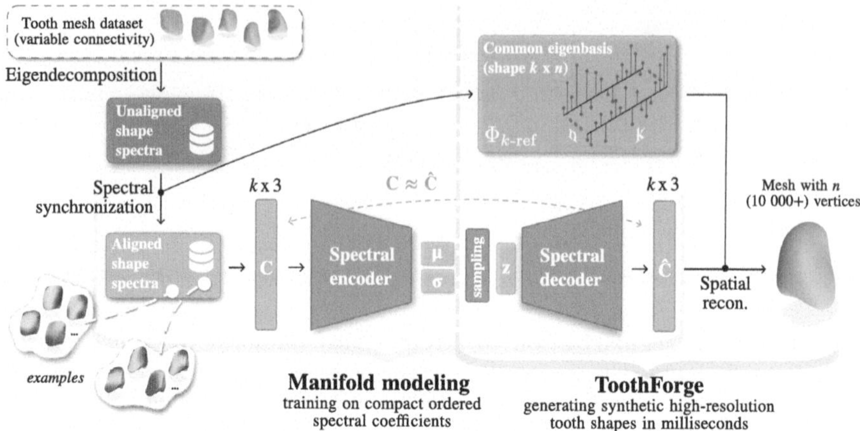

Fig. 4. Method outline. Our framework for generating synthetic shapes of teeth via sampling on a latent manifold. Such manifold is modeled using spectral *synchronized* (spectrally aligned to common basis) coefficients of tooth shapes, denoted as C. For novel data sampling using ToothForge, two ingredients are necessary: decoder weights for inferring novel modal coefficient $\hat{C}$ and common eigenbasis $\mathbf{\Phi}_{k-\text{ref}}$ to project it to spatial domain. Synthetising novel tooth shape with n (10 000+) vertices is performed in order of miliseconds thanks to operating in the spectral domain.

low-scale medical shape datasets, it becomes especially challenging to model a smooth latent manifold. Small datasets lack the diversity needed to cover the full space of possible shapes, leading to *gaps* in the learned latent space.

We modify the encoder function e_θ defined in Sect. 2.2 by introducing a stochastic component, mapping synchronized spectral coefficients into vectors of mean μ and variance Σ, representing a distribution: $e_\theta(z|R\mathbf{C}_k) \sim \mathcal{N}(\mu, \Sigma)$. Notice that the coefficients are aligned with their corresponding spectral alignment matrix so all shapes are represented on a common basis. The decoder $d_\gamma(\hat{\mathbf{C}}_k|z)$ remains unchanged, but since the network is trained on aligned spectra, the reconstructed coefficients $\hat{\mathbf{C}}_k$ follow the same trend. The objective is a combination of the reconstruction loss and β-weighted Kullback-Leibler (KL) divergence, balancing between reconstruction fidelity and distribution approximation:

$$\mathcal{L} = \|R\mathbf{C}_k - \hat{\mathbf{C}}_k\|^2 + \beta \text{KL}(e_\theta(z|R\mathbf{C}_k)|p(z)). \tag{7}$$

Positive values of β puts pressure on the bottleneck to match the prior $p(z)$. The effect of this regularization on teeth spectral generation is visualized in Fig. 3.

Post-training, two ingredients are necessary to generate new data: trained decoder weights ϕ and common eigenbase $\mathbf{\Phi}_{k-\text{ref}}$. Novel vertex matrix $\mathbf{V}_{\text{syn}} = d_\gamma(z_r)\mathbf{\Phi}_{k-\text{ref}} = \hat{\mathbf{C}}_k\mathbf{\Phi}_{k-\text{ref}}$ is coupled with the edges $\mathcal{E}$ of the reference mesh $\mathcal{M}_{\text{ref}}$ to form a realistic high-resolution synthetic mesh. The visual summary of the whole framework can be seen in Fig. 4.

3 Experimental Setup

Dataset. We conducted our experiments on a private dataset of 430 tooth shapes provided by industrial partner, approved by an ethics committee. It is divided into three categories: incisors, premolars and molars. The canine class was excluded from the experiments due to insufficient data availability. Separate models were trained for each category. The samples are fully anonymized, containing only the mesh data without additional information. Data were divided by 80–20 ratio into training/validation and testing.

Evaluation Metrics. To compare two ordered coefficient sets of equal size, $C_1 \subseteq \mathbb{R}^3$ and $C_2 \subseteq \mathbb{R}^3$, we calculate the distance as $d_{\text{MSE}}(C_1, C_2) = \|C_1 - C_2\|^2$. For unordered vertex sets $S_1 \subseteq \mathbb{R}^3$ and $S_2 \subseteq \mathbb{R}^3$ we use the *Chamfer distance* defined as $d_{\text{CD}}(S_1, S_2) = \sum_{x \in S_1} \min_{y \in S_2} \|x - y\|_2^2 + \sum_{y \in S_2} \min_{x \in S_1} \|x - y\|_2^2$. To evaluate how well the distribution of synthesized samples P approximates the real distribution G, we employ a *Minimum Matching Distance* (*MMD*) metric, calculated using either d_{MSE} or d_{CD} as structural distance. MMD quantifies *fidelity* of P with respect to G. Each sample in G is matched to the closest sample in P, and the average distance is computed. Lower MMD scores indicate that the samples in P closely resemble those in G, reflecting higher fidelity.

Implementation Details. We utilize a compact model with 5-stage encoder and decoder with hidden feature sizes of 32, 64, 128, 256 and 512. Training was carried out on a single Tesla T4 with 16 GB of VRAM spanned over approximately 2 h. The model was optimized by AdamW optimizer with an initial learning rate of 1e−4, dynamically changed using cosine annealing restarts each 10 000 iterations. The value of β in β-VAE changed within the range of 0 to 0.05 using cyclical annealing scheduler. We fix the value of k to 256 when generating truncated spectral coefficients. Latent size is fixed to 16.

4 Results

We report the key quantitative results in Table 1. Reconstruction metrics are the average distance errors computed from all test cases for each tooth class. To evaluate the quality of synthesized shapes, we randomly sample from $\mathcal{N}(0, I)$, decode, and reconstruct 1 000 latent vectors and use training samples as the ground truth distribution. The measured metrics remain consistently low across all tooth classes, demonstrating the learned representation's ability to generalize to unseen shapes while providing reliable reconstructions. We also report the average time of generating 1 000 novel tooth shapes. It takes approximately 1 millisecond to perform both spectral decoder inference and projection to the spatial domain, highlighting ToothForge's capability to enhance downstream dental tasks by synthesizing novel shapes without significant additional overhead. The text follows with qualitative and comparative analysis.

Table 1. Quantitative evaluation of generated shapes using ToothForge. Metrics are the average scores across all test cases for given tooth class. MMD is calculated from 1 000 randomly sampled and reconstructed latent vectors. Measured times include the decoder's forward pass and spatial reconstruction without batching optimizations.

Tooth class	$d_{\text{MSE-spectral}} \downarrow$	$d_{\text{MSE-spatial}} \downarrow$	MMD ↓	Shape generation time (ms)↓
Incisors	0.08737	0.00211	0.00754	0.71 ± 0.05 (10 623 vertices)
Premolars	0.09764	0.00209	0.00716	0.75 ± 0.06 (11 960 vertices)
Molars	0.03225	0.00104	0.00325	0.77 ± 0.03 (11 671 vertices)

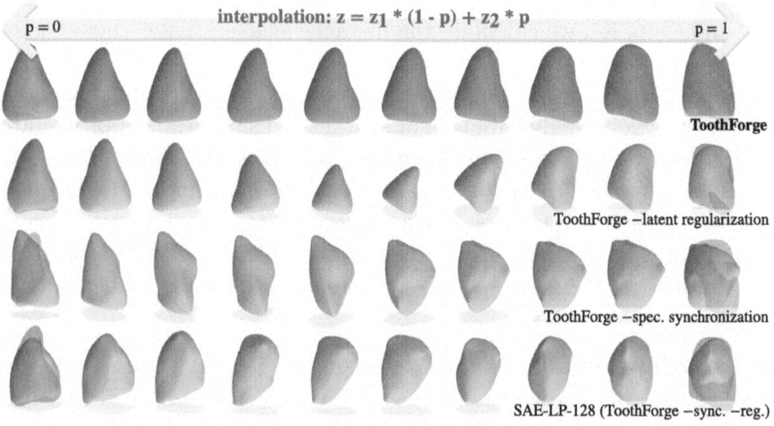

Fig. 5. Synthetic data generation via interpolation in latent space. Individual rows show linear interpolation of latent codes for various framework setups. Two incisor samples from the test set were chosen (leftmost and rightmost), encoded into latent representations z_1 and z_2, sampled along the line $z = z_1 * (1 - p) + z_2 * p$, decoded and finally reconstructed using common basis. Ground truth shapes overlay leftmost and rightmost reconstructions.

Effects of Spectral Synchronization and Latent Regularization. We conduct a qualitative assessment to validate the design choices outlined in Sect. 2.3 and Sect. 2.4 by training on unaligned spectral coefficients and omitting latent space regularization. In the absence of both components, the overall framework design closely resembles SAE-LP-128, as discussed in Sect. 2.2 and is currently the state-of-the-art in spectral generative modeling [14,15]. The results in Fig. 5 demonstrate the importance of these components in scenarios with sparse data and various mesh connectivities. Throughout the interpolation sequence, artifacts from incorrectly reconstructed harmonic coefficients are mitigated, leading to smoother and more realistic shapes while avoiding shrinkage.

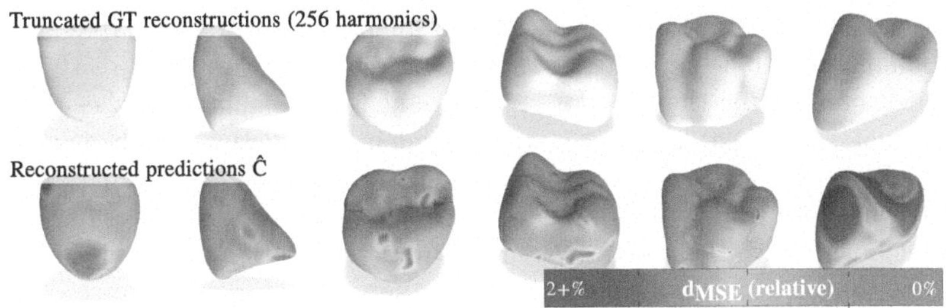

Fig. 6. Reconstructions of unseen tooth shapes. The predictions accurately capture the overall tooth shape. The reconstructions may appear smoothed at times (rightmost premolar). This is due to inaccuracies in predicting high-frequency components.

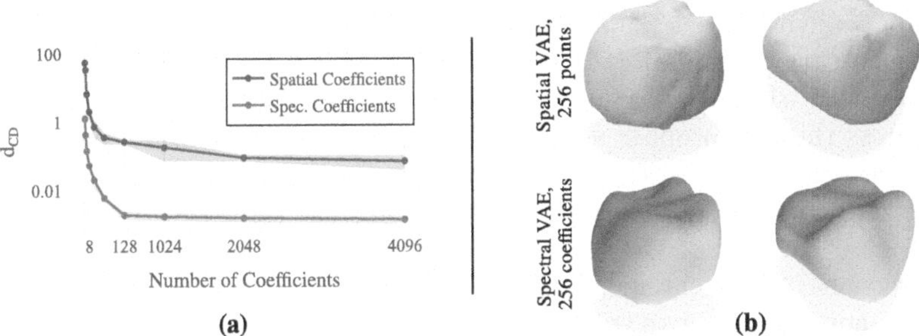

Fig. 7. Spectral vs. spatial coordinates in tooth analysis. In **(a)**, Chamfer Distances (d_{CD}) for spatial coefficients are computed by sampling a given number of points, reconstructing the mesh using Poisson reconstruction, and then calculating the d_{CD} with the original mesh. The logarithmic y-axis highlights this difference. Both methods converge to low error levels as the number of coefficients increases, but the spectral approach remains consistently more efficient. Values are scaled by 10. **(b)** depicts reconstruction capability of autoencoders when processing 256 coordinates.

Reconstruction Fidelity. We demonstrate various reconstruction results in Fig. 6. We use the network to encode and decode the spectral coefficients of the test set. The predictions effectively capture the overall shape of the tooth. Reconstructions occasionally appear overly smooth, which is attributed to prediction errors in the high-frequency coefficients.

Spectral vs. Spatial: A Comparative Analysis. Spectral coefficients provide a compact representation of 3D data by transforming spatial coordinates into the frequency domain. We highlight this compactness for teeth in the graph in Fig. 7. The graph compares Chamfer distances for spatial and spectral coefficients as a function of the number of coefficients used for reconstruction of all molar cases in our dataset. Spectral coefficients exhibit errors lower than that of spatial coefficients, demonstrating their superior compactness in representing geometry. Spectral coefficients also offer natural ordering that simplifies pooling operations since it allows for hierarchical processing directly along this order, eliminating the need for expensive nearest-neighbor searches in high-dimensional spaces [20,26]. Reconstructing dental structures with a spectral autoencoder yields smoother but more plausible results, unlike the spatial counterpart, which often adds high-frequency noise and misses key features like cusps. While smoothing can help eliminate noise from the spatial output, incorporating more anatomical details would require a substantial increase in resolution, making the spectral approach more efficient in on-the-fly generation.

5 Conclusion

In this work, we presented *ToothForge*, a spectral approach to generate novel 3D teeth in real time. The main motivation is to address the sparsity of dental shape datasets, so synthesized shapes increase the accuracy of downstream tasks in digital dentistry [11,24,28]. It is important that the method introduces minimal time overhead when used as an augmentation during the training of these tasks. To address this, we propose a compact autoencoder that learns from spectral embeddings, as mesh spectra are known to provide an effective shape representation [21]. However, generating shape spectra comes with the instability of the decomposed harmonics. We propose modeling the latent manifold on *synchronized* versions of frequential embeddings. Spectra of all data samples are aligned to a common basis prior to the training procedure, effectively eliminating biases introduced by the decomposition instability. More importantly, we eliminate the assumption that all shapes in the training dataset must share a common fixed connectivity, a factor imposed by previous works [14,15]. Our work extends its application to scenarios where guaranteeing consistent connectivity across shapes is unrealistic. We evaluated our framework using a private dataset of real dental crowns. It takes less than a millisecond to generate a new tooth shape with more than 10 000 vertices using a single GPU. Reconstruction accuracy is consistently higher when training on aligned embeddings, supported by qualitative analysis. Synthetic shapes closely resemble ground truth shapes, achieving the average MMD value of 0.00598 across tooth classes. We demonstrate that spatial networks perform poorly compared to spectral frameworks given the same size of input features. Using 256 spectral coefficients is enough to generate realistic teeth, whereas the same number of spatial coefficients generates noisy meshes that lack important anatomical features such as molar cusps.

In summary, our method, *ToothForge*, provides a tool for synthesizing tooth shapes, introducing a new strategy for data augmentation in dental shape analysis tasks. This has the potential to significantly enhance their accuracy with minimal computational cost. Future research will investigate the disentanglement properties of the manifold, specifically how different latent dimensions represent independent and interpretable anatomical features. This disentanglement would allow for more precise control over features such as cusp sizes or groove depths in patient-specific crowns. By navigating the disentangled manifold, these anatomical features could be modified in real-time, paving the way for broader applications in the daily practice of dental technicians and beyond the field of dentistry.

References

1. Achlioptas, P., Diamanti, O., Mitliagkas, I., Guibas, L.: Learning representations and generative models for 3D point clouds. In: International Conference on Machine Learning (ICML) (2018)
2. Adams, J., Elhabian, S.: Point2SSM: learning morphological variations of anatomies from point clouds. In: International Conference on Learning Representations (ICLR) (2024)
3. Agus, M., Gobbetti, E., Pintore, G., Cali, C., Schneider, J.: WISH: efficient 3D biological shape classification through Willmore flow and spherical harmonics decomposition. In: Conference on Computer Vision and Pattern Recognition Workshops (CVPR) (2020)
4. Babiloni, F., Maggioni, M., Tanay, T., Deng, J., Leonardis, A., Zafeiriou, S.: Adaptive spiral layers for efficient 3D representation learning on meshes. In: International Conference on Computer Vision (ICCV) (2023)
5. Bronstein, M.M., Bruna, J., Cohen, T., Veličković, P.: Geometric deep learning: grids, groups, graphs, geodesics, and gauges (2021)
6. Fu, H., Li, C., Liu, X., Gao, J., Celikyilmaz, A., Carin, L.: Cyclical annealing schedule: a simple approach to mitigating KL vanishing. In: Conference of the North American Chapter of the Association for Computational Linguistics: Human Language Technologies (2019)
7. de la Fuente, N., Majó, M., Luzko, I., Córdova, H., Fernández-Esparrach, G., Bernal, J.: Enhancing image classification in small and unbalanced datasets through synthetic data augmentation. In: Workshop on Clinical Image-Based Procedures (MICCAI CLIP) (2024)
8. Gopinath, K., Desrosiers, C., Lombaert, H.: Graph convolutions on spectral embeddings for cortical surface parcellation. Med. Image Anal. (MedIA) (2019)
9. Ha, S., Lyu, I.: SPHARM-Net: spherical harmonics-based convolution for cortical parcellation. IEEE Trans. Med. Imaging (2022)
10. Hanocka, R., Hertz, A., Fish, N., Giryes, R., Fleishman, S., Cohen-Or, D.: MeshCNN: a network with an edge. ACM Trans. Graph. (2019)
11. Hosseinimanesh, G., Ghadiri, F., Guibault, F., Cheriet, F., Keren, J.: From mesh completion to AI designed crown. In: Medical Image Computing and Computer Assisted Intervention (MICCAI) (2023)
12. Hu, S.M., et al.: Subdivision-based mesh convolution networks. ACM Trans. Graph. (2022)
13. Le, T., Bui, G., Duan, Y.: A multi-view recurrent neural network for 3D mesh segmentation. Comput. Graph. (2017)
14. Lemeunier, C., Denis, F., Lavoué, G., Dupont, F.: Representation learning of 3D meshes using an autoencoder in the spectral domain. Comput. Graph. (2022)
15. Lemeunier, C., Denis, F., Lavoué, G., Dupont, F.: SpecTrHuMS: spectral transformer for human mesh sequence learning. Comput. Graph. (2023)
16. Marin, R., Rampini, A., Castellani, U., Rodolà, E., Ovsjanikov, M., Melzi, S.: Spectral shape recovery and analysis via data-driven connections. Int. J. Comput. Vis. (IJCV) (2021)

17. Meyer, M., Desbrun, M., Schröder, P., Barr, A.H.: Discrete differential-geometry operators for triangulated 2-manifolds. In: Visualization and Mathematics III (2003)
18. Qi, C.R., Su, H., Nießner, M., Dai, A., Yan, M., Guibas, L.J.: Volumetric and multi-view CNNs for object classification on 3D data. In: Conference on Computer Vision and Pattern Recognition (CVPR) (2016)
19. Qi, C.R., Su, H., Mo, K., Guibas, L.J.: PointNet: deep learning on point sets for 3D classification and segmentation. In: Conference on Computer Vision and Pattern Recognition (CVPR) (2017)
20. Qi, C.R., Yi, L., Su, H., Guibas, L.J.: PointNet++: deep hierarchical feature learning on point sets in a metric space. In: Conference on Neural Information Processing Systems (NeurIPS) (2017)
21. Reuter, M., Biasotti, S., Giorgi, D., Patanè, G., Spagnuolo, M.: Discrete Laplace–Beltrami operators for shape analysis and segmentation. Comput. Graph. (2009)
22. Styner, M., et al.: Framework for the statistical shape analysis of brain structures using SPHARM-PDM. Insight J. (2006)
23. Su, H., Maji, S., Kalogerakis, E., Learned-Miller, E.: Multi-view convolutional neural networks for 3D shape recognition. In: International Conference on Computer Vision (ICCV) (2015)
24. Tian, S., et al.: A dual discriminator adversarial learning approach for dental occlusal surface reconstruction. J. Healthcare Eng. (2022)
25. Wang, P.S., Liu, Y., Guo, Y.X., Sun, C.Y., Tong, X.: O-CNN: octree-based convolutional neural networks for 3D shape analysis. ACM Trans. Graph. (2017)
26. Wang, Y., Sun, Y., Liu, Z., Sarma, S.E., Bronstein, M.M., Solomon, J.M.: Dynamic graph CNN for learning on point clouds. ACM Trans. Graph. (2019)
27. Wu, X., et al.: Point transformer V3: simpler, faster, stronger. In: Conference on Computer Vision and Pattern Recognition (CVPR) (2024)
28. Yang, S., et al.: DCrownFormer: morphology-aware point-to-mesh generation transformer for dental crown prosthesis from 3D scan data of antagonist and preparation teeth. In: Medical Image Computing and Computer Assisted Intervention (MICCAI) (2024)
29. Zhao, H., Jiang, L., Jia, J., Torr, P., Koltun, V.: Point transformer. In: International Conference on Computer Vision (ICCV) (2021)

LEDA: Log-Euclidean Diffeomorphism Autoencoder for Efficient Statistical Analysis of Diffeomorphisms

Krithika Iyer[1,2](✉), Shireen Elhabian[1,2], and Sarang Joshi[1,3]

[1] Scientific Computing and Imaging Institute, University of Utah, Salt Lake City, UT, USA
{krithika.iyer,sarang.joshi}@utah.edu, shireen@sci.utah.edu

[2] Kahlert School of Computing, University of Utah, Salt Lake City, UT, USA

[3] Biomedical Engineering Department, University of Utah, Salt Lake City, UT, USA

Abstract. Image registration is a core task in computational anatomy that establishes correspondences between images. Invertible deformable registration, which computes a deformation field and handles complex, non-linear transformation, is essential for tracking anatomical variations, especially in neuroimaging applications where inter-subject differences and longitudinal changes are key. Analyzing the deformation fields is challenging due to their non-linearity, limiting statistical analysis. However, traditional approaches for analyzing deformation fields are computationally expensive, sensitive to initialization, and prone to numerical errors, especially when the deformation is far from the identity. To address these limitations, we propose the Log-Euclidean Diffeomorphism Autoencoder (LEDA), an innovative framework designed to compute the principal logarithm of deformation fields by efficiently predicting consecutive square roots. LEDA operates within a linearized latent space that adheres to the diffeomorphisms group action laws, enhancing our model's robustness and applicability. We also introduce a loss function to enforce inverse consistency, ensuring accurate latent representations of deformation fields. Extensive experiments with the OASIS-1 dataset demonstrate the effectiveness of LEDA in accurately modeling and analyzing complex non-linear deformations while maintaining inverse consistency. Additionally, we evaluate its ability to capture and incorporate clinical variables, enhancing its relevance for clinical applications.

Keywords: deformable image registration · manifold statistics · non-rigid registration · diffeomorphisms · shape population statistics · log-euclidean statistics

1 Introduction

The link between the form and function of anatomies is critical in understanding morphological variations to effectively diagnose diseases, plan procedures,

I. Oguz et al. (Eds.): IPMI 2025, LNCS 15830, pp. 327–341, 2026.
https://doi.org/10.1007/978-3-031-96625-5_22

and establish treatment methods [7]. Historically, observational studies were used to analyze anatomical variations [1]. However, increasing accessibility of medical imaging technologies led to the availability of high-resolution in-vivo functional and structural imaging, providing a deeper understanding of organs. Consequently, computational anatomy has emerged as a critical tool to model, analyze, and quantify the variability of anatomical structures across individuals or populations [2]. Image registration is a fundamental component of computational anatomy that introduces voxel-level/spatial correspondences. The voxel-level spatial correspondences transform a large dataset of images onto a standard coordinate frame to facilitate detailed morphological analysis. Image registration is widely used in applications such as neuroimaging studies, adaptive radiotherapy planning, and the development of population-specific atlases [12,40,44].

There are two types of image transformation: rigid and non-rigid (or deformable) registration [20,43]. Images are aligned using simple translations and rotations in rigid image registration, which does not account for interior deformations. This method benefits solid structures like bones, where the relative spatial connections between image components stay constant. It is also effective in scenarios where no significant deformation is expected, such as intra-subject alignment of brain scans taken close in time, dental imaging, or preoperative and postoperative comparisons in orthopedics. In contrast, non-rigid registration accommodates complex, non-linear changes, which are essential for accurately aligning pre- and post-treatment images (e.g., in oncology), tracking progressive changes over time due to patient movement or disease progression, and constructing detailed anatomical atlases that reflect individual variability [18]. Non-rigid methods can be broadly classified into parametric and non-parametric approaches [29]. Parametric methods use models such as B-splines or radial basis functions to represent the transformation in a structured and computationally efficient manner. Non-parametric methods estimate the deformation field directly from the data without assuming a specific functional form. This flexibility allows non-parametric methods to accommodate complex, non-linear anatomical variations, making them beneficial for highly deformable structures.

A smooth and invertible mapping that preserves the continuity and topology of anatomical structures is called a *diffeomorphic* transform. These transformations ensure that no regions overlap or fold, making them ideal for capturing biologically plausible deformations. Methods for estimating deformation fields in computational anatomy span traditional mathematical approaches and modern deep learning techniques. Traditional methods, such as Large Deformation Diffeomorphic Metric Mapping (LDDMM) [11,25,30] and optical flow [45], focus on producing biologically plausible, smooth transformations. Techniques like Direct Deformation Estimation (DDE) [14] improve precision by directly computing deformation gradients. Deep learning models [10,17,41] leverage neural networks to efficiently learn complex deformation fields, enabling scalable and near real-time registration of large datasets. These advancements make deep learning approaches indispensable in clinical and research applications. However, ensuring inverse consistency—the symmetry and reversibility of transformations—is critical for reliable results, particularly in bidirectional or longitudinal studies. Loss functions intended to improve the consistency and robustness of learned transformations were incorporated into models such as GradICON [41] and ICON [24].

Analyzing deformation fields is critical for detecting anatomical differences. While spatial correspondence is necessary, it cannot account for the complexities of these variations, necessitating the use of statistical tools to interpret the transformations [8,16]. However, conventional tools face significant challenges due to the complex, non-linear nature of diffeomorphic transforms [38]. The set of smooth, invertible transformations collectively forms the diffeomorphisms group of a manifold $\mathcal{M}$, denoted as $\mathrm{Diff}(\mathcal{M})$. This group is infinite-dimensional when $\mathcal{M}$ is not zero-dimensional $\dim(\mathcal{M}) > 0$ and simultaneously exhibits the structure of a Fréchet manifold and a Fréchet Lie group [39]. This duality arises because group operations—composition and inversion—are smooth, and the tangent space at the identity of this group corresponds to vector fields on $\mathcal{M}$. The manifold structure of deformation fields introduces significant challenges. Unlike Euclidean spaces, where linear statistics such as means or Principal Component Analysis (PCA) are well-defined, these operations lack meaningful interpretation on curved manifolds like $\mathrm{Diff}(\mathcal{M})$. Adding two deformation fields or directly averaging them in Euclidean terms fails to preserve the smoothness, invertibility, or geometric significance of the transformations. Such operations have no anatomical or mathematical relevance within the diffeomorphic framework.

Several methods such as Principal Geodesic Analysis (PGA) and its variants [21,46,47], Fréchet means [33], and geodesic regression [22] have been developed to address the challenges of statistical estimation on manifolds. However, applying these methods to Fréchet Lie groups of diffeomorphisms requires significant adaptation due to the unique complexities of their structure. Adding a Riemannian metric to a Lie group to convert it into a Riemannian manifold is non-trivial, as not all Lie groups naturally possess such metrics. For example, a bi-invariant metric is a type of Riemannian metric that remains unchanged under left and right multiplication by group elements. They simplify computations, provide consistent geometric interpretations, and are particularly useful for analyzing symmetry. Compact Lie groups (e.g., $\mathrm{SO}(n)$, the group of rotations) have bi-invariant metrics; however, they are generally absent for non-compact Fréchet Lie groups such as $\mathrm{Diff}(\mathcal{M})$. This necessitates alternative meaningful metrics for non-compact groups to capture complex relationships between deformation fields in $\mathrm{Diff}(\mathcal{M})$ and ensure anatomically meaningful models.

The Log-Euclidean framework [6] leverages the group structure of $\mathrm{Diff}(\mathcal{M})$ instead of working directly on the non-linear manifold. The corresponding Lie algebra, represented by smooth vector fields, provides a linearized space for analyzing transformations. This approach simplifies computations like geodesic distances and statistical averaging, which are challenging in the Riemannian setting due to non-linearity and curvature. The *principal logarithm* maps group elements to their counterparts in the Lie algebra, enabling the representation of smooth, invertible transformations in a linear vector field space. This representation facilitates efficient computation of distances between transformations, offering an alternative to Riemannian approaches. However, optimization-based methods for estimating logarithms, such as the non-linear inverse scaling and square rooting algorithm [6,26,31], face challenges, including high computational cost, sensitivity to initialization, and susceptibility to noise. Moreover, these methods typically operate independently on each deformation field and do not consider the inverse consistency of transformations- a property crucial for ensuring reliable analyses in the Log-Euclidean framework [36].

These drawbacks underscore the need for more robust and efficient approaches to estimating principal logarithms. To address these limitations, we propose *Log-Euclidean Diffeomorphism Autoencoder (LEDA)*, an innovative framework designed to compute the principal logarithm of deformation fields by efficiently predicting the consecutive square roots of deformation fields. The framework facilitates statistical analysis within a linearized latent space that respects the group action laws of the diffeomorphism group. By appropriately mapping composition in the data space to scaling in the latent space, the framework enables the application of vector-space-based statistical methods, enhancing the robustness and applicability of statistical analysis for non-linear deformations. Furthermore, we introduce a loss function to enforce inverse consistency constraints, ensuring the latent representations accurately capture the properties of the deformation fields.

2 Related Work

The study of anatomical variability has led to the development of several computational methodologies for capturing complex, nonlinear changes inherent in biological structures. Diffeomorphism-based image registration is a method that helps capture nonlinear geometrical deformation in a population of images. Quantitatively comparing nonlinear registration algorithms necessitates computing global statistics about the deformation fields and is closely related to how diffeomorphisms are parameterized [5].

Marsland and Twining proposed using geodesic interpolating splines (GIS) [15] and polyharmonic clamped plate splines for low-dimensional representation of warps to enable statistical analysis. However, these methods are computationally expensive and unsuitable for complex invertible transformations frequently used in medical imaging [35]. Pennec and Fillard developed a Riemannian geometry framework for statistical analysis on nonlinear spaces, particularly for anatomical structures [38]. Similarly, Principal Geodesic Analysis (PGA) [21] proposed by Fletcher et al. is designed explicitly for Riemannian manifolds and effectively utilizes geodesics to define principal components of underlying nonlinear manifolds. However, these approaches rely heavily on the choice of Riemannian metric, which may lack clear anatomical interpretation and entail significant computational overhead. Several studies have modeled diffeomorphisms as flows using time-varying velocity fields. Vaillant et al. proposed using the space of initial momentum as a linear representation of the nonlinear diffeomorphic shape space [42]. While advantageous, it faces challenges such as high computational costs, limited interpretability of momentum representations, sensitivity to initialization, and difficulty handling large deformations beyond finite dimensional landmark matching.

More recently, Hinkle et al. proposed diffeomorphic autoencoders for Large Deformation Diffeomorphic Metric Mapping (LDDMM) specifically for atlas building [27], integrating momentum fields into diffeomorphisms through vector field flows governed by the Euler-Poincaré equation. Similarly, Bône et al. introduce diffeomorphic autoencoder [13] to simplify the shape analysis by using the vector momentum formulation of LDDMM. The approach leverages the EPDiff equation to ensure the transformations follow optimal paths between shapes. Although LDDMM supports statistical analysis of transformations by uniquely encoding shape by vectors normal to the outline of the template, they remain computationally expensive, limiting scalability to large datasets.

In summary, while existing methods, including deep learning approaches, have advanced the modeling of anatomical variability using coordinate transformations, challenges remain in balancing scalability, computational efficiency, and interpretability. There is a continued need for robust frameworks that accurately capture complex transformations while ensuring efficiency and consistency.

3 Background

Here, we provide a brief overview of the theory and notations required for the Log-Euclidean framework. For a more detailed discussion, please refer to Vincent Arsigny [5] (Chaps. 2 and 8).

Log-Euclidean Framework: A diffeomorphism $\phi : \mathbb{R}^D \to \mathbb{R}^D$ ($D = 2$ for 2D images and $D = 3$ for volumetric images) is a smooth, invertible mapping with a smooth inverse ϕ^{-1}. The diffeomorphism ϕ maps every point $\mathbf{x}$ in the original D-dimensional grid to a new location defined as $\phi(\mathbf{x}) = \mathbf{x} + \mathbf{u}(\mathbf{x})$, where $\mathbf{u}(\mathbf{x})$ is the displacement field, while preserving topology and ensuring that no overlaps or folds occur.

Diffeomorphic transformations can be constructed by composing multiple small transformations. Specifically, ϕ can be expressed as $\phi(\mathbf{x}) = (\mathbf{x} + \epsilon\mathbf{v}_1(\mathbf{x})) \circ (\mathbf{x} + \epsilon\mathbf{v}_2(\mathbf{x})) \circ \cdots \circ (\mathbf{x} + \epsilon\mathbf{v}_n(\mathbf{x}))$, where each term $(\mathbf{x} + \epsilon\mathbf{v}_i(\mathbf{x}))$ represents a small deformation controlled by the magnitude $\epsilon \in \mathbb{R}^+$ and $\mathbf{v}_i(\mathbf{x}) : \mathbb{R}^D \to \mathbb{R}^D$ is a smooth, bounded vector field. Each small deformation $(\mathbf{x} + \epsilon\mathbf{v}_i(\mathbf{x}))$ is close to the identity when ϵ is sufficiently small, ensuring smoothness and invertibility. The resulting full transformation ϕ, remains diffeomorphic and smooth, though it may deviate from the identity depending on the cumulative effect of the vector fields $\mathbf{v}_i(\mathbf{x})$.

The space of diffeomorphisms, $\mathrm{Diff}(\mathcal{M})$, forms a Lie group under composition, with the identity map serving as the group's identity element; i.e., $\phi \circ \mathbf{id} = \mathbf{id} \circ \phi = \phi, \forall \phi \in \mathrm{Diff}(\mathcal{M})$. The associated Lie algebra consists of smooth vector fields $\mathbf{v}(\mathbf{x}) : \mathbb{R}^D \to \mathbb{R}^D$, which generate diffeomorphisms through their flows. These flows are described by one-parameter subgroups,

$\{\phi_t\}_{t\in\mathbb{R}}$, a continuous family of diffeomorphisms satisfying the group property: $\phi_s \circ \phi_t = \phi_{s+t} \forall s, t \in \mathbb{R}$, with $\phi_0 = \mathbf{id}$ as the identity map. The flows generated by these vector fields are governed by the ordinary differential equation (ODE) that describes how a point $\mathbf{x}$ is transported by the vector field $\mathbf{v}$ over time t starting from the identity map at $t = 0$:

$$\frac{d\phi_t(\mathbf{x})}{dt} = \mathbf{v}(\phi_t(\mathbf{x})), \quad \phi_0(\mathbf{x}) = \mathbf{x}. \tag{1}$$

The exponential map denoted as $\exp(\mathbf{v})$, establishes a connection between the Lie algebra and the Lie group by generating a one-parameter subgroup $\{\phi_t\}_{t\in\mathbb{R}}$ from a vector field $\mathbf{v}$. The logarithm map $\log(\phi)$ serves as the local inverse of the exponential map, allowing us to represent a diffeomorphism ϕ in terms of its generating vector field. Specifically, $\log(\phi) = \mathbf{v}$ and $\exp(\mathbf{v}) = \phi$.

Logarithm Estimation: To compute the logarithm map $\log(\phi)$, a non-linear inverse scaling and square rooting algorithm has been proposed [6]. This approach begins by selecting a scaling factor 2^N, which defines the number of successive square root operations needed to be computed. These operations iteratively transform ϕ into a version that is closer to the identity map $\mathbf{id}$, where the logarithm map is more accurately approximated. Once N^{th} square root ϕ^{-2^N} is obtained, the logarithm of ϕ is estimated as:

$$\log(\phi) = 2^N \log(\phi^{-2^N}) \quad \text{where} \quad \log(\phi^{-2^N}) = \phi^{-2^N} - \mathbf{id} \tag{2}$$

However, this algorithm has several drawbacks. The repeated square root operations are computationally expensive, especially for high-dimensional data or when ϕ is far from the identity. Furthermore, the algorithm is sensitive to initialization, and the iterative process can accumulate numerical errors, reducing the accuracy of non-identity diffeomorphisms.

4 Methods

4.1 LEDA: Log-Euclidean Diffeomorphism Autoencoder

We introduce the Log-Euclidean Diffeomorphism Autoencoder (LEDA), a novel approach to efficiently estimate N successive square roots of the input deformation field. Such an iterative decomposition progressively reduces the deformation field closer to the identity, offering a computationally efficient and accurate approximation of the logarithm map, even for complex, high-dimensional deformations. The LEDA architecture (Fig. 1.b) includes an encoder $f_\gamma(\phi)$ that maps the input ϕ to a low-dimensional latent representation $\mathbf{z} \in \mathbb{R}^L$. The decoder function g_θ reconstructs the square roots of the deformation field, ensuring a consistent mapping between scaling in the latent space and composition in the deformation space. It predicts the nth root of the deformation field using scaled versions of the latent representation $\mathbf{z}$. Mathematically, this can be expressed as:

$$f_\gamma(\phi) = \mathbf{z} \qquad g_\theta(\mathbf{z}/m) = \phi^{1/m} \quad \text{where } m = 2^n \ \ \forall n \in \{0, 1, \ldots N\} \tag{3}$$

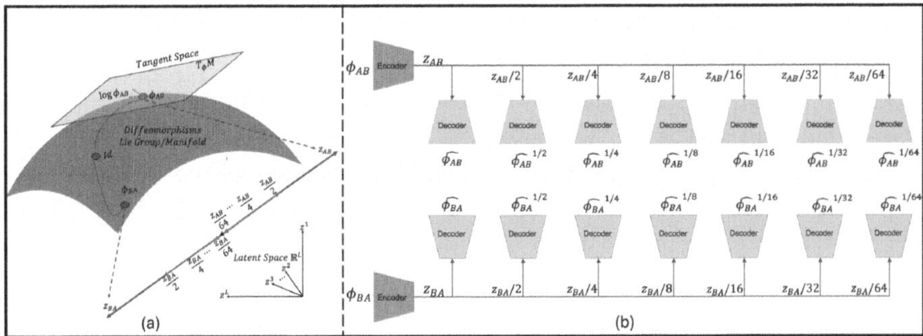

Fig. 1. (a) **Mathematical framework for diffeomorphic transformations**: Shows the relationship between the Euclidean vector space (tangent space $T_{\phi}M$), the diffeomorphic group (Lie group/manifold), and the latent space ($\mathbb{R}^L$). Here, A and B represent two coordinate spaces or imaging domains between which the deformations occurs. Transformations (ϕ_{AB}, ϕ_{BA}) and the logarithmic map ($\log(\phi_{AB})$), alongside their projections into LEDA's latent space ($\mathbf{z}_{AB}, \mathbf{z}_{BA}$), demonstrating inverse consistency and the mapping of composition in the data space to scaling in the latent space. (b) **LEDA architecture**

The successive square roots estimation framework allows us to approximate the logarithm map using Eq. 2. The remaining $N-1$ estimated roots contribute to establishing a direct relationship between latent space operations and deformation field compositions, enabling intuitive manipulation of complex spatial transformations. The LEDA framework achieves its objectives through three core criteria incorporated into its design and loss function:

1. Faithful Reconstruction: The estimated roots must accurately reconstruct the original deformation field when composed a specified number of times i.e., if ϕ^{-m} is the predicted root at stage n where $m = 2^n$, then ϕ^{-m} composed m times should yield the original deformation field ϕ.

2. Inverse consistency: The estimated roots should maintain inverse consistency, i.e., $\phi_{AB}(\phi_{BA}(\mathbf{x})) = \mathbf{x}$ for all $\mathbf{x}$. Here, A and B represent two coordinate spaces or imaging domains between which the deformations occur. To facilitate this, we model LEDA as a **Siamese** [32] autoencoder, that processes the forward ϕ_{AB} and inverse ϕ_{BA} fields simultaneously.

3. Inverse consistency in latent space: Representing the forward deformation ϕ_{AB} by $\mathbf{z}_{AB}$ and the inverse field ϕ_{BA} by $\mathbf{z}_{BA}$, inverse consistency in the latent space requires that $\mathbf{z}_{AB} = -\mathbf{z}_{BA}$, ensuring the latent representations of a deformation field and its inverse are equal in magnitude but opposite in direction.

These contributions establish a computationally efficient framework for analyzing and manipulating diffeomorphic transformations while preserving their key structural properties.

4.2 LEDA: Architecture and Loss

The LEDA processes pairs of deformation fields $(\phi_{AB,k}, \phi_{BA,k})$ from the dataset $\mathcal{D} = \{(\phi_{AB,k}, \phi_{BA,k}) : k = 1, 2, \ldots, K\}$, where $\phi_{AB,k}$ represents a forward deformation field and $\phi_{BA,k}$ is its corresponding inverse deformation field. The Siamese nature of the LEDA is illustrated in Fig. 1.b. It simultaneously processes one pair of deformation fields $(\phi_{AB,k}, \phi_{BA,k})$, using networks that share weights, as indicated by identical coloring in the figure. Although the decoder architecture in Fig. 1.b is unrolled to show the scaling of the latent space and corresponding root estimation, the implementation uses a single shared decoder network to perform these tasks. LEDA framework is implemented using 2D convolutional layers and fully connected layers for the encoder and decoder. However, it can be straightforwardly extended to 3D.

To define the loss functions used by the model, we introduce the notation $C_m(\phi)$ to indicate the composition of ϕ with itself m times:

$$C_m(\phi) = \underbrace{\phi \circ \phi \circ \cdots \circ \phi}_{m \text{ times}} \tag{4}$$

To ensure the model satisfies the criteria described in Sect. 4.1, the training objective incorporates three loss terms:

Reconstruction Loss: Ensures accurate reconstruction of the original deformation field ϕ from its predicted roots.

$$\mathcal{L}_{rec} = \sum_{k=1}^{K} \sum_{n=0,m=2^n}^{N} \left\{ \left\| C_m(\widehat{\phi}^{-m}_{AB,k}) - \phi_{AB,k} \right\|^2 + \left\| C_m(\widehat{\phi}^{-m}_{BA,k}) - \phi_{BA,k} \right\|^2 \right\} \tag{5}$$

where ϕ is the ground truth deformation field and $\widehat{\phi}^{-m}$ is the output of the decoder at the $n-$th stage, i.e., after scaling the latent representation by $1/2^n$.

Inverse Consistency Loss: We use the approximate inverse consistency loss [24] proposed for image registration models to enforce inverse consistency for each estimated root.

$$\mathcal{L}_{inv} = \sum_{k=1}^{K} \sum_{n=0,m=2^n}^{N} \left\{ \left\| \widehat{\phi}^{-m}_{AB,k} \circ \widehat{\phi}^{-m}_{BA,k} - \mathbf{id} \right\|_2^2 + \left\| \widehat{\phi}^{-m}_{BA,k} \circ \widehat{\phi}^{-m}_{AB,k} - \mathbf{id} \right\|_2^2 \right\} \tag{6}$$

Latent Inverse Consistency Loss: Enforces latent inverse consistency using cosine similarity Θ_k and magnitude constraints:

$$\mathcal{L}_{linv} = \sum_{k=1}^{K} \left\{ \frac{1 + \cos(\Theta_k)}{2} + \left\| \mathbf{z}_{AB,k} + \mathbf{z}_{BA,k} \right\|^2 \right\} \tag{7}$$

where Θ_k representing the cosine similarity between $\mathbf{z}_{AB,k}$ and $\mathbf{z}_{BA,k}$.

The total loss function is given as: $\mathcal{L} = \alpha_{rec}\mathcal{L}_{rec} + \alpha_{inv}\mathcal{L}_{inv} + \alpha_{linv}\mathcal{L}_{linv}$ where α's represent the weight of each term. By explicitly linking latent space

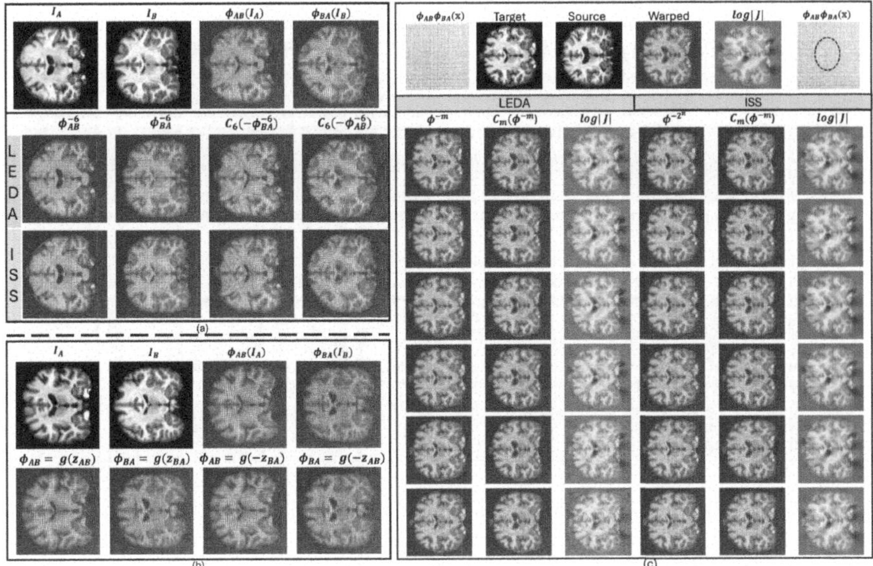

Fig. 2. (a) Validation of Small Deformation Field Assumption (b) Validation of Latent Inverse Consistency (c) Comparison of Square Root Estimation: LEDA (left) and ISS (right)

operations to deformation field compositions, LEDA offers a robust and intuitive framework for manipulating complex spatial transformations. Its emphasis on inverse consistency ensures reliability, making it well-suited for applications requiring efficient diffeomorphic transformations.

5 Experiments

Dataset: In this paper, we use the OASIS-1 dataset [34], a widely-used neuroimaging resource containing 3D brain MRI scans. It includes multiple T1-weighted scans per subject, with 100 subjects over 60 diagnosed with very mild to moderate Alzheimer's disease (AD). We use 2D coronal slices of the scans [28] and resize them to 160×160. Deformation fields are generated by training a 2D GradICON model [41] on the 2D coronal slices dataset, producing 85,078 pairs of deformation fields from 413 2D images.

Baseline: We compare the logarithm maps and square root estimated from LEDA with the non-linear inverse scaling and squaring (ISS) proposed by Arsigny [6] to assess their ability to recover the original deformation. The authors use the closed from gradient updates, but we use PyTorch to estimate the gradients for square root estimation updates to simplify the process.

Results: Figure 2.c shows a qualitative comparison of both methods. Notably, ISS is highly sensitive to initialization values, requiring a specific initializa-

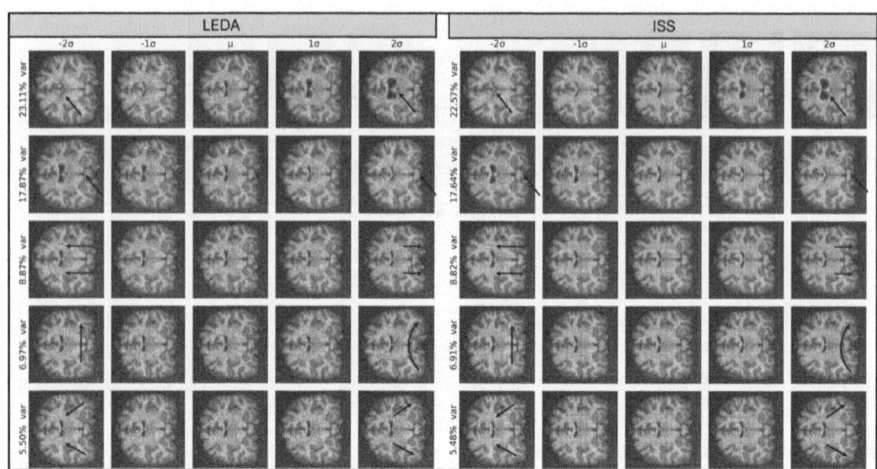

Fig. 3. PCA Modes of Logarithm Maps LEDA (left) and ISS (right): Variations can be associated with structural changes seen in AD, including ventricular expansion and hippocampal atrophy.

tion of $\frac{\phi}{2}$. We tested multiple initializations and present results from the best-performing one. Both methods produce similar square root estimates, with higher roots approaching the identity transformation progressively (top to bottom). The log-determinant of the Jacobian reveals local deformations, where red indicates contraction and blue indicates expansion, and both methods show consistent patterns. However, ISS struggles with inverse consistency, as evidenced by the grid plots in the first row. While LEDA fully recovers the identity grid, ISS leaves residual deformations, highlighting LEDA's superior robustness in maintaining inverse consistency. Moreover, estimating the square roots using ISS for a single deformation field takes **2.3** s (given the PyTorch implementation), whereas performing inference via the trained LEDA to estimate all the roots for a pair of deformation fields takes **only 0.02** s. This substantial difference in computation time becomes even more pronounced when dealing with high-resolution deformation fields, underscoring the scalability and efficiency of the proposed framework.

We further verify the estimated logarithm maps to test their consistency with the small deformation field assumption, which states that forward displacement fields $u_{AB}(\mathbf{x})$ should approximate the negation of inverse displacement fields $u_{BA}(\mathbf{x})$. Using the 2^6th root, we negate the forward displacement field $\phi^{-2^6}(\mathbf{x})$, compose it appropriately, and recover the inverse field. Similarly, negating the inverse transform and composing it recovers the forward field. Figure 2.a demonstrates that roots estimated by both methods satisfy this assumption, validating the accuracy of the logarithm maps and ensuring expected small deformation model behavior.

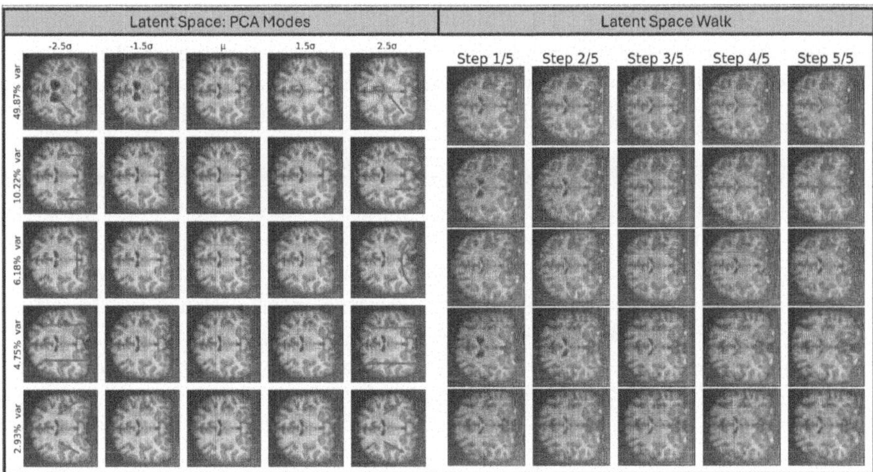

Fig. 4. (a) PCA Modes of Latent Representations: Variations show structural changes associated with AD, including ventricular expansion and hippocampal atrophy. (b) Latent Space Walk: Random walk demonstrates smooth, continuous transitions through anatomical variations.

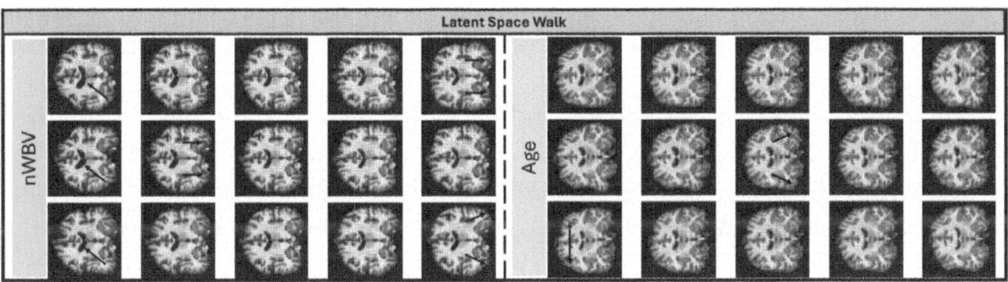

Fig. 5. Latent Walk Along Latent Dimensions Predictive of Age and nWBV

To assess inverse consistency within the latent space of LEDA, we negate the latent representation of forward and inverse displacement fields and decode them using LEDA. Figure 2.b shows that the LEDA accurately decodes the negated latent representation into the corresponding inverse field, confirming inverse consistency. This validation is crucial as it demonstrates the LEDA's ability to represent and preserve relationships between forward and inverse transformations accurately.

We performed PCA on the logarithm maps of deformation fields predicted by both methods to analyze dominant modes of anatomical variation. Figure 3 displays the identified modes aligning with known neuroanatomical changes in AD. The first mode (~23% variance) captures large-scale atrophy patterns, including ventricular expansion and surrounding tissue reduction, particularly in the medial temporal lobe, consistent with AD pathology [4,37]. The second mode (~18%) highlights hippocampal atrophy and adjacent gray matter loss, reflecting early-to-moderate AD stages [4]. This mode exhibits slight asymmetry, suggesting individual disease progression differences. The third mode (∼9%) reveals bilateral ventricular expansion with cortical thinning, while subsequent modes (∼6 – 5%) capture localized cortical and subcortical changes, such as finer tissue loss.

PCA was also conducted on the latent space of the LEDA, with the modes of variation shown in Fig. 4.a. The latent space modes closely align with the logarithm map PCA results, capturing clinically consistent changes such as ventricular expansion, hippocampus atrophy in early modes, and localized cortical thinning [9]. Figure 4.b illustrates progressive interpolations in the latent space, visualized over five steps. These smooth and structured transitions effectively preserve anatomical coherence while capturing key variations, confirming the model's ability to generate realistic deformations and interpolate between anatomical states.

Moreover, low-dimensional latent space representation is crucial for capturing essential features while reducing computational complexity, enabling more efficient analysis of high-dimensional neuroimaging data. We employed a linear regression framework to analyze deformation fields based on their latent presentations. The process involves (1) selecting a reference image B of a healthy young adult, (2) computing deformation fields ϕ_{AB} for all samples A excluding B, and (3) fitting linear regression models using the latent representations as independent variables and clinical variables from the OASIS dataset as dependent variables. This approach aims to uncover relationships between structural changes and clinical characteristics. The clinical variables included are age, Normalized Whole Brain Volume (nWBV), Mini-Mental State Examination (MMSE), Clinical Dementia Rating (CDR), estimated Total Intracranial Volume (eTIV), and Atlas Scaling Factor (ASF). Our analysis indicated that the latent dimensions were most predictive of age and nWBV, with r-scores of 0.71 and 0.78, respectively. Figure 5 illustrates the latent walk along the top three directions predictive of age and nWBV based on linear regression coefficients. Changes associated with age reflect overall brain shape and size, with minimal lateral ventricular expansion, consistent with the literature on age-related brain changes [19,23]. In contrast, dimensions indicative of nWBV show significant alterations in ventricle shape and size, aligning with findings that decreased brain volume correlates with ventricular expansion in neurodegenerative conditions [3].

6 Conclusion

In this work, we address the challenges of analyzing non-linear deformation fields in image registration by introducing a novel framework called Log-Euclidean Diffeomorphism Autoencoder (LEDA), designed to compute the principal logarithm of deformation fields by efficiently predicting the consecutive square roots of deformation fields. Extensive evaluations show LEDA's effectiveness in estimating logarithm maps that capture clinically relevant anatomical variations. LEDA's latent space can robustly link deformation fields to clinical variables, offering valuable insights into disease progression. With its efficiency and accuracy, LEDA opens avenues for efficient analysis of deformation fields, enabling more precise neuroimaging and medical applications.

Acknowledgement. This work was supported by the National Institutes of Health under grant numbers NIH-R01CA259686 (Sarang Joshi) and NIH-R01DE032366 (Shireen Elhabian).

References

1. Alraddadi, A.: Literature review of anatomical variations: clinical significance, identification approach, and teaching strategies. Cureus **13**(4) (2021)
2. Ambellan, F., Lamecker, H., von Tycowicz, C., Zachow, S.: Statistical shape models: understanding and mastering variation in anatomy. In: Rea, P.M. (ed.) Biomedical Visualisation. AEMB, vol. 1156, pp. 67–84. Springer, Cham (2019). https://doi.org/10.1007/978-3-030-19385-0_5
3. Apostolova, L.G., et al.: Ventricular enlargement and its clinical correlates in the imaging cohort from the ADCS MCI donepezil/vitamin E study. Alzheimer Dis. Assoc. Disord. **27**(2), 174–181 (2013)
4. Apostolova, L.G., et al.: Hippocampal atrophy and ventricular enlargement in normal aging, mild cognitive impairment (MCI), and Alzheimer disease. Alzheimer Dis. Assoc. Disord. **26**(1), 17–27 (2012)
5. Arsigny, V.: Processing data in lie groups: an algebraic approach. Ph.D. thesis (2006)
6. Arsigny, V., Commowick, O., Pennec, X., Ayache, N.: A Log-Euclidean framework for statistics on diffeomorphisms. In: Larsen, R., Nielsen, M., Sporring, J. (eds.) MICCAI 2006. LNCS, vol. 4190, pp. 924–931. Springer, Heidelberg (2006). https://doi.org/10.1007/11866565_113
7. Asghar, A., Patra, A., Naaz, S., Kumar, R., Babu, C.R., Singh, B.: Investigating the integration of anatomical variabilities into medical education as a potential strategy for mitigating surgical errors. J. Anat. Soc. India **73**(1), 70–81 (2024)
8. Avants, B.B., Epstein, C.L., Gee, J.C.: Geodesic image normalization and temporal parameterization in the space of diffeomorphisms. In: Yang, G.-Z., Jiang, T.Z., Shen, D., Gu, L., Yang, J. (eds.) MIAR 2006. LNCS, vol. 4091, pp. 9–16. Springer, Heidelberg (2006). https://doi.org/10.1007/11812715_2
9. Bakkour, A., Morris, J.C., Dickerson, B.C.: The cortical signature of prodromal AD: regional thinning predicts mild AD dementia. Neurology **72**(12), 1048–1055 (2009)
10. Balakrishnan, G., Zhao, A., Sabuncu, M.R., Guttag, J., Dalca, A.V.: VoxelMorph: a learning framework for deformable medical image registration. IEEE Trans. Med. Imaging **38**(8), 1788–1800 (2019)
11. Beg, M.F., Miller, M.I., Trouvé, A., Younes, L.: Computing large deformation metric mappings via geodesic flows of diffeomorphisms. Int. J. Comput. Vision **61**, 139–157 (2005)
12. Binte Alam, S., Nii, M., Shimizu, A., Kobashi, S.: Spatiotemporal statistical shape model for temporal shape change analysis of adult brain. Current Med. Imaging **16**(5), 499–506 (2020)

13. Bône, A., Louis, M., Colliot, O., Durrleman, S.: Learning low-dimensional representations of shape data sets with diffeomorphic autoencoders. In: Chung, A., Gee, J.C., Yushkevich, P.A., Bao, S. (eds.) IPMI 2019. LNCS, vol. 11492, pp. 195–207. Springer, Cham (2019). https://doi.org/10.1007/978-3-030-20351-1_15
14. Boyle, J.J., et al.: Regularization-free strain mapping in three dimensions, with application to cardiac ultrasound. J. Biomech. Eng. **141**(1), 011010 (2019)
15. Camion, V., Younes, L.: Geodesic interpolating splines. In: Figueiredo, M., Zerubia, J., Jain, A.K. (eds.) EMMCVPR 2001. LNCS, vol. 2134, pp. 513–527. Springer, Heidelberg (2001). https://doi.org/10.1007/3-540-44745-8_34
16. Charon, N.: Analysis of geometric and functional shapes with extensions of currents: applications to registration and atlas estimation. Ph.D. thesis, École normale supérieure de Cachan-ENS Cachan (2013)
17. Chen, J., Frey, E.C., He, Y., Segars, W.P., Li, Y., Du, Y.: TransMorph: transformer for unsupervised medical image registration. Med. Image Anal. **82**, 102615 (2022)
18. Crum, W.R., Hartkens, T., Hill, D.: Non-rigid image registration: theory and practice. Br. J. Radiol. **77**(suppl_2), S140–S153 (2004)
19. Currà, A., et al.: The ventricular system enlarges abnormally in the seventies, earlier in men, and first in the frontal horn: a study based on more than 3,000 scans. Front. Aging Neurosci. **11**, 294 (2019)
20. Deng, B., Yao, Y., Dyke, R.M., Zhang, J.: A survey of non-rigid 3D registration. Comput. Graph. Forum **41**, 559–589 (2022)
21. Fletcher, P.T., Lu, C., Pizer, S.M., Joshi, S.: Principal geodesic analysis for the study of nonlinear statistics of shape. IEEE Trans. Med. Imaging **23**(8), 995–1005 (2004)
22. Fletcher, T.: Geodesic regression on Riemannian manifolds. In: Proceedings of the Third International Workshop on Mathematical Foundations of Computational Anatomy-Geometrical and Statistical Methods for Modelling Biological Shape Variability, pp. 75–86 (2011)
23. Fujita, S., et al.: Characterization of brain volume changes in aging individuals with normal cognition using serial magnetic resonance imaging. JAMA Netw. Open **6**(6), e2318153–e2318153 (2023)
24. Greer, H., Kwitt, R., Vialard, F.X., Niethammer, M.: ICON: learning regular maps through inverse consistency. In: Proceedings of the IEEE/CVF International Conference on Computer Vision, pp. 3396–3405 (2021)
25. Hernandez, M., Julvez, U.R.: Insights into traditional large deformation diffeomorphic metric mapping and unsupervised deep-learning for diffeomorphic registration and their evaluation. Comput. Biol. Med. **178**, 108761 (2024)
26. Higham, N.J.: The scaling and squaring method for the matrix exponential revisited. SIAM J. Matrix Anal. Appl. **26**(4), 1179–1193 (2005)
27. Hinkle, J., Womble, D., Yoon, H.J.: Diffeomorphic autoencoders for LDDMM atlas building (2018)
28. Hoopes, A., Hoffmann, M., Fischl, B., Guttag, J., Dalca, A.V.: HyperMorph: amortized hyperparameter learning for image registration. In: Feragen, A., Sommer, S., Schnabel, J., Nielsen, M. (eds.) IPMI 2021. LNCS, vol. 12729, pp. 3–17. Springer, Cham (2021). https://doi.org/10.1007/978-3-030-78191-0_1
29. Hu, W.R., Xie, Y., Li, L., Zhang, W.S.: A TV-L 1 based nonrigid image registration by coupling parametric and non-parametric transformation. Int. J. Autom. Comput. **12**(5), 467–481 (2015)
30. Joshi, S.C., Miller, M.I.: Landmark matching via large deformation diffeomorphisms. IEEE Trans. Image Process. **9**(8), 1357–1370 (2000)

31. Kenney, C., Laub, A.J.: Condition estimates for matrix functions. SIAM J. Matrix Anal. Appl. **10**(2), 191–209 (1989)
32. Koch, G., Zemel, R., Salakhutdinov, R., et al.: Siamese neural networks for one-shot image recognition. In: ICML Deep Learning Workshop, Lille, vol. 2, pp. 1–30 (2015)
33. Le, H., Kume, A.: The Fréchet mean shape and the shape of the means. Adv. Appl. Probab. **32**(1), 101–113 (2000)
34. Marcus, D.S., Wang, T.H., Parker, J., Csernansky, J.G., Morris, J.C., Buckner, R.L.: Open access series of imaging studies (OASIS): cross-sectional MRI data in young, middle aged, nondemented, and demented older adults. J. Cogn. Neurosci. **19**(9), 1498–1507 (2007)
35. Marsland, S., Twining, C.J.: Constructing diffeomorphic representations for the groupwise analysis of nonrigid registrations of medical images. IEEE Trans. Med. Imaging **23**(8), 1006–1020 (2004)
36. Miller, M.I., Arguillère, S., Tward, D.J., Younes, L.: Computational anatomy and diffeomorphometry: a dynamical systems model of neuroanatomy in the soft condensed matter continuum. Wiley Interdisc. Revi. Syst. Biol. Med. **10**(6), e1425 (2018)
37. Nestor, S.M., et al.: Ventricular enlargement as a possible measure of Alzheimer's disease progression validated using the Alzheimer's disease neuroimaging initiative database. Brain **131**(9), 2443–2454 (2008)
38. Pennec, X., Fillard, P.: Statistical computing on non-linear spaces for computational anatomy. In: Handbook of Biomedical Imaging: Methodologies and Clinical Research, pp. 147–168 (2015)
39. Schmid, R.: Infinite dimensional lie groups with applications to mathematical physics (2004)
40. Suganyadevi, S., Seethalakshmi, V., Balasamy, K.: A review on deep learning in medical image analysis. Int. J. Multimedia Inf. Retrieval **11**(1), 19–38 (2022)
41. Tian, L., et al.: GradiCON: approximate diffeomorphisms via gradient inverse consistency. In: Proceedings of the IEEE/CVF Conference on Computer Vision and Pattern Recognition, pp. 18084–18094 (2023)
42. Vaillant, M., Miller, M.I., Younes, L., Trouvé, A.: Statistics on diffeomorphisms via tangent space representations. Neuroimage **23**, S161–S169 (2004)
43. Yaniv, Z.: Rigid registration. In: Image-Guided Interventions: Technology and Applications, pp. 159–192 (2008)
44. Zachiu, C., et al.: Anatomically-adaptive multi-modal image registration for image-guided external-beam radiotherapy. Phys. Med. Biol. **65**(21), 215028 (2020)
45. Zhai, M., Xiang, X., Lv, N., Kong, X.: Optical flow and scene flow estimation: a survey. Pattern Recogn. **114**, 107861 (2021)
46. Zhang, M., Fletcher, P.T.: Bayesian principal geodesic analysis for estimating intrinsic diffeomorphic image variability. Med. Image Anal. **25**(1), 37–44 (2015)
47. Zhang, M., Fletcher, T.: Probabilistic principal geodesic analysis. In: Advances in Neural Information Processing Systems, vol. 26 (2013)

CoRLD: Contrastive Representation Learning of Deformable Shapes in Images

Tonmoy Hossain[1](✉) and Miaomiao Zhang[1,2]

[1] Computer Science, University of Virginia, Charlottesville, VA, USA
pwg7jb@virginia.edu

[2] Electrical and Computer Engineering, University of Virginia, Charlottesville, VA, USA

Abstract. Deformable shape representations, parameterized by deformations relative to a given template, have proven effective for improved image analysis tasks. However, their broader applicability is hindered by two major challenges. First, existing methods mainly rely on a known template during testing, which is impractical and limits flexibility. Second, they often struggle to capture fine-grained, voxel-level distinctions between similar shapes (e.g., anatomical variations among healthy individuals, those with mild cognitive impairment, and diseased states). To address these limitations, we propose a novel framework - Contrastive Representation Learning of Deformable shapes (CoRLD) in learned deformation spaces and demonstrate its effectiveness in the context of image classification. Our CoRLD leverages a class-aware contrastive supervised learning objective in latent deformation spaces, promoting proximity among representations of similar classes while ensuring separation of dissimilar groups. In contrast to previous deep learning networks that require a reference image as input to predict deformation changes, our approach eliminates this dependency. Instead, template images are utilized solely as ground truth in the loss function during the training process, making our model more flexible and generalizable to a wide range of medical applications. We validate CoRLD on diverse datasets, including real brain magnetic resonance imaging (MRIs) and adrenal shapes derived from computed tomography (CT) scans. Experimental results show that our model effectively extracts deformable shape features, which can be easily integrated with existing classifiers to substantially boost the classification accuracy. Our code is available at GitHub.

1 Introduction

Deformable shape features have demonstrated their effectiveness in various image analysis tasks, including image classification [13,16,31], segmentation [21,38], and object recognition [12,28]. Existing methods have studied different representations of geometric shapes, such as landmarks [8,11], point clouds [1], and medial axes [27]. However, these techniques often overlook the interior structures of objects, limiting their ability to capture the intricate details of complex

I. Oguz et al. (Eds.): IPMI 2025, LNCS 15830, pp. 342–357, 2026.
https://doi.org/10.1007/978-3-031-96625-5_23

objects in images. In contrast, deformation-based shape representations (e.g., elastic deformations or fluid flows) focus on detailed shape information from images [10,30]. These methods assume that many objects in generic classes can be represented as deformed versions of an ideal template, enabling a transformation that reflects geometric changes and captures fine-grained shape details.

Recent advances in deep learning have significantly expanded the capabilities of deformation-based shape representation learning [4,15,32]. These methods leverage two main categories of parametrization: stationary velocity fields (SVF) [2] and large deformation diffeomorphic metric mapping (LDDMM) [7]. The former offers a computationally efficient parametrization, widely adopted across deep learning architectures [6,9,23]. LDDMM, on the other hand, provides a mathematically elegant approach to model complex large deformations with well-defined distance metrics in deformation spaces [7,17,34]. Both methods have been heavily integrated into networks, with primary applications focused on image alignment and registration [6,37], a task they perform well but do not fully explore the power of deformation-based representations. Recent efforts have developed a deep learning framework that integrates image intensity features with geometric transformations in unified spaces [15,32]; demonstrating its effectiveness in substantially improving the performance of classifiers. However, these methods still face challenges. They often require reference images during inference, which can be impractical and limit their flexibility in dynamic and diverse real-world scenarios. Additionally, they struggle to capture fine-grained voxel-level differences between similar shapes, such as distinguishing between healthy individuals and those with mild cognitive impairment.

To address these issues, this paper introduces a novel framework, Contrastive Representation Learning of Deformable shapes (CoRLD), in learned deformation spaces. We demonstrate its effectiveness in image classification by integrating the learned representations with image intensity and texture features. Inspired by recent works in contrastive learning that aim to capture fine-grained structural and semantic differences [5,22], our model CoRLD employs a class-aware contrastive supervised learning objective in latent deformation spaces. This promotes similarity among representations of similar classes while ensuring clear separation between dissimilar groups. Additionally, CoRLD decouples the template from the network input, using it exclusively as a guidance in the training process. The contributions of our proposed model are threefold:

(i) Develop a novel model, CoRLD, that for the first time learns class-aware contrastive shape features in the latent space of geometric deformations.
(ii) Effectively eliminate the requirement of a template image to learn deformable shape representations.
(iii) Our model is easily adaptable to be integrated with various feature extractors, serving as a plug-and-play enhancement module to boost the performance of classification tasks.

We validate the effectiveness of our model in diverse multi-class/binary image classification tasks on real brain MRIs [19] and adrenal shapes derived from

CTs [36]. Experimental results show that our model outperforms the state-of-the-art, achieving improved performance across all tasks without requiring the reference images during inference time.

2 Background: Deformable Registration to Derive Shape Representations from Images

This section briefly reviews the concept of deformable image registration [2,7,39], which is commonly used to derive deformation-based shape representations from images [20,40]. Based on the premise that objects in many generic classes can be described as deformed versions of an ideal template image, descriptors in this class arise naturally by matching the template to an input image while preserving topology. The resulting transformation can be viewed as geometric shapes that capture the variations between individual images and the template. In highly sensitive domains such as medical imaging, it is critical for transformations to be diffeomorphisms (i.e., smooth, bijective mappings with smooth inverses) in order to preserve the topological structures of objects.

Let $\Omega = \mathbb{R}^d/\mathbb{Z}^d$ be a d-dimensional torus domain with periodic boundary conditions. Given a reference image S and a target image T on the torus domain Ω $(S(x), T(x) : x \in \Omega \to \mathbb{R})$, a diffeomorphic transformation, ϕ_t, for $t \in [0,1]$ can be defined as a flow over time to deform a template image to match a target image by a composite function, $S \circ \phi_t^{-1}$, where $\circ$ denotes an interpolation operator. Such a transformation is typically parameterized by time-dependent velocity fields under the LDDMM [7], or SVF that remains constant over time [2]. While we employ SVF for implementation in this paper, our framework is easily applicable to the other.

For a stationary velocity field v, the diffeomorphisms, $\{\phi_t\}$, are generated as solutions to the equation:

$$\frac{d\phi_t}{dt} = v \circ \phi_t, \text{ s.t. } \phi_0 = x. \tag{1}$$

The solution of Eq. (1) is identified as an exponential map using a scaling and squaring scheme [2]. The velocity field, v, is often used as representations of diffeomorphisms due to its nice properties of linearity [2,26].

The objective function to estimate velocity fields from a given pair of template and target images can be formulated as

$$E(v) = \frac{1}{\sigma^2}\text{Dist}(S \circ \phi_1^{-1}(v), T) + \text{Reg}(v), \text{s.t. Eq. (1)}, \tag{2}$$

where σ^2 is a noise variance and $\text{Reg}(\cdot) = \|\nabla v\|$ serves as a regularization term ensuring the smoothness of the transformation fields. The $\text{Dist}(\cdot,\cdot)$ is a distance function that measures the dissimilarity between images, i.e., sum-of-squared differences [7], normalized cross correlation [3], and mutual information [33].

3 Our Method: CoRLD

This section presents CoRLD, a novel representation learning algorithm that, for the first time, learns class-aware contrastive shape features in the latent space of image deformations. We further demonstrate the effectiveness of these learned contrastive shape features by integrating them with image features, resulting in a boosted classifier aiming to deliver robust and improved network predictions. The details of our network architecture are introduced below (Fig. 1).

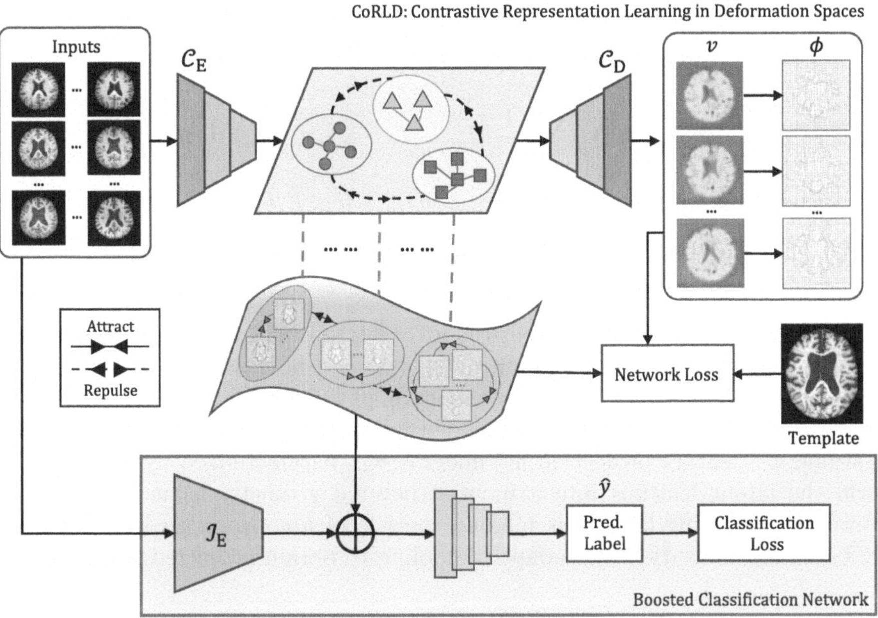

Fig. 1. An overview of our proposed model CoRLD.

3.1 Network Architecture

Given a number of C image classes, there exists a number of $N_c, c \in \{1, ..., C\}$ images in each class. With a group of training images $\{I_{nc}\}_{n=1,c=1}^{N_c,C}$ and their associated class labels $\{y_{nc}\}$, we define the training data as $X = \{(I_{nc}, y_{nc})\}_{c=1}^{C}$, where $I \in \mathbb{R}^{H \times W \times D}$ with $H \times W \times D$ being the image dimension. Let $\mathcal{C}_\text{E}$ denote our L-layer encoder network, parameterized by θ_e. The representation at layer $l \in \{1, \ldots, L\}$ is defined as

$$\mathbf{E}_l(I_{nc}; \theta_e) = g\big(\mathbf{K}_l * \mathbf{E}_{l-1}(I_{nc}; \theta_e) + \mathbf{b}_l\big),$$

where $g(\cdot)$ is a non-linear activation function, $\mathbf{K}_l$ represents the learnable convolutional filters, $*$ denotes the convolution operation, $\mathbf{b}_l$ is the bias term for layer l, and $\mathbf{E}_0 = I_{nc}$ is the input image. Following, the encoded latent representation z_{nc} is given by the output of the L-th layer, i.e., $z_{nc} = \mathbf{E}_L(I_{nc}; \theta_e) = g\big(\mathbf{K}_L * \mathbf{E}_{L-1}(I_{nc}; \theta_e) + \mathbf{b}_L\big)$, which are decoded to get velocity field v_{nc} by a decoder $\mathcal{C}_{\mathrm{D}}$, parameterized by θ_d.

In contrast to existing methods to predict geometric deformations [6,32,41], our model CoRLD eliminates the dependence on the template image required as an input during the learning process. In particular, CoRLD directly encodes the input images I_{nc} and predicts the latent representations of transformations. The template image is used exclusively in the loss function, serving as ground truth to guide the network in learning geometric deformations. Analogous to Eq. (2), the objective function to predict the velocity fields can then be defined as

$$\mathcal{L}_{\text{shape}}(\theta_e, \theta_d) = \sum_{n=1}^{N_c} \sum_{c=1}^{C} \frac{1}{\sigma^2} \| I_{nc} - T_c \circ \phi_{nc}(v_{nc}(I_{nc}; \theta_e, \theta_d)) \|_2^2 \\ + \delta \, \| \nabla v_{nc}(I_{nc}; \theta_e, \theta_d) \| + \text{reg}(\theta_e, \theta_d), \text{ s.t. Eq. (1)}, \tag{3}$$

where reg$(\cdot)$ is a regularity term on the network parameters and δ are the weighting parameter.

Class-Aware Contrastive Representation Learning in Deformation Spaces. We introduce a class-aware supervised contrastive objective in the latent space to enhance the discriminative power of learned representations across classes. Starting from the encoded latent feature representation z_{nc}, we first adopt a feature projection module $\mathcal{T}(.; \theta_s)$, parameterized by θ_s, to transform the latent features into a more structured geometric space suitable for contrastive learning [22]. This module consists of a convolution layer followed by batch normalization and adaptive pooling to obtain projected features, i.e., $\tilde{z}_{nc} := \mathcal{T}(z_{nc}; \theta_s)$.

Now, let $\mathbf{z}^i (:= \tilde{z}^i_{nc}, \forall i)$ and $\mathbf{z}^p (:= \tilde{z}^p_{nc}, \forall i)$ represent the latent features of samples i and p, respectively. To guide the contrastive objective, we define the following:

- A *positive sample set* $P(i) = \{p \mid y^i = y^p \text{ and } i \neq p\}$, representing all samples whose labels matching the label of sample i, excluding i itself;
- A *candidate sample set* $A(i) = \{a \mid y^i \neq y^a\}$, representing all samples in the batch except sample i itself.

Under these definitions, we are ready to define our class-aware contrastive loss in the latent deformation spaces, formulated as

$$\mathcal{L}_{\text{CSR}}(\theta_e, \theta_s) = -\sum_{i \in I} \sum_{p \in P(i)} \frac{1}{|P(i)|} \log \left(\frac{\exp\left(\text{sim}(\mathbf{z}^i, \mathbf{z}^p)/\tau\right)}{\sum_{a \in A(i)} \exp\left(\text{sim}(\mathbf{z}^i, \mathbf{z}^a)/\tau\right)} \right), \tag{4}$$

where sim$(\mathbf{z}^i, \mathbf{z}^p)$ denotes the cosine similarity between feature vectors $\mathbf{z}^i$ and $\mathbf{z}^p$, and $\tau > 0$ is a temperature parameter that controls the concentration of the similarity distribution. This supervised contrastive objective encourages the model to group semantically similar features closer together while separating dissimilar features; hence enabling the learning of semantically meaningful and discriminative representations in the latent deformation space.

Network Loss. Defining $\Theta(\theta_e, \theta_s, \theta_d)$, we are now ready to formulate the total loss function of our CoRLD network as

$$\mathcal{L}_{\text{CoRLD}}(\Theta) = \mathcal{L}_{shape}(I_{nc}, T_c; \Theta) + \beta\, \mathcal{L}_{\text{CSR}}(\mathcal{T}(I_{nc}; \theta_e, \theta_s)) + \text{reg}(\Theta), \tag{5}$$

where $\text{reg}(\cdot)$ is a regularity term on the network parameters, and β are the weighting parameters.

3.2 Boosted Classifier with Contrastive Shape Representations

We validate the effectiveness of our proposed CoRLD model on image classification tasks by training a classifier using latent features from images and learned contrastive geometric shape spaces. Let $\mathcal{I}_{\text{E}}$ denote the image encoder network parameterized by θ_{IE} extracting intensity features. We then integrate learned contrastive shape features from latent spaces with the image features to train a boosted classifier, parameterized by θ_{clf}. For each input image I_{nc}, this classifier predicts the class label $\hat{y}_{nc}$ with respect to the ground truth label y_{nc}. While we use a non-parameterized feature concatenation module for shape

Algorithm 1: Two-step CoRLD training with boosted image classifier.

Input : A group of N input images, class labels y_{nc}, a number of iterations $r_{\text{CoRLD}}/r_{\text{clf}}$, and stopping thresholds $\epsilon_{\text{CoRLD}}/\epsilon_{\text{clf}}$.

Output: Latent contrastive shape features $\tilde{z}_{nc}$, initial velocity fields v_{nc}, and classification labels $\hat{y}_{nc}$.

`/* Train CoRLD */`

repeat
  for $i = 1$ to r_{CoRLD} **do**
    Predict latent contrastive shape features, initial velocity fields v_{nc} and derive the deformation field ϕ_{nc};
    Optimize the CoRLD network loss $\mathcal{L}_{\text{CoRLD}}$ in Eq. (5);
  end
until $|\Delta \mathcal{L}_{\text{CoRLD}}| < \epsilon_{\text{CoRLD}}$;

`/* Train boosted image classification network */`

repeat
  for $i = 1$ to r_{clf} **do**
    Integrate intensity features and learned contrastive shape features derived from image feature extractor $\mathcal{I}_{\text{E}}$ and CoRLD $(\mathcal{C}_{\text{E}}, \mathcal{C}_{\text{D}})$;
    Optimize the boosted classification loss $\mathcal{L}_{\text{clf}}$ in Eq. (6);
  end
until $|\Delta \mathcal{L}_{\text{clf}}| < \epsilon_{\text{clf}}$;

integration, other advanced fusion methods can easily be incorporated. Optimized over a cross-entropy loss to train this boosted classifier, parameterized by $\Theta_{\text{clf}}(\theta_{\text{clf}}, \theta_{\text{IE}})$, and denoting γ is a weighting parameter, we are now ready to define the classification loss function as

$$\mathcal{L}_{\text{clf}}(I_{nc}; \Theta_{\text{clf}}) = \gamma \sum_{n=1}^{N_c} \sum_{c=1}^{C} -y_{nc} \cdot \log \hat{y}_{nc}(\Theta_{\text{clf}}) + \text{reg}(\Theta_{\text{clf}}). \tag{6}$$

Network Optimization. We design a two-step optimization module: first, CoRLD is trained to predict contrastive features in the latent geometric space. Next, a boosted classifier is trained using features from an integrated space, combining image and learned shape features. A summary of our two-step training of CoRLD with boosted classification task is presented in Algorithm 1.

4 Experiments and Evaluations

We validate the effectiveness of our model CoRLD on diverse datasets, including real brain MRI scans capturing complex neurological structures and adrenal shapes derived from CT scans reflecting the variability and complexity of soft tissue. Examples of the experimental datasets are shown in Fig. 2.

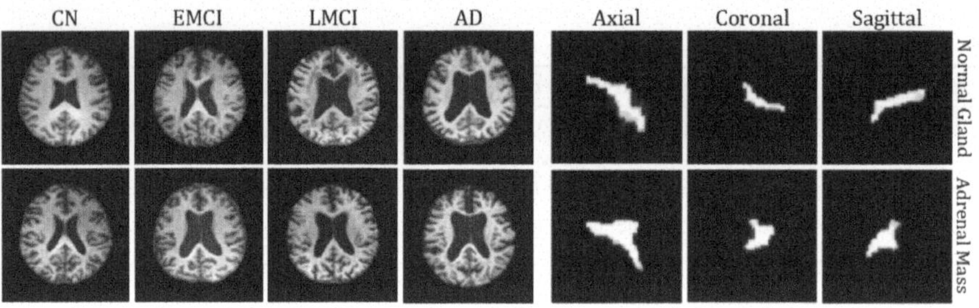

Fig. 2. Left to Right: Examples of brain MRI slices across four diagnostic groups (CN, EMCI, LMCI, AD) vs 3D adrenal shapes derived from CTs visualized in three anatomical planes (Axial, Coronal, and Sagittal).

2D Brain MRIs. We include axial views of 2219 public T1-weighted brain MRIs from the Alzheimer's Disease Neuroimaging Initiative (ADNI) [19]. All subjects ranged in age from 50 to 100, covering cognitively normals (CN: 497), patients affected by Alzheimer's disease (AD: 368), and individuals with early and late mild cognitive impairment (EMCI: 733/LMCI: 621), respectively. All MRIs were preprocessed to be the size of 128×128, and underwent skull-stripping, intensity min-max normalization, bias-field correction, and affine registration [29].

3D Adrenal Shapes. We select 1584 left and right real 3D adrenal gland shapes derived from CT scans, representing 792 patients from the AdrenalMNIST3D data repository [36]. This dataset is specifically collected to identify the presence of adrenal mass differentiating from normal adrenal glands. All images underwent affine registration and normalization of intensity with the size of $64 \times 64 \times 64$, with isotropic voxels of $1\,\text{mm}^3$.

4.1 Experiments

We evaluate CoRLD based on two perspectives: (i) assessing the quality of latent representations by measuring the classification performance and (ii) demonstrating the effectiveness of our learned contrastive representation model in downstream image classification tasks. We validate the effectiveness of our proposed model by comparing it with GeoSIC [32], a deep network that learns deformable shapes in a deformation space and various intensity-backed network backbones.

Evaluation of Learned Contrastive Shape Representations. We evaluate the quality of the learned representations in our model CoRLD by ablating different components (the presence of template images during inference and the contrastive objective) through classification tasks across all datasets. We train a three-layer fully connected classifier with ReLU and dropout on the learned latent geometric features and report performance metrics such as accuracy, precision, and F1-score. We also measure the training time (per epoch) for each ablation setting to assess the computational efficiency of the representation learning model itself, excluding the classifier training time, for fair evaluation.

To further validate our template-free strategy, we qualitatively assess the deformation quality through visualizations of deformed images, error maps between deformed and target images, predicted velocities, and deformation fields.

Evaluation of CoRLD in Downstream Classification Tasks. We further demonstrate the effectiveness of CoRLD over all baselines by comparing their learned latent representations integrated into the downstream image classification tasks. Here, we compare CoRLD with Geo-SIC and various image feature extractor backbones, including ResNet [14], DenseNet [18], ConvNext [25], and ResNeXt [35]. To evaluate performance, we report classification accuracy, precision, F1-score, sensitivity, specificity, and area under the curve (AUC) scores.

Robustness to Input Perturbations. We demonstrate the robustness of CoRLD to variations in image intensity by performing a brief experiment on all datasets under ResNet and DenseNet backbones where we add different scales of universal adversarial noises and compare CoRLD with all baselines.

Evaluation of Contrastive Temperature and Template Strategies. We analyze the impact of contrastive temperature τ on model performance through classification tasks for our proposed model CoRLD on 2D brain MRIs. Increasing τ evaluates the sensitivity of the model to similarity margins in latent representations. Besides, we validate our model's flexibility by comparing single versus multi-template training settings to investigate the model behavior under different template configurations while preserving template-free inference across both 2D brain and 3D adrenal classifications.

Parameter Setting. We set the noise variance $\sigma = 0.01$ and batch size of 512 and 16 for the 2D brain MRI and 3D adrenal shape experiments, respectively. We split the dataset into 70%, 15%, and 15% for training, validation, and testing, respectively. For network training, we utilize the cosine annealing learning rate scheduler that starts with a learning rate of $\eta = 1e^{-3}$. We extensively evaluate various configurations of the weighting parameters (δ, β, and γ) to analyze the network's convergence behavior and stability characteristics, with empirical results demonstrating optimal performance at $\delta = 0.1$, $\beta = 0.1$, and $\gamma = 1.0$. We train all the models with Adam optimizer [24] and obtain the best validation performance until convergence. The training and prediction for all methods are conducted on a 40 GB NVIDIA A100 Tensor Core GPU.

5 Results

Table 1 presents an ablation study of CoRLD under different settings. The results show that contrastive learning improves performance in both template and template-free settings, achieving 79.58/84.90% and 83.78/85.23% accuracies on both 2D brain and 3D adrenal datasets, respectively. Notably, the template-free variant with contrastive learning achieves optimal performance, demonstrating that template dependency can be removed in inference time while maintaining robust shape-based classification. CoRLD also maintains efficiency with training times of 1.13 s and 67.53 s per epoch on brain MRIs and adrenal shapes, respectively, with minimal computational overhead compared to non-contrastive methods, while providing substantial performance gains.

Table 1. Ablation studies examining the impact of template dependency and contrastive objective on CoRLD's classification performance using only shape features. The *Time* metric reflects the training time for the representation learning model per epoch, excluding the classifier.

Objective		2D Brain MRIs				3D Adrenal Shapes			
Template	$\mathcal{L}_{CSR}$	Accuracy	Precision	F1-sc.	Time (s)	Accuracy	Precision	F1-sc.	Time (s)
Yes	No	78.98	78.92	78.91	1.066	82.22	82.67	80.38	68.65
Yes	Yes	79.58	79.51	79.41	1.086	84.90	83.08	82.21	72.40
No	No	80.78	80.91	80.68	1.064	83.89	82.43	80.72	64.64
No	**Yes**	**83.78**	**84.16**	**83.86**	1.132	**85.23**	**84.23**	**82.42**	67.53

Figure 3 visualizes the qualitative comparisons between template-guided and our template-free approaches. Both models achieve comparable performance, evidenced by nearly identical deformed outputs with negligible differences. The transformation fields demonstrate similar smoothness and topological properties

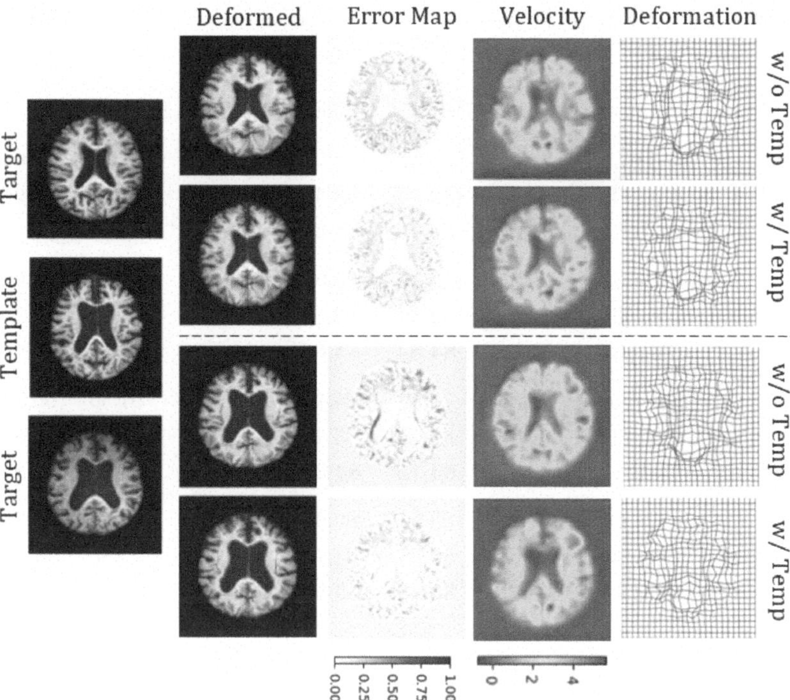

Fig. 3. Left to right: Visual comparison of the deformed template, its error map with the target, velocity in colormap, and deformation field for template-guided (w/ Temp) and CoRLD (w/o Temp) models.

across both approaches, indicating our model can predict anatomically plausible deformations without requiring explicit template guidance.

Table 2 reports the classification performances of CoRLD against established baselines across four different backbone architectures, evaluating the effectiveness of learned geometric and intensity features. Our model consistently achieves state-of-the-art results, outperforming all baselines by a margin of over 1–2% across all network backbones. These extensive analyses yield two significant insights: (i) the models leveraging both intensity and shape features (Geo-SIC and CoRLD) substantially outperform classifiers trained solely on intensity features, highlighting the complementary nature of geometric information in downstream tasks, specifically image classification, and (ii) CoRLD's template-free approach to learning contrastive shape features and integrating them with intensity features showcases its effectiveness. The consistent improvements across diverse network backbones validate the robustness and generalizability of our method, establishing its superiority over traditional image-only approaches and existing shape-aware frameworks.

Table 3 demonstrates the classification performance of CoRLD on 3D adrenal shapes across multiple network backbones. Our method consistently achieves

Table 2. Classification performances (%) comparison on real brain MRI dataset between CoRLD vs. other baselines.

Backbone	Model	Accuracy	Precision	F1-Score	Sensitivity	Specificity	AUC
ResNet	Image Only	80.28	80.41	81.92	80.28	93.89	95.08
	Geo-SIC	82.58	82.92	82.58	82.58	94.17	**96.93**
	CoRLD	**84.68**	**84.75**	**84.67**	**84.68**	**94.83**	96.85
DenseNet	Image Only	79.54	79.53	79.07	79.54	92.57	95.92
	Geo-SIC	82.28	82.40	82.16	82.28	93.91	96.38
	CoRLD	**84.68**	**84.84**	**84.69**	**84.68**	**94.89**	**96.90**
ConvNext	Image Only	75.08	75.59	74.79	75.08	91.55	90.96
	Geo-SIC	83.48	83.63	83.47	83.48	94.36	96.85
	CoRLD	**84.38**	**84.78**	**84.38**	**84.38**	**94.65**	**96.89**
ResNext	Image Only	81.98	83.53	82.09	81.98	93.86	95.93
	Geo-SIC	84.68	85.28	84.79	84.68	94.86	**96.70**
	CoRLD	**85.59**	**85.65**	**85.56**	**85.59**	**95.12**	96.61

state-of-the-art results across all evaluation metrics. The integration of geometric shape features shows marked improvement over intensity-only approaches, while our template-free contrastive shape learning strategy significantly outperforms existing baselines. These results, consistent with our brain MRI experiments, further validate the generalizability and effectiveness of CoRLD's shape-aware learning framework in 3D medical image classification tasks.

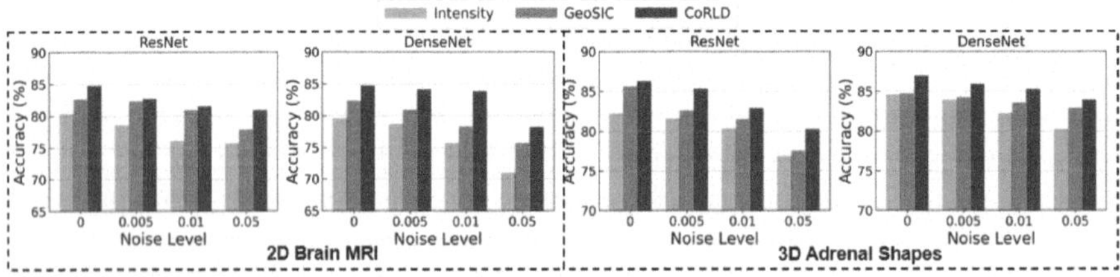

Fig. 4. Classification accuracy comparison across all models, including CoRLD, under ResNet and DenseNet backbones for 2D brain MRI (left panel) and 3D adrenal shape (right panel) datasets at different scales of adversarial noise levels.

Table 3. Classification performances (%) comparison on real adrenal shape dataset between CoRLD vs. other baselines.

Backbone	Model	Accuracy	Precision	F1-Score	Sensitivity	Specificity	AUC
ResNet	Image Only	82.21	80.96	80.02	82.21	90.39	86.51
	Geo-SIC	85.58	85.16	84.50	85.58	92.14	89.31
	CoRLD	**86.24**	**86.90**	**85.39**	**86.24**	**93.01**	**90.28**
DenseNet	Image Only	84.56	83.98	82.89	84.56	90.39	88.08
	Geo-SIC	84.69	83.67	84.61	84.69	93.45	89.30
	CoRLD	**86.91**	**86.44**	**86.03**	**86.91**	**94.76**	**89.67**
ConvNext	Image Only	74.83	71.80	70.83	74.83	81.27	70.39
	Geo-SIC	76.51	71.92	72.04	76.51	83.89	71.74
	CoRLD	**77.85**	**74.65**	**74.22**	**77.85**	**84.32**	**73.72**
ResNext	Image Only	78.52	75.96	76.20	78.52	87.75	83.97
	Geo-SIC	80.54	78.65	77.81	80.54	88.21	82.38
	CoRLD	**81.88**	**80.55**	**80.72**	**81.88**	**90.39**	**83.04**

Figure 4 visualizes the robustness analysis of three approaches (Intensity, GeoSIC, and CoRLD) against increasing adversarial noise perturbations σ (ranging from 0 to 0.05) across ResNet and DenseNet architectures on 2D brain MRI and 3D adrenal shapes datasets. For brain MRI classification, CoRLD maintains consistently superior accuracy under increasing noise levels, particularly at $\sigma = 0.05$ where it outperforms baseline methods by a significant margin (>4% higher accuracy). The 3D Adrenal Shapes results demonstrate a similar trend, though with smaller performance gaps, where CoRLD exhibits better resilience specifically in the DenseNet backbone. While all methods show expected performance degradation with increasing noise levels, CoRLD's integrated contrastive learning approach demonstrates stronger robustness across both datasets and architectures, validating its effectiveness under adversarial conditions.

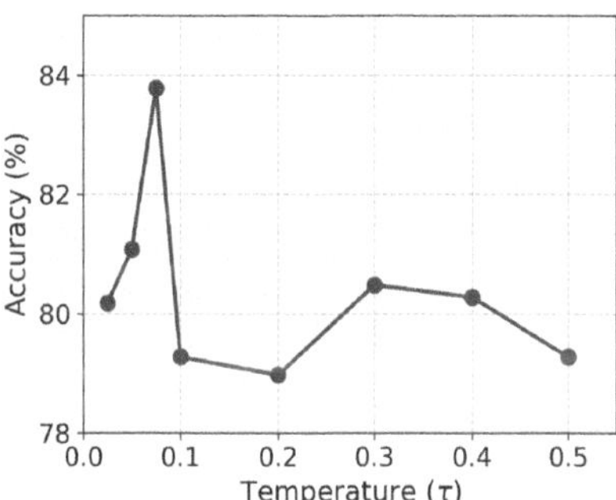

Fig. 5. Effect of the temperature parameter (τ) on network performance for the 2D brain dataset.

Figure 5 demonstrates the temperature parameter τ controls the concentration of the $\mathcal{L}_{\text{CSR}}$ loss distribution in the latent geometric space, with optimal feature discrimination at $\tau = 0.75$ from well-separated positive/negative pairs, while higher τ values over-smooth the distribution. Table 4 demonstrates multi-template training achieves superior classification by capturing class-specific geometric variations and anatomical patterns, compared to single-template which is constrained to one reference geometry.

Table 4. Accuracy (%) comparison between single vs multi-template CoRLD model on all datasets.

Template	2D Brains	3D Adrenals
Single	83.78	85.23
Multi	85.29	86.24

6 Conclusion and Discussion

In this paper, we present a novel geometric representation learning framework, CoRLD, which learns shape features through contrastive image deformations in latent space. Our approach eliminates the need for template-based geometric analysis by directly learning diffeomorphic transformations from input images through supervised contrastive optimization. Extensive evaluations on real 2D brain MRIs and 3D adrenal CT shapes demonstrate CoRLD's superior performance in classification tasks, highlighting the effectiveness of template-free geometric feature learning in medical imaging.

Building upon CoRLD's promising results, we can advance this framework in several compelling directions: i) extending our contrastive learning strategy to unsupervised settings, and eliminating the need for paired shape annotations in scenarios with limited labeled data, ii) integrating uncertainty estimation into the learned geometric representations, providing confidence measures for anatomical variations in clinical applications, and iii) adapting the proposed framework to handle multi-modal geometric features, enabling robust shape analysis across different imaging protocols while maintaining anatomical consistency.

Acknowledgements. This work was supported by NSF CAREER Grant 2239977.

References

1. Achlioptas, P., Diamanti, O., Mitliagkas, I., Guibas, L.: Learning representations and generative models for 3D point clouds. In: International Conference on Machine Learning, pp. 40–49 (2018)
2. Arsigny, V., Commowick, O., Pennec, X., Ayache, N.: A Log-Euclidean framework for statistics on diffeomorphisms. In: Larsen, R., Nielsen, M., Sporring, J. (eds.) MICCAI 2006. LNCS, vol. 4190, pp. 924–931. Springer, Heidelberg (2006). https://doi.org/10.1007/11866565_113
3. Avants, B.B., Epstein, C.L., Grossman, M., Gee, J.C.: Symmetric diffeomorphic image registration with cross-correlation: evaluating automated labeling of elderly and neurodegenerative brain. Med. Image Anal. **12**(1), 26–41 (2008)

4. Azad, R., et al.: Beyond self-attention: deformable large kernel attention for medical image segmentation. In: Proceedings of the IEEE/CVF Winter Conference on Applications of Computer Vision, pp. 1287–1297 (2024)
5. Azizi, S., et al.: Big self-supervised models advance medical image classification. In: Proceedings of the IEEE/CVF International Conference on Computer Vision, pp. 3478–3488 (2021)
6. Balakrishnan, G., Zhao, A., Sabuncu, M.R., Guttag, J., Dalca, A.V.: VoxelMorph: a learning framework for deformable medical image registration. IEEE Trans. Med. Imaging **38**(8), 1788–1800 (2019)
7. Beg, M.F., Miller, M.I., Trouvé, A., Younes, L.: Computing large deformation metric mappings via geodesic flows of diffeomorphisms. Int. J. Comput. Vision **61**, 139–157 (2005)
8. Bhalodia, R., Elhabian, S.Y., Kavan, L., Whitaker, R.T.: DeepSSM: a deep learning framework for statistical shape modeling from raw images. In: Shape in Medical Imaging: International Workshop, ShapeMI 2018, Held in Conjunction with MICCAI 2018, Granada, Spain, 20 September 2018, Proceedings, pp. 244–257. Springer (2018)
9. Chen, J., Frey, E.C., He, Y., Segars, W.P., Li, Y., Du, Y.: TransMorph: transformer for unsupervised medical image registration. Med. Image Anal. **82**, 102615 (2022)
10. Christensen, G.E., Rabbitt, R.D., Miller, M.I., et al.: A deformable neuroanatomy textbook based on viscous fluid mechanics. In: 27th Annual Conference on Information Sciences and Systems, pp. 211–216. Citeseer (1993)
11. Cootes, T.F., Taylor, C.J., Cooper, D.H., Graham, J.: Active shape models-their training and application. Comput. Vis. Image Underst. **61**(1), 38–59 (1995)
12. Deng, J., Xu, D., Li, W., Duan, L.: Harmonious teacher for cross-domain object detection. In: Proceedings of the IEEE/CVF Conference on Computer Vision and Pattern Recognition, pp. 23829–23838 (2023)
13. Hao, F., He, F., Liu, L., Wu, F., Tao, D., Cheng, J.: Class-aware patch embedding adaptation for few-shot image classification. In: Proceedings of the IEEE/CVF International Conference on Computer Vision, pp. 18905–18915 (2023)
14. He, K., Zhang, X., Ren, S., Sun, J.: Deep residual learning for image recognition. In: Proceedings of the IEEE Conference on Computer Vision and Pattern Recognition, pp. 770–778 (2016)
15. Hossain, T., Ma, J., Li, J., Zhang, M.: Invariant shape representation learning for image classification. arXiv preprint arXiv:2411.12201 (2024)
16. Hossain, T., Shishir, F.S., Ashraf, M., Al Nasim, M.A., Shah, F.M.: Brain tumor detection using convolutional neural network. In: 2019 1st International Conference on Advances in Science, Engineering and Robotics Technology (ICASERT), pp. 1–6. IEEE (2019)
17. Hossain, T., Zhang, M.: MGAug: multimodal geometric augmentation in latent spaces of image deformations. Med. Image Anal., 103540 (2025)
18. Huang, G., Liu, Z., Van Der Maaten, L., Weinberger, K.Q.: Densely connected convolutional networks. In: Proceedings of the IEEE Conference on Computer Vision and Pattern Recognition, pp. 4700–4708 (2017)
19. Jack, C.R., Jr., et al.: The Alzheimer's disease neuroimaging initiative (ADNI): MRI methods. J. Magn. Reson. Imaging **27**(4), 685–691 (2008)
20. Joshi, S., Davis, B., Jomier, M., Gerig, G.: Unbiased diffeomorphic atlas construction for computational anatomy. Neuroimage **23**, S151–S160 (2004)
21. Ke, T.-W., Mo, S., Stella, X.Y.: Learning hierarchical image segmentation for recognition and by recognition. In: The Twelfth International Conference on Learning Representations (2023)

22. Khosla, P., et al.: Supervised contrastive learning. In: Advances in Neural Information Processing Systems, vol. 33, pp. 18661–18673 (2020)
23. Kim, B., Han, I., Ye, J.C.: DiffuseMorph: unsupervised deformable image registration using diffusion model. In: European Conference on Computer Vision, pp. 347–364. Springer (2022)
24. Kingma, D.P.: Adam: a method for stochastic optimization. arXiv preprint arXiv:1412.6980 (2014)
25. Liu, Z., Mao, H., Wu, C.-Y., Feichtenhofer, C., Darrell, T., Xie, S.: A convnet for the 2020s. In: Proceedings of the IEEE/CVF Conference on Computer Vision and Pattern Recognition, pp. 11976–11986 (2022)
26. Mok, T., Chung, A.: Conditional deformable image registration with convolutional neural network. In: de Bruijne, M., et al. (eds.) MICCAI 2021. LNCS, vol. 12904, pp. 35–45. Springer, Cham (2021). https://doi.org/10.1007/978-3-030-87202-1_4
27. Pizer, S.M., Fritsch, D.S., Yushkevich, P.A., Johnson, V.E., Chaney, E.L.: Segmentation, registration, and measurement of shape variation via image object shape. IEEE Trans. Med. Imaging **18**(10), 851–865 (1999)
28. Pu, Y., et al.: Rank-DETR for high quality object detection. In: Advances in Neural Information Processing Systems, vol. 36 (2024)
29. Reuter, M., Schmansky, N.J., Rosas, H.D., Fischl, B.: Within-subject template estimation for unbiased longitudinal image analysis. Neuroimage **61**(4), 1402–1418 (2012)
30. Rueckert, D., Frangi, A.F., Schnabel, J.A.: Automatic construction of 3-D statistical deformation models of the brain using nonrigid registration. IEEE Trans. Med. Imaging **22**(8), 1014–1025 (2003)
31. Vilas, M.G., Schaumlöffel, T., Roig, G.: Analyzing vision transformers for image classification in class embedding space. In: Advances in Neural Information Processing Systems, vol. 36 (2024)
32. Wang, J., Zhang, M.: Geo-SIC: learning deformable geometric shapes in deep image classifiers. In: Advances in Neural Information Processing Systems, vol. 35, pp. 27994–28007 (2022)
33. Wells, W.M., III., Viola, P., Atsumi, H., Nakajima, S., Kikinis, R.: Multi-modal volume registration by maximization of mutual information. Med. Image Anal. **1**(1), 35–51 (1996)
34. Wu, N., Zhang, M.: NeurEPDiff: neural operators to predict geodesics in deformation spaces. In: International Conference on Information Processing in Medical Imaging, pp. 588–600. Springer (2023)
35. Xie, S., Girshick, R., Dollár, P., Tu, Z., He, K.: Aggregated residual transformations for deep neural networks. In: Proceedings of the IEEE Conference on Computer Vision and Pattern Recognition, pp. 1492–1500 (2017)
36. Yang, J., et al.: MedMNIST V2-a large-scale lightweight benchmark for 2D and 3D biomedical image classification. Sci. Data **10**(1), 41 (2023)
37. Yang, X., Kwitt, R., Niethammer, M.: Fast predictive image registration. In: Deep Learning and Data Labeling for Medical Applications: First International Workshop, LABELS 2016, and Second International Workshop, DLMIA 2016, Held in Conjunction with MICCAI 2016, Athens, Greece, 21 October 2016, Proceedings, vol. 1, pp. 48–57. Springer (2016)
38. You, C., et al.: Rethinking semi-supervised medical image segmentation: a variance-reduction perspective. In: Advances in Neural Information Processing Systems, vol. 36 (2024)

39. Zhang, M., et al.: Frequency diffeomorphisms for efficient image registration. In: Niethammer, M., et al. (eds.) IPMI 2017. LNCS, vol. 10265, pp. 559–570. Springer, Cham (2017). https://doi.org/10.1007/978-3-319-59050-9_44
40. Zhang, M., Singh, N., Fletcher, P.T.: Bayesian estimation of regularization and atlas building in diffeomorphic image registration. In: Gee, J.C., Joshi, S., Pohl, K.M., Wells, W.M., Zöllei, L. (eds.) IPMI 2013. LNCS, vol. 7917, pp. 37–48. Springer, Heidelberg (2013). https://doi.org/10.1007/978-3-642-38868-2_4
41. Zhao, A., Balakrishnan, G., Durand, F., Guttag, J.V., Dalca, A.V.: Data augmentation using learned transformations for one-shot medical image segmentation. In: Proceedings of the IEEE/CVF Conference on Computer Vision and Pattern Recognition, pp. 8543–8553 (2019)

Time-Series Image Analysis

4DRGS: 4D Radiative Gaussian Splatting for Efficient 3D Vessel Reconstruction from Sparse-View Dynamic DSA Images

Zhentao Liu[1], Ruyi Zha[2], Huangxuan Zhao[3], Hongdong Li[2], and Zhiming Cui[1(✉)]

[1] School of Biomedical Engineering and State Key Laboratory of Advanced Medical Materials and Devices, ShanghaiTech University, Shanghai, China
cuizhm@shanghaitech.edu.cn
[2] School of Computing, Australian National University, Canberra, Australia
[3] School of Computer Science, Wuhan University, Wuhan, China

Abstract. Reconstructing 3D vessel structures from sparse-view dynamic digital subtraction angiography (DSA) images enables accurate medical assessment while reducing radiation exposure. Existing methods often produce suboptimal results or require excessive computation time. In this work, we propose 4D radiative Gaussian splatting (4DRGS) to achieve high-quality reconstruction efficiently. In detail, we represent the vessels with 4D radiative Gaussian kernels. Each kernel has time-invariant geometry parameters, including position, rotation, and scale, to model static vessel structures. The time-dependent central attenuation of each kernel is predicted from a compact neural network to capture the temporal varying response of contrast agent flow. We splat these Gaussian kernels to synthesize DSA images via X-ray rasterization and optimize the model with real captured ones. The final 3D vessel volume is voxelized from the well-trained kernels. Moreover, we introduce accumulated attenuation pruning and bounded scaling activation to improve reconstruction quality. Extensive experiments on real-world patient data demonstrate that 4DRGS achieves impressive results in 5 min training, which is 32× faster than the state-of-the-art method. This underscores the potential of 4DRGS for real-world clinics.

Keywords: Sparse-View DSA Reconstruction · Gaussian Splatting

1 Introduction

Digital subtraction angiography (DSA) is a widely recognized gold standard for diagnosing vascular diseases, such as arteriovenous malformation, arteriovenous fistula, and intracranial aneurysms [13,22–24]. As shown in Fig. 1(a), DSA imaging involves two rotational cone-beam X-ray scans: the mask run, performed before the contrast agent injection, and the fill run, performed after the

Z. Liu and R. Zha—Equal contribution.

I. Oguz et al. (Eds.): IPMI 2025, LNCS 15830, pp. 361–374, 2026.
https://doi.org/10.1007/978-3-031-96625-5_24

injection. Subtracting X-ray images in the fill run from those in the mask run yields 2D DSA images, which highlight blood flow marked by the contrast agent while removing background tissues. However, significant vessel overlap in DSA images hinders accurate anatomical assessment. Therefore, reconstructing 3D vessel structures from DSA images is essential for clear visualization to support medical diagnosis. Existing DSA systems typically capture hundreds of images for precise reconstruction [33] based on the Feldkamp-Davis-Kress (FDK) algorithm [6,8], exposing patients and radiographers to significant radiation. In this work, we aim to achieve high-quality reconstruction efficiently with dozens of images to reduce radiation exposure.

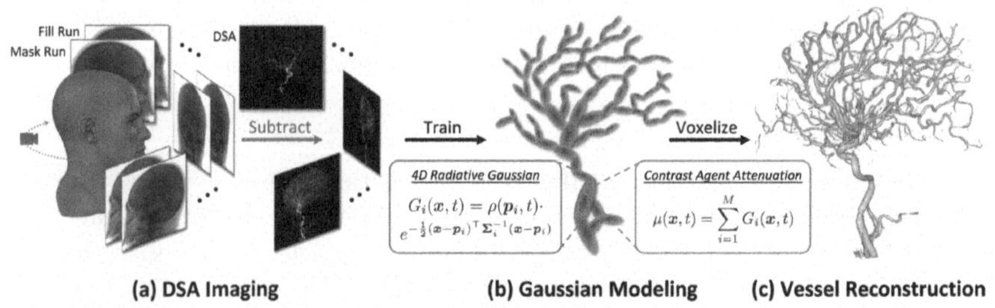

Fig. 1. Overview of DSA imaging and vessel reconstruction. (a) DSA images are generated by subtracting fill-run X-ray images from their mask-run counterparts. (b) We model vessels as a set of 4D radiative Gaussians. (c) The final 3D vessel volume is reconstructed via attenuation voxelization.

Sparse-view DSA reconstruction is a challenging task for two reasons. First, as observed from Fig. 1(a), each DSA image captures a distinct blood state as the contrast agent gradually flows through the vessels. Static computed tomography (CT) reconstruction algorithms [1,5,7,8,21,25,29–31] often fail under such dynamic scenarios. Second, sparse-view reconstruction is highly ill-posed. Commonly used algorithms such as FDK [6,8] would produce severe artifacts when measurements are insufficient. Although recent neural radiance fields (NeRF) [17]-based method VPAL [15] addresses these challenges well, it takes hours to process a single case and is thus impractical for real-world usage.

Our work is inspired by 3D Gaussian splatting (3DGS) [11], which represents scenes using explicit kernels and employs rasterization for RGB image rendering. Compared with NeRF-based methods [15] which model the entire space using neural networks, the kernel-based representation allows us to bypass large empty regions in backgrounds and focus on sparse vessel structures. Furthermore, differentiable rasterization offers faster rendering and training than volume rendering [15], making it well-suited for time-sensitive DSA reconstruction. Existing 3DGS-based solutions in X-ray imaging address static CT reconstruction [14,30], X-ray image synthesis [4,9], and DSA image synthesis [32]. However, no prior work has applied Gaussian splatting to DSA reconstruction.

In this paper, we introduce 4D radiative Gaussian splatting (4DRGS), the first Gaussian splatting-based framework for efficient 3D vessel reconstruction from sparse-view dynamic DSA images. A key observation is that vessels maintain static structures during scanning process, while their attenuation varies over time due to the contrast agent flow. Therefore, we represent vessels as a set of 4D radiative Gaussian kernels (Fig. 1(b)), where each kernel acts as a local Gaussian-shaped time-varying attenuation distribution. The static vessel structures are modeled with time-invariant geometry parameters, including position, rotation, and scale. To mimic the temporal attenuation changes, we use a compact neural network to predict each kernel's central attenuation based on its position and the given timestamp. We splat these Gaussian kernels to synthesize DSA images via X-ray rasterization [30] and optimize them by minimizing the disparities with real captured images. After training, 3D vessel volume is reconstructed via attenuation voxelization (Fig. 1(c)) [30]. Two innovations are further introduced to enhance reconstruction quality: (1) accumulated attenuation pruning to remove non-vessel kernels and (2) bounded scaling activation to reduce needle artifacts. Experiments on real-world data demonstrate our method's effectiveness for both 3D vessel reconstruction and 2D DSA image synthesis. Notably, 4DRGS achieves impressive results in 5 min and converges in 13 min, offering a speedup of 32× and 12× compared to the state-of-the-art (SOTA) method VPAL [15].

In summary, our contributions are threefold. First, we propose 4DRGS, the first Gaussian splatting-based framework for efficient 3D vessel reconstruction. Second, we develop 4D radiative Gaussian kernels for DSA imaging and introduce key innovations to improve reconstruction quality. Finally, our method achieves SOTA results within minutes, demonstrating its potential for real-world usage.

2 Preliminary

DSA Imaging. As depicted in Fig. 1(a), DSA imaging involves two scans: one before injecting the contrast agent (mask run) and one after (fill run). Both scans follow the same procedure, where a cone-beam X-ray machine rotates around the patient and captures 2D X-ray images at equal angular intervals. Denote X-ray images from the mask run as $\{\mathbf{I}_j^m \in \mathbb{R}^{H\times W}\}_{j=1}^T$ and their counterparts from the fill run as $\{\mathbf{I}_j^f \in \mathbb{R}^{H\times W}\}_{j=1}^T$, where j, T, and $H \times W$ are frame index, total number of frames, and image resolution, respectively. The DSA images $\{\mathbf{I}_j \in \mathbb{R}^{H\times W}\}_{j=1}^T$ can be generated through logarithmic subtraction: $\mathbf{I}_j = \ln(\mathbf{I}_j^m) - \ln(\mathbf{I}_j^f)$. In this way, DSA images highlight dynamic blood flows while removing non-relevant tissues, delivering useful insights for vascular disease diagnosis.

Sparse-View DSA Reconstruction. We define the timestamp t_j of the j-th DSA image as $t_j = \frac{j}{T}$, which indicates its capture order. The complete set of frame data is represented as $\{\mathbf{I}_j, t_j\}_{j=1}^T$. The goal of sparse-view DSA reconstruction is to recover an attenuation volume representing vessel structures with a uniformly sampled subset $\{\mathbf{I}_{j_k}, t_{j_k}\}_{k=1}^N$, where $N < T$ and $j_k = \lfloor (k-1)\cdot \frac{T}{N} \rfloor + 1$.

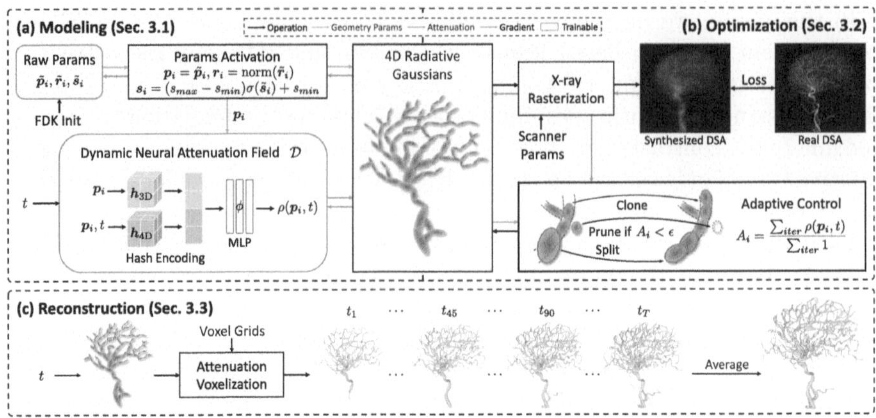

Fig. 2. The overall pipeline of 4DRGS. We model vessels as a set of 4D radiative Gaussian kernels (Sect. 3.1) and optimize them with image losses (Sect. 3.2). 3D vessel volume is reconstructed via attenuation voxelization (Sect. 3.3).

Radiative Gaussian Splatting. R^2-Gaussian [30] is the first work to leverage 3DGS for static CT reconstruction. It represents the scanned scene using 3D radiative Gaussian kernels, each defined by the central attenuation, position, rotation, and scale. It develops differentiable X-ray rasterization and voxelization to produce X-ray images and attenuation volumes in real time. Building on R^2-Gaussian, we introduce 4DRGS designated for dynamic DSA imaging, enabling high-quality and efficient vessel reconstruction.

3 4D Radiative Gaussian Splatting

The overall pipeline of 4DRGS is shown in Fig. 2. First, we present 4D radiative Gaussian kernels as DSA scene representation in Sect. 3.1. Our 4D kernels extend the static 3D kernels in [30] by modeling time-varying contrast agent flow and incorporating improved activation modules. Next, in Sect. 3.2, we outline the training process and introduce accumulated attenuation pruning, which is designed for DSA imaging to improve reconstruction quality. Finally, we detail the process of extracting 3D vessels from the trained kernels in Sect. 3.3.

3.1 4D Radiative Gaussian Modeling

DSA Scene Representation. As shown in Fig. 1(b) and Fig. 2(a), we model vessels with a set of 4D radiative Gaussian kernels $\mathbb{G} = \{G_i\}_{i=1}^{M}$, where M is the number of kernels. A key observation is that vessels maintain static structures during the scanning process, but their attenuation varies over time due to the flowing contrast agents. Based on this, we define each kernel with two types of attributes including geometry parameters and central attenuation. Time-invariant geometry parameters describe static vessel structures. These parameters include position $\boldsymbol{p}_i \in \mathbb{R}^3$ for kernel location, rotation quaternion $\boldsymbol{r}_i \in \mathbb{R}^4$

for kernel orientation, and scale vector $\boldsymbol{s}_i \in \mathbb{R}^3_+$ for kernel size. The rotation quaternion $\boldsymbol{r}_i$ and scale vector $\boldsymbol{s}_i$ can be further converted into rotation matrix $\mathbf{R}_i \in \mathrm{SO}(3)$ and diagonal scale matrix $\mathbf{S}_i \in \mathbb{R}^{3\times 3}$, respectively, yielding the Gaussian covariance matrix $\boldsymbol{\Sigma}_i = \mathbf{R}_i\mathbf{S}_i\mathbf{S}_i^\top\mathbf{R}_i^\top \in \mathbb{R}^{3\times 3}$ [11]. The central attenuation $\rho(\boldsymbol{p}_i, t) \in \mathbb{R}_{\geq 0}$ is time-dependent to capture attenuation changes caused by the contrast agent flow. Finally, the Gaussian-shaped kernel response $G_i(\boldsymbol{x}, t) \in \mathbb{R}_{\geq 0}$ for any spatial point $\boldsymbol{x} \in \mathbb{R}^3$ and timestamp $t \in \mathbb{R}$ is defined as:

$$G_i(\boldsymbol{x}, t) = \rho(\boldsymbol{p}_i, t) \cdot \exp\left(-\frac{1}{2}(\boldsymbol{x} - \boldsymbol{p}_i)^\top \boldsymbol{\Sigma}_i^{-1}(\boldsymbol{x} - \boldsymbol{p}_i)\right). \tag{1}$$

The overall contrast agent attenuation value $\mu(\boldsymbol{x}, t) \in \mathbb{R}_{\geq 0}$ is computed as the sum of responses from all kernels:

$$\mu(\boldsymbol{x}, t) = \sum_{i=1}^{M} G_i(\boldsymbol{x}, t). \tag{2}$$

Bounded Scaling Activation. Geometry parameters $\boldsymbol{p}_i$, $\boldsymbol{r}_i$, and $\boldsymbol{s}_i$ are activated from their optimizable raw counterparts $\tilde{\boldsymbol{p}}_i \in \mathbb{R}^3$, $\tilde{\boldsymbol{r}}_i \in \mathbb{R}^4$, and $\tilde{\boldsymbol{s}}_i \in \mathbb{R}^3$, respectively, to ensure they remain within valid ranges. Following 3DGS [11], we activate positions with $\boldsymbol{p}_i = \tilde{\boldsymbol{p}}_i$ and rotation quaternions with $\boldsymbol{r}_i = \mathrm{norm}(\tilde{\boldsymbol{r}}_i)$, where $\mathrm{norm}(\cdot)$ is the normalization operation. Regarding scale vectors, we avoid using exponential activation in [11]. Because its unbounded positive output range would cause elongated kernels, resulting in needle artifacts [11]. In DSA imaging scenario, kernel sizes should remain small to model fine vessel details. Motivated by this structure prior, we adopt bounded scaling activation similar to [28]:

$$\boldsymbol{s}_i = (s_{max} - s_{min})\,\sigma(\tilde{\boldsymbol{s}}_i) + s_{min}, \tag{3}$$

where $\sigma(\cdot)$ is the sigmoid activation, and $s_{min}, s_{max} \in \mathbb{R}_+$ are the minimum and maximum scale bounds, respectively. As a result, elements of $\boldsymbol{s}_i$ are constrained in the range of (s_{min}, s_{max}), effectively mitigating needle artifacts and enhancing reconstruction quality.

Dynamic Neural Attenuation Field. To model attenuation changes in Gaussian kernels, we introduce a compact neural network $\mathcal{D}$ dubbed dynamic neural attenuation field (DNAF). It takes kernel position and timestamp as input, and predicts the central attenuation: $\mathcal{D} : (\boldsymbol{p}_i, t) \in \mathbb{R}^3 \times \mathbb{R} \rightarrow \rho \in \mathbb{R}_{\geq 0}$. To enhance spatial-temporal expressiveness, we adopt the 3D hash encoding $\boldsymbol{h}_{\mathrm{3D}}$ [18] for time-invariant features and the 4D hash encoding $\boldsymbol{h}_{\mathrm{4D}}$ [20] for time-variant features. A subsequent decoding multi-layer perceptron (MLP) ϕ then maps these features to attenuation values, activated by rectified linear unit (ReLU) [19]. Mathematically, $\rho(\boldsymbol{p}_i, t)$ is formulated as:

$$\rho(\boldsymbol{p}_i, t) = \phi\left(\boldsymbol{h}_{\mathrm{3D}}(\boldsymbol{p}_i) \oplus \boldsymbol{h}_{\mathrm{4D}}(\boldsymbol{p}_i, t)\right), \tag{4}$$

where $\oplus$ denotes concatenation operation.

Model Initialization. Inspired by R^2-Gaussian [30], we initialize geometry parameters of kernels from a low-quality volume reconstructed by FDK [8]. Specifically, we sample M non-empty voxels with an attenuation threshold δ, and assign their locations as initial raw positions $\{\tilde{\boldsymbol{p}}_i\}_{i=1}^M$. We set the activated scales $\{\boldsymbol{s}_i\}_{i=1}^M$ as the nearest neighbor distances of positions, and compute raw scales $\{\tilde{\boldsymbol{s}}_i\}_{i=1}^M$ with the inverse of Eq. (3). We set raw rotations as $\{\tilde{\boldsymbol{r}}_i = [1,0,0,0]^\top\}_{i=1}^M$. Network parameters in DNAF are randomly initialized.

3.2 Model Optimization

DSA Image Synthesis. Consider an X-ray path $\boldsymbol{l}(a) = \boldsymbol{o} + a\boldsymbol{d} \in \mathbb{R}^3$ sampled from the j_k-th frame ray set $\mathbb{L}_{j_k} \subset \mathbb{R}^3$ for training, where $\boldsymbol{o} \in \mathbb{R}^3$, $a \in \mathbb{R}_{\geq 0}$, and $\boldsymbol{d} \in \mathbb{R}^3$ are source position, length variable, and ray direction, respectively. Based on the Beer-Lambert law [10], the corresponding synthesized DSA pixel value $\hat{I} \in \mathbb{R}_{\geq 0}$ is obtained by integrating the contrast agent attenuation μ (Eq. (2)) along the ray path:

$$\hat{I}(\boldsymbol{l}, t_{j_k}) = \int_{a_n}^{a_f} \mu\left(\boldsymbol{l}(a), t_{j_k}\right) \mathrm{d}a, \tag{5}$$

where $[a_n, a_f]$ is the path bound. The complete synthesized DSA image $\hat{\mathbf{I}}(t_{j_k}) \in \mathbb{R}^{H \times W}$ is then obtained by compositing the pixel values from the entire ray set $\mathbb{L}_{j_k}$: $\hat{\mathbf{I}}(t_{j_k}) = \{\hat{I}(\boldsymbol{l}, t_{j_k})\}_{\boldsymbol{l} \in \mathbb{L}_{j_k}}$. Such a synthesizing process can be efficiently achieved via X-ray rasterization [30], which splats Gaussian kernels onto the image plane and parallelly computes ray integrations.

Training Objective. We optimize our model in Sect. 3.1 by minimizing the L1 loss $\mathcal{L}_1$ and D-SSIM loss $\mathcal{L}_{ssim}$ [26] between the synthesized DSA images $\hat{\mathbf{I}}(t_{j_k})$ and real captured ones $\mathbf{I}_{j_k}$. To mitigate overfitting on training frames and improve temporal consistency, we follow [15] to incorporate temporal perturbation into the loss function. The overall loss function is formulated as:

$$\mathcal{L} = (1-\lambda_{ssim})\mathcal{L}_1\left(\hat{\mathbf{I}}(t_{j_k}+\tau), \mathbf{I}_{j_k}\right) + \lambda_{ssim}\mathcal{L}_{ssim}\left(\hat{\mathbf{I}}(t_{j_k}+\tau), \mathbf{I}_{j_k}\right), \tau \sim \mathcal{N}(0, w^2), \tag{6}$$

where $w = t_{j_k+1} - t_{j_k}$ is the standard deviation of temporal Gaussian noise τ.

Accumulated Attenuation Pruning. During training, we refine the kernel distribution to better match the target vessel geometry via adaptive control mechanism including densification and pruning. The densification strategy is the same as in 3DGS [11] and R^2-Gaussian [30]. If a kernel's position gradient exceeds a predefined threshold, it suggests that this kernel does not accurately represent the underlying area and requires densification. As illustrated in Fig. 2(b), small Gaussians (indicated by light blue circles) in under-reconstructed regions are densified through cloning, while large Gaussians (dark blue circles) in over-reconstructed regions are densified by splitting.

Table 1. Configurations of DSA images and reconstructed volumes used in experiment.

Cases	Image resolution	Pixel size (mm)	Volume resolution	Voxel size (mm)
#7	1240×960	0.3144×0.3173	512×512×400	0.4768^3
#10, #11, #13	960×960	0.3236×0.3198	512×512×506	0.3802^3
#15	1240×960	0.3081×0.3070	512×512×395	0.4663^3
Others	1240×960	0.3219×0.3208	512×512×395	0.4881^3

R^2-Gaussian [30] prunes empty kernels with central attenuation values below a predefined threshold. However, this approach is unsuitable for our 4D kernels in DSA imaging, as their attenuation values vary over time. Using current timestamp attenuation as the pruning criterion may mistakenly remove vessel kernels that are not marked by the contrast agent at that moment. TOGS [32] proposes a random pruning strategy, but it would also prune vessel kernels, resulting in degraded reconstruction. To overcome this limitation, we introduce accumulated attenuation pruning. The accumulated attenuation A_i of a kernel is defined as:

$$A_i = \frac{\sum_{iter} \rho(\boldsymbol{p}_i, t)}{\sum_{iter} 1}, \tag{7}$$

where $\sum_{iter}$ denotes the summation over training iterations between neighboring pruning operations. If the accumulated attenuation remains consistently small, i.e., $A_i < \epsilon$, it means this kernel has been negligibly marked by contrast agent during training. This suggests the kernel belongs to backgrounds rather than vessels, and therefore should be pruned. In this way, we precisely prune empty kernels and retain useful ones, eventually improving reconstruction quality.

3.3 Vessel Reconstruction

Given any timestamp t, we leverage CUDA-based voxelization in [30] to efficiently query an attenuation volume $\mathbf{V}(t) = \{\mu(\boldsymbol{x}, t)\}_{\boldsymbol{x} \in \mathbb{X}}$. Here, $\mathbb{X} \subset \mathbb{R}^3$ is the set of voxel grids defined by the target volume's resolution and spacing. The final 3D vessel volume is obtained by averaging attenuation volumes across all timestamps $\{t_j\}_{j=1}^{T}$ in the complete set of DSA data (Fig. 2(c)): $\overline{\mathbf{V}} = \frac{1}{T}\sum_{j=1}^{T} \mathbf{V}(t_j)$.

4 Experiments

4.1 Experimental Settings

Dataset. In this study, we collected data from 15 real-world patient cases using the Siemens AXIOM-Artis DSA scanning system, whose source-to-object and source-to-detector distances are 750mm and 1200mm, respectively. For each case, the system captured 133 mask-fill X-ray image pairs, evenly distributed across a rotational range of 198°C. Additionally, the system provided vessel volumes reconstructed by its inbuilt FDK algorithm [6]. Detailed configurations of DSA images and reconstructed volumes are listed in Table 1. Notably, although the provided volumes are not entirely accurate, we treat them as references for evaluating 3D reconstruction. We subsampled 30, 40, 50, and 60 views from the complete set as four training scenarios, and left the remaining views for evaluating 2D image synthesis.

Implementation Details. DNAF comprises two hash encoders $\boldsymbol{h}_{3\mathrm{D}}$ and $\boldsymbol{h}_{4\mathrm{D}}$ followed by a two-layer MLP with a width of 64. $\boldsymbol{h}_{3\mathrm{D}}$ consists of 12 hash levels. The hash table size is 2^{19} with every entry storing a trainable 2-dimensional feature vector. The base resolution is set to 8, increasing by a factor of 1.45 per level. $\boldsymbol{h}_{4\mathrm{D}}$ uses a similar configuration, except the base resolution is set to 2, with a growth factor of 1.4. The rasterization and voxelization modules are borrowed from R^2-Gaussian [30]. We optimize our model with Adam optimizer [12] for 30k iterations. Adaptive control runs from 600 to 15k iterations, adjusting every 200 iterations with a gradient threshold of 0.0001. The threshold for accumulated attenuation pruning is $\epsilon = 1e-6$. A fast version is also provided, where the model trains for 10k iterations, and adaptive control stops at 5k iterations. The learning rates for position, rotation, scale, and DNAF start from 0.0001, 0.001, 0.005, and 0.001, respectively, and exponentially decay to 0.1 of the initial ones. There is also a weight decay factor of $5e-5$ for parameters in DNAF. s_{min} and s_{max} are set to 0.1 and 10 times of the voxel spacing, respectively. We initialize $M = 30$k kernels with threshold $\delta = 0.016$. The loss weight is $\lambda_{ssim} = 0.2$.

Evaluation Metrics. We evaluated both 3D vessel reconstruction and 2D DSA image synthesis. For 3D reconstruction, we did not directly compare the reconstructed volumes with the reference volumes due to the unknown data calibration issue. Instead, we evaluated vessel surfaces, which are easier to align and compare. Specifically, we used marching cubes [16] to extract meshes from volumes, with attenuation thresholds of 0.025 for the reference volumes and 0.008 for the algorithm reconstructed volumes. The reference and reconstructed meshes were then aligned using iterative closest point (ICP) [2]. After that, we computed the Chamfer distance (CD) and Hausdorff distance (HD) between the aligned meshes, both measured in millimeters. For 2D image synthesis evaluation, we calculated the peak signal-to-noise ratio (PSNR) and structural similarity index measure (SSIM) [26] between the synthesized and ground truth DSA images in the test set. We also recorded the running time as an efficiency metric.

Competing Methods. We compared 4DRGS with four methods: FDK with hann filtering [6,8], 3DGS-based CT reconstruction method R^2-Gaussian [30], 3DGS-based DSA image synthesis method TOGS [32], and current SOTA NeRF-based method VPAL [15]. FDK was implemented based on TIGRE-toolbox [3], while others were adapted from their source codes. FDK and R^2-Gaussian are static methods, so we directly evaluated their output volumes. TOGS, VPAL, and our 4DRGS are dynamic methods, and we computed the average of generated volumes across timestamps as the final output, as described in Sect. 3.3. TOGS does not support direct volume reconstruction. As a workaround, we uniformly rendered 720 views in a full circle [27] using TOGS and then reconstructed

Table 2. Quantitative results on 3D vessel reconstruction and 2D DSA image synthesis. The best performance is shown in bold, while the second best is underlined.

Views	Method	CD (mm) ↓	HD (mm) ↓	PSNR (dB) ↑	SSIM ↑	Time ↓
30	FDK	32.63±5.90	90.24±13.61	-	-	**1.06s±0.20s**
	R^2-Gaussian	8.34±1.93	34.56±8.30	28.61±1.22	0.803±0.053	11m55s±20s
	TOGS	5.24±1.36	18.66±11.08	34.17±1.79	0.803±0.072	8m46s±22s
	VPAL	1.79±0.51	4.07±1.79	34.32±1.84	0.819±0.068	2h36m±52s
	Ours (10k)	1.88±0.36	4.05±1.22	34.96±1.73	0.867±0.053	4m31s±9s
	Ours (30k)	**1.72±0.29**	**3.44±0.85**	**35.07±1.65**	**0.869±0.052**	12m38s±37s
40	FDK	30.61±6.12	87.23±15.10	-	-	**1.36s±0.24s**
	R^2-Gaussian	8.13±1.68	33.91±7.16	28.63±1.31	0.804±0.054	11m50s±27s
	TOGS	5.04±1.35	18.19±9.80	34.43±1.83	0.803±0.075	8m49s±39s
	VPAL	1.69±0.40	3.78±1.28	35.05±2.33	0.824±0.071	2h38m±3m54s
	Ours (10k)	1.81±0.30	3.89±1.00	35.50±1.76	0.871±0.053	4m40s±10s
	Ours (30k)	**1.67±0.27**	**3.15±0.80**	**35.69±1.77**	**0.874±0.051**	12m33s±28s
50	FDK	28.80±6.64	84.36±16.57	-	-	**1.62s±0.33s**
	R^2-Gaussian	8.07±1.77	34.16±8.62	28.70±1.30	0.807±0.054	12m2s±25s
	TOGS	5.01±1.50	18.25±11.37	34.69±1.95	0.808±0.080	8m50s±38s
	VPAL	**1.58±0.19**	3.59±1.05	35.91±1.83	0.832±0.070	2h44m±8m30s
	Ours (10k)	1.78±0.38	3.89±1.00	35.85±1.82	0.875±0.052	4m43s±13s
	Ours (30k)	1.67±0.29	**3.20±0.82**	**36.13±1.86**	**0.879±0.050**	12m56s±29s
60	FDK	27.04±7.36	82.33±18.08	-	-	**1.89s±0.32s**
	R^2-Gaussian	7.95±1.60	33.72±7.42	28.75±1.26	0.808±0.053	12m5s±18s
	TOGS	4.96±1.24	18.16±11.10	34.73±1.82	0.812±0.067	9m17s±34s
	VPAL	1.75±0.68	3.94±2.17	35.85±2.73	0.830±0.074	2h42m±7m28s
	Ours (10k)	1.91±0.52	3.70±0.90	36.06±1.83	0.876±0.052	4m38s±9s
	Ours (30k)	**1.63±0.26**	**3.27±0.83**	**36.33±1.87**	**0.880±0.051**	12m52s±47s

the volume with FDK [6] for each timestamp. All experiments were conducted on a single RTX3090 GPU.

4.2 Experimental Results

Quantitative Evaluation. Table 2 presents a quantitative comparison of different methods. Our method achieves SOTA 2D and 3D performance in most scenarios, with only a slightly poorer CD metric than VPAL at 50 input views. In terms of efficiency, our method converges in 13 min, offering a speedup of over 12× compared to VPAL, which requires around 160 min for training. Remarkably, our fast version, trained for 10k iterations, achieves comparable performance as VPAL in 5 min, which is 32× faster. Our efficiency arises from two reasons. First, we use Gaussian kernels to model only vascular structures, avoiding unnecessary computations for empty backgrounds. In contrast, VPAL employs MLPs to model entire scanning scenes. Second, highly parallelized rasterization is inherently faster than VPAL's volume rendering.

Qualitative Evaluation. A qualitative comparison of 3D vessel reconstruction is shown in Fig. 3. Two static methods, FDK and R^2-Gaussian, exhibit severe streaky artifacts and noises, highlighting the necessity of dynamic flow modeling. TOGS generates blurry reconstructions because it trivially adapts RGB rasterization from original 3DGS to DSA imaging without considering X-ray formation principles. While VPAL recovers complete vessel structures, it still fails to capture some fine details. In contrast, our method effectively preserves intricate

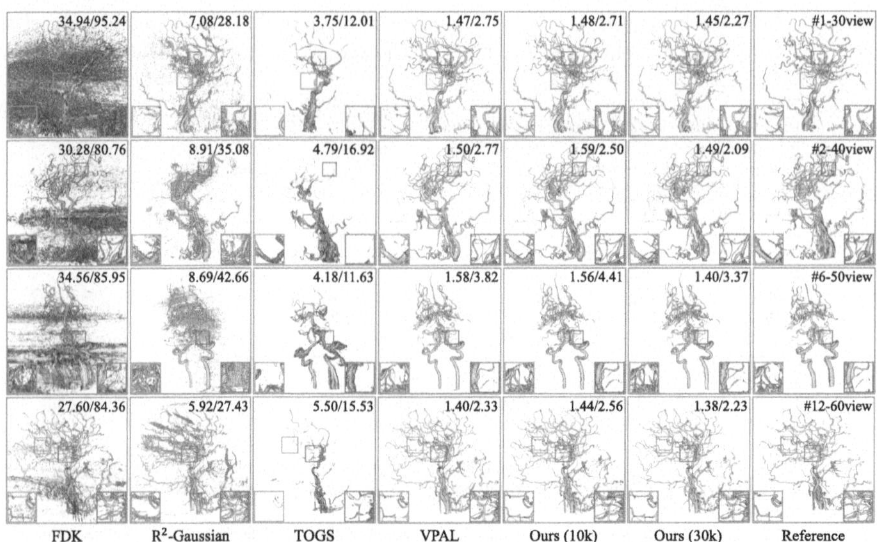

Fig. 3. 3D vessel reconstruction of different methods with CD(mm)/HD(mm) values shown at the top right of each image.

vessel details, demonstrating the superiority of our kernel-based representation compared to the pure MLP-based representation used in VPAL.

Figure 4 presents a qualitative comparison of 2D DSA image synthesis on test frames. R^2-Gaussian loses significant vessel structures due to its inability to model dynamic DSA sequences. TOGS performs better but vessel details remain under-reconstructed. Because its limited temporal modeling hinders accurate recovery of contrast agent dynamics. VPAL renders complete structures but lacks fine details, and suffers from noise and blurriness. In contrast, our method captures fine details while minimizing artifacts and noise, showcasing its effectiveness in vessel structure representation.

4.3 Ablation Study

In this section, we conduct ablation studies to validate our two innovations: accumulated attenuation pruning and bounded scaling activation. Accumulated pruning is compared to random pruning in TOGS and threshold pruning in R^2-Gaussian. We follow the original papers to set the random pruning proportion to 8% and the threshold pruning attenuation to $1e-6$. Bounded activation is compared with exponential activation in vanilla 3DGS. All ablation experiments are conducted with 30 input views for 30k training iterations.

Quantitative results in Table 3 show that random pruning and threshold pruning lead to significant degradation of both 3D and 2D quality. Additionally, kernel numbers tend to increase under these strategies. Exponential activation causes a minor reduction in 3D quality, and it slightly extends the training time. In Fig. 5(a), we provide 3D vessel reconstruction results at the top

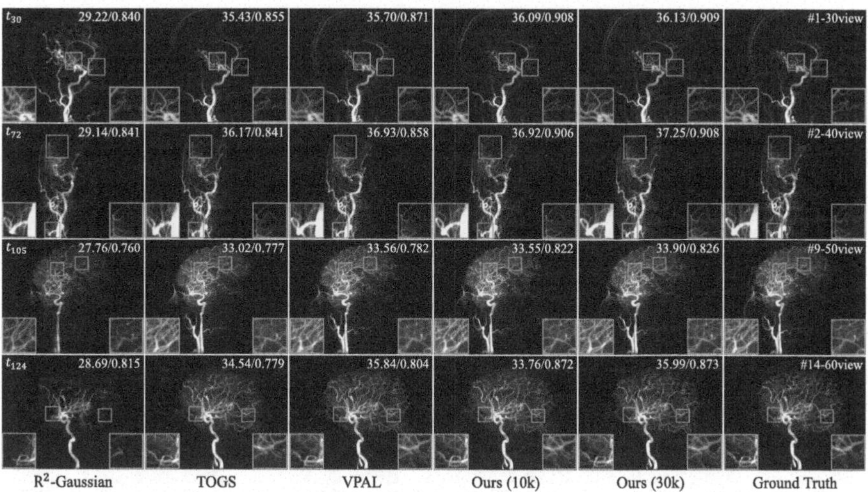

Fig. 4. 2D DSA image synthesis of different methods at test frames. PSNR(dB)/SSIM values averaged over the test set are shown at the top right of each image.

Table 3. Quantitative results of ablation study. **RP** stands for random pruning, **TP** stands for threshold pruning, **Exp** stands for exponential scaling activation, and **Gau.** stands for number of Gaussians.

	CD (mm) ↓	HD (mm) ↓	PSNR (dB) ↑	SSIM ↑	Time ↓	Gau.
RP	6.79±1.89	28.84±11.59	34.64±1.60	0.866±0.052	12m27s±40s	93k±27k
TP	5.46±1.85	28.41±13.42	34.76±1.61	0.866±0.052	12m44s±36s	76k±29k
Exp	2.06±0.41	4.04±0.94	35.05±1.68	0.870±0.052	13m5s±41s	61k±19k
Ours	1.72±0.29	3.44±0.85	35.07±1.65	0.869±0.052	12m38s±37s	61k±19k

row and the corresponding sagittal slices at the bottom row. Random pruning and threshold pruning introduce considerable noisy artifacts because they incorrectly prune kernels that belong to vessels. In contrast, our accumulated pruning method precisely removes non-vessel Gaussians while preserving vessel-related ones, enabling high-quality reconstruction with limited noise. Exponential activation produces needle artifacts as highlighted by the red arrows. These artifacts stem from elongated Gaussian kernels caused by the unbounded positive range of exponential activation. In contrast, our bounded scaling activation addresses this issue by constraining the Gaussian kernel size within a suitable range. In Fig. 5(b), we provide 2D DSA image synthesis results at test frames. Consistent with 3D results, random pruning and threshold pruning exhibit noticeable quality degradation with noise, while exponential activation introduces needle artifacts. Overall, our full model achieves the best results both quantitatively and qualitatively, verifying the effectiveness of accumulated attenuation pruning and bounded scaling activation.

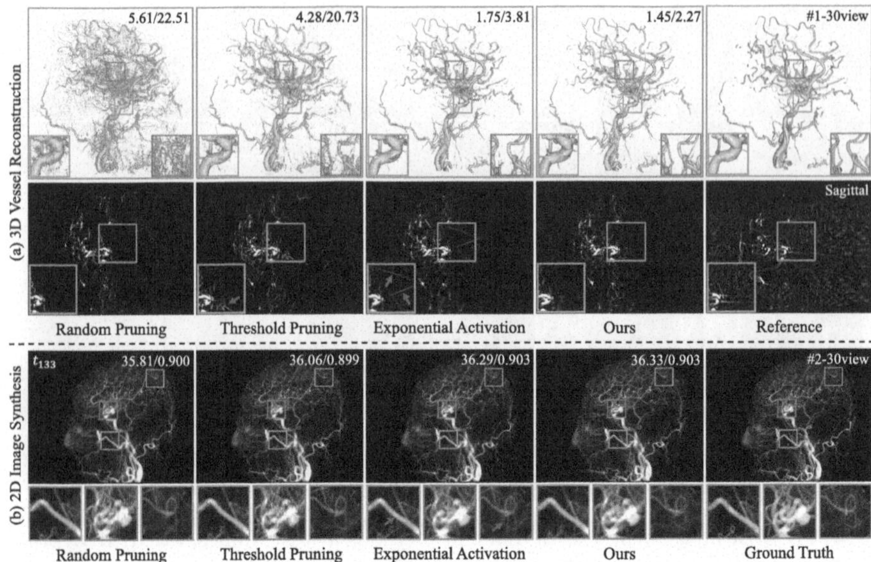

Fig. 5. Qualitative results of ablation study. (a) 3D vessel reconstruction. Top row: 3D visualization with CD(mm)/HD(mm) values shown at the top right of each image. Bottom row: sagittal slice of reconstructed volume. (b) 2D DSA image synthesis at test frame. PSNR(dB)/SSIM values averaged over the test set are shown at the top right of each image.

5 Discussion and Conclusion

Discussion. While our method achieves SOTA results, it has some limitations. For instance, we assume no patient movement during scanning, though minor motion may occur in practice. Additionally, we have not considered calibration errors in scanner geometry and imaging noises in DSA scanning process. Moreover, our dataset is a small private set of multi-view cerebral DSA scans. However, single- or bi-plane DSA is more common especially for other body parts like coronary arteries. Our method would fail on such data due to limited views or unmodeled cardiac motion. We leave these aspects for future work.

Conclusion. In this work, we present 4DRGS, the first Gaussian splatting-based framework for sparse-view DSA reconstruction. 4DRGS represents vessels using 4D radiative Gaussian kernels, which effectively model static vessel structures and time-varying attenuation changes. We train these kernels with image losses and extract the target vessel volume from them. Two innovations are proposed to enhance reconstruction quality, including accumulated attenuation pruning and bounded scaling activation. Extensive experiments demonstrate the superiority of our method in both 3D vessel reconstruction and 2D DSA image synthesis. Remarkably, our method achieves impressive results within minutes, highlighting its potential for clinical applications.

Acknowledgments.. This work was supported in part by NSFC grants (No. 6230012077), and HPC Platform of ShanghaiTech University. Shanghai Municipal Central Guided Local Science and Technology Development Fund Project no: YDZX20233100001001.

References

1. Andersen, A.H., Kak, A.C.: Simultaneous algebraic reconstruction technique (SART): a superior implementation of the art algorithm. Ultrason. Imaging **6**(1), 81–94 (1984)
2. Besl, P., McKay, N.D.: A method for registration of 3-d shapes. IEEE Trans. Pattern Anal. Mach. Intell. **14**(2), 239–256 (1992). https://doi.org/10.1109/34.121791
3. Biguri, A., Dosanjh, M., Hancock, S., Soleimani, M.: Tigre: a MATLAB-GPU toolbox for CBCT image reconstruction. Biomed. Phys. Eng. Express **2**(5), 055010 (2016)
4. Cai, Y., et al.: Radiative gaussian splatting for efficient x-ray novel view synthesis. In: Leonardis, A., Ricci, E., Roth, S., Russakovsky, O., Sattler, T., Varol, G. (eds.) Computer Vision. ECCV 2024. LNCS, vol. 15059, pp. 283–299. Springer, Cham (2025). https://doi.org/10.1007/978-3-031-73232-4_16
5. Cai, Y., Wang, J., Yuille, A., Zhou, Z., Wang, A.: Structure-aware sparse-view x-ray 3d reconstruction. In: CVPR (2024)
6. Fahrig, R., Fox, A., Lownie, S., Holdsworth, D.: Use of a c-arm system to generate true three-dimensional computed rotational angiograms: preliminary in vitro and in vivo results. Am. J. Neuroradiol. **18**(8), 1507–1514 (1997)
7. Fang, Y., et al.: SNAF: Sparse-view CBCT reconstruction with neural attenuation fields. arXiv preprint arXiv:2211.17048 (2022)
8. Feldkamp, L.A., Davis, L.C., Kress, J.W.: Practical cone-beam algorithm. JOSA A **1**(6), 612–619 (1984)
9. Gao, Z., Planche, B., Zheng, M., Chen, X., Chen, T., Wu, Z.: DDGS-CT: direction-disentangled gaussian splatting for realistic volume rendering. arXiv preprint arXiv:2406.02518 (2024)
10. Kak, A.C., Slaney, M.: Principles of computerized tomographic imaging. SIAM (2001)
11. Kerbl, B., Kopanas, G., Leimkühler, T., Drettakis, G.: 3d gaussian splatting for real-time radiance field rendering. ACM Trans. Graph. **42**(4), 139–1 (2023)
12. Kingma, D.P., Ba, J.: Adam: a method for stochastic optimization. In: International Conference on Learning Representations (ICLR) (2015)
13. Lang, S., et al.: 4d DSA for dynamic visualization of cerebral vasculature: a single-center experience in 26 cases. Am. J. Neuroradiol. **38**(6), 1169–1176 (2017)
14. Li, Y., Fu, X., Zhao, S., Jin, R., Zhou, S.K.: Sparse-view CT reconstruction with 3d gaussian volumetric representation. arXiv preprint arXiv:2312.15676 (2023)
15. Liu, Z., et al.: 3d vessel reconstruction from sparse-view dynamic DSA images via vessel probability guided attenuation learning. arXiv preprint arXiv:2405.10705 (2024)
16. Lorensen, W.E., Cline, H.E.: Marching cubes: a high resolution 3d surface construction algorithm. In: Seminal Graphics: Pioneering Efforts that Shaped the Field, pp. 347–353 (1998)

17. Mildenhall, B., Srinivasan, P.P., Tancik, M., Barron, J.T., Ramamoorthi, R., Ng, R.: NERF: representing scenes as neural radiance fields for view synthesis. Commun. ACM **65**(1), 99–106 (2021)
18. Müller, T., Evans, A., Schied, C., Keller, A.: Instant neural graphics primitives with a multiresolution hash encoding. ACM Trans. Graph. (ToG) **41**(4), 1–15 (2022)
19. Nair, V., Hinton, G.E.: Rectified linear units improve restricted Boltzmann machines. In: Proceedings of the 27th International Conference on Machine Learning (ICML-2010), pp. 807–814 (2010)
20. Park, S., Son, M., Jang, S., Ahn, Y.C., Kim, J.Y., Kang, N.: Temporal interpolation is all you need for dynamic neural radiance fields. In: Proceedings of the IEEE/CVF Conference on Computer Vision and Pattern Recognition, pp. 4212–4221 (2023)
21. Rückert, D., Wang, Y., Li, R., Idoughi, R., Heidrich, W.: NeAT: neural adaptive tomography. ACM Trans. Graph. (TOG) **41**(4), 1–13 (2022)
22. Ruedinger, K., Schafer, S., Speidel, M., Strother, C.: 4D-DSA: development and current neurovascular applications. Am. J. Neuroradiol. **42**(2), 214–220 (2021)
23. Sandoval-Garcia, C., Royalty, K., Aagaard-Kienitz, B., Schafer, S., Yang, P., Strother, C.: A comparison of 4d DSA with 2d and 3d DSA in the analysis of normal vascular structures in a canine model. Am. J. Neuroradiol. **36**(10), 1959–1963 (2015)
24. Sandoval-Garcia, C., et al.: 4D DSA a new technique for arteriovenous malformation evaluation: a feasibility study. J. Neurointerv. Surg. **8**, 300–304 (2015)
25. Sidky, E.Y., Pan, X.: Image reconstruction in circular cone-beam computed tomography by constrained, total-variation minimization. Phys. Med. Biol. **53**(17), 4777 (2008)
26. Wang, Z., Bovik, A.C., Sheikh, H.R., Simoncelli, E.P.: Image quality assessment: from error visibility to structural similarity. IEEE Trans. Image Process. **13**(4), 600–612 (2004)
27. Wu, Q., Feng, R., Wei, H., Yu, J., Zhang, Y.: Self-supervised coordinate projection network for sparse-view computed tomography. IEEE Trans. Comput. Imaging **9**, 517–524 (2023)
28. Xu, Y., et al.: GRM: large gaussian reconstruction model for efficient 3d reconstruction and generation. arXiv preprint arXiv:2403.14621 (2024)
29. Zang, G., Idoughi, R., Li, R., Wonka, P., Heidrich, W.: Intratomo: self-supervised learning-based tomography via sinogram synthesis and prediction. In: Proceedings of the IEEE/CVF International Conference on Computer Vision, pp. 1960–1970 (2021)
30. Zha, R., Lin, T.J., Cai, Y., Cao, J., Zhang, Y., Li, H.: R^2-gaussian: rectifying radiative gaussian splatting for tomographic reconstruction. arXiv preprint arXiv:2405.20693 (2024)
31. Zha, R., Zhang, Y., Li, H.: NAF: neural attenuation fields for sparse-view CBCT reconstruction. In: Wang, L., Dou, Q., Fletcher, P.T., Speidel, S., Li, S. (eds.) Medical Image Computing and Computer Assisted Intervention. MICCAI 2022. LNCS, vol. 13436, pp. 442–452. Springer, Cham (2022). https://doi.org/10.1007/978-3-031-16446-0_42
32. Zhang, S., et al.: Togs: Gaussian splatting with temporal opacity offset for real-time 4D DSA rendering. arXiv preprint arXiv:2403.19586 (2024)
33. Zhao, H., et al.: Self-supervised learning enables 3d digital subtraction angiography reconstruction from ultra-sparse 2d projection views: a multicenter study. Cell Rep. Med. **3**(10), 1–39 (2022)

Brightness-Invariant Tracking Estimation in Tagged MRI

Zhangxing Bian[1(✉)], Shuwen Wei[1], Xiao Liang[2], Yuan-Chiao Lu[3,4], Samuel W. Remedios[1], Fangxu Xing[5], Jonghye Woo[5], Dzung L. Pham[1,3], Aaron Carass[1], Philip V. Bayly[6], Jiachen Zhuo[2], Ahmed Alshareef[7], and Jerry L. Prince[1]

[1] Johns Hopkins University, Baltimore, MD, USA
zbian4@jhu.edu
[2] University of Maryland School of Medicine, Baltimore, MD, USA
[3] Uniformed Services University of the Health Sciences, Bethesda, MD, USA
[4] The Henry M. Jackson Foundation for the Advancement of Military Medicine, Inc., Bethesda, MD, USA
[5] Massachusetts General Hospital and Harvard Medical School, Boston, MA, USA
[6] Washington University, St. Louis, MO, USA
[7] University of South Carolina, Columbia, SC, USA

Abstract. Magnetic resonance (MR) tagging is an imaging technique for noninvasively tracking tissue motion in vivo by creating a visible pattern of magnetization saturation (tags) that deforms with the tissue. Due to longitudinal relaxation and progression to steady-state, the tags and tissue brightnesses change over time, which makes tracking with optical flow methods error-prone. Although Fourier methods can alleviate these problems, they are also sensitive to brightness changes as well as spectral spreading due to motion. To address these problems, we introduce the brightness-invariant tracking estimation (BRITE) technique for tagged MRI. BRITE disentangles the anatomy from the tag pattern in the observed tagged image sequence and simultaneously estimates the Lagrangian motion. The inherent ill-posedness of this problem is addressed by leveraging the expressive power of denoising diffusion probabilistic models to represent the probabilistic distribution of the underlying anatomy and the flexibility of physics-informed neural networks to estimate biologically-plausible motion. A set of tagged MR images of a gel phantom was acquired with various tag periods and imaging flip angles to demonstrate the impact of brightness variations and to validate our method. The results show that BRITE achieves more accurate motion and strain estimates as compared to other state of the art methods, while also being resistant to tag fading.

Keywords: MR tagging · Spectral overlap · Motion tracking · Strain

1 Introduction

Motion tracking using magnetic resonance imaging (MRI) is used to understand tissue deformations and organ biomechanics. Applications include analyz-

I. Oguz et al. (Eds.): IPMI 2025, LNCS 15830, pp. 375–389, 2026.
https://doi.org/10.1007/978-3-031-96625-5_25

ing cardiac motion [15], investigating musculoskeletal dynamics [19], studying tongue movements during speech [20,23,32], quantifying brain deformation during mild head impact [7,16], and evaluating liver strain [17]. Several imaging strategies including SPAMM (Spatial Modulation of Magnetization) tagging [6], DENSE (Displacement Encoding with Stimulated Echoes) [2], and SENC (Strain ENCoded) imaging [22] have been developed to aid this endeavor. Among these, SPAMM tagging has been widely adopted in clinical imaging due to its low specific absorption rate and simplicity [15] in application and interpretation. SPAMM modulates the MR signal with periodic patterns, known as tags, that deform with the tissue. This technique enables visual inspection and computational quantification of in vivo tissue deformation and strain.

Methods for automated tissue-motion tracking of tagged MR images fall into two categories: (1) Fourier-domain and (2) image-domain analyses. Among the Fourier-domain methods, HARP (Harmonic Phase) analysis [21] has emerged as a popular tag tracking method for SPAMM and related sequences [15]. The HARP technique and its descendants [9,18,30] can achieve sub-pixel accuracy. Although image-domain methods using relaxed brightness constancy assumptions have been proposed [11,14,24], recent work has shown that HARP is more accurate [8,15]. Despite HARP's superiority, its accuracy remains limited because it depends on extracting the harmonic phase using a band-pass filter. The resultant harmonic peaks are never perfectly isolated for the following reasons: (1) Tag fading due to T1 relaxation and steady-state progression [8] causes an increase in the central peak and a decrease in the harmonic peaks; (2) Large tag periods bring the harmonic peaks closer together; and (3) Non-rigid tissue motion deforms the tag lines, which broadens the harmonic peaks. In addition, object rotation can cause the spectral peak to rotate entirely outside the band-pass region in Fourier space. Figure 1 shows some of these effects on data from a silicone gel phantom. Figure 2 shows the impact of tag fading and spectral overlap on HARP.

The acquisition of additional tagged MR images has been used to avoid this problem. The complementary SPAMM (CSPAMM) [12] method suppresses the central peak by combining two SPAMM acquisitions; to yield both horizontal and vertical tags, four acquisitions are required. The TruHARP [1] method uses five SPAMM acquisitions to produce completely isolated vertical and horizontal spectral peaks. Neither method is clinically feasible due to the additional data acquisition. Current practice in MR tagging is to use two SPAMM acquisitions, one horizontal and one vertical, or a single acquisition of grid tags. In this paper, we address the spectral overlap problem in the presence of brightness changes when considering horizontal or vertical SPAMM tagged image sequences.

To achieve better accuracy from tagged MRI in the presence of brightness changes, we propose the brightness-invariant tracking estimation (BRITE) method. The key innovation is to disentangle the anatomical signal (characterized by the central Fourier-domain peak) and the tag patterns (characterized by the harmonic peaks) given the observed tagged image sequence, while simultaneously estimating motion. Our main contributions are as follows: **(1)** We analyze

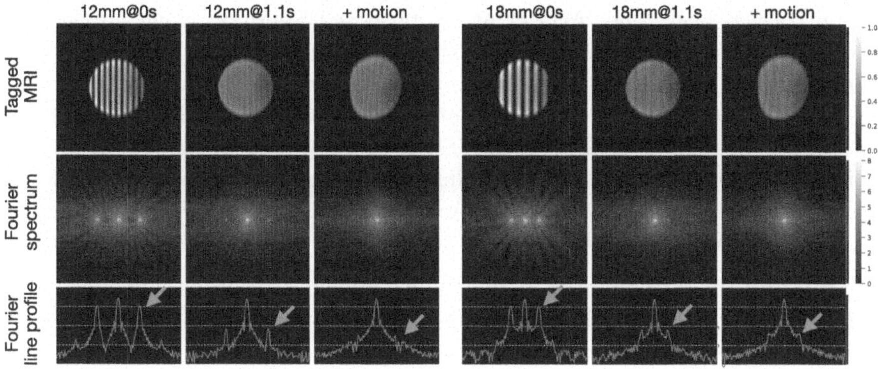

Fig. 1. Top row: Tagged MRI. **Middle Row:** Their corresponding Fourier spectrum. **Bottom Row:** The midline profiles extracted from the Fourier spectra. The column names "**Xmm@Ys**" denote a tag period of X mm acquired at Y s following tagging.

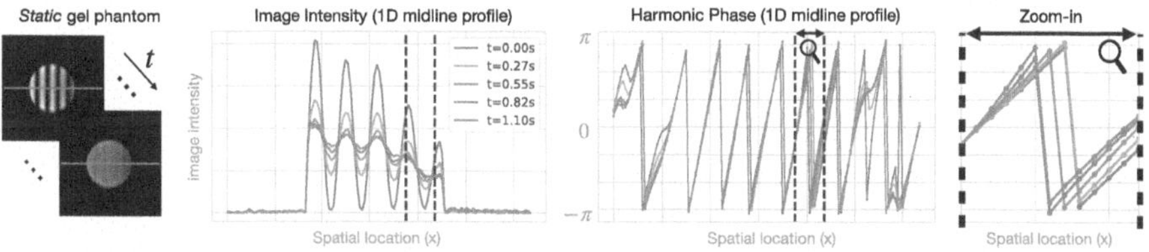

Fig. 2. Tagged MRI of a *static* phantom. Harmonic phases are distorted over time due to tag fading and spectral overlap, resulting in erroneous non-zero motion estimation.

spectral overlap and tag fading in tagged MRI and quantify their impacts on motion tracking. **(2)** We introduce BRITE to disentangle anatomy from tags and account for tag fading while estimating Lagrangian motion. **(3)** Our physics-informed neural network (PINN) is adapted for Lagrangian motion tracking with sequential images. **(4)** We validate BRITE using SPAMM-tagged MRI data of a silicone gel phantom acquired with various tag periods and imaging flip angles.

2 Related Work

Fourier Domain Approaches. Fourier-based methods have long been used to extract motion information, typically via band-pass filtering. As early as 1996, Zhang et al. [33] extracted paired harmonic peaks to reconstruct and track horizontal and vertical tag lines. The HARP method [21] tries to isolate a single harmonic peak with a band-pass filter. The inverse Fourier transform generates a phase image which allows for tracking of a tissue's phase—a material property of the tissue—over time. Discontinuities in the phase image can cause interpolation issues which have been tackled with various approaches [9,18,30]. Qian et al. [25,26] used convolution with Gabor filters to extract tag lines that can be tracked over time. Arts et al. [5] proposed SinMod, which applies two skewed band-pass filters—one emphasizing low frequencies and the other high frequencies—to isolate a single harmonic peak. From the resulting pair of complex images, the displacement is estimated as the phase difference between two time points divided by local frequency. SinMod demonstrates good accuracy and robustness, especially at later time points where tag contrast diminishes. All of these approaches rely on band-pass filters applied to inherently overlapping spectra, which can reduce motion-tracking accuracy.

Image Domain Approaches. Prince and McVeigh [24] proposed the variable brightness optical flow (VBOF) method which relaxes the brightness constancy assumption by explicitly modeling tag fading through the imaging equation. The approach requires knowledge of the longitudinal relaxation time (T1), transverse relaxation time (T2), and spin density of the tissue. Subsequent work [14] used a local linear intensity model, but imposed restrictions on the scaling field throughout the time course. Dougherty et al. [11] proposed an alternative method known as registration and change visualization (RCV), which applied Laplacian filtering to remove local offset and enhance edges. The underlying assumptions of RCV are viable only when a pair of images have minimal brightness variation. Template-matching strategies have also been proposed [3,13] that have drawbacks due to large tag deformations [13], tag fading [3,13], and require prior knowledge of the tissue and sequence parameters [3].

In the deep learning (DL) era, Ye et al. [31] introduced DeepTag, an unsupervised DL-based registration method that uses normalized cross-correlation as the similarity measure to directly register *raw* tagged images without addressing tag fading. While DeepTag can achieve rapid inference, it requires a sufficient number of tagged images from the target domain during training, thereby limiting its general applicability. More recently, Bian et al. [8] analyzed various commonly used image similarity measures for training DL-based registrations on *raw* tagged images, and showed that tag fading remains a significant challenge.

Inspired by some of this previous work, BRITE disentangles the anatomy and tag patterns throughout the time course. Unlike VBOF and template matching approaches, BRITE does not rely on prior knowledge of the imaging parameters; different from RCV, BRITE learns both the amplitude and offset of the tag pattern, and can handle large brightness variations; and unlike Deeptag, BRITE does not require tagged data to be collected for training.

A Unified Perspective. Fourier- and image-domain approaches are fundamentally two sides of the same coin, related by the Fourier transform and its inverse. Although spectral overlap effects are more directly observed in the Fourier domain, they ultimately influence the methods applied to either domain.

3 Methodology

From a communication theory perspective, MR tagging can be partially viewed as the amplitude modulation of the anatomical signal with a *low-frequency* carrier. This causes significant spectral overlap, making the reconstruction of the

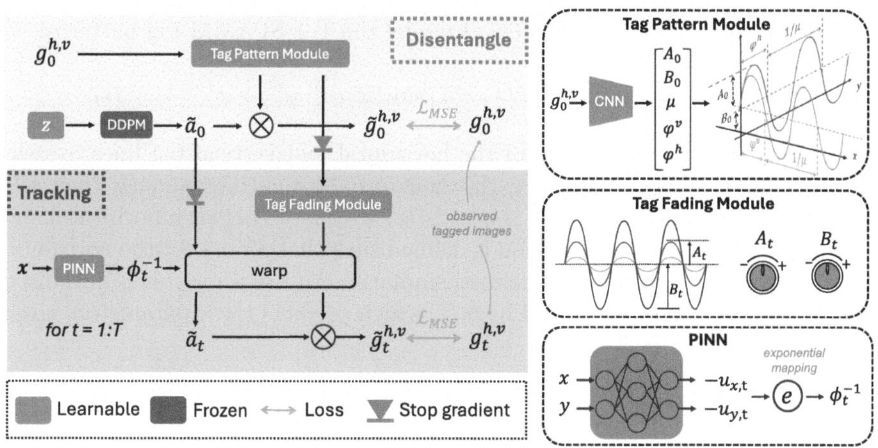

Fig. 3. Overview of the Proposed BRITE Framework. The pipeline consists of two stages: disentangling and tracking. At $t = 0$, the "Tag Pattern Module" and a pretrained DDPM model are used to separate the anatomical image $\tilde{a}_0$ from the tag patterns $\tilde{p}_0^{h,v}$. Subsequently, for each time $t > 0$, the PINN estimates the Lagrangian motion while the "Tag Fading Module" captures tag fading.

signal from the observed data an ill-posed problem. In our task, separating the harmonic and central spectral components (i.e., eliminating spectral overlap) is further complicated by tag fading and tissue motion. In other words, from an image-domain perspective, we aim to simultaneously estimate the underlying anatomy, the fading tag pattern, and the tissue motion over time.

We first introduce the following notation. In 2D, g denotes a tagged image, a an anatomical image, and p a tag pattern. Superscripts h, v indicate horizontal and vertical directions, respectively, and subscripts t indicate time frame (e.g., g_t^v denotes the vertically tagged image at time frame t). When an equation applies to both directions, we write $\{h, v\}$ in the superscript for brevity. The spatial variables (x, y) are omitted whenever it does not cause ambiguity.

Figure 3 shows an overview of the two key parts of BRITE, the disentanglement and the tracking. Disentanglement estimates both the tag patterns and the underlying anatomical signal without tags. Tracking estimates tissue motion across time, incorporating the effects of tag fading at each frame. Ultimately, this provides Lagrangian motion fields relative to the reference frame and yields separated anatomical and tag patterns that deform over time.

Disentangling Tagged Images. The forward model for MR tagging at $t = 0$ can be approximated by,

$$g_0^{\{h,v\}} = a_0 \otimes p_0^{\{h,v\}},$$

where g_0^h and g_0^v are the observed tagged images with horizontal and vertical tags, respectively, $a_0(x, y)$ represents the unknown anatomical image at $t = 0$, and $\otimes$ denotes the element-wise product.

The functions $p_0^h(x, y)$ and $p_0^v(x, y)$ model the 1–1 SPAMM tag patterns,

$$p_0^{\{h,v\}}(x, y\,; A_0, B_0, \mu, \varphi^{\{h,v\}}) = A_0 \sin(2\pi\mu x^{\delta^v} y^{\delta^h} + \varphi^{\{h,v\}}) + B_0,$$

where φ^h, φ^v denote the phases of the horizontal and vertical tag lines, respectively. μ is their spatial frequency, A_0 their initial amplitude, and B_0 their DC offset; these three parameters (μ, A_0, and B_0) are shared between horizontal and vertical tags. The indicator function is defined such that $\delta^h = 1$ for the horizontal direction and 0 otherwise; δ^v is defined similarly. We use a *randomly*-initialized CNN encoder $f_{\theta_{\text{tag}}}$ parameterized by θ_{tag}, which predicts these parameters given the observed tagged images,

$$\{A_0, B_0, \mu, \varphi^h, \varphi^v\} = f_{\theta_{\text{tag}}} := f_{\theta_{\text{tag}}}(\texttt{cat}(g_0^h, g_0^v))\,,$$

where `cat` is channel-wise concatenation. Empirically, we found using this CNN-based prediction strategy helps optimization and convergence compared with directly optimizing parameters, which may be explained by deep image priors [29]. We estimate the tag frequency μ rather than fixing it as a constant (e.g., read from the pulse sequence) to account for potential magnetic gradient imperfections.

To estimate the unknown anatomy a_0, we adopt a compressed sensing framework [10]. We use a denoising diffusion probabilistic model (DDPM) $\mathcal{G}(z)$, pre-trained on anatomical MR images (e.g., cine or T1-weighted) to generate realistic candidate anatomies, which confines the search region for this ill-posed inverse problem. Solving

$$z^*, \theta_{\text{tag}}^* = \arg\min_{z, \theta_{\text{tag}}} \sum_{i \in \{h,v\}} \left\| g_0^i - \mathcal{G}(z) \otimes p_0^i(; f_{\theta_{\text{tag}}}) \right\|_2^2 \tag{1}$$

yields the estimated anatomy $\tilde{a}_0$ and tag patterns $\tilde{p}_0^h, \tilde{p}_0^v$ that jointly explain the observed tagged images, where

$$\tilde{a}_0 = \mathcal{G}(z^*) \;\; \text{and} \;\; \tilde{p}_0^{\{h,v\}} = p_0^{\{h,v\}}\left(x, y\,; f_{\theta_{\text{tag}}^*}\right).$$

Tracking. Having disentangled the initial frame, we now address subsequent time frames from $t = 1$ to $t = T$. Tag patterns fade and deform over time, and the anatomy should deform in the same way. Our objective is to estimate tissue motion and evolving tag parameters for each t, ultimately providing a Lagrangian motion field referenced to $t = 0$.

We assume a deformation field ϕ_t that maps points from the reference frame ($t = 0$) to their positions at time t. We adapt a PINN [27] framework for tracking. Specifically, a multilayer perception $f_{\theta_{\text{pinn},t}}$ parameterized by $\theta_{\text{pinn},t}$ takes coordinates (x, y) as input and outputs a stationary velocity field $\mathbf{u}_t(x, y)$,

$$\mathbf{u}_t(x, y) = f_{\theta_{\text{pinn},t}}(x, y)\,. \tag{2}$$

We then integrate this velocity field and its negation via an fast exponential mapping [4], also known as scaling and squaring, to obtain a biologically plausible diffeomorphic deformation and its inverse,

$$\phi_t(x,y) = \exp(\mathbf{u}_t)(x,y)\,, \quad \phi_t^{-1}(x,y) = \exp(-\mathbf{u}_t)(x,y)\,. \tag{3}$$

The *inverse* mapping allows us to warp the estimated anatomy to time t as

$$\tilde{a}_t = \tilde{a}_0 \circ \phi_t^{-1}. \tag{4}$$

To account for tag fading, we introduce a "Tag Fading Module" that has two learnable parameters A_t and B_t that model the time-varying amplitude and offset. The faded tag patterns (before considering deformations) are,

$$\hat{p}_t^{\{h,v\}}(x,y\,;A_t,B_t) = A_t \sin\left(2\pi\tilde{\mu}x^{\delta^v}y^{\delta^h} + \tilde{\varphi}^{\{h,v\}}\right) + B_t\,.$$

Note that the tag frequency $\tilde{\mu}$ and phase $\tilde{\varphi}^{\{h,v\}}$ are carried over from the solution to Equation (1). Then incorporating the deformation gives the *faded* and *deformed* tag pattern at time t as,

$$\tilde{p}_t^{\{h,v\}} = \hat{p}_t^{\{h,v\}} \circ \phi_t^{-1}\,. \tag{5}$$

Combining (4) and (5), the reconstructed tagged images at time t are,

$$\tilde{g}_t^{\{h,v\}} = \tilde{a}_t \otimes \tilde{p}_t^{\{h,v\}}\,.$$

We jointly optimize $\theta_{\text{pinn},t}$ and $\{A_t, B_t\}$ to minimize the reconstruction error between $\tilde{g}_t^{h,v}$ and the observed $g_t^{h,v}$ at time t,

$$\{A_t^*, B_t^*\}, \theta_{\text{pinn},t}^* = \mathop{\arg\min}_{\theta_{\text{pinn},t},\{A_t,B_t\}} \sum_{i\in\{h,v\}} \left\|\tilde{g}_t^i - g_t^i\right\|_2^2\,. \tag{6}$$

Once $\theta_{\text{pinn},t}^*$ is obtained, the optimal Lagrangian motion field ϕ_t and its inverse ϕ_t^{-1} are computed from Equations (2) and (3). To solve the motion at $t+1$, we initialize $\theta_{\text{pinn},t+1}$ with $\theta_{\text{pinn},t}^*$ and then optimize Equation (6). This is an important feature of BRITE, as it prevents tag-jumping when the tissue's Lagrangian motion exceeds half of the tag period. Furthermore, unlike methods that compose motion between every pair of adjacent frames, BRITE avoids accumulating drifting errors. Notably, no smoothness penalty is required for motion estimation.

A brief delay (usually a few milliseconds) exists between tag preparation and the start of imaging, during which tissue may move and deform the tag lines. BRITE handles this by simply tracking from the first timeframe ($t = 0$), so that the estimated ϕ_0^{-1} captures any pre-imaging deformation. Then for $t > 0$, we substitute $\tilde{a}_0 \leftarrow \tilde{a}_0 \circ \phi_0^{-1}$ and $\hat{p}_t^{\{h,v\}} \leftarrow \hat{p}_t^{\{h,v\}} \circ \phi_0^{-1}$ into Equations (4) and (5), and optimization of (6) proceeds as usual. We use this strategy for all experiments.

Implementation Details. The function $f_{\theta_{\text{tag}}}$ is a shallow 4-layer ResNet with basic residual blocks. It accepts a two-channel input and outputs residual values for the tag-pattern parameters, which are added to user-specified initial values. We initialize A_0 and B_0 at 0.45 and 0.55, respectively, both φ^h and φ^v at 2π, and μ at the read from pulse setting. The DDPM model uses a UNet backbone with three resolution levels (channels = (64, 64, 64)) and is pretrained on a synthetic dataset of random ovals (see Fig. 4(b) for examples). A DDIM [28] scheduler with 20 steps is employed to sample the anatomy from the latent code z. The function $f_{\theta_{\text{pinn},t}}$ is a fully connected network of three hidden layers, each with 128 neurons. The number of steps for scaling and squaring is set to 7. For the tag fading module, (A_t, B_t) are both initialized to 0.5. The optimization of Equation (1) runs for 600 iterations. Equation (6) runs for 2,000 iterations per time frame. All learnable parameters are trained with the Adam optimizer, using a learning rate of 1×10^{-2} for z, 1×10^{-4} for $f_{\theta_{\text{tag}}}$ and $f_{\theta_{\text{pinn},t}}$, and 5×10^{-2} for (A_t, B_t). BRITE processes a 100-frame tagged MRI in 14 min (8.4 s/frame), with a peak GPU memory usage of 5 GB on an Nvidia 4070 GPU.

4 Experimental Setup

Datasets. We acquired tagged MRI data from a static cylindrical gel phantom (made from Sylgard 527) using a 1–1 SPAMM tagging pulse followed by a Siemens FLASH imaging sequence. The total tagging angle was 90°. The duration was 1.1 s with an 11 ms temporal resolution, yielding 100 time frames. The imaging parameters were TR=3.67 ms, TE=1.63 ms, and a $128 \times 128 \times 1$ voxel matrix with a $2 \times 2 \times 10$ mm resolution. Eight tagged sequences were acquired with tag periods of 9, 12, 18, and 26 mm and imaging flip angles of 5° and 10°. The choice of a 5° flip angle approximated the Ernst angle, assuming $T_1 = 900$ ms. Simulated rigid rotation (N=1) and non-rigid deformations (N=20) were applied to each of the acquired tagged sequences (N=8) to provide ground-truth motion fields for evaluation. For non-rigid deformations, random displacements were generated at predefined control points and interpolated pixel-wise in between using B-splines.

Evaluation Metrics. We report the end-point error (EPE), $\texttt{EPE}(x, y) = \|\mathbf{d}_{\text{gt}}(x, y) - \mathbf{d}_{\text{est}}(x, y)\|_2$, the magnitude of the difference between the ground-truth and estimated displacement fields. We also report the maximum principal strain (MPS), which captures the largest principal strain experienced by the tissue. MPS reflects largest directional deformation which is of interest in biomechanical analyses. The MPS error (eMPS), defined as $\texttt{eMPS}(x, y) = |\texttt{MPS}_{\text{gt}}(x, y) - \texttt{MPS}_{\text{est}}(x, y)|$, quantifies how close the ground truth and estimated strain fields are. A Lagrangian reference frame was employed, with the first time frame being the reference.

Methods for Comparison. We compared BRITE with five other methods. We used the *HARP* software provided by the original authors. The band-pass filter radius was set to half of the tagging frequency. The source code of *SinMod* is not publicly available and thus we implemented it following the original publication. The size of the squared cosine kernel used for smoothing was set to 15 and the exponent for quality measurement was set to 8. For *SyN + raw* and *SyN + DRIMET*, we used the open-source SyN[1] and the DRIMET[2] implementation. For both variants of *SyN*, we performed a grid search over different image similarity metrics (`CC` and `MeanSquares`), various window sizes for `CC`, and two motion-tracking strategies: (1) registering adjacent frame pairs and composing transformations, and (2) initializing the transformation with the previous time step's result and registering directly to the reference image. For *SyN + raw*, where raw tagged images are input, `CC` with a window size of 4 outperforms the default of 2. For *SyN + DRIMET*, the `MeanSquares` metric yields the best results. Both the raw and DRIMET inputs are multivariate: raw has two images (horizontal and vertical), while DRIMET has four (two per orientation), so we used SyN's multivariate mode. The composition strategy works best for both variants of *SyN*, and we report the best performance achieved under these parameter settings. We used the open-source algorithm *Deeptag*[3], which requires in-domain tagged data for training prior to testing. We trained Deeptag on synthetic oval-shaped tagged images, which are subject to tag-fading and simulated deformations (following the same procedure as the test set). Because Deeptag was originally designed for grid-tag patterns, we doubled the number of channels in the network's first layer to accommodate both horizontal and vertical tagged images.

5 Results and Discussion

A closer look at BRITE. Figure 4 provides a detailed view of how each component of BRITE operates on a tagged MRI sequence undergoing tissue deformation and tag fading. Figure 4(a) shows the input sequence, where the contrast of the horizontal and vertical tags diminishes over time. Figure 4(c) illustrates the disentanglement step, which involves jointly optimizing the latent code z and tag-pattern as per Equation (1). During optimization, the disentanglement loss decreases and tag pattern module converges. Notably, even though the DDPM model is trained only on synthetic oval-shaped images (shown in Fig. 4(b)), it successfully estimates the phantom's underlying anatomy from the real tagged data. As new time frames are introduced sequentially, the tag fading module and PINN work in tandem. In Fig. 4(d), the tag fading module updates the amplitude and DC offset parameters (A_t, B_t), quickly adapting to changes in tag contrast caused by T1 relaxation and progression towards steady-state. Figure 4(e) demonstrates motion estimation using the PINN, where the Lagrangian and diffeomorphic deformation are estimated throughout the sequence.

Evaluation: Non-rigid Deformation. We applied simulated non-rigid deformation fields to each tagged sequence acquired with varying imaging flip angles (FA) and tag periods (TP). Specifically, we considered FA $\in \{5°, 10°\}$ and TP $\in \{9, 12, 18, 26$ mm$\}$. Figure 5 summarizes the performance of all methods. Figure 6 shows qualitative results. BRITE generally outperforms all other

[1] https://github.com/ANTsX/ANTs.

[2] https://github.com/jasonbian97/DRIMET-tagged-MRI.

[3] https://github.com/DeepTag/cardiac_tagging_motion_estimation.

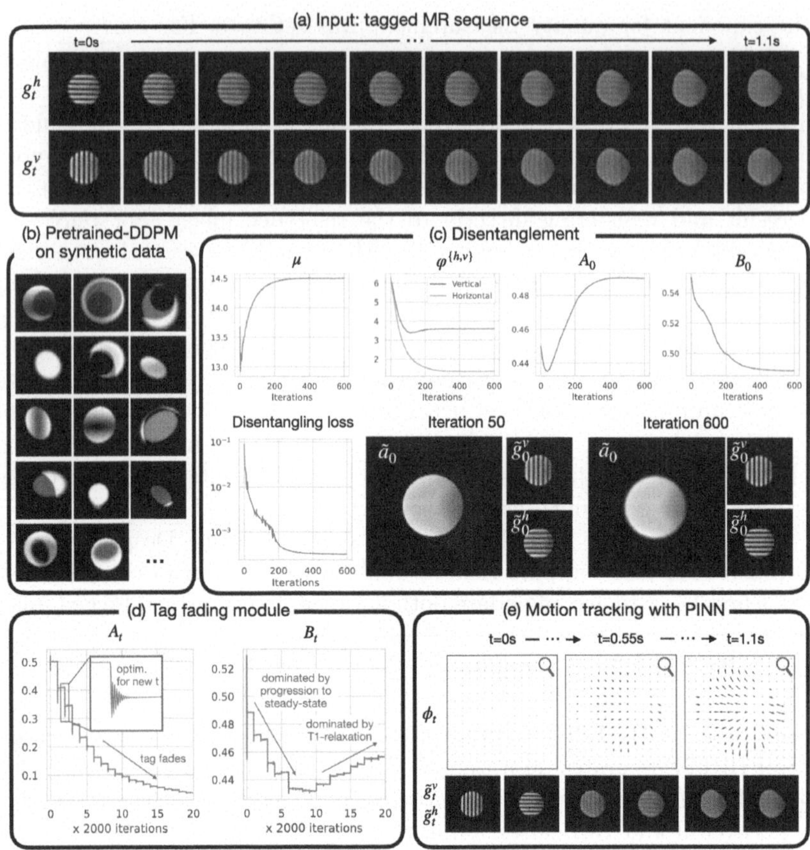

Fig. 4. (a) Input tagged MR sequence with horizontal and vertical tags. (b) Synthetic oval-shaped images used for training the DDPM. (c) Disentanglement process: shown are the optimization trajectories of $\mu, \varphi^h, \varphi^v, A_0, B_0$, the disentanglement loss, estimated anatomy $\tilde{a}_0$, and reconstructed tagged images $\tilde{g}_0$ at 50 and 600 iterations. (d) Tag fading module optimization. Every 5th frame (total of 20 frames over 1.1s) is used for optimization and display clarity. (e) Estimated Lagrangian motion fields.

methods in terms of *both* motion (EPE) and strain estimation accuracy (eMPS). SinMod generally ranks second but its accuracy degrades for larger tag periods, where spectral overlap becomes more pronounced. Similar performance drops are observed in HARP-based approaches, including HARP itself and DRIMET, as they rely on Fourier-domain filtering, which is susceptible to spectral overlap as tags fade and the period increases. Methods that operate directly on raw tagged images experience difficulties when tags fade, especially with smaller tag periods, where the denser tag pattern has a steeper sinusoidal profile. A possible explanation is that, as tags fade, the algorithm must "compress" (or "stretch", depending on which image is fixed) the unfaded portion to maximize matching with the faded portion, creating erroneous motion. A steeper sinusoidal profile

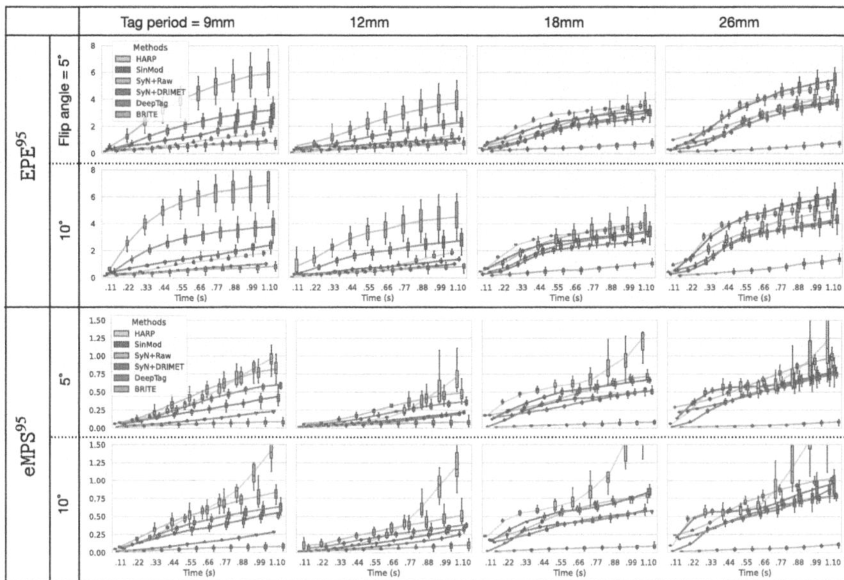

Fig. 5. Non-rigid Deformation Evaluation. Results are shown for EPE (top two rows) and eMPS (bottom two rows) at different time points. Each column represents a different tag period, and each row corresponds to a flip angle.

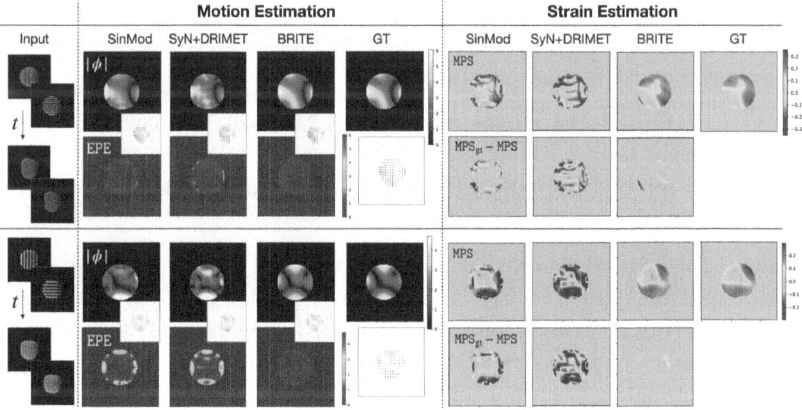

Fig. 6. Qualitative Results under Non-Rigid Deformations. Two examples are shown with FA = 5°, TP = 18mm on the top and FA = 10°, TP = 12mm on the bottom. The left column displays the input images at $t = 0$s and 1.1s. The Motion Estimation panel shows the displacement fields and its magnitude at 1.1s, and EPE. The Strain Estimation panel shows the MPS fields and the corresponding errors. Only the well-performing methods are shown due to the limited space.

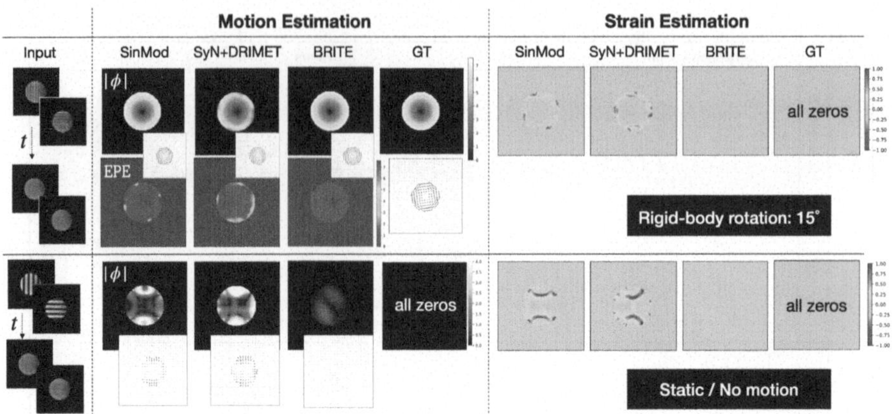

Fig. 7. Qualitative Results under Rigid-Body Rotations and No Motion. The top row has FA = 10°, TP = 9mm for the 15° rigid-body rotation, and the bottom row shows FA = 10°, TP = 26mm for a static, no-motion scenario.

exaggerates this error. Comparing results across flip angles, we see that all methods generally perform better at 5° than at 10°. This likely reflects the fact that 5° is closer to the Ernst angle, thus providing higher SNR and longer-lasting tag patterns for tracking.

The smoothness of deformation fields can significantly affect the MPS strain metric. For instance, HARP tracks each material point independently, making its strain measurements more noise-prone. In contrast, SinMod uses a squared cosine kernel for smoothing, while DRIMET, DeepTag, SyN, and BRITE apply regularization (e.g., smoothness terms or velocity-field integration) to produce smoother fields and less extreme strain values.

Evaluation: Rigid-body Rotation and No Motion. All methods are evaluated on a 15° rigid-body rotation and static scenario. BRITE generally performs best among all methods. Quantitative results are omitted due to limited space. Qualitative examples are shown in Fig. 7.

Summary. All methods show declining performance over time as the deformation grows and tags fade. Larger deformations broaden spectral peaks. Fading tags reduce the tag contrast needed by intensity-based methods and exacerbate spectral overlap effects in Fourier-based methods. BRITE suffers *less* from these issues by disentangling the anatomy from the faded tags and applying its tag fading module independently at each time frame, ensuring a robust adaptation to variable brightness conditions.

Limitations and Future Work. Our current approach assumes a sinusoidal tag pattern (1–1 SPAMM or CSPAMM). Future work will accommodate higher-order SPAMM and grid-tagging by modifying the tag model. Although we only demonstrated our method's performance in 2D, extending to 3D is straightforward. Future work will validate BRITE on human organs.

6 Conclusion

We analyzed the phenomenon of tag fading and spectral overlap in tagged MRI and quantified their impacts on various motion tracking techniques. We introduced BRITE, a tracking approach that disentangles anatomy from tags and accounts for tag fading while estimating Lagrangian motion. We validated BRITE using SPAMM-tagged MR images of a silicone gel phantom acquired with various tag periods and imaging flip angles, demonstrating that BRITE is resistant to tag fading and provides more accurate motion and strain estimates compared to other state-of-the-art methods.

Acknowledgment. This study was partially supported by the National Institutes of Health (NIH) grants U01NS112120, R01NS136056, R01DC018511, and National Science Foundation (NSF) Graduate Research Fellowship (Grant No. DGE-1746891, Remedios).

Under a license agreement between Myocardial Solutions and the Johns Hopkins University, Dr. Jerry L Prince and Johns Hopkins University are entitled to royalty distributions related to the HARP technology discussed in this publication. This arrangement has been reviewed and approved by the Johns Hopkins University in accordance with its conflict of interest policies.

The opinions and assertions expressed herein are those of the authors and do not reflect the official policy or position of the Uniformed Services University, the Henry M. Jackson Foundation for the Advancement of Military Medicine, or the Department of Defense.

References

1. Agarwal, H.K., Prince, J.L., Abd-Elmoniem, K.Z.: Total removal of unwanted harmonic peaks (TruHARP) MRI for single breath-hold high-resolution myocardial motion and strain quantification. Magn. Reson. Med. **64**(2), 574–585 (2010)
2. Aletras, A.H., Ding, S., Balaban, R.S., Wen, H.: DENSE: displacement encoding with stimulated echoes in cardiac functional MRI. J. Magn. Reson. **137**(1), 247 (1999)
3. Amini, A., Curwen, R., Constable, R.T., Gore, J.C.: MR physics-based snake tracking and dense deformations from tagged cardiac images. In: American Association for Artificial Intelligence (AAAI) Spring Symposium Series. In: Applications of Computer Vision in Medical Image Processing, pp. 126–129 (1994)
4. Arsigny, V., Commowick, O., Pennec, X., Ayache, N.: A log-euclidean framework for statistics on diffeomorphisms. In: Larsen, R., Nielsen, M., Sporring, J. (eds.) Medical Image Computing and Computer-Assisted Intervention. MICCAI 2006. LNCS, vol. 4190, pp. 924–931. Springer, Berlin, Heidelberg (2006). https://doi.org/10.1007/11866565_113
5. Arts, T., Prinzen, F.W., Delhaas, T., Milles, J.R., Rossi, A.C., Clarysse, P.: Mapping displacement and deformation of the heart with local sine-wave modeling. IEEE Trans. Med. Imaging **29**(5), 1114–1123 (2010)
6. Axel, L., Dougherty, L.: Heart wall motion: improved method of spatial modulation of magnetization for MR imaging. Radiology **172**(2), 349–350 (1989)

7. Bayly, P.V., et al.: MR imaging of human brain mechanics in vivo: new measurements and applications to the development of computational models of brain injury. Ann. Biomed. Eng. **49**, 2677–2692 (2021)
8. Bian, Z., et al.: Is registering raw tagged-MR enough for strain estimation in the era of deep learning? In: Medical Imaging 2024: Image Processing, vol. 12926, pp. 79–85. SPIE (2024)
9. Bian, Z., et al.: DRIMET: deep registration-based 3d incompressible motion estimation in tagged-MRI with application to the tongue. In: Medical Imaging with Deep Learning, pp. 134–150. PMLR (2024)
10. Bora, A., Jalal, A., Price, E., Dimakis, A.G.: Compressed sensing using generative models. In: International Conference on Machine Learning, pp. 537–546. PMLR (2017)
11. Dougherty, L., Asmuth, J.C., Blom, A.S., Axel, L., Kumar, R.: Validation of an optical flow method for tag displacement estimation. IEEE Trans. Med. Imaging **18**(4), 359–363 (1999)
12. Fischer, S.E., McKinnon, G., Maier, S., Boesiger, P.: Improved myocardial tagging contrast. Magn. Reson. Med. **30**(2), 191–200 (1993)
13. Fisher, D.J.: Automatic tracking of cardiac wall motion using magnetic resonance markers. The University of Iowa (1990)
14. Gupta, S.N., Prince, J.L.: On variable brightness optical flow for tagged MRI. In: Information Processing in Medical Imaging, vol. 3, pp. 323–334. Kluwer Dordrecht (1995)
15. Ibrahim, E.: Myocardial tagging by cardiovascular magnetic resonance: evolution of techniques-pulse sequences, analysis algorithms, and applications. J. Cardiovasc. Magn. Reson. **13**(1), 36 (2011)
16. Knutsen, A.K., et al.: Improved measurement of brain deformation during mild head acceleration using a novel tagged MRI sequence. J. Biomech. **47**(14), 3475–3481 (2014)
17. Mannelli, L., et al.: Assessment of the liver strain among cirrhotic and normal livers using tagged MRI. J. Magn. Reson. Imaging **36**(6), 1490–1495 (2012)
18. Mella, H., et al.: HARP-I: a harmonic phase interpolation method for the estimation of motion from tagged MR images. IEEE Trans. Med. Imaging **40**(4), 1240–1252 (2021)
19. Moerman, K.M., Sprengers, A.M., Simms, C.K., Lamerichs, R.M., Stoker, J., Nederveen, A.J.: Validation of continuously tagged MRI for the measurement of dynamic 3D skeletal muscle tissue deformation. Med. Phys. **39**(4), 1793–1810 (2012)
20. Niitsu, M., Kumada, M., Campeau, N.G., Niimi, S., Riederer, S.J., Itai, Y.: Tongue displacement: visualization with rapid tagged magnetization-prepared MR imaging. Radiology **191**(2), 578–580 (1994)
21. Osman, N.F., Kerwin, W.S., McVeigh, E.R., Prince, J.L.: Cardiac motion tracking using CINE harmonic phase (HARP) magnetic resonance imaging. Magn. Reson. Med. **42**(6), 1048–1060 (1999)
22. Osman, N.F., Sampath, S., Atalar, E., Prince, J.L.: Imaging longitudinal cardiac strain on short-axis images using strain-encoded MRI. Magn. Reson. Med. **46**(2), 324–334 (2001)
23. Parthasarathy, V., Prince, J.L., Stone, M., Murano, E.Z., NessAiver, M.: Measuring tongue motion from tagged cine-MRI using harmonic phase (HARP) processing. J. Acoust. Soc. Am. **121**(1), 491–504 (2007)
24. Prince, J.L., McVeigh, E.R.: Motion estimation from tagged MR image sequences. IEEE Trans. Med. Imaging **11**(2), 238–249 (1992)

25. Qian, Z., Metaxas, D.N., Axel, L.: Extraction and tracking of MRI tagging sheets using a 3D Gabor filter bank. In: 2006 International Conference of the IEEE Engineering in Medicine and Biology Society, pp. 711–714. IEEE (2006)
26. Qian, Z., Montillo, A., Metaxas, D.N., Axel, L.: Segmenting cardiac MRI tagging lines using Gabor filter banks. In: Proceedings of the 25th Annual International Conference of the IEEE Engineering in Medicine and Biology Society (IEEE Cat. No. 03CH37439), vol. 1, pp. 630–633. IEEE (2003)
27. Raissi, M., Perdikaris, P., Karniadakis, G.E.: Physics-informed neural networks: a deep learning framework for solving forward and inverse problems involving nonlinear partial differential equations. J. Comput. Phys. **378**, 686–707 (2019)
28. Song, J., Meng, C., Ermon, S.: Denoising diffusion implicit models. arXiv preprint arXiv:2010.02502 (2020)
29. Ulyanov, D., Vedaldi, A., Lempitsky, V.: Deep image prior. In: IEEE Conference on Computer Vision and Pattern Recognition (CVPR), pp. 9446–9454 (2018)
30. Xing, F., et al.: Phase vector incompressible registration algorithm for motion estimation from tagged magnetic resonance images. IEEE Trans. Med. Imaging **36**(10), 2116–2128 (2017)
31. Ye, M., et al.: DeepTag: an unsupervised deep learning method for motion tracking on cardiac tagging magnetic resonance images. In: IEEE Conference on Computer Vision and Pattern Recognition (CVPR), pp. 7261–7271 (2021)
32. Yu, J., et al.: New starting point registration method for tagged MRI tongue motion estimation. In: Proceedings of SPIE Medical Imaging (SPIE-MI 2023), San Diego, CA, February 19 – 23, 2023, p. 1246429 (2023)
33. Zhang, S., et al.: A Fourier based algorithm for tracking SPAMM tags in gated magnetic resonance cardiac images. Med. Phys. **23**(8), 1359–1369 (1996)

SafeTriage: Facial Video De-identification for Privacy-Preserving Stroke Triage

Tongan Cai[1], Haomiao Ni[2], Wenchao Ma[1], Yuan Xue[3], Qian Ma[1], Rachel Leicht[4], Kelvin Wong[4], John Volpi[4], Stephen T.C. Wong[4], James Z. Wang[1], and Sharon X. Huang[1(✉)]

[1] The Pennsylvania State University, University Park, PA, USA
suh972@psu.edu
[2] The University of Memphis, Memphis, TN, USA
[3] The Ohio State University, Columbus, OH, USA
[4] Houston Methodist Hospital, Houston, TX, USA

Abstract. Effective stroke triage in emergency settings often relies on clinicians' ability to identify subtle abnormalities in facial muscle coordination. While recent AI models have shown promise in detecting such patterns from patient facial videos, their reliance on real patient data raises significant ethical and privacy challenges—especially when training robust and generalizable models across institutions. To address these concerns, we propose *SafeTriage*, a novel method designed to de-identify patient facial videos while preserving essential motion cues crucial for stroke diagnosis. SafeTriage leverages a pretrained video motion transfer (VMT) model to map the motion characteristics of real patient faces onto synthetic identities. This approach retains diagnostically relevant facial dynamics without revealing the patients' identities. To mitigate the distribution shift between normal population pre-training videos and patient population test videos, we introduce a conditional generative model for visual prompt tuning, which adapts the input space of the VMT model to ensure accurate motion transfer without needing to fine-tune the VMT model backbone. Comprehensive evaluation, including quantitative metrics and clinical expert assessments, demonstrates that SafeTriage-produced synthetic videos effectively preserve stroke-relevant facial patterns, enabling reliable AI-based triage. Our evaluations also show that SafeTriage provides robust privacy protection while maintaining diagnostic accuracy, offering a secure and ethically sound foundation for data sharing and AI-driven clinical analysis in neurological disorders.

1 Introduction

Stroke is a leading cause of disability and mortality worldwide [16]. Timely detection and intervention significantly enhance survival rates and the quality of life for stroke patients. In emergency room (ER) triage, potential strokes are often identified by observing subtle abnormalities in oral-facial movements. However,

T. Cai, H. Ni, and W. Ma—These authors contributed equally to this work.

I. Oguz et al. (Eds.): IPMI 2025, LNCS 15830, pp. 390–404, 2026.
https://doi.org/10.1007/978-3-031-96625-5_26

the shortage of experienced neurologists [20] and the nuanced nature of these indicators [32] undermine the reliability of triage in critical stroke situations.

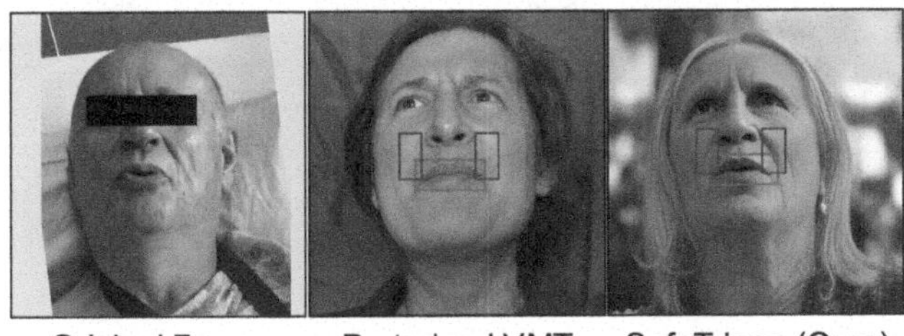

Fig. 1. Illustration of the advantages of the SafeTriage framework. A direct application of a pretrained video motion transfer (VMT) model, trained with normal population faces, fails to accurately capture subtle motions (e.g., lip movements highlighted by the red box) and facial asymmetries (e.g., asymmetries highlighted by the blue boxes) in patient faces. In contrast, SafeTriage overcomes these limitations through visual prompt tuning (VPT) to effectively adapt the pretrained VMT model for patient faces. (Color figure online)

Recent advances in machine intelligence have shown promise in identifying neurological disorders from visual and audio inputs. For example, Cai *et al.* [2] introduced *DeepStroke*, an effective stroke triage framework tailored for ER environments, utilizing video and speech data for stroke detection. Zhuang *et al.* [38] developed a method for evaluating facial weakness from videos. Despite the progress, the success of existing stroke triage methods and AI-assisted models heavily depends on access to datasets of real patient videos. Privacy and ethical considerations, however, severely limit sharing these sensitive data across institutions. As a result, models are often trained on limited, in-house datasets, hindering their robustness, scalability, and generalizability.

In the broader domain of video analysis, various strategies have been explored for privacy protection and de-identification. Dave *et al.* [6] introduced a self-supervised framework for action recognition that employs a minimax optimization strategy. This framework aims to minimize the cost of action recognition while maximizing privacy through a contrastive self-supervised loss. Although it effectively removes privacy attributes, the actions in the generated videos are only recognizable by the system's modules and not by humans. Xia *et al.* [31] developed a text-guided sign language video anonymization technique using a large-scale diffusion model [24], which relies on text inputs for guidance—inputs that are often unavailable in patient facial videos. Meanwhile, approaches in medical imaging—such as using Generative Adversarial Networks (GAN) [9] for de-identification of skin lesion images [1], MRI results [27], and Chest X-Rays [21]—have seen some success. Translating these strategies to clinical videos is less straightforward. Hou *et al.* [13] utilized landmark-guided face morphing on seizure patient videos, and Flouty *et al.* [8] applied facial blurring in surgical

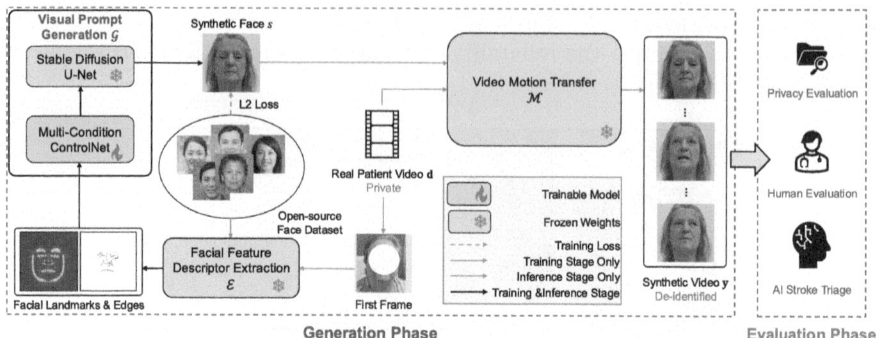

Fig. 2. Overview of the proposed SafeTriage framework, which comprises two main phases: the generation phase and the evaluation phase. The generation phase is based on a *frozen* video motion transfer model $\mathcal{M}$. Given a private video of a real patient $\mathbf{d}$, SafeTriage introduces a visual prompt generation model, $\mathcal{G}$, to generate a synthetic face, s, with facial asymmetries and head pose aligned to the first frame of $\mathbf{d}$. Then $\mathcal{M}$ is employed to generate a de-identified synthetic video, $\mathbf{y}$, given the conditionally-generated input s and real patient video $\mathbf{d}$. In the evaluation phase, the diagnostic utility and privacy preservation of the synthetic video $\mathbf{y}$ are assessed. It is important to note that the training of $\mathcal{G}$ only uses publicly available, non-private face data.

settings. While these methods obscure subject identities, they fail to preserve the critical facial motions necessary for video-based facial analysis, thus hindering large-scale data sharing essential for advancing AI-assisted medical diagnosis and intervention. Zhu *et al.* [37] recently proposed a generative approach for facial de-identification and medical information preservation, but their method relies on a conditional GAN model trained from scratch and uses faces of healthy volunteers, which may limit its scalability and adaptability to general applications.

To address the persistent challenges of data scarcity and to facilitate the sharing of patient facial videos, we present *SafeTriage*, a novel framework that de-identifies patient facial videos by retargeting them onto synthetic appearances while preserving the critical facial motion patterns needed for accurate stroke diagnosis. SafeTriage leverages a pretrained video motion transfer (VMT) model to map patient-specific motions onto synthetic faces, generating videos that combine synthetic appearances with pathological motion dynamics. Directly applying the VMT model to patient data, however, introduces a significant domain gap, as the model is originally trained on videos of healthy individuals (see Fig. 1). To mitigate this, we draw inspiration from recent advancements in visual prompt tuning (VPT) [15] and introduce a conditional generation module that adapts the input space of the VMT model. This module reduces distributional shifts between training and testing while keeping the model backbone frozen. Unlike traditional model fine-tuning, our approach requires training only a small number of parameters in the introduced generative model, thereby largely preserving the prior knowledge embedded in the pretrained VMT model. Comprehensive experiments and evaluation demonstrate that SafeTriage can generate high-quality, de-identified synthetic videos that effectively retain stroke-relevant diagnostic features, achieving an optimal balance between privacy protection and diagnostic accuracy.

2 Methodology

Figure 2 presents an overview of the proposed SafeTriage framework for privacy-preserving stroke triage. The framework comprises two phases: a generation phase and an evaluation phase. In the generation phase, de-identified synthetic videos are generated while preserving the diagnostic motion characteristics of real patient videos. This is achieved through two main components: (1) visual prompt generation and (2) video motion transfer (VMT). To address the limited availability of patient video data, we leverage a large-scale pretrained VMT model as the foundation model of our framework. To minimize computational cost and avoid using privacy-sensitive patient data for model training, rather than fine-tuning the VMT network using patient data, we introduce a conditional generation module to produce a visual prompt (i.e., a synthetic subject image) that has the subject head pose and facial edge features aligned to the first frame of a driving patient video. Using the conditionally generated subject images can effectively help the VMT model adapt to a new patient distribution while keeping its backbone parameters frozen. In the evaluation phase, we assess the synthetic videos in three aspects: (1) human evaluation to ascertain visual realism and the preservation of diagnostic patterns; (2) privacy evaluation to measure the effectiveness of de-identification; and (3) performance assessment of an AI-based stroke triage model with synthetic videos to verify that diagnostic accuracy is maintained. Details are presented in the following sections.

2.1 The Generation Phase

Visual Prompt Generation. Considering the limited availability of patient videos and the privacy concerns associated with using such data for model training, our framework leverages a VMT foundation model, $\mathcal{M}$, pretrained on publicly available, large-scale datasets of in-the-wild facial videos of primarily healthy individuals. However, directly applying this pretrained VMT model to patient video motion transfer results in degraded performance due to the distributional shift between healthy and patient cohorts (see Fig. 1). Traditional parameter fine-tuning methods, such as full fine-tuning, partial fine-tuning (e.g., updating a subset of parameters [33]), and additive fine-tuning (e.g., using external modules like adapters [4,5,14]), can address domain gaps but often incur notable computational costs and, in the context of patient facial videos, pose an additional risk of privacy leakage. To minimize training overhead and the risk of privacy leakage, we adopt an approach inspired by visual prompt tuning (VPT) [15]. Similar to text prompts in large language models (LLMs), the input subject image and driving video for the VMT foundation model can be viewed as visual prompts. The subject image is typically a face synthesized by

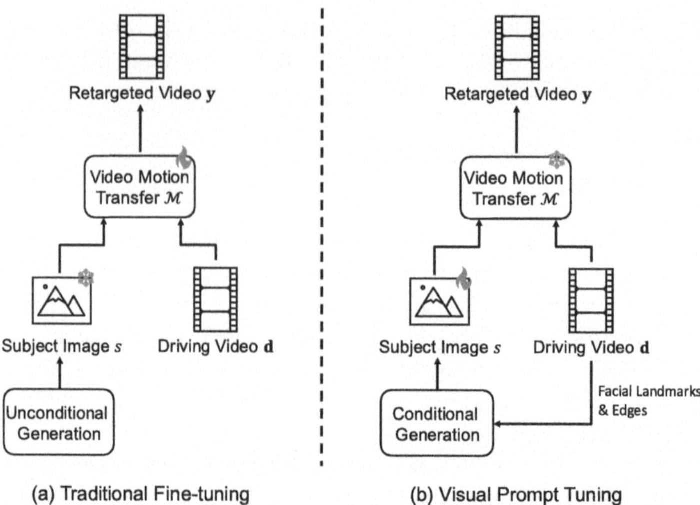

Fig. 3. Illustration of Visual Prompt Tuning. Given a subject image s and a driving video $\mathbf{d}$, the video motion transfer model $\mathcal{M}$ generates a retargeted video $\mathbf{y}$ with the appearance of the subject image and motion in the driving video. Similar to text prompts in large language models (LLMs), s and $\mathbf{d}$ can be viewed as *visual* input prompts for the model $\mathcal{M}$. To adapt $\mathcal{M}$ to a new distribution, unlike traditional methods that fine-tune the model backbone, we employ an efficient and effective visual prompt engineering technique that uses a conditional generation process to generate the visual prompt s that is specifically *tuned* to be aligned with the first frame of the driving video $\mathbf{d}$, enabling $\mathcal{M}$ to perform high-quality transfer of out-of-distribution motion.

a generative AI model with a pseudo-identity (i.e., a non-real identity), which will be later used to conceal the real identity of patients. As shown in Fig. 3, our proposed VPT approach differs from the traditional fine-tuning method by freezing the VMT model backbone while using a conditionally generated subject image prompt specifically tuned to align with privacy-safe features of the driving patient video.

To achieve this, we employ a conditional facial generative model $\mathcal{G}$ to synthesize a pseudo-identity face image s that has its head pose and facial features (such as facial asymmetries) aligned with the real patient video d. Specifically, a pretrained facial feature descriptor model $\mathcal{E}$ is used to extract facial edges e and landmarks m from the first frame of the patient video d. The extracted facial edges are mainly in the eye, mouth and cheek regions to capture potential pathological features such as asymmetries (see the edge feature example shown in Fig. 2). The facial edge and landmark features serve as conditions for $\mathcal{G}$ to generate the visual prompt s, which also ensures alignment of the head pose between s and the first frame of d. To maintain the quality of the generated visual prompts, we filter out low-quality synthetic images using an off-the-shelf image quality assessment tool. As a general framework, we can use various model backbones to implement the model $\mathcal{G}$. Here we choose ControlNet [35] for its high performance and efficiency. To avoid identity leakage, we train the model $\mathcal{G}$ exclusively with publicly available face data, ensuring that no private patient data is used during the training process. More details can be found in Section 3.2.

Video Motion Transfer. Let s be a pseudo-identity face image, referred to as the subject image. Let $\mathbf{d} = \langle d_1, d_2, \cdots, d_N \rangle$ be a video of a patient speaking, such as describing the "cookie theft" picture commonly used for rating aphasia severity in the NIH Stroke Scale. $\mathbf{d}$, consisting of N individual frames $d_i, i = 1, \ldots, N$, is used as the driving video for motion transfer from the patient to the pseudo-identity subject. We utilize a pretrained VMT model $\mathcal{M}$ to generate an output video $\mathbf{y} = \langle y_1, y_2, \cdots, y_N \rangle$, where the identity in y_i's is inherited from the synthetic face s and the motions are derived from the original patient video d_i's. A typical warp-based VMT network [22,28] consists of three main stages: (1) semantic correspondence learning, (2) deformation flow modeling, and (3) warping and inpainting, with inpainting often handled by a generator. Specifically, the VMT model $\mathcal{M}$ first identifies semantic correspondences (i.e., locations in multiple images that share the same semantic meaning [12]) between s and d_i by extracting keypoints (either 2D [28] or 3D [30]) or regions [29] in the feature space. Using these correspondences, $\mathcal{M}$ predicts the dense deformation flow, aligning the pose of the subject in s with the pose in d_i via a warp operation. Due to potential occlusions, deformation flow alone may not be sufficient for generation as occluded parts cannot be recovered by image warping. Therefore, a generator to inpaint occluded areas is employed in $\mathcal{M}$ to produce the final complete and realistic output. This motion transfer process ensures the de-identification of patient videos while preserving the essential motion patterns for downstream tasks. For implementation, various VMT model backbones can be used for motion transfer; we discuss our specific choice in Sect. 3.2.

2.2 The Evaluation Phase

Human Expert Evaluation. We conduct a human evaluation to assess the *visual realism* and *diagnostic utility* of the generated videos $\mathbf{y}$. First, qualified visual content analysts evaluate the visual realism of each synthetic video by responding to the prompt: "*Do you think this video is realistic?*". The analysts choose from the following options "A. *Very realistic,*" "B. *Moderately realistic,*" or "C. *Not realistic at all.*" Only those synthetic videos unanimously rated as "Very realistic" proceed to a subsequent clinical review to validate the preservation of diagnostic patterns. During the clinical review, we perform a paired user study in which qualified clinicians compare each synthetic video generated by our model with its corresponding real patient video. The clinicians evaluate each pair of videos by answering the question: "*Do you think the diagnostic patterns between these two videos are consistent?*". Clinicians respond with either "Yes" or "No".

Privacy Evaluation. We validate the privacy-protection ability of the SafeTriage method by comparing the facial embedding vectors of real and synthetic videos generated with a pretrained face recognition network VGG-Face [23]. The VGG-Face model achieves 98.9% recognition accuracy on the commonly-used benchmark FaceNet [25] while being computationally efficient. For each patient-to-synthetic retargeting case, we randomly select pairs of two frames and calculate the similarity between their facial embedding vectors. The two frames are either from the same real video of a patient, or one from the real video and the other from the corresponding synthetic video. We demonstrate the change in identity similarity score distribution before and after applying the SafeTriage framework.

AI-Assisted Stroke Triage. An additional quantitative evaluation is conducted by comparing stroke classification results using real patient videos vs. generated synthetic videos, by an AI-based stroke triage model. We adopt the *DeepStroke* [2] model, which was developed for the task of stroke triage using patient facial videos. All the subjects used for training and testing the *DeepStroke* model are patients visiting the ER and only some of them have a stroke confirmed by diffusion-weighted MRI. The *DeepStroke* network backbone is a temporal-averaging ResNet-34 [11] that treats all frames in one video as having the same label. The network uses the adjacent-frame difference vectors as input and performs a binary classification task (i.e., stroke vs. non-stroke) guided by a binary cross-entropy loss. The frame-level logits are stacked and averaged to generate the case-level prediction.

To ensure robust evaluation and address the limited dataset size, we adopt the five-fold cross-validation approach to train and test the stroke triage model. Our experiments are designed to assess the preservation of diagnostic facial motion features by comparing the testing performance on real patient videos with that on their corresponding synthetic videos, using identical models trained on real data. Additionally, to explore the potential of our generated videos in facilitating data sharing and the development of future large AI models, we also examine the outcomes when synthetic videos are used for training and real or synthetic videos are used for validation within the cross-validation framework.

3 Experiments

3.1 Dataset and Metrics

Dataset. We acquired a clinical dataset for this IRB-approved study from Houston Methodist Hospital. Patients suspected of stroke are selected during their ER visits without consideration of race or sex, ensuring equity and diversity. Patients are asked to perform two speech tasks while being video-recorded at a resolution of 1920×1200 at 30 frames per second (fps). The videos are collected "in the wild" without imposing restrictions on the patients' positions, allowing for recordings of patients lying in bed, sitting, or standing, with varied backgrounds and lighting conditions. Each video is paired with ground truth verification from diffusion-weighted MRI scans to determine stroke presence. Our study's cohort includes 113 patients, with 66 diagnosed with stroke and 47 identified as non-stroke but presenting other conditions based on MRI. On average, these videos have a length of 1,895 frames (about 63.16 s). Following prior work [2], we construct a binary classification task, omitting stroke subtypes for simplicity.

Evaluation Metrics. For the human evaluation described in Sect. 2.2, to quantitatively analyze the survey response results, we assign scores to the options provided: 3 points to option A, 2 points to option B, and 1 point to option C. Similarly, for responses regarding the preservation of diagnostic motion features in the generated videos, we assign 1 point for a "Yes" response and 0 point for a "No" response during the clinician review. For both evaluations, we calculate and report the Fleiss's kappa agreement score [7] to measure inter-rater agreement and the mean scores to provide an overall quantitative measure of performance.

For the privacy protection evaluation, we use the cosine similarity (CSIM) [34] between the aforementioned identity embedding vectors to measure identify distance and visualize the identity distance distributions for real-real video frame pairs and real-synthetic video frame pairs, respectively.

For the evaluation of diagnostic utility with an AI stroke triage model, we use commonly referenced metrics, including accuracy, specificity, sensitivity, F1 score, and area under the ROC curve (AUC), to assess the effect of SafeTriage on the model's stroke classification performance. We first establish the baseline of training the *DeepStroke* triage model using real videos and testing on real videos. We then measure performance change from the baseline when testing the model on synthetic videos, or when both training and testing the model using synthetic videos. Furthermore, we measure the absolute change in prediction class probabilities and report the Mean Squared Error (MSE) between the prediction result (before applying softmax) from the real video baseline and the corresponding predictions from other train-test schemes involving synthetic videos.

3.2 Model Implementation

As mentioned in Sect. 2.1, we use a conditional synthetic face generation model $\mathcal{G}$ for visual prompt tuning. Two types of facial feature descriptors, landmark heatmaps and edge maps, are used as the conditional inputs for model $\mathcal{G}$. The landmark heatmaps capture the pose of the patient in the driving video, and generating a synthetic face with matching pose helps the VMT model better retarget facial motions. Moreover, existing VMTs, pretrained primarily on normal faces, largely ignore pathological facial appearance features during the retargeting process—such as facial asymmetries and wrinkles (e.g., asymmetrically positioned eyes, eyebrows, or mouth corners)—which causes their direct application in transferring patient facial motion to produce subpar results with reduced clinical utility. To address this, we use edge maps as an additional condition in visual prompt tuning to retain such features.

In practice, we first use `mmpose`[1] to detect facial landmarks and generate corresponding landmark heatmaps. For edges, since the edges of interest are around facial organs such as eyes, nose, and mouth, we utilize the detected landmarks to define bounding boxes around the interested facial organs and apply the Canny edge detector [3] within these bounding boxes to generate facial edge maps.

We implement the conditional generative model $\mathcal{G}$ with ControlNet [35], a state-of-the-art computationally efficient model that reuses a frozen large pre-trained text-to-image model, Stable Diffusion [24], and introduces only a small number of trainable parameters to enable new conditioning controls. To train this multi-condition ControlNet model, we collected 2,000 human face images from the publicly available FFHQ dataset [19]. We resized the images to a resolution of 256×256 to accelerate training. The model was trained for 100 epochs with a batch size of 2, using two NVIDIA RTX 6000 GPUs. After training the model, multiple candidate synthetic faces are generated using the model given each set of conditions, and a public image quality assessment tool[2] is applied to filter out low-quality generated images. A randomly chosen high-quality synthetic face is then further enhanced using CodeFormer [36] and upsampled to 512×512 to be used as the visual prompt s.

We subsequently preprocess the original patient video to obtain the driving video $\mathbf{d}$ for motion retargeting by 1) eliminating constant camera roll, 2) square-cropping the face regions with sufficient borders, and 3) resizing to 512×512. Given the subject image s and the driving video $\mathbf{d}$, we implement the video motion transfer model $\mathcal{M}$ with LivePortrait [10], which achieves superior generalization due to large-scale training and advanced network architectures, while incorporating specialized modules for precise eye and lip retargeting. We use a specific version[3] of LivePortrait with default settings, as it produces more stable videos in our experiments.

For AI-assisted stroke triage, we use a stand-alone video module of *DeepStroke*, without multi-modal fusion or adversarial training. The backbone is a ResNet-34 pretrained on the FairFace dataset [17]. The model parameters are frozen except for the last residual block and the output fully connected layer. The cross-validation experiments have a common manual seed to ensure the same fold-wise training/validation data points.

3.3 Result Analysis

Results of Human Evaluation. Four independent raters with computer vision expertise participated in evaluating the visual realism of 113 synthetic videos. The overall mean quality score is 2.55 with a standard deviation of 0.65, indicating very good to excellent visual realism since the highest possible score is 3.0. The raters achieved a Fleiss's kappa of 0.608. Among the 113 synthetic

[1] https://github.com/open-mmlab/mmpose.
[2] https://github.com/LAION-AI/aesthetic-predictor.
[3] https://github.com/KwaiVGI/LivePortrait/commit7fda9.

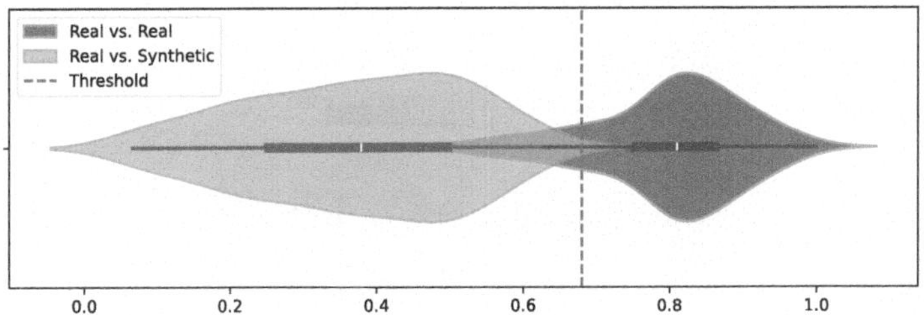

Fig. 4. Privacy protection evaluation results. The distributions of facial embedding similarity scores are plotted in the violin plot. *Real vs. Real* refers to scores between two random frames that come from the same real video, and *Real vs. Synthetic* refers to scores between two random frames, one from a real patient video and the other from the corresponding synthetic video. The outlines of the violin shapes are estimated distributions, and a thicker band indicates a higher density. The thicker horizontal lines indicate quarter quantile ranges. The red dashed line at 0.68 is the default face verification threshold in the VGG-Face model [26].

videos, 38 are rated as "Very realistic" (with score 3.0) by all raters and subsequently selected for further clinician review. A qualified clinician assessed the diagnostic patterns in these selected videos. The mean evaluation score is 0.76, indicating that the majority of retargeted synthetic videos preserved the diagnostic patterns present in the original videos. Some examples of generated videos are also provided in our GitHub repository.[4]

Results of Privacy Evaluation. The result of privacy protection evaluation is shown in Fig. 4. The violin plot illustrates the similarity score distribution of two random frames. The pairs are either from the same real video or consist of one from a real video and the other from the corresponding synthetic video. The similarity scores for the pairs from the same real videos are very high overall since they are essentially frames of the same person. When applying SafeTriage and calculating the similarity scores for the real-synthetic pairs, the scores drop to the below 0.6 range, indicating that the synthetic video represents a different identity from the corresponding real video. This result confirms that the generated synthetic videos have minimal privacy leakage, even though some facial features, such as landmarks and edges, are used as conditions for the visual prompt tuning model.

Results of AI Stroke Triage. Table 1 shows the stroke triage results of SafeTriage-generated synthetic videos using the *DeepStroke* AI triage model. Note that the evaluation of the proposed SafeTriage framework involves the performance of the *DeepStroke* video-only module under several train-test schemes: 1. trained and tested with generated videos (Syn-Syn), 2. trained with generated videos and tested with real videos (Syn-Real), and 3. trained with real videos and

[4] https://github.com/Wenchao-M/SafeTriage-Supplementary.

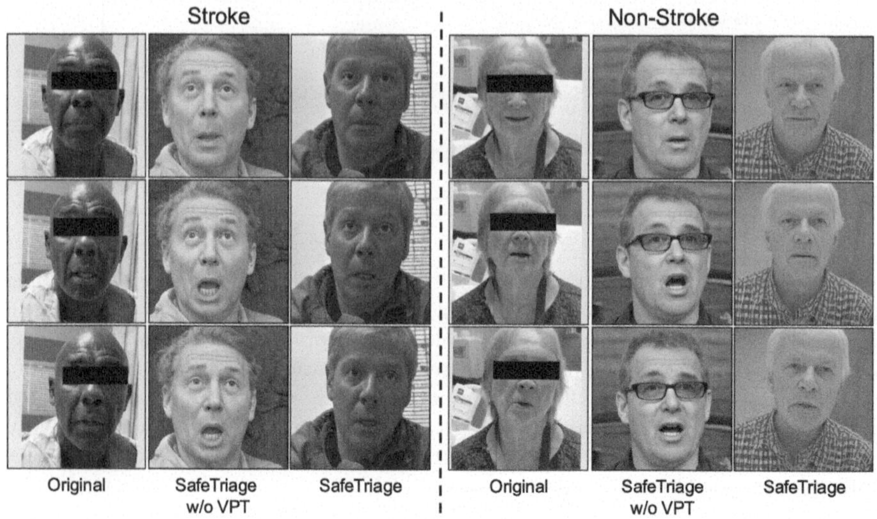

Fig. 5. Qualitative ablation study demonstrating the effectiveness of the visual prompt tuning (VPT) module in SafeTriage. The left block shows SafeTriage without vs. with VPT comparison results for a patient with a stroke. The right block shows the comparison for a patient without a stroke. In each block, the columns display the 1st, 200th, and 400th frames at a resolution of 512×512.

tested with generated videos (Real-Syn). Meanwhile, the baseline refers to the model's performance when both training and testing are conducted using real videos (Real-Real). To measure the level of agreement in class prediction probabilities, we also report the mean-squared error (MSE) between the probability score vector from the real video baseline and those of other train-test schemes involving synthetic videos. Furthermore, to demonstrate the effectiveness of our proposed visual prompt tuning (VPT) method, we conduct an ablation study by generating another set of synthetic videos using synthetic faces from an unconditional model (StyleGAN3-FFHQ) [18], referring to it as the w/o VPT model.

The evaluation schemes are designed to assess the potential use cases of the SafeTriage framework. We anticipate that the most common use case is when synthetic videos are used to train AI models, which are subsequently applied to real patient diagnosis, corresponding to the Syn-Real scheme. As shown in Table 1, SafeTriage achieves performance very close to the baseline, with or without VPT. The full SafeTriage framework with VPT gives results that are the closest to the baseline in terms of classification metrics and MSE. Besides the quantitative ablation study regarding the visual prompt tuning model, we also present in Fig. 5 a qualitative comparison of results with vs. without VPT.

Another potential use case can involve using synthetic videos for privacy-preserving diagnosis, where the diagnosis model is trained with videos of real patients who have already consented to data usage. This corresponds to the Real-Syn scheme. This scheme, as shown in the Table 1, results in a slightly lower

Table 1. Stroke triage results. *Acc.*, *Spec.*, and *Sens.*, represent accuracy, specificity, and sensitivity, respectively. Triage refers to human triage performance, which is a general statistic from the ER department of the data provider. Baseline, w/o VPT, and SafeTriage (ours) results are from the DeepStroke AI triage model. The AI triage performance statistics are adjusted so that the sensitivity level roughly matches with that of human triage.

Method	Train	Test	Acc.(%)↑	Spec.(%)↑	Sens.(%)↑	F1↑	AUC↑	**MSE↓**
Triage	-	-	64.03	45.71	70.19	-	-	-
Baseline	Real	Real	62.10	50.44	71.10	0.6869	0.6885	0
w/o VPT	Syn	Real	61.83	49.33	71.10	0.6741	0.6730	0.1386
	Real	Syn	55.62	35.11	70.77	0.5708	0.5526	0.1358
	Syn	Syn	54.04	28.44	71.65	0.6241	0.5883	0.1479
SafeTriage(Ours)	Syn	Real	62.06	49.78	71.21	0.6757	0.6800	0.0875
	Real	Syn	61.06	46.81	71.21	0.6811	0.6421	0.1058
	Syn	Syn	54.98	32.22	71.43	0.6443	0.5869	0.1017

performance compared to the baseline. We contemplate that this performance drop could be attributed to the more diverse backgrounds in synthetic videos, whereas the real videos have more uniform hospital-setting backgrounds.

The Syn-Syn scheme, in which the AI triage model is both trained and tested on synthetic videos, showed a significant performance drop. One possible reason is that artifacts in the synthetic videos, resulting from imperfections in the conditional generative model and video transfer model, may have amplified effects in the same-domain (Syn-Syn) scheme compared to the diminished effects in cross-domain models (Syn-Real and Real-Syn). Future efforts could explore potential solutions to this issue.

4 Conclusion and Discussion

In this study, we have introduced SafeTriage, a novel facial video de-identification framework designed for privacy-preserving stroke triage. SafeTriage generates synthetic videos that anonymize patient identities while retaining crucial diagnostic motion patterns. By integrating visual prompt tuning and a large pretrained video motion transfer model, our approach enables zero-shot, privacy-preserving video synthesis. Through comprehensive evaluation, SafeTriage has been shown to effectively protect patient privacy while maintaining clinical utility. We believe that SafeTriage represents a major step forward in addressing challenges associated with sharing privacy-sensitive clinical facial video data, laying the groundwork for enabling the training of large-scale AI foundation models for screening and diagnosing neurological conditions, without compromising patient privacy.

Despite its contributions, SafeTriage has certain limitations. First, its generation quality is inherently constrained by the capabilities of the pretrained VMT model. Since our framework is general in terms of which VMT model to use, we expect its performance to improve as more powerful VMT foundation models become available in the future. Second, the current SafeTriage framework focuses solely on visual data. However, the value of multi-modal diagnosis, incorporating corresponding audio data, is well-established. Future work will integrate audio de-identification to enable the secure sharing of both video and audio data. Additionally, due to time constraints, our clinical evaluation is limited to diagnostic pattern assessment in synthetic videos. In the future, we will conduct broader clinical studies to assess their real-world applicability. Finally, we will explore improved privacy evaluation metrics, such as video-based similarity measures, to better assess identity preservation across frames.

Acknowledgments. Research reported in this publication was supported in part by the National Institute of Neurological Disorders and Stroke of the National Institutes of Health under award number R01NS140292, the T.T. and W.F. Chao Foundation, and the John S. Dunn Foundation. Additionally, part of the experiments were supported by computing resources from the iTiger GPU cluster [1]https://itiger-cluster.github.io/, funded by NSF award CNS-2318210 and partially by generous contributions from the College of Arts and Sciences and Information Technology Services at the University of Memphis.

References

1. Bissoto, A., Valle, E., Avila, S.: Gan-based data augmentation and anonymization for skin-lesion analysis: A critical review. In: Proceedings of the IEEE/CVF Conference on Computer Vision and Pattern Recognition (CVPR) Workshops, pp. 1847–1856 (June 2021)
2. Cai, T., et al.: Deepstroke: an efficient stroke screening framework for emergency rooms with multimodal adversarial deep learning. Med. Image Anal. **80**, 102522 (2022)
3. Canny, J.F.: A computational approach to edge detection. IEEE Trans. Pattern Anal. Mach. Intell. **8**(6), 679–698 (1986)
4. Chen, H., et al.: Conv-adapter: Exploring parameter efficient transfer learning for convnets. In: Proceedings of the IEEE/CVF Conference on Computer Vision and Pattern Recognition, pp. 1551–1561 (2024)
5. Chen, S., et al.: Adaptformer: adapting vision transformers for scalable visual recognition. Adv. Neural. Inf. Process. Syst. **35**, 16664–16678 (2022)
6. Dave, I.R., Chen, C., Shah, M.: Spact: Self-supervised privacy preservation for action recognition. In: Proceedings of the IEEE/CVF Conference on Computer Vision and Pattern Recognition, pp. 20164–20173 (2022)
7. Fleiss, J.L., Levin, B., Paik, M.C., et al.: The measurement of interrater agreement. Statistical methods for rates and proportions **2**(212–236), 22–23 (1981)
8. Flouty, E., Zisimopoulos, O., Stoyanov, D.: Faceoff: anonymizing videos in the operating rooms. In: OR 2.0 Context-Aware Operating Theaters, Computer Assisted Robotic Endoscopy, Clinical Image-Based Procedures, and Skin Image Analysis, pp. 30–38. Springer International Publishing, Cham (2018)
9. Goodfellow, I., et al.: Generative adversarial nets. Advances in neural information processing systems **27** (2014)

10. Guo, J., et al.: Liveportrait: Efficient portrait animation with stitching and retargeting control. arXiv preprint arXiv:2407.03168 (2024)
11. He, K., Zhang, X., Ren, S., Sun, J.: Deep residual learning for image recognition. In: Proceedings of the IEEE Conference on Computer Vision and Pattern Recognition, pp. 770–778 (2016)
12. Hedlin, E., et al.: Unsupervised semantic correspondence using stable diffusion. In: Advances in Neural Information Processing Systems, vol. 36 (2024)
13. Hou, J.C., et al.: Artificial intelligence-based face transformation in patient seizure videos for privacy protection. Mayo Clinic Proceedings: Digital Health **1**(4), 619–628 (2023)
14. Houlsby, N., et al.: Parameter-efficient transfer learning for NLP. In: International Conference on Machine Learning, pp. 2790–2799. PMLR (2019)
15. Jia, M., et al.: Visual prompt tuning. In: European Conference on Computer Vision, pp. 709–727. Springer (2022)
16. Johnson, W., Onuma, O., Owolabi, M., Sachdev, S.: Stroke: a global response is needed. Bull. World Health Organ. **94**(9), 634 (2016)
17. Karkkainen, K., Joo, J.: FairFace: Face attribute dataset for balanced race, gender, and age for bias measurement and mitigation. In: Proceedings of the IEEE/CVF Winter Conference on Applications of Computer Vision, pp. 1548–1558 (2021)
18. Karras, T., et al.: Alias-free generative adversarial networks. Adv. Neural. Inf. Process. Syst. **34**, 852–863 (2021)
19. Karras, T., Laine, S., Aila, T.: A style-based generator architecture for generative adversarial networks. In: Proceedings of the IEEE/CVF Conference on Computer Vision and Pattern Recognition, pp. 4401–4410 (2019)
20. Leira, E.C., Kaskie, B., Froehler, M.T., Adams, H.P., Jr.: The growing shortage of vascular neurologists in the era of health reform: planning is brain! Stroke **44**(3), 822–827 (2013)
21. Montenegro, H., Cardoso, J.S.: Anonymizing medical case-based explanations through disentanglement. Med. Image Anal. **95**, 103209 (2024)
22. Ni, H., Liu, J., Xue, Y., Huang, S.X.: 3d-aware talking-head video motion transfer. In: Proceedings of the IEEE/CVF Winter Conference on Applications of Computer Vision, pp. 4954–4964 (2024)
23. Parkhi, O., Vedaldi, A., Zisserman, A.: Deep face recognition. In: BMVC 2015-Proceedings of the British Machine Vision Conference 2015. British Machine Vision Association (2015)
24. Rombach, R., Blattmann, A., Lorenz, D., Esser, P., Ommer, B.: High-resolution image synthesis with latent diffusion models. In: Proceedings of the IEEE/CVF Conference on Computer Vision and Pattern Recognition, pp. 10684–10695 (2022)
25. Schroff, F., Kalenichenko, D., Philbin, J.: Facenet: A unified embedding for face recognition and clustering. In: Proceedings of the IEEE Conference on Computer Vision and Pattern Recognition, pp. 815–823 (2015)
26. Serengil, S.I., Ozpinar, A.: Lightface: A hybrid deep face recognition framework. In: 2020 Innovations in Intelligent Systems and Applications Conference (ASYU), pp. 23–27. IEEE (2020)
27. Shin, H.C., et al.: Medical image synthesis for data augmentation and anonymization using generative adversarial networks. In: Simulation and Synthesis in Medical Imaging: Third International Workshop, SASHIMI 2018, Held in Conjunction with MICCAI 2018, Granada, Spain, September 16, 2018, Proceedings 3, pp. 1–11. Springer (2018)

28. Siarohin, A., Lathuilière, S., Tulyakov, S., Ricci, E., Sebe, N.: First order motion model for image animation. In: Advances in Neural Information Processing Systems, vol. 32 (2019)
29. Siarohin, A., Woodford, O.J., Ren, J., Chai, M., Tulyakov, S.: Motion representations for articulated animation. In: Proceedings of the IEEE/CVF Conference on Computer Vision and Pattern Recognition, pp. 13653–13662 (2021)
30. Wang, T.C., Mallya, A., Liu, M.Y.: One-shot free-view neural talking-head synthesis for video conferencing. In: Proceedings of the IEEE/CVF Conference on Computer Vision and Pattern Recognition, pp. 10039–10049 (2021)
31. Xia, Z., Neidle, C., Metaxas, D.N.: Diffslva: Harnessing diffusion models for sign language video anonymization. arXiv preprint arXiv:2311.16060 (2023)
32. Yu, M., et al.: Toward rapid stroke diagnosis with multimodal deep learning. In: Martel, A.L., Abolmaesumi, P., Stoyanov, D., Mateus, D., Zuluaga, M.A., Zhou, S.K., Racoceanu, D., Joskowicz, L. (eds.) MICCAI 2020. LNCS, vol. 12263, pp. 616–626. Springer, Cham (2020). https://doi.org/10.1007/978-3-030-59716-0_59
33. Zaken, E.B., Ravfogel, S., Goldberg, Y.: Bitfit: Simple parameter-efficient fine-tuning for transformer-based masked language-models. arXiv preprint arXiv:2106.10199 (2021)
34. Zakharov, E., Shysheya, A., Burkov, E., Lempitsky, V.: Few-shot adversarial learning of realistic neural talking head models. In: Proceedings of the IEEE/CVF International Conference on Computer Vision pp. 9459–9468 (2019)
35. Zhang, L., Rao, A., Agrawala, M.: Adding conditional control to text-to-image diffusion models. In: Proceedings of the IEEE/CVF International Conference on Computer Vision, pp. 3836–3847 (2023)
36. Zhou, S., Chan, K., Li, C., Loy, C.C.: Towards robust blind face restoration with codebook lookup transformer. Adv. Neural. Inf. Process. Syst. **35**, 30599–30611 (2022)
37. Zhu, B., Zhang, C., Sui, Y., Li, L.: Facemotionpreserve: a generative approach for facial de-identification and medical information preservation. Sci. Rep. **14**(1), 17275 (2024)
38. Zhuang, Y., et al.: Video-based facial weakness analysis. IEEE Trans. Biomed. Eng. **68**(9), 2698–2705 (2021)

Author Index

I. Oguz et al. (Eds.): IPMI 2025, LNCS 15830, pp. 405–408, 2026.
https://doi.org/10.1007/978-3-031-96625-5

The manufacturer's authorised representative in the EU is Springer Nature Customer Service Centre GmbH, Europaplatz 3, 69115 Heidelberg, Germany. If you have any concerns regarding our products, please contact ProductSafety@springernature.com

Printed and bound by CPI Group (UK) Ltd, Croydon, CR0 4YY

29/04/2026

02099511-0005